Advances in Indian Earthquake Engineering and Seismology

M. L. Sharma • Manish Shrikhande • H. R. Wason
Editors

Advances in Indian Earthquake Engineering and Seismology

Contributions in Honour of Jai Krishna

Springer

Editors
M. L. Sharma
Department of Earthquake Engineering
Indian Institute of Technology Roorkee
Roorkee, Uttarakhand, India

Manish Shrikhande
Department of Earthquake Engineering
Indian Institute of Technology Roorkee
Roorkee, Uttarakhand, India

H. R. Wason
Department of Earthquake Engineering
Indian Institute of Technology Roorkee
Roorkee, Uttarakhand, India

ISBN 978-3-030-08310-6 ISBN 978-3-319-76855-7 (eBook)
https://doi.org/10.1007/978-3-319-76855-7

Printed on acid-free paper

This Springer imprint is published by the registered company Springer International Publishing AG part of Springer Nature.
The registered company address is: Gewerbestrasse 11, 6330 Cham, Switzerland

Preface

Earthquake engineering education in India started with the introduction of a post-graduate course on 'Structural Dynamics' in 1958 by Prof. Jai Krishna at the then University of Roorkee (now IIT Roorkee) with active support from Prof. D.E. Hudson of California Institute of Technology, USA. The first seminar on earthquake engineering was organized in February 1959 to focus the attention of engineers and technologists on the various aspects of this vital engineering problem, with particular emphasis for India which has about 60% of its area vulnerable to earthquakes. With the growing interest in this field amongst a large number of engineers and seismologists, Indian Society of Earthquake Technology was established in 1962 with Prof. Jai Krishna as the founder President. The society has since been working in close association with the Department of Earthquake Engineering, Indian Institute of Technology Roorkee, Roorkee. Under the dynamic leadership of Prof. Jai Krishna, the subject developed at a very rapid rate and Earthquake Engineering Symposia became a regular four yearly event with sizeable national and international participation. It has been only due to his stewardship throughout his life that there is an increasing awareness among the engineers in India to adopt earthquake resistant design practices. Prof. Jai Krishna remained active in popularizing earthquake engineering education and research in India until his death on August 27, 1999. The President of India conferred Padma Bhushan (the third highest civilian honour) on him in 1972 for his meritorious services to the nation. He was also posthumously honoured as 'Legend of Earthquake Engineering' by IAEE in 2008 during the 14th WCEE in Beijing, China.

The Indian Society of Earthquake Technology (ISET) in its Golden Jubilee year decided to bring out a special commemorative state-of-the-art volume entitled *Advances in Indian Earthquake Engineering and Seismology* to mark the 100th birth anniversary of the late Prof. Jai Krishna featuring contributed chapters from leading domain experts. This compilation presents the current state and trends of earthquake engineering practice ranging from engineering seismology to earthquake resistant design and construction for disaster risk reduction.

We begin with a short biography of Prof. Jai Krishna and a note on growth and contributions by ISET during the last 50 years. Various aspects in engineering seismology, the evolution of methodologies and the state-of-the-art techniques used to accomplish various tasks in engineering seismology are dealt with in Chap. 1. The developments in instrumentation seismology in India since the first seismological observatory of the country set up at Alipore (Kolkata) by India Meteorological Department (IMD) in 1898 are discussed in Chap. 2. The evolution of strong motion instrumentation programme in India is detailed in Chap. 3. Conversion of magnitude scales for use in seismic hazard assessment is discussed in Chap. 4, while tsunami hazard assessment is covered in Chap. 5. Chapter 6 marks the beginning of the Geotechnical Earthquake Engineering section with a discussion of recent developments in the areas of liquefaction, stability of rigid retaining structures, shallow foundations and pile foundations. A state-of-the-art review of two important problems in geotechnical earthquake engineering, namely, seismic analysis and design of retaining walls and shallow foundations, is presented in Chap. 7. Various aspects of liquefaction hazard mitigation of loose saturated sands with a spectrum of ground engineering methods are presented in Chap. 8. The recent advances in soil dynamics, especially in the area of geotechnical earthquake engineering towards evaluating the local site effects, dynamic soil properties, different field and laboratory tests required, various site classification schemes and different methods to evaluate the surface level ground motion, are covered in Chap. 9. The section on Structural Dynamics begins with a discussion of earthquake resistant design of bridges in Chap. 10. This is followed by a discussion of principles of earthquake engineering and earthquake resistant design of reinforced concrete and masonry buildings in Chaps. 11 and 12 respectively. Developments in seismic analysis procedures and the displacement-based design are addressed in Chap. 13. A review of developments in the earthquake resistant design during the past 100 years and shape of a futuristic code that would give more freedom and flexibility to the owners, designers and builders and improve the performance of a structure under a severe seismic loading is presented in Chap. 14. In Chap. 15, application of system identification techniques in structural dynamics for extraction of modal properties is discussed. A review of the response spectrum superposition methods to estimate the maximum amplitudes of various response quantities at different levels of multi-degree-of-freedom structures for earthquake resistant design applications is presented in Chap. 16. The contours of complex and vast problem of mainstreaming disaster risk reduction in building sector in India are explored in Chap. 17. It is analysed that the task of mainstreaming could be accomplished only with concerted effort and cooperation from all stakeholders of the society. The recent trends in disaster mitigation and management and details of 'Vulnerability Atlas of India' which to date is the only major document existing on damage risk to housing stock in India with respect to natural hazards, i.e. earthquake, wind and cyclone, and flood, are discussed in Chap. 18. The socio-economic issues in disaster risk reduction in local buildings are discussed in Chap. 19.

Roorkee, Uttarakhand, India

M. L. Sharma
Manish Shrikhande
H. R. Wason

Prologue 1: Jai Krishna

1 Preamble

Eminent educationist, researcher and academic administrator Jai Krishna was credited with pioneering teaching and research in the field of earthquake engineering in India. A doyen of structural and earthquake engineering, he created a band of dedicated engineers and scientists for mitigating earthquake disasters. In his passing away on August 27, 1999, the University of Roorkee (now Indian Institute of Technology Roorkee) lost a gifted and motivating teacher, an imaginative and original researcher and above all an intelligent and warm person. For the institution, where he spent nearly six decades of dedicated and devoted work, it would not be easy to fill the big void thus created.

2 Family Background and Early Education

Jai Krishna was born on Feb 14, 1912, in the district town of Muzaffarnagar in west Uttar Pradesh where his father Pyarey Lal had moved in 1896 at the age of 16 years following the sudden death of his own father in their village nearby. Pyarey Lal used to work in the local court. Jai Krishna coming from this middle class family environment did his schooling in Muzaffarnagar and excelled in it. At that stage he moved to Meerut for higher education and obtained his B.Sc. from Meerut College, a constituent college of Agra University, topping the list of successful candidates in B.A./B.Sc. The year 1932 was the turning point in Jai Krishna's life, as simultaneously with his graduation, he got through the prestigious entrance examination of Thomason College of Civil Engineering at Roorkee, topping the list of successful candidates. He obtained his C.E. (Honours) Diploma from Roorkee in 1935 and soon joined UP Public Works Department which used to be the preferred choice of most young men in those days, but the job did not suit his talents. After a period of about 4 years, he returned to Roorkee in 1939 to serve his alma mater—to start with as a lecturer and later in various higher positions eventually rising to the coveted position of vice chancellor in 1971 to become the first teacher of the institution to achieve this rare distinction.

In 1937, at the age of 25, Jai Krishna married Shashi Prabha, a charming homely girl from Meerut, and the happy couple was blessed with a son, Prem (1938), and two daughters—Veena (1939) and Bimla (1941). The family was soon amongst the elite of the town and later became synonymous with the University of Roorkee due to their deep association with the institution during the six decades which followed. Being one in the faculty from the days of the British, Jai Krishna played a leading role in institution building, maintaining the discipline and traditions of the Thomason College with great zeal and enthusiasm. He and his wife were doing morning walks through the campus and nearby areas of Bengal Engineers Group and Centre regularly and religiously until about 1984 when Shashi Prabha was unfortunately taken ill. The couple was, however, lucky to have lived happily all through until Shashi Prabha passed away about 9 months earlier than her beloved husband.

3 Higher Education and Career

At Thomason College and later at the University of Roorkee, Jai Krishna taught Civil Engineering and was one of the best and most revered teachers on the campus. He was a trendsetter and innovative in his approach to explain fundamentals of the subject to his students. He became a Reader in 1954 and Associate Professor in 1957.

Jai Krishna, under a British Council programme, went to England in 1952 to pursue doctoral work at the University of London which he completed in 1954. Later, he spent 6 months at California Institute of Technology, Pasadena, during 1956–1957 and worked on earthquake engineering problems with G.W. Housner and D.E. Hudson who were the leading experts in the subject. It was here that Jai Krishna developed a vision for similar work in India and on his return initiated research studies in earthquake engineering while introducing an optional course in Structural Dynamics at the post-graduate level. A seminar on Earthquake Engineering was also organized in February 1959 which brought civil engineers, geologists and seismologists on a common platform to focus attention of all concerned, particularly civil engineers, on the serious problem faced by the sub-continent.

The progress of earthquake engineering studies at the Civil Engineering Department was soon recognized, and a School of Research and Training in Earthquake Engineering was set up in 1960 with Jai Krishna as Professor and Director. This 'School' earned a name for itself in the national and international arena with Jai Krishna's all-round contributions in research and developmental activities inspiring and motivating his younger colleagues. This eventually resulted in the establishment of a full-fledged academic department of the University of Roorkee in 1971 through funding from the Government of Uttar Pradesh/University Grants Commission. Though Jai Krishna rose to become vice chancellor of the university, he continued to work with his colleagues and guided them on various issues related to earthquake engineering research. Jai Krishna's most significant contribution to the university as its vice chancellor was in expanding and galvanizing teaching and research activities. The university, under his dynamic leadership, formulated the fifth 5-year plan proposals for diversifying some of the major fields of activities into specialities for the first time and got these sanctioned by UGC for implementation. This gave a big boost to research in frontier areas in science and technology relevant to the futuristic demands of the profession. After his retirement as vice chancellor, Jai Krishna was appointed as Professor Emeritus and continued to pursue his technical work and review the progress of various projects being done by some of his colleagues. He remained in close touch with the activities and research work till his very end.

4 Scientific Contributions

During his study tour to CALTECH in 1956–1957, Jai Krishna prepared an excellent paper highlighting the various aspects of earthquake engineering problem in India, which became the launching pad for studies and research work relevant to all engineering structures, ranging from common man's house to multi-storey buildings, dams, bridges etc. He brought in new and original ideas in the design of earthquake resistant structures, which led to the evolution of economic design and construction practices. Some of his important research contributions are as follows:

1. He initiated analytical and experimental studies on the behaviour of brick masonry structures under vertical and lateral loads with a view to assess their vulnerability in earthquake regions, and to evolve and incorporate appropriate measures to strengthen such systems. This was extremely topical and relevant to India as large number of casualties resulted from collapse of common man's housing. The results of Jai Krishna's work found wide acceptability not only in India but in several other countries facing this problem. The innovative concepts of incorporating simple and economic methods of improving seismic resistance of masonry buildings have found a place in building codes in India and elsewhere. As a matter of fact, the International Association for Earthquake Engineering (IAEE) also published a manual on non-engineered construction, which was largely based on the work carried out initially by Jai Krishna and later by his colleagues under his guidance.
2. Realizing that there was almost no ground motion data available in the country, Jai Krishna in the early 1960s initiated work in design, fabrication and installation of accelerographs and structural response recorders in seismic areas of the country. This work motivated a number of scientists and other technical staff to implement this very ambitious project to collect strong motion data from within

the country for guidance in earthquake resistant design of structures. This naturally was to be a long-term programme of continuing nature and was later supported by the Department of Science and Technology, Government of India. This activity has now picked up momentum and has begun to yield very useful data base which would go a long way in guiding engineers to adopt appropriate attenuation relationships in seismic design criteria.

3. Jai Krishna's approach to research investigations was typically one of supplementing analytical results with experimental tests. He fully realized the limitations of either types of studies but always emphasized on ensuring that the analytical results were realistic and that they are supported by real behaviour. For this purpose, he along with his co-workers designed and fabricated shake tables on which models of structures could be constructed and tested for dynamic (or shock) loading. His outstanding contribution on the development of facilities for laboratory and field testing led to independence from foreign equipment and consultancy on such problems and helped in import substitution.
4. Jai Krishna introduced the concept of iso-acceleration lines indicative of seismic forces in epicentral tracts for study of distribution and attenuation of energy from earthquake sources.
5. Design of water tanks and staging in epicentral areas was of great concern to Jai Krishna since these are structures of post-earthquake importance and needed most in fighting fires following an earthquake. Dynamic characteristics of a number of existing water tanks were studied through extensive field testing programme and results were used in seismic analysis and design. The efficacy of providing diagonal bracings in the staging of certain types of water tanks was also examined from the viewpoint of improving their seismic performance.

5 Professional Contributions

As mentioned earlier, Jai Krishna was a brilliant teacher highly respected for clarity of his thoughts and expression and one who could effectively communicate with students with utmost ease. He believed in 'teaching not only the subject but also the student'. With rich experience of teaching he authored a treatise on *Plain and Reinforced Concrete* in two volumes (with his colleague O.P. Jain) in 1951. These books bearing the character of the excellent teaching talents of the authors were widely acclaimed as the best resource on the subject and used by students and field engineers all over the country. These books have been updated and revised several times now and are amongst the most popular references even today. Later in 1976, Jai Krishna authored another textbook on Earthquake Engineering (with his colleague A.R. Chandrasekaran) for helping the students with the much needed inputs for providing the necessary background and insight to understand the codal provisions for design of civil engineering structures in seismic regions.

Jai Krishna had a leading role in the field of standardization. He was the prime motivating force behind the formulation of Indian Standards for earthquake resistant design of structures brought out by Indian Standards Institution (now Bureau of Indian Standards) under his chairmanship. These standards have been extensively used during the past four decades and are some of the most comprehensive set of provisions which are constantly reviewed and updated as new inputs become available through research and data recordings.

Jai Krishna advised on seismic aspects related to a number of important major projects in the country like Bhakra and Beas, Yamuna, Maneri Bhali, Obra, Tenughat, Koyna, Bhatsa, Narmada, Brahmaputra and Tehri. He also led, for the first time in the country, seismic studies for feasibility of a Nuclear Power Plant at Narora and eventually the detailed dynamic analysis of various structures related to this power plant. Subsequently, work on a number of other nuclear, hydro and thermal power plants was also carried out fully through know-how thus developed under his leadership.

In the words of G.W. Housner he was ‘a man of great personal integrity and very influential in mitigating earthquake hazards and promoting public safety during earthquakes, both nationally and internationally’.

Jai Krishna’s approach to teaching and research was always very innovative. He had the distinction of introducing the first courses in Soil Mechanics (1948) and Structural Dynamics (1958) in the engineering syllabi in the country. His role as institution builder was highlighted when he was invited by UNESCO as a consultant to help developing a School of Earthquake Engineering and Seismology at Skopje in Yugoslavia for a year during 1967–1968. He was earlier a visiting lecturer at the International Institute of Seismology and Earthquake Engineering at Tokyo in 1965.

Jai Krishna with his vast experience of academic activities was chosen as Chairman of a high powered committee to review the working and for improving the facilities at Regional Engineering Colleges in India. He was also a member of management boards of several universities, Indian Institutes of Technology and Institute of Science, Bangalore.

Committed and devoted to the cause and objectives of disaster mitigation programmes all over the seismic world, Jai Krishna was an active member of UNESCO consultative committee on earthquake engineering and was also the President of Inter-governmental Meeting of UNESCO on the subject in 1976. Within the country, he was a member of several Expert Committees on river valley projects notably Bhakra, Brahmaputra, Narmada, Bhatsa, Koyna and Tehri.

Jai Krishna led an Indian delegation to the USSR in 1972 for exchange of information on seismic design practices particularly for high dams. He also represented the Government of India at the US-China Bilateral Workshop on Earthquake Engineering at Harbin (China) in 1982.

6 Association with National and International Bodies

Jai Krishna was Fellow of Indian National Science Academy and Indian National Academy of Engineering. He was Honorary Fellow of American Academy of Engineering, International Association of Earthquake Engineering, New Zealand Society of Earthquake Engineering, Third World Academy of Sciences and the Institution of Engineers (India).

Jai Krishna was the founder president of Indian Society of Earthquake Engineering and Indian National Academy of Engineering. He also held the office of the President of the Institution of Engineers (India) during 1974–1975 and was elected President of International Association of Earthquake Engineering for the period 1977–1980 at the conclusion of Sixth World Conference on Earthquake Engineering held at New Delhi in January 1977 of which he was the Organizing Chairman.

7 Prizes, Awards and Honours

Starting from Agra University Merit Scholarship and Gold Medal for standing first amongst undergraduate students (Science/Arts) in 1932, Jai Krishna continued to excel in studies and passed his C.E. (Hons) winning Cautley Gold Medal, Thomason prize and Callcott Reilly Memorial Gold Medal from Thomason College. Later, in 1957–1958, he was awarded Railway Board's first prize for his classical paper highlighting 'Earthquake Engineering Problems in India' by the Institution of Engineers (India). He also won the Khosla Research Prize of the University of Roorkee in 1963. His contributions in the subject were recognized by the Council of Scientific and Industrial Research, by way of Sir Shanti Swarup Bhatnagar Award for the year 1966. Jai Krishna also received the National Design Award for 1971 from the Institution of Engineers (India) and K.L. Moudgill Award for 1972 from Indian Standards Institution. The high point of recognition came in 1972 when he was conferred Padma Bhushan by the President of India for his meritorious services to the nation.

Jai Krishna was highly respected and patiently heard not only in India but also by the international community. He was cited and honoured by International Award at the Ninth World Conference on Earthquake Engineering at Tokyo in 1988 by Japan Society of Disaster prevention. In 1997, the Indian National Academy of Engineering honoured him for his Life Time Contributions in Engineering.

Jai Krishna was also honoured and awarded honorary degrees of Doctor of Science and Doctor of Engineering by Agra University and Roorkee University, respectively.

8 Extracurricular Activities

As a student Jai Krishna's activities were not confined to classrooms or laboratories. He was a very active person on tennis courts and would sweat it out with his friends over there for hours whenever he found time. He was equally at home in Billiards and played regularly in the Engineers Students Club at Thomason College and used to pick up the 'cue' even during his visits to the club later on various ceremonial occasions. He also did horse riding as part of studentship and thoroughly enjoyed the same. Jai Krishna had a passion for 'contract bridge' and would not miss an opportunity to have a 'sitting' whenever his friends visited Roorkee and he used to call some colleagues if the number was falling short. He was like many others keen to watch cricket telecasts particularly when India were in it—but would prefer to do something more useful if the 'going was not good'.

With his attributes in teaching and interest in games and sports, Jai Krishna was the natural choice to be asked to become President of Recreation (later President of University Sports Association), in 1957 to organize and manage its activities for the students who had to compulsorily opt for two of these for proficiency and for which performance was evaluated to be included in their marks sheets. He continued this work with great enthusiasm for almost 10 years before handing it over to a younger colleague. Earlier, Jai Krishna also looked after the Engineer Students Mess and the Club as President.

9 Motivation to Colleagues

Jai Krishna's outstanding ability to lead from the front was the hallmark of his success as researcher, administrator and organizer. His capacity to deal with some of the most difficult situations was so great that he never allowed such events to affect the younger staff. He motivated the younger faculty into working with existing facilities and equipment while simultaneously striving to procure the best available. He used to often say that 'best is the biggest enemy of good' and this used to go rather well even with those who were apprehensive of getting any worthwhile results with the available facilities.

Jai Krishna had tremendous sense of humour. A young colleague, soon after joining the university, enquired of him about the 'working hours' in the department. Jai Krishna very coolly told the youngster 'we are very liberal in this regard—come any time before 8 a.m. and go any time after 5 p.m.'. The young man quickly took the hint and moulded himself to the culture expected of him.

10 On Personal Side

Jai Krishna had an excellent personal library at home which housed large number of books and journals on earthquake engineering and allied subjects. All these were open and available to researchers for reference all the time even when he was not at home. He had wished that all this rich collection be donated to Earthquake Engineering Department for the benefit of staff and students. This was indeed done soon after his death by his illustrious son, Prem Krishna, himself a distinguished structural engineer who, like his father, served Roorkee University for over 32 years and is presently the Railway Board's Chair Professor in Bridge Engineering.

Extremely fond of mangoes, Jai Krishna enjoyed his breakfast, lunch and dinner with these when the season was on. In the bungalow where he lived for almost 25 years, there were about 12 trees of the choicest variety of 'langra' which used to yield thousands of mangoes. The family used to organize mango parties for friends, colleagues and their families to share the fruits of the bungalow. Older generations of alumni who visit the campus invariably make nostalgic references to this feature at Jai Krishna's bungalow which has now been dismantled to house the imposing new building of the Department of Management Studies.

11 Last Days

Shaken and heartbroken after the passing away of his wife in November 1998, Jai Krishna tried to busy himself in reading books/journals related to science and technology and also on philosophy. Unfortunately, however, he was taken ill in June 1999 when on a visit with his daughter Bimla and son-in-law Indu, both doctors of long standing at S.N. Medical College, Agra. Unable to recover from prostate problems in spite of the best care, he desired that he be taken to Roorkee, where after a few days the inevitable occurred.

12 Conclusion

Summing up Jai Krishna's life, a quote from his obituary would be appropriate. 'Guru and mentor to generations of earthquake engineers, Jai Krishna epitomized the human spirit of enquiry and unflinching loyalty and devotion to academic pursuits, so very essential for development of knowledge'. The obituary makes reference to the existence of Earthquake Engineering Department at the University of Roorkee, 'as a testimony to the creative genius of Jai Krishna who was not only instrumental in its creation, but also nurtured it till he breathed his last. His life and work will guide the future generations of engineers inspiring them in the path of creative pursuit of profession to enrich human endeavour for knowledge.'

13 Acknowledgments

Information regarding the early life of Jai Krishna was obtained from his son Prem Krishna. All other information is based on the author's close association with him for over 37 years at the Department of Earthquake Engineering.

Help and support received from Prof. & Head, Department of Earthquake Engineering, Indian Institute of Technology Roorkee, is gratefully acknowledged.

Papers Published

1. "Reinforced Concrete Circular Slabs", Journal of Institution of Engineers (India), Vol. XXXII No.4, June 1952.
2. "Economic Aspects of the Design of Reinforced Concrete Retaining Walls" (Part 1 & 2), Civil Engineering and Public Works Review, Vol. 49, Nos. 573 and 574, March 1954 and April 1954.
3. "The Beam Strength of the Reinforced Concrete Cylindrical Shell", (Part I & II), Civil Engineering and Public Works Review, Vol. 49, Nos. 578 & 579, Aug. 1954 and Sept. 1954, (Co-author: O.P. Jain).
4. "New Equation for R.C. Retaining Walls", Journal of National Building Organization, Vol. 2, No.1–2, 1956.
5. "Photogrammetry Applied to Structural Engineering", Civil Engineering and Public Works Review, Vol. 52, No. 615, Sept. 1957.
6. "Design of Prestressed Continuous Beams", Indian Concrete Journal, Vol. 31, No. 7, July 1957, (Co-author: A.S. Arya).
7. "Prestressing Continuous Beams", Proc. Symposium on Prestressed Concrete, Indian Construction News, Calcutta, Aug. 1958. (Co-author: A.S. Arya).
8. "Slabs Constrained on Three Edges and Subjected to Hydrostatic Pressure", JOURNAL OF Institution of Engineers (India), Vol. XXXVIII, No. 9, Pt. I, May 1958.
9. "Earthquake Engineering Problems in India", Journal of Institution of Engineers (India), Vol. XXXIX No. 1 Pt. I, Sept. 1958.
10. "Aerodynamic Stability of Non-uniform Cantilevers", University Research Journal, Roorkee, Vol. 1, No.1, 1958.
11. "Secondary Effects in the Design of Multistoreyed Frames", Proc. Symposium on Multistoreyed Building, NBO, Bombay, 1958.
12. "Seismic Zoning of India", Proc. Seminar on Earthquake Engineering, University of Roorkee, Feb. 1959.
13. "Aseismic Design Structures", Proc. Seminar on Earthquake Engineering, University of Roorkee, Roorkee, Feb. 1959.
14. "Seismic Data for the Design of Structures", Proc. Second World Conference on Earthquake Engineering, Tokyo, 1960.
15. "Kapkote Earthquake of Dec. 28, 1958", Proc. Second World Conference on Earthquake Engineering, Tokyo, 1960. (Co-authors: R.S. Mithal, A.R. Chandrasekaran and S.K. Manglik).
16. "Earthquake Resistant Design of Earth Dams", Proc. Second Symposium on Earthquake Engineering, University of Roorkee, Roorkee, 1962.
17. "Building Construction in Seismic Zones of India", Proc. Second Symposium on Earthquake Engineering, University of Roorkee, Roorkee, 1962. (Co-author: A.S. Arya).

18. "Design of Structural Response Recorders", Proc. Second Symposium on Earthquake Engineering, University of Roorkee, Roorkee, 1962. (Co-author: A.R. Chandrasekaran).
19. "Design of Shock Vibration Tables", Proc. Second Symposium on Earthquake Engineering, University of Roorkee, Roorkee, 1962. (Co-author: A.R. Chandrasekaran).
20. "Earthquake Resistant Design of Earth Dams", Journal of Institution of Engineers (India). Vol. XLIV No. I, Pt. CI, Sept. 1963.
21. "Economic Considerations in the Design of Concrete Structural Elements", JOURNAL OF Institution of Engineers (India), Vol. XLIV, No. 5, Pt. C13, Jan. 1964. (Co-author: J.L. Ajmani).
22. "Problems of Earthquakes", Bull. Indian Society of Earthquake Technology, Vol. 1, No. 1, 1964.
23. "Badgam Earthquake of Sept. 2, 1963", Bull. Indian Society of Earthquake Technology, Vol. 1, No. 1, 1964 (Co-author: L.S. Srivastava, S.E. Hasan, R.S. Mithal and A.R. Chandrasekran).
24. "Earthquake Engineering Makes Progress", Bull. Indian Society of Earthquake Technology, Vol. 1, No. 2, 1964.
25. "Earthquake Resistant Design of an Elevated Water Tower", Bull. Indian Society of Earthquake Technology, Vol. 1, No. 1, 1964. (Co-author: A.R. Chandrasekran).
26. "Water Towers in Seismic Zones", Proc. 3rd World Conference on Earthquake Engineering, New Zealand, 1965. (Co-author: A.R. Chandrasekran).
27. "Engineering Aspects of Badgam Earthquake", Proc. Third World Conference on Earthquake Engineering, New Zealand, 1965.
28. "Structural Response Recorders", Proc. Third World Conference on Earthquake Engineering, New Zealand. (Co-author: A.R. Chandrasekran).
29. "Strengthening of Brick Buildings against Earthquake Forces", Proc. Third World Conference on Earthquake Engineering, 1965. (Co-author: Brijesh Chandra).
30. "Earth Dams Subjected to Earthquakes", Proc. Third World Conference on Earthquake Engineering, 1965. (Co-author: S. Prakash).
31. "Earthquake Resistant Design of Buildings", Journal of Institution of Engineers (India) Vol. XLV No. 7, Pt. C14, March 1965. (Co-author: A.S. Arya).
32. "Earthquake Engineering Makes Progress", Bulletin of Indian Society of Earthquake Technology, Vol. II, No. 2, 1965.
33. "Protection Against Earthquake Damage", Publication and Information Directorate, CSIR, 1966. (Co-author: Brijesh Chandra).
34. "Importance of Earthquake Studies as part of Investigations for Power Projects", Bull. of Indian Society of Earthquake Technology, Vol. III, No. 2, May 1966. (Co-author: L.S. Srivastava).
35. "Basic Principles Underlying Seismic Design of Bridges", Proc. Third Symposium on Earthquake Engineering, University of Roorkee, Roorkee. 1966.
36. "Seismological Instrumentation", Proceedings of the Seminar on Geophysical Investigations in the Peninsular Shield, Hyderabad, 1966.
37. "Damage in Kashmir due to June 6, 1966, Hindukush Earthquake", Proc. Third Symposium on Earthquake Engineering, University of Roorkee, Roorkee 1966. (Co-author: A.S. Arya).
38. "Behaviour of Load Boaring Brick Shear Walls during Earthquakes", Proc. Third Symposium on Earthquake Engineering, University of Roorkee, Roorkee, 1966. (Co-authors: Brijesh Chandra and S.K. Kanungo).
39. "Behaviour of Earth Dam Models under Seismic Loading", Proc. Third Symposium on Earthquake Engineering, University of Roorkee, Roorkee, 1966. (Co-author: S. Prakash).
40. "Study of Liquefaction of Obra Dam Sands", Journal of Institution of Engineers (India), Vol. XLVIII, No. 1, Pt. CI 1, Sept. 1967 (Special), (Co-authors: S. Prakash, J.N. Mathur and M.K. Gupta).
41. "Earthquake Resistant Design of Buildings—Lessons from Recent Earthquakes", Bulletin of Indian Society of Earthquake Technology, Vol. 4, No. 2, April 1967.
42. "Pavement Performance as a Basis of Design", Journal of Highway Division, Proc. American Society of Civil Engineers, No. HW 2, Nov. 1967, (Co-author: N.K. Vaswani).

43. "Blast Tests at Obra Dam Site", Journal of Institution of Engineers (India), Vol. XLVIII, No. 9, Pt. C15, May 1968.
44. "Philosophy of Aseismic Design", Bulletin Indian Society of Earthquake Technology, Vol. 6, No. 1, March 1969.
45. "Strengthening of Brick Buildings in Seismic Zones", Proc. Fourth World Conference on Earthquake Engineering, Chile, 1969, (Co-author: Brijesh Chandra).
46. "Ground Acceleration During Earthquakes", Bulletin of Indian Society of Earthquake Technology, Vol. 6, No. 1, March (Co-author: Brijesh Chandra).
47. "A Study of Earth—Dam Models under Shock Loading", Proc. Fourth World Conference on Earthquake Engineering, Chile, 1969, (Co-author: S. Prakash and S.K. Thakkar).
48. "Distribution of the Maximum Intensity of Forces in the Koyna Earthquake of December 11, 1967", Earthquake Engineering Studies, School of Research and Training in Earthquake Engineering, Roorkee, 1969. (Co-authors: A.S. Arya and K. Kumar).
49. "Analysis of Koyna Accelerogram of Dec. 11, 1967", Bulletin of Seismological Society of America, Vol. 59, No. 4, Aug. 1969. (Co-authors: A.R. Chandrasekran and S.S. Saini).
50. "Characteristics of a Strong Ground Motion Record as Applicable to Design Factors", Bulletin of Indian Society of Earthquake Technology, Vol. 6, No. 2, June 1969). (Co-author: S.S. Saini).
51. "Problems Underlying the Code of Practice for Earthquake Resistant Design of Structures", Bulletin Indian Society of Earthquake Technology, Vol. VII, No. 2, 1970.
52. "Importance of Isoforce Lines of an Earthquake with Special Reference to Koyna Earthquake of Dec. 11, 1967", Proc. Fourth Symposium on Earthquake Engineering, Vol. 1, pp. 1–14, 1970, (Co-authors: A.S. Arya and K. Kumar).
53. "Strong Motion Data from Koyna Earthquakes", Proc. Fourth Symposium on Earthquake Engineering Vol. II, 1970. (Co-authors: A.R. Chandrasekran and P.N. Agarwal).
54. "Earthquake Stress Analysis of Koyna Dam Motion", Proc. 4th Symposium on Earthquake Engineering Vol. I, 1970. (Co-authors: S.S. Saini and A.R. Chandrasekran).
55. "Earthquake Engineering Problem in Himalayas", Symposium on Engineering in Himalayas, Institution of Engineers (India), July 1970. (Co-author: B. Chandra).
56. "Distribution of the Maximum Ground Acceleration in the Broach Earthquake on March 23, 1970", Earthquake Engineering Studies, School of Research and Training in Earthquake Engineering, Roorkee, 1970. (Co-authors: A.S. Arya and K. Kumar).
57. "Earthquake Engineering Problems in Asia", Build International, 1972, (Co-author: Brijesh Chandra).
58. "Determination of Isoacceleration Lines by Sliding and Overturning of Objects", Proc. Fifth World Conference on Earthquake Engineering, Rome, 1973. (Co-authors: A.S. Arya and K. Kumar)
59. "Influence of Bridge Bearings on Aseismic Design of Simply Supported Bridge Spans", Proc. Fifth World Conference on Earthquake Engineering, Rome, 1973. (Co-author: A.R. Chandrasekran).
60. "Feasibility Studies for an Atomic Power Plant on Alluvial Soil in Seismic Zone", Proc. Fifth World Conference on Earthquake Engineering, Rome, 1973. (Co-authors: A.S. Arya, A.R. Chandrasekran, Brijesh Chandra, S.K. Thakkar and P. Nandkumaran).
61. "Overturning of Top Profile of Koyna Dam During Severe Ground Motion", International Journal of Earthquake Engineering and Structural Dynamics, Vol. 2, No. 3, Jan–March 1974. (Co-author: S.S. Saini).
62. "Dynamic Earth Pressures Distribution Behind Flexible Retaining Walls", Indian Geotechnical Journal, Vol. 4, No. 3 , July 1974. (Co-authors: S. Prakash and P. Nandkumaran).
63. "Seismic Data for Structural Design", Presented at Second Iranian Congress of Civil Engineering, Shiraj, Iran, May 1976. (Co-author: Brijesh Chandra).
64. "Aseismic Building Construction in Developing Countries", Presented at Second Iranian Congress of Civil Engineering, Shiraj. Iran, May 1976. (Co-author: Brijesh Chandra).

65. "Problems in Earthquake Resistant Design", Proceedings Sixth World Conference on Earthquake Engineering, New Delhi, January 1977, pp. 1927–34 (Co-author: A.R. Chandrasekran).
66. "Energy Approach to Earthquake Resistant Design", Proc. 6th World Conference on Earthquake Engineering, New Delhi, Jan. 1977, pp. 1848–53. (Co-authors: R.Y. Soni and Brijesh Chandra).
67. "Seismic Data for Design Structures", CBIP Golden Jubilee Commemorative Volume, March 1977, pp. 132–38. (Co-authors: Brijesh Chandra and V.N. Singh).
68. "Earthquake Consideration in Design of Embankment Dams Design and Engineering", Institution of Engineers (India) Karnatak Centre Nos. 2 & 3, 1977. (Co-authors: P. Nandhkumaran).
69. "Effect of Seismic Environment on Design of Major Projects", Inaugural Talk, American National Conference, San Francisco, 1979.
70. "Earthquake Prediction", Technical Note, Bulletin of Indian Society of Earthquake Technology, Vol. 17, No. 2, June 1980.
71. "Experimental Work as an Aid to Earthquake Resistant Design", Proc. Seventh World Conference on Earthquake Engineering, Istanbul. Vol. 6 pp. 151–57, Sept. 1980.
72. "Case Study of Koyna Dam", Bulletin of Indian Society of Earthquake Technology, Vol. 17, No. 3, Sept. 1980.
73. "Future of Earthquake Engineering", Bulletin of Indian Society of Earthquake Technology, Vol. 17, No. 4, Dec. 1980.
74. "Effects of Earthquakes on Earthen Structures", International Conference on Geotechnical Engineering. St. Louis, 1981.
75. "Earthen Buildings in Seismic Areas of India", International Workshop on Earthen Buildings in Seismic Areas, Albuquerque, May 1981. (Co-author: Brijesh Chandra).
76. "Disaster Mitigation", Publication of Central Building of Research Institute, Roorkee, June 1982.
77. "Earthquake Design Criteria", Presented at US—PRC Bilateral Workshop on Earthquake Engineering, Harbin, China, Aug. 1982. (Co-authors: Brijesh Chandra and D.K. Paul).
78. "Uncertainities in Earthquake Resistant Design", Eight World Conference on Earthquake Engineering, San Francisco, 1984. (Co-author: Brijesh Chandra).
79. "Influence of Seismic Environment on Design of Major Structures", Invited Luncheon Address, Eight World Conference on Earthquake Engineering, San Francisco, 1984.
80. "Neotectonics in South Asia", International Symposium on Neotectonics in South Asia, Survey of India, Dehradun, 1986.
81. "Importance of Seismological Studies in India", National Meet on Earthquake Mechanism and Mitigation, Department of Science and Technology, New Delhi, 1986.
82. "River Valley Projects, Earthquakes and Environment", Khosla Lecture, Institution of Engineers (India), Jan. 1989.

Prologue 2: ISET Growth and Contributions During Last 50 Years

1 Preamble

The Indian Society of Earthquake Technology (ISET) came into being to provide a proper forum to scientists and engineers drawn from amongst Civil Engineers, Electro-Mechanical Engineers, Engineering Seismologists and Engineering Geologists who had a common interest related to earthquake engineering. The Indian Society of Earthquake Technology was established on 12th November, 1962, during the second symposium on Earthquake Engineering held at the University of Roorkee with the late Prof. Jai Krishna as the founding President. The headquarters of the Society was decided to be at Roorkee and the Society was registered in December 1964 under the Societies Registration Act 1961. A list of past and current Officers of the Society and Executive Committee Members is given in Annexure I.

2 Aims and Objectives of the Society

The aims and objectives of ISET are mainly directed towards promoting research, development and awareness work in the field of earthquake technology as follows and further as detailed in its bye-laws:

1. To provide necessary forum for scientists and engineers of various specialization to come together and exchange ideas on the problems of earthquake technology.
2. To disseminate knowledge in the field of earthquake technology dealing with scientific and engineering aspects.
3. To honour pioneering and meritorious contributions in the field of earthquake technology.

3 ISET Journal

An important activity of the ISET since 1964 has been bringing out a periodical entitled *Bulletin of the Indian Society of Earthquake Technology* containing technical papers in the areas of structural dynamics, soil dynamics, seismology and other allied fields. A very useful feature of this publication has been the inclusion of the seismological notes to cover the earthquake activity in India and its neighbourhood with input from India Meteorological Department, New Delhi. This feature continued till the year 1993 after which it was discontinued for some administrative reasons. This Bulletin continued as a half-yearly publication till 1966 and started a quarterly publication from 1967. The name of the Bulletin was subsequently changed to *ISET JOURNAL OF EARTHQUAKE TECHNOLOGY* from March, 1998. So far 506 research papers have been published up to the year 2009. In addition to the journal, the Society also publishes a quarterly newsletter.

It has also published a number of books that further its aims and objectives. In particular, the Society has published *Catalogue of Earthquakes in India and Neighbourhood* in the year 1983, compiled by Arun Bapat, Ms. R.C. Kulkarni and S.K. Guha, and reprinted the *Manual of Earthquake Resistant Non-engineered Construction*, a publication of IAEE.

To keep the ISET in the forefront, the standard of the ISET Journal and its other publications will have to be improved with the passage of time.

4 Membership

In spite of the specialized and professional nature of the Society, its membership has steadily grown over the years as is depicted in Fig. 1. The Society has a good national and international exposure through its life members, institutional members and Local Chapters. The ISET membership has gradually increased, the trend governed by the growing number of life members, whereas the yearly membership has declined. The current total membership of the Society stands over 1670, which includes Honorary Fellows (06), Life Fellows (46), Life Members (1536), Annual Members (30) and Institution Members (53). There are 55 I.E. members from abroad. Besides the members, there are 30 subscribers to the ISET Journal.

5 ISET Local Chapters

The ISET activities are spread all over the country through ISET Local Chapters. The Roorkee Chapter was established in 1983 with Prof. Shamsher Prakash (the then Director, CBRI) as the Chairman and Dr. H.R. Wason as the Secretary. Soon after its formation, it organized an International Symposium on ‘Creation of awareness about

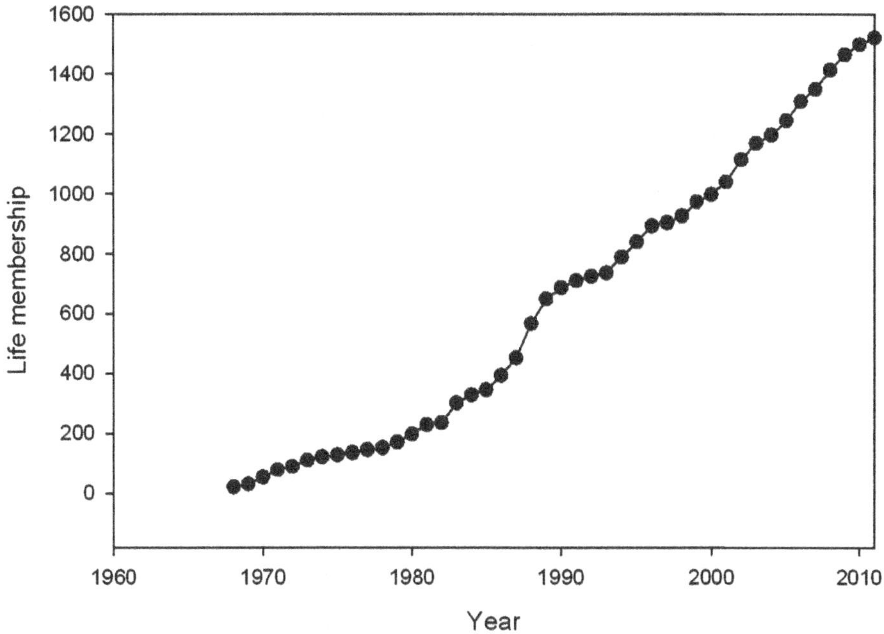

Fig. 1 ISET life membership

earthquake hazards and mitigation of seismic risks' in November 1984. The Society has currently 11 Local Chapters, namely at Amravati, Bangaluru, Chennai, Delhi, Guwahati, Jorhat, Kolkata, Mumbai, Nagpur, Pune and Roorkee, which have been set up to widen the activities of the Society by way of direct involvement of the members located in the respective neighbouring areas. These Chapters organize activities towards creation of earthquake awareness among general public and earthquake engineering education in furtherance of the aims and objectives of the parent body. The affairs of Chapters are managed by a Managing Committee. The names of current Chairmen and Secretaries of the Local Chapters are given in Annexure II.

6 ISET Annual Lectures

ISET holds an Annual Lecture along with Annual General Meeting every year, which is delivered by an eminent engineer/scientist in the field of earthquake technology. This lecture is also printed in one of the forthcoming issues of the ISET Journal. Thirty-two Annual Lectures have been delivered since 1979. The 2012 lecture was delivered by Dr. Shailesh Nayak, Secretary, Ministry of Earth

Sciences, Government of India, New Delhi, at Jamia Millia Islamia, organized by the ISET Delhi Chapter. A list of ISET Annual Lectures along with the names of eminent speakers is given in Annexure III.

7 ISET Awards

In an effort to encourage research by providing suitable incentives and to give recognition to noteworthy research papers published in the ISET Journal and also the proceedings of symposia sponsored by it, the Society has instituted several awards as follows:

'ISET Jai Krishna Award'—Certificate and award money for the best paper in the publications of the Society for block years 1985–1988, 1989–1992 and so on.

'ISET B.N. Gupta Award'—Certificate and award money for the best paper in Seismology and Seismotectonics published in four yearly blocks of 2001–2004, 2005–2008 etc. in the ISET Journal.

'ISET V.H. Joshi Award'—Certificate and award money for the best paper in Soil Dynamics published in four yearly blocks 1997–2000, 2001–2004, 2005–2008 etc. in the ISET Journal.

'ISET Hanumantacharaya Joshi Award'—Certificate and award money for the best paper in Structural Dynamics published in four yearly blocks 1997–2000, 2001–2004, 2005–2008 etc. in the ISET Journal.

'ISET Arbinda Mukhopadhyay Award'—Certificate and award money for the best paper dealing with 'Theoretical Aspects of Earthquake Technology' published in four yearly blocks 1997–2000, 2001–2004, 2005–2008 etc. in the ISET Journal.

Further, in reference to one of the objectives of the Society to honour distinguished persons for their outstanding contributions, the following awards have also been instituted.

'ISET Trifunac Award'—Citation and award money for significant contributions in 'Strong Motion Earthquake Studies', a 4-yearly award.

'ISET Shamsher Prakash Award'—Shawl, plaque and award money for significant contributions in 'Geotechnical Earthquake Engineering', a 4-yearly award.

A research award for recognizing best Ph.D. thesis work in India has also been recently instituted.

'ISET B.K. Maheshwari Research Award'—Citation and award money for best Ph.D. thesis in 'Soil-Structure Interaction in India', a 4-yearly National Research Award.

8 Organization of Symposia/Workshops in Liaison with the Department of Earthquake Engineering, IITR

The Society has maintained close liaison with the Department of Earthquake Engineering, IIT Roorkee (formerly School of Research and Training in Earthquake Engineering). Ever since its formation, the Society has been actively associated with the Department of Earthquake Engineering, in organizing the 4-yearly Symposia on Earthquake Engineering. The Society has organized several symposia/workshops and sponsored special courses during the last 50 years in the field of earthquake technology, which has been a very major activity of the Society. ISET also played a most pivotal role in holding of the 6th World Conference on Earthquake Engineering in January 1977 at New Delhi. It is also engaged in various activities directed towards creation of awareness about earthquakes and earthquake technology amongst general public as well as scientific/engineering communities.

The Bhuj earthquake of January 26, 2001, with a magnitude 6.9 was the largest earthquake to occur in urban India since its independence. As such, the Society organized a workshop on 'Recent Earthquakes of Chamoli and Bhuj' at Roorkee during May 24–26, 2001, jointly with the Department of Earthquake Engineering, IIT Roorkee. This workshop had a participation of 98 delegates including representatives from 47 organizations and a total of 60 technical presentations were made covering different themes during the workshop. The last 4-yearly Symposium '14th Symposium on Earthquake Engineering' was co-organized by the Society jointly with the Department of Earthquake Engineering, Indian Institute of Technology, Roorkee, during December 17–19, 2010.

9 Liaison with the International Association for Earthquake Engineering (IAEE)

ISET is a founder member of the International Association for Earthquake Engineering (IAEE) and represents India on this world forum dealing with various aspects of earthquake engineering. The Society is represented by the President as the National Delegate and the Vice President as the Deputy National Delegate in the General Assembly of Delegates of the IAEE. The Society is keeping close liaison with IAEE and has actively participated in its activities. Professor Jai Krishna has served the IAEE as Vice President (1973–1977) and then its President (1977–1980). In view of his distinguished service to earthquake engineering, the late Prof. Jai Krishna was elected as an Honorary Member of the IAEE. In addition to Prof. Jai Krishna, other ISET members to have served on the Executive Committee of the IAEE as Directors (or in other capacities) include Prof. A.S. Arya, Prof. Brijesh Chandra Mathur and Prof. P. N. Agrawal.

ISET Honorary Fellow the late Prof. Jai Krishna was elected as one of the Legends of Earthquake Engineering in the meeting of IAEE held during 14th

World Conference on Earthquake Engineering at Beijing, P.R. China. During the 15th World Conference on Earthquake Engineering recently held at Lisbon, Portugal, during September 24–28, 2012, Prof. Sudhir Jain has been elected as President of IAEE (2014–2018 term) and Prof. A.S. Arya has been nominated as an Honorary Member of the IAEE in recognition of their distinguished contributions to earthquake engineering. Prof. Sudhir Jain has also been on the E.C. of IAEE as its Executive Vice President (2008–2012) and as Director earlier to it.

10 National Role

The Executive Committee of ISET also acts as the Indian National Committee on Earthquake Engineering and has contributed to the policy making by way of its representation on various committees of Planning Commission, DST, UGC, INSA, AICTE, BIS, etc. The Society has contributed to the formulation of various Indian Standard Codes of Practice related to earthquake engineering. The President of ISET is represented on the relevant committees responsible for drafting these codes.

ISET has been continuously alive to the changing needs of the community around and its members have been appropriately formulating its policies and action plans. The introduction of earthquake engineering in undergraduate syllabus, training of people in good construction practice evolving safe and economical design for mass housing, preparation of monographs on topics of current national importance and involvement of construction industry in its activities may be notable aspects needing greater attention in the near future.

The Society has succeeded in creating interest among the engineers and scientists community in better understanding of the science of earthquake phenomena, earthquake resistant design, vulnerability assessment and earthquake risk reduction. I am sure due to popularization of earthquake technology and earthquake awareness, the future earthquakes will cause less and less damage than they would have caused had this forum not come into being. It is my hope that more and more engineers will put into practice the new knowledge that is being generated in this field in India and elsewhere. The relevance of the Society will be appreciated increasingly with time among different disciplines directly or indirectly related with earthquake technology.

It is a matter of great pleasure that the Society has blossomed through the first 50 years and I do hope it will grow further at an accelerated rate and its activities shall be much more meaningful to the needs of the general public in the coming years.

11 Annexure I

List of Officers and Executive Committee Members of ISET

Term	President	Vice president	Secretary	Editor	Members
1962–1965	Jai Krishna	Tandon, A.N.	Chandrasekaran, A.R.	Shamsher Prakash	Gupta, S.K., Joshi, R.N., Kapila, I.P., Krishna-Swamy, V.S., Mithal, R.S., Patel, V.J., and Visvesvaraya, H.C.
1965–1966	Jai Krishna	Krishnaswamy, V.S.	Arya, A.S.	Agarwal, S.L.	Chandrasekaran, A.R., Hari Narain, Shamsher Prakash, Tandon, A.N., and Visvesvaraya, H.C. Indian Institute of Technology Delhi, Regional Research Laboratory, Jorhat, Assam
1966–1967	Jai Krishna	Krishnaswamy, V.S.	Arya, A.S.	Agarwal, S.L. (till Aug. 1966) Srivastava, L.S. (from Sept. 1966 onwards)	Chandrasekaran, A.R., Hari Narain, Mathur, B.C., Shamsher Prakash, Tandon, A.N., and Visvesvaraya, H.C. Indian Institute of Technology Delhi, Regional Research Laboratory, Jorhat, Assam
1967–1968	Krishnaswamy, V.S.	Mithal, R.S.	Arya, A.S.	Srivastava, L.S.	Chandrasekaran, A.R., Chetty, S.M.K., Gupta, Y.P., Narain, J., Shamsher Prakash, and Sandhawaliya, P.S. Sahu Cement Service, New Delhi Gammon India Ltd., Bombay
1968–1969	Tandon, A.N.	Jai Krishna	Shamsher Prakash	Srivastava, L.S.	Arya, A.S., Chandrasekaran, A.R., Joshi, R.N., Krishnaswamy, V.S., Mithal, R.S., and Jagdish Narain Sahu Cement Service, New Delhi Bhakra and Beas Designs Orgn., New Delhi
1969–1970	Mane, P.M.	Jai Krishna	Shamsher Prakash	Srivastava, L.S.	Arya, A.S., Chandrasekaran, A.R., Mithal, R.S., Hari Narain, Jagdish Narain, and Tandon, A.N. Central Board of Irrigation and Power, New Delhi Army Head Quarters, Engineer-in-chief's Branch, New Delhi
1970–1971	Murti, N.G.K.	Jai Krishna	Srivastava, L.S. (up to Aug. 1970) Shamsher Prakash (Sept. 1970 onwards)	Chandrasekaran, A.R.	Arya, A.S., Bodhe, J.G., Chandra, Satish, Das, V.C., Rao, V.V.S., and Sandhawalia, P.S. Central Water and Power Commission, New Delhi Hindustan Construction Company, Bombay

(continued)

Term	President	Vice president	Secretary	Editor	Members
1971–1972	Murti, N.G.K.	Garg, S.P.	Shamsher Prakash	Chandrasekaran, A.R.	Agrawal, P.N., Arya, A.S., Das, V.C., Desai, G.N., Chandra, Satish, and Mallick, D.V. Central Board of Irrigation and Power, New Delhi Central Water and Power Commission, New Delhi
1972–1973	Bodhe, J.G.	Arya, A.S.	Srivastava, L.S.	Chandrasekaran, A.R.	Brijesh Chandra, Das, Y.C., Guha, S.K., Iyeer, I.S. R., Gopal Ranjan, and Shamsher Prakash Central Water and Power Commission, New Delhi India Meteorological Dept., New Delhi
1973–1974	Bodhe, J.G.	Arya, A.S.	Mathur, B.C.	Thakkar, S.K.	Guha, S.K., Iyenger, K.T.S., Nayak, G.C., Nigam, N.C., Ramesh, C.K., and Saini, S.S. Central Designs Organisation, Bombay Central Water and Power Commission, New Delhi
1974–1975	Murti, N.G.K.	Arya, A.S.	Mathur, B.C.	Thakkar, S.K.	Anand Prakash, Bodhe, J.G., Chandrasekaran, A.R., Nayak, G.C., Saini, S.S., and Shamsher Prakash Central Designs Organisation, Bombay Indian Meteorological Dept., New Delhi
1975–1976	Murthy, Y.K.	Chandrasekaran, A.R.	Mathur, B.C.	Thakkar, S.K.	Anand Prakash, Gopal Ranjan, Nandakumaran, P., Shamsher Prakash, and Srivastava, L.S. Central Designs Organisation, Bombay Hindustan Construction Company, Bombay Central Public Works Department, New Delhi
1976–1977	Murthy, Y.K.	Arya, A.S.	Nandakumaran, P.	Singh, V.N.	Agarwal, P.N., Agrawal, S.K., Bakare, V.D., Gopal Ranjan, Nene, R.L., and Sharda, S.C. Indian Meteorological Department, New Delhi Department of Atomic Energy, Bombay
1977–1978	Vaish, V.R.	Arya, A.S.	Nandakumaran, P.	Singh, V.N.	Agarwal, P.N., Agrawal, S.K., Bhatia, K.G., Gopal Ranjan, Radhakrishnan, R., Ramachandran, V., Sharda, S.C., and Sen Sharma, S.B. Central Building Research Institute, Roorkee

1978–1979	Vaish, V.R.	Arya, A.S.	Nandakumaran, P.	Singh, V.N.	Agarwal, P.N., Agrawal, S.K., Bhatia, K.G., Gopal Ranjan, Radhakrishnan, R., Ramachandran, V., and Sharda, S.C. Central Building Research Institute, Roorkee Central Water Commission, New Delhi
1979–1981	Arya, A.S.	Bhatia, K.G.	Thakkar, S.K.	Agrawal, R.C.	Bakare, V.D., Bodhe, J.G., Gupta, M.K., Gupta, S.P., Harkauli, A.N., Murthy, A.V.S.S., Sivaram, B., Swami, M.L., and Vaish, V.R. Geological Survey of India, New Delhi India Meteorological Department, New Delhi Central Water Commission, New Delhi
1981–1983	Arya, A.S.	Nigam, N.C.	Thakkar, S.K.	Agrawal, R.C.	Bakare, V.D., Bapat, A., Chandrasekaran, A.R., Desai, G.N., Gupta, A.K., Gupta, V.K., and Vaish, V.R. India Meteorological Department, New Delhi Geological Survey of India, New Delhi Central Water Commission, New Delhi Central Design Organisation, Nasik, Maharashtra
1983–1985	Singh, Pritam	Agrawal, P.N.	Gupta, Satyendra, P.	Joshi, V.H. (up to March 1984) Lavania, B.V.K. (March 1984 onwards)	Arya, A.S., Gupta, Arvind Kumar, Gupta, Y.P., Joshi, V.H., Krishna, Prem, Mukhopadhyay, Arbinda, Munirudrappa, N., Wason, H.R., and Khurana, K.K. Central Building Research Institute, Roorkee Bharat Heavy Electrical Ltd., Hyderabad
1985–1987	Shamsher Prakash	Krishna Prem	Gupta, M.K.	Wason, H.R.	Agrawal, R.C., Bapat, A., Kaushik, S.K., Munirudrappa, N., Singh, Pritam, and Singh, V.N. Department of Earthquake Engg., UOR, Roorkee Nuclear Power Board, Dept. of Atomic Energy Govt. of India, Colaba, Bombay India Meteorological Department New Delhi, NEEPC, Shillong

(continued)

Term	President	Vice president	Secretary	Editor	Members
1987–1989	Ramachandran, V.	Joshi, V.H.	Agarwal, R.C. (up to 1989) Gupta, M.K. (onwards)	Dubey, A.K	Basavanna, B.M., Gupta, M.K., Gupta, V.K., Jain, A.K., Kaushik, S.K., Padale, J.G., Paul, D.K., and Thakkar, S.K. Central Building Research Institute, Roorkee National Thermal Power Corporation, New Delhi India Meteorological Department, New Delhi Central Water Commission, New Delhi Department of Earthquake Engg., UOR, Roorkee Geological Survey of India, Calcutta
1989–1991	Chummar, A.V.	Kaushik, S.K.	Gupta, V.K.	Pandey, A.D.	Chandra Prakash, Gupta, M.K., Jain, P.K., Khare, P.S., Pachauri, A.K., Paul, D.K., and Thakkar, S.K. Irrigation Research Institute, Roorkee Jorhat Engineering College, Jorhat, Assam India Meteorological Department, New Delhi Central Water Commission, New Delhi Department of Earthquake Engg., UOR, Roorkee Geological Survey of India, Calcutta
1991–1993	Agrawal, P.N.	Kaushik, S.K.	Gupta, V.K.	Dr. Pankaj	Gupta, M.K., Kumar, Ashok, Dube, R.N., Jain, A.K., Mittal, Satyendra, Wason, H.R., Bose, P., Rastogi, B.K., and Kalita, U.C. Department of Earthquake Engg., UOR, Roorkee National Building Organisation, New Delhi Bharat Heavy Electricals Ltd., Hyderabad Department of Atomic Energy, Bombay Central Water Commission, New Delhi India Meteorological Department, New Delhi Geological Survey of India, Dehradun Chairmen of all Local Chapters
1993–1995	Chandrasekaran, A.R.	Gupta, M.K.	Kumar, Ashok	Dr. Pankaj	Gupta, S.P., Jain, P.K., Kaushik, S.K., Mittal, Satyendra, Prakash, Chandra, Saxena, N.K., Sharma, Gokul and Singh, Amar Irrigation Research Institute, Roorkee

					Central Building Research Institute, Roorkee Nuclear Power Corporation, Bombay Immediate Past Secretary India Meteorological Department, New Delhi Central Water Commission, New Delhi Geological Survey of India, Calcutta Department of Earthquake Engg., UOR, Roorkee
1995–1997	Bhatia, K.G.	Paul, D.K.	Dr. Pankaj	Sarkar, I.	Bajaj, R.S., Basu, S., Chandra, Brijesh, Gupta, M.K., Kumar, Ashwani, Munirudrappa, N., Singh, A.K., and Sinvhal, A. Development Consultants Limited, Calcutta Maharashtra Engg. Research Institute, Nasik Nuclear Power Corporation, Bombay India Meteorological Department, New Delhi Central Water Commission, New Delhi Department of Earthquake Engg., UOR, Roorkee Geological Survey of India, Calcutta Chairmen of all Local Chapters
1997–1999	Nigam, N.C.	Paul, D.K.	Wason, H.R.	Gupta, Vinay K.	Bose, P.R., Gupta, Vinay K., Gupta, M.K., Gupta, V.K., Gupta, Sushil, Kumar, Ashok, Sinha, Ravi, Singh, A.K., and Thakkar, S.K. Central Building Research Institute, Roorkee Corporate R&D, BHEL, Hyderabad Nuclear Power Corporation, Colaba, Bombay Metallurgical & Engg. Consultants, Ranchi Regional Research Lab., Jorhat, Assam India Meteorological Department, New Delhi Central Water Commission, New Delhi Department of Earthquake Engineering, UOR, Roorkee Immediate Past Secretary, ISET Geological Survey of India, Calcutta

(continued)

Term	President	Vice president	Secretary	Editor	Members
1999–2001	Joshi, V.H.	Gupta, Y.P.	Wason, H.R.	Gupta, Vinay K.	Garg, K.G., Gupta, V.K., Ingle, R.K., Mittal, Satyendra, Mukerjee, Shayamal, Pachauri, A.K., Prakash, Vipul, Saxena, N.K., Sinha, Ravi, and Suryawanshi, C.S. Ministry of Railways, New Delhi Narmada Project Designs (Dam & Power House), Vadodara H.P., P.W.D., Shimla North Eastern Electric Power Corp. Ltd., Shillong Geology & Mining Industries Dept., Mizoram India Meteorological Department, New Delhi Central Water Commission, New Delhi Department of Earthquake Engineering, UOR, Roorkee Immediate Past Secretary, ISET Geological Survey of India, Calcutta
2001–2003	Joshi, V.H.	Gupta, M.K.	Prakash, Vipul	Gupta, Vinay K.	Bansal, Brijesh K., Bose, P.R., Garg, K.G., Mathur, Ashok K., Mittal, Satyendra, Mukerjee, Shyamal, Ramasamy, G., Sharma, M.L., and Sinhval, Amita Ministry of Railways, New Delhi Narmada Project Designs (Dam & Power House), Vadodara H.P., P.W.D., Shimla North Eastern Electric Power Corp. Ltd., Shillong Geology & Mining Industries Dept., Mizoram India Meteorological Department, New Delhi Central Water Commission, New Delhi Department of Earthquake Engineering, IITR Immediate Past Secretary, ISET Geological Survey of India, Calcutta

2003–2005	Thakkar, S.K.	Gupta, Vinay K.	Prakash, Vipul	Gupta, Vinay K.	Agarwal, Shailesh K., Bhatnagar, N.K., Goyal, Alok, Gupta, Sushil, Gupta, V.K., Mukerjee, Shyamal, Rai, Durgesh C., and Singh, D.P. Development Consultants Pvt. Ltd., Kolkata Ministry of Railways, New Delhi India Meteorological Department, New Delhi Central Water Commission, New Delhi Department of Earthquake Engineering, IITR Immediate Past Secretary, ISET Geological Survey of India, Kolkata
2005–2007	Thakkar, S.K.	Gupta, Vinay K.	Singhal, N.C.	Gupta, Vinay K.	Manohar, C.S., Saxena, A.K., Chakraborty Subrata, Saxena, N.K., Naithani, K.C., Aterkar Sandesh, Bhatnagar, N.K., and Muthumani, K.M. International Institute of Information Technology, Hyderabad Tandon Consultants Private Limited, New Delhi India Meteorological Department, New Delhi Central Water Commission, New Delhi Department of Earthquake Engineering, IITR Immediate Past Secretary, ISET Geological Survey of India, Kolkata
2007–2009	Paul, D.K.	Wason, H.R.	Maheshwari, B.K.	Gupta, Vinay K.	Dange, A.P., Gupta, Sushil, Gupta, Vinay K., Joshi, G.C., Naithani, K.C., Ramana, G.V., Sankarasubramanian, G., Sharma, M.L., and Shrikhande, M. Govt. College of Engineering, Amravati Nuclear Power Corporation, Mumbai Geological Survey of India, Kolkata India Meteorological Department, New Delhi Central Water Commission, New Delhi Department of Earthquake Engineering, IITR Immediate Past Secretary, ISET

(continued)

Term	President	Vice president	Secretary	Editor	Members
2009–2011	Paul, D.K.	Wason, H.R.	Maheshwari, B.K.	Gupta, Vinay K.	Gupta, Sushil, Gupta, Vinay, K., Gupta, Y.K., Hegde, M.N., Joshi, G.C., Naithani, K.C., Patil, S.U., and Singh, S.K. Disaster Mitigation & Management Centre, Dehradun European Centre for Training and Research in Earthquake Engineering, Pavia-Italy Geological Survey of India, Kolkata Indian Meteorological Department, New Delhi, Central Water Commission, New Delhi Department of Earthquake Engineering, IITR Immediate Past Secretary, ISET
2011–2013	Wason, H.R.	Sharma, M.L.	Shrikhande, Manish	Gupta, Vinay K	Gupta, Y.K., Hira, B.N., Khan, P.K., Khose, V.N., Kumar, Ratnesh, Pandey, A.K., and Sankarasubramanian, G. Geological Survey of India, Kolkata India Meteorological Department, New Delhi, Central Water Commission, New Delhi Department of Earthquake Engineering, IITR Immediate Past Secretary, ISET

12 Annexure II

List of Chairmen and Secretaries of Local Chapters

Chapter	Chairman	Secretary
Delhi	Prof. Khalid Moin Chairman, ISET Delhi Chapter Department of Civil Engg. Faculty of Engineering & Tech. Jamia Millia Islamia, Jamia Nagar New Delhi—110025	Dr. Rehan Ahmad Khan Secretary, ISET Delhi Chapter Department of Civil Engg. F/O Engg. and Technology Jamia Millia Islamia New Delhi—110025
Mumbai	Sri S.P. Bagli Vice President (Business Development) Chairman, ISET Mumbai Chapter Shapoorji Pallonji & Co. Ltd. SP Centre, 41/44, Minoo Desai Marg Colaba, Mumbai—400005	Sri Arun Samota Dy. Chief Engineer (Business Development) Secretary, ISET Mumbai Chapter Nuclear Power Corporation of India Ltd. A-1 Block, Nabhikiya Urja Bhavan Anushakti Nagar, Mumbai—400094
Kolkata	*VACANT*	Mr. Archan Kumar Sarkar Secretary, ISET Kolkata Chapter Fd-61/4, Sali Lake City Kolkata—700091
Chennai	Dr. R. Radhakrishan Chairman, ISET Chennai Chapter Retd. Prof. of Structural Engineering Indian Institute of Technology Chennai—600036 (*Present address is not known*)	Dr. A.R. Shanthakumar Secretary, ISET Chennai Chapter 22, Prakash Mudali Street, T. Nagar Chennai—600017
Roorkee	Prof. Krishen Kumar Chairman, ISET Roorkee Chapter Flat 3b, Block 2, Janues Residences 1, Moti Lal Gupta Road Barisha, Calcutta—700008, West Bengal	Dr. M.L. Sharma Professor Secretary, ISET Roorkee Chapter Department of Earthquake Engg. IIT Roorkee, Roorkee—247667
Bangalore	Prof. V. Devraj Chairman, ISET Bangalore Chapter Faculty of Engineering-Civil, UVCE Bangalore University, Jnana Bharathi Bangalore—560056	Dr. Sunil Mohana Murthy #136, 2nd Floor, 5th Main Hanumanthnagar Bengaluru—560019
Jorhat	Mr. R.A. Swamy Chairman, ISET Jorhat Chapter Civil Engineering Department Jorhat Engineering College Jorhat—785001	Mr. Bhagwati Prasad Malpani Secretary, ISET Jorhat Chapter Marwari Petty Jorhat—785001 (Assam)
Pune	Er. S.G. Shirke Chartered Engineer Director General, Walmi (Retd.) Chairman, ISET Pune Chapter 'Kaumudi', Dnyaneshwar Society Pune—411009	Mr. J.G. Padale Senior Research Officer CWPRS (Retd.) Secretary, ISET Pune Chapter 810, Second Floor, Bhawani Peth Pune—411042

(continued)

Chapter	Chairman	Secretary
Amravati	Prof. Atmaram P. Dange Chairman, ISET Amravati Chapter Narhari Nagar Rajapeth, Amravati—444606	Er. Pravin Vyankatrao Khandve Secretary, ISET Amravati Chapter Ravi Nagar Amravati Behind Hanuman Temple Opposite Mahajan House AT.PO. TQ. Dist-Amravati—444605
Nagpur	Dr. S.S. Kulkarni Chairman, ISET Nagpur Chapter 26, Visvesvaraya National Institute of Technology, Nagpur—440011 (*Present address is not known*)	Mr. R.K. Ingle Secretary, ISET Nagpur Chapter Department of Applied Mechanics Visvesvaraya National Institute of Technology Nagpur—440011
Guwahati	Prof. Gokul Sharma Chairman, ISET Guwahati Chapter C/O Hudco Regional Office Guwahati Opp. P.N. Bank, R.G. Barua Road Ganeshguri, Guwahati—781005, Assam	Er. Jyotirindra N. Khataniar Consultant Engineer Secretary, ISET Guwahati Chapter Srishtie, 5th Floor Eureka Tower Chandmari, Guwahati—781003, Assam

13 Annexure III

List of ISET Annual Lectures and Eminent Speakers

Sl. no.	Year	Speaker	Topic
1.	1979	V.S. Krishnaswamy	A message
2.	1981	Y.K. Murthy	Mitigation from earthquake hazards from dam failures
3.	1982	Jai Krishna	Earthquake disasters
4.	1983	H.B. Seed	Recent developments in the evaluation of soil liquefaction
5.	1984	A.S. Arya	Earthquake disaster mitigation programme
6.	1985	B.M. Hukku	On seismic and neotectonic activity around river valley projects
7.	1986	A.K. Chopra	Earthquake analysis design and safety of gravity dams
8.	1987	Shamsher Prakash	Analysis and design of pile foundations under dynamic load
9.	1988	A.R. Chandrasekaran	Strong motion array and ground motion characteristics of Shillong Region
10.	1989	Haresh C. Shah	Recent developments in seismic risk analysis challenges for the coming international decade for natural hazard reduction
11.	1990	R.D. Hanson	Supplemental damping for improved seismic performance of building
12.	1991	N.C. Nigam	Probabilistic approach to earthquake engineering emergence of a practical theory
13.	1992	Harsh K. Gupta	Artificial water reservoir and earthquake

(continued)

Sl. no.	Year	Speaker	Topic
14.	1993	P. Bormann	Seismotectonics and seismic hazard of Germany in the European context
15.	1994	M.P. Singh	Structural control for seismic protection of structures
16.	1995	Anil Kakodkar	Some issue in a seismic design of nuclear structures and components
17.	1996	H.N. Srivastava	Recent developments in seismicity associated with reservoirs
18.	1997	R.N. Iyengar	Strong motion: analysis, modeling and applications
19.	1998	B.L. Jatana	Fail safe large dams in earthquake prone Himalayan region
20.	2000	V.S. Ramamurthy	Earthquake technology in the new millennium
21.	2001	Ravi Shanker	Seismotectonics of Kutch Rift Basin and its bearing on the Himalayan seismicity
22.	2002	T.K. Datta	A state-of-the-art review on active control of structures
23.	2003	M.D. Trifunac	70th anniversary of biot spectrum
24.	2004	Bruce A. Bolt	Seismic input motions for nonlinear structural analysis
25.	2005	Mahesh Tandon	Economical design of earthquake resistant bridges
26.	2006	N. Lakshmanan	Seismic evaluation and retrofitting of buildings and structures
27.	2006	Ahmed Elgamal	Nonlinear modeling of large-scale ground-foundation-structure seismic response
28.	2008	K.G. Bhatia	Foundations for industrial machines and earthquake effects
29.	2009	S.K. Singh	Estimation of earthquake ground motion in Mexico City and Delhi, two mega cities
30.	2010	K.V. Subramanian	Evolution of seismic design of structures, systems and components of nuclear power plants
31.	2010	W.D. Liam Finn	Evolution in selection and application of ground motions for design
32.	2012	Shailesh Nayak	Tsunami warning system

Contents

Part I
Engineering Seismology

Chapter 1
Engineering Seismology

M. L. Sharma

1 Introduction

Seismology is the branch of geophysics which deals with the occurrence of earthquakes and related phenomena on the planet earth. Seismology, as a science of earthquakes, includes the direct and inverse problems dealing with earthquake source, medium characteristics, data recording and interpretation of data. The outcome is the modeling and source characterization in terms of its location, geometry and potential, characterization of medium in terms of geometry, attenuation and other rheological properties and quantification of size of earthquakes in terms of magnitude and energy, modeling earthquake occurrence and study the earth structure etc. Seismology, in general, can be divided into four sections namely: observational seismology, theoretical seismology, strong motion seismology and engineering seismology.

Observational seismology deals with earthquake recording, interpretation of the records and analysis of the data obtained. This part includes the theory of sensors, seismographs, installation requirements, data acquisition and processing of data. Various types of arrays varying in time (long and short durations) and space (worldwide, regional and local) with a large range of instrumentation (broadband to short period seismometers, accelerographs, tiltmeters, strain meters etc.) and specific objectives come under this portion of seismology.

Theoretical seismology consists of mathematical treatment of the problem of wave propagation, medium characteristics and source modeling. Theoretical seismology is based on the principles of the mechanics of continuous media. Earth is assumed to be an imperfectly elastic, heterogeneous, anisotropic body of approximate elliptical shape in which vibrations are produced by earthquakes for which

M. L. Sharma (✉)
Department of Earthquake Engineering, IIT Roorkee, Roorkee, Uttarakhand, India

M. L. Sharma et al. (eds.), *Advances in Indian Earthquake Engineering and Seismology*, https://doi.org/10.1007/978-3-319-76855-7_1

mathematical analysis is carried out seeking adequate solutions for the wave equation.

Strong motion seismology deals with the measurement, processing, interpretation and estimation of strong ground motion generated by potentially damaging earthquakes. The data recorded are essential to understand the high frequency nature of seismogenic failure process, the nature of radiation from the source, its propagation through the medium in near field conditions, all of which have a first order effect on the seismic loading applied to the physical environment. The main objective of this branch of seismology is to improve the scientific understanding of the physical process that controls strong shaking and to develop reliable estimates of strong ground motion for the reduction of loss of life and property during future probable earthquakes through earthquake resistant design and retrofitting.

Engineering seismology, in addition to being part of seismology, is an integral part of earthquake engineering, a specialized branch of civil engineering (structural engineering in particular) concerned with the protection of the built environment against the potentially destructive effects of earthquakes. Engineering seismology bridges the gap between seismology and earthquake engineering by providing the inputs to an earthquake engineer for earthquake resistant designs. Starting from the knowledge of the earthquake source, its physical processes, wave propagation in the media and its characteristics, instrumentation and records, their processing and interpretation and size estimation are generally used in engineering seismology to generate intermediate results like seismicity and its temporal and spatial variation, seismotectonic modeling, attenuation characteristics, etc. (Sharma et al., 2009; Sharma, 1998). In turn, these intermediate results are then used to estimate the seismic hazard in terms of peak values of acceleration/velocity/displacement or the whole response spectra due to a seismogenic source (scenario earthquake or deterministic methodology) or combination of sources (probabilistic approach). The generation of synthetic strong motion time history using recorded motion and/or spectrum compatible time histories is the next step in engineering seismology which is further used in earthquake engineering for earthquake resistant designs. Since the prediction of ground motion at a site involve a complex phenomenon of earthquake source mechanism, wave propagation, medium characteristics, instrumentation and local site effects, it is necessary to have a sound knowledge of the other branches of seismology as described above to predict complete and realistic ground motion.

Once planners and developers have taken decisions regarding the location of civil engineering projects, or once people have begun to settle in an area, the level of exposure to natural hazards specially earthquakes is determined. In general, seismic hazard cannot be altered, hence the key to mitigating seismic risk levels lies in the reduction of vulnerability, or stated another way, through the provision of earthquake resistance. The civil engineer requires quantitative information on the nature and likelihood of the expected earthquake hazards in order to provide effective earthquake protection. This has to be accomplished by defining the hazard not in terms of the effects a single earthquake event may produce at the site, but a synthesis of the potential effects of many possible earthquake scenarios and quantitative definitions of the particular effects that may be expected to occur with different specified frequencies. Thus, the essence of Engineering Seismology may be to provide quantitative assessments of earthquake hazards. When such exercise is carried out for an area of interest and micro zones of specific ground motion are prepared, it leads to seismic microzonation. The present chapter deals with various aspects in engineering seismology, the evolution of methodologies and the state of the art techniques used to accomplish various tasks in engineering seismology. It is not possible to look into every aspect of engineering seismology but some of the important ones have been discussed in the present chapter.

2 Engineering Seismology-Basics

Earthquakes are caused by sudden rupture on geological faults, the slip on the fault rupture ranging from a few centimeters for moderate to many meters for large events. The energy radiated from earthquake occurrence creates a number of effects that are potentially threatening to the built environment and society at large. Seismic risk is the resultant of the convolution of three factors: seismic hazard, exposure and vulnerability. Seismic hazard refers to the effects of earthquakes that can cause damage in the built environment, such as the primary effects of ground shaking or ground rupture or secondary effects such as soil liquefaction or landslides. Exposure refers to the population, buildings, installations and infrastructure encountered at the location where earthquake effects could occur. Vulnerability represents the likelihood of damage being sustained by a structure when exposed to a particular earthquake effect. Such effects are termed as seismic hazard and Engineering Seismology, in the broadest sense, is concerned with assessing, the likelihood and characteristics of these hazards and their possible impact in a given region (seismic microzonation) or at a given site of interest.

In most cases the hazard is directly related to the shaking of the ground caused by the passage of seismic waves. The main focus of engineering seismology is the assessment of the hazard of ground shaking along with the primary effects which are due to the source itself. The subject as such covers the whole spectrum of studies starting from the sources – the cause, the medium and the recordings to the interpretation and direct and indirect problems and their solutions. Further, the amplification of ground motion, an important subject in engineering seismology, deals with earthquake ground motion that can be amplified by local site effects. Topographic features such as ridges and/or basins can cause amplification of the shaking and soft soil deposits tend to increase the amplitude of the shaking with respect to rock sites. The basin edge effect in amplifying the ground motion has been experienced in past many earthquakes. The shaking can induce secondary geotechnical hazards by causing failure of the ground. In mountainous or hilly areas, earthquakes frequently trigger landslides, which can significantly compound the losses. In areas where saturated sandy soils are encountered, the ground shaking can induce liquefaction through the generation of high pore water pressures leading to reduced effective stress and a significant loss of shear strength, which in turn leads to the sinking of buildings into the ground and lateral spreading on river banks and along coasts. The assessment of landslide and liquefaction hazard involves evaluating the susceptibility of slopes and soil deposits and determining the expected level of earthquake ground motion.

One of the important secondary effects include tsunamis which are caused due to surface fault ruptures in the ocean floor and seismic sea waves are generated by the sudden displacement of the surface of the sea that travel with very high speeds. As the waves approach the shore and the water depth decreases, the amplitude of the waves increases to maintain the momentum, reaching heights of up to 30 m. As the waves impact on low lying coastal areas, the destruction can be almost total. Seismic hazard assessment in terms of strong ground- motion is the activity that defines objectives for the most part of Engineering Seismology in addition to many of the specific topics which are sometimes closer to science i.e., seismology and on the other hand engineering, i.e., earthquake engineering.

3 Seismic Hazard Assessment

The Seismic Hazard Assessment (SHA) refers to estimation of some measure of the strong earthquake ground motion expected to occur at a selected site which is necessarily used for the purpose of evolving earthquake resistant design of a new structure or for estimating the safety of an existing structure of importance at that site. It is used to describe the severity of ground motion at a site (Anderson and Trifunac, 1978), regardless of the consequences, while the risk refers to the consequences (Jordanovski et al., 1991). Two basic methodologies used for SHA are the Deterministic Seismic Hazard Assessment (DSHA) and the Probabilistic Seismic Hazard Assessment (PSHA) approaches.

There are few basic steps required in seismic hazard assessment (both DSHA and PSHA). The earthquake catalogues which is the basic input to both approaches has to be treated for homogenization of earthquake size (Das et al., 2012a), completeness in size, time and space (Sharma, 2003), declustring of events for making the data mutually exclusive and collectively exhaustive (Sharma and Lindolhm, 2012), identification of seismogenic sources by seismotectonic modeling of the region around the site of interest and estimating the strong ground motion using empirical/semi empirical/theoretical and/or analytical modeling. Moment magnitude is the most appropriate measure of earthquake size because it is directly related to the seismic moment of the earthquake, and thus to the fault area and the average displacement on the fault. When earthquake occurrence rates are instead based on historical seismicity in seismic hazard analyses, other magnitude measures are sometimes used because the seismic moments of the historical earthquakes may be poorly known. In such cases, empirical equations relating other measures of size of earthquakes to the moment magnitude are used (Das et al., 2012b; Wason et al., 2012).

Various models are used to describe the size and the temporal variations of earthquake occurrence. For size mainly the Gutenberg-Richter (G-R) relationship is considered along with some other models like different form of G-R relationship or power laws and the characteristic earthquake model. The most commonly used model which provides a simple framework for evaluating probabilities of events is described by a Poisson model. The Poisson process yields values of random variables describing the number of occurrence of a particular event during a given time interval and/or in a specified spatial region. A number of models that account for prior seismicity have also been proposed in the past (Anagnos and Kiremidjian, 1988). Nonhomogeneous Poisson models (e.g., Vere-Jones and Ozaki, 1982) allow the annual rates of exceedance to vary with time. Renewal models (Esteva, 1970; Hagiwara, 1974; Savy et al., 1980; Kiremidjian and Anagnos, 1984; Cornell and Winterstein, 1988) use arrival-time distributions other than exponential to allow the hazard rate to increase with time since the last event; Gamma and Weibull distributions are the most common. Time- predictable models specify a distribution of the time to the next earthquake that depends on the magnitude of the most recent earthquake; slip-predictable models consider the distribution of earthquake magnitude to depend on the time since the most recent earthquake. Markov models incorporate a type of memory that describes the chances that a process moves from some past "state" to a particular future "state." The time for which the process stays in a particular "state" before moving to another "state" is exponentially distributed. Semi-Markov models are not restricted to exponential distribution. Both Markov models (Vere-Jones and Devies, 1966; Vagliente, 1973; Veneziano and Cornell, 1974; Nishioka and Shah, 1980) and semi-Markov models (Patwardhan et al., 1980; Cluff et al., 1980; Coppersmith, 1981; Guagenti-Grandori and Molina, 1984) have been used in seismic hazard analysis. Extreme value statistics are sometimes used when the data on lower magnitudes is not available from a region. The seismic hazard is estimated using Gumbel's distribution in such cases.

A primary issue of critical relevance for the seismic hazard assessment is the choice between a deterministic and a probabilistic approach to the representation of ground shaking and to damage estimation. For ease of reference, the two approaches are briefly outlined and discussed below.

4 Deterministic Seismic Hazard Assessment

The deterministic approach develops a particular earthquake scenario upon which a ground-motion hazard evaluation is based. The scenario consists of the postulated occurrence of an earthquake of a specified size occurring at a specified location, typically a seismically active fault. The standard deterministic approach consists of the first step as the identification of the locations and characteristics of all significant earthquake sources that might affect the site/zone of interest. The seismic potential is quantified by assigning to each source a significant earthquake that can be the maximum historical event known from that source (Maximum Probable Earthquake, or MPE), or the maximum earthquake that appears capable of occurring under the known tectonic framework (Maximum Credible Earthquake, or MCE). The shortest distance from the source to the sites in the zone of interest is generally assumed, in the absence of other information. Many definitions of the distance measures from source to site exist. The second step in DSHA consists of the selection of the attenuation relations enabling one to estimate the ground motion within the zone of interest as a function of earthquake magnitude, source-to-site distance, and local ground conditions (Sharma, 1998). The final step is the definition of the controlling earthquake (i.e., the earthquake that is expected to produce the strongest level of shaking). The selection is made comparing, through the attenuation relation, the effects produced at the site by the combination of earthquake distance and magnitude. The resultant earthquake will define the seismic scenario.

However, depending on the historical seismicity of the region, variations with respect to this standard procedure are possible. Thus, it may be desirable to adopt more than one scenario earthquake, for instance, a destructive event (such as the MPE) and a less severe, but damaging one with a higher likelihood of occurrence. In earthquake engineering generally the MCE and the DBE (Design Basis Earthquake) are preferred. At the end of the process, the deterministic analysis gives as a result one or more scenarios that represent the worst situations expected, but it does not provide any information about their occurrence in time. The deterministic approach proposes design for the so-called "scenario" earthquakes (MCE). The underlying philosophy is that "scenario" earthquake of the seismic source is scientifically reasonable and is expected to produce the most severe strong ground motion at the site. One of the most important issues is the reliable estimation of the MCE, for each of the identified seismic source zones, on the basis of the available data on past earthquakes and the seismotectonic and geological features of source zones. However, the delineation of the source zones, the estimation of MCE magnitudes, as well as the estimation of the corresponding ground motion at a certain distance from the earthquake source, are commonly associated with large uncertainties.

5 Probabilistic Seismic Hazard Assessment

The probabilistic approach is a typical application of Probabilistic Seismic Hazard Analysis (PSHA), whose basic features have been established by Cornell (1968). It consists of four basic steps, some of which are partly coincident with those of the deterministic approach, namely (i) Seismic source zone identification, (ii) Probabilistic characterization of source zone activity, (iii) Selection of attenuation relation, and (iv) Integration over the whole range of magnitudes and distances for each source zone in order to obtain, for each site of interest, the probabilistic hazard values in the form of a cumulative distribution for the ground-motion parameters of interest. Many seismic hazard studies carried out report various variations in the methodology of seismic hazard assessment (Herbindu et al.,

2012a, 2012b; Wason et al., 2012; Joshi et al., 2012; Tripathi et al., 2012; Sharma et al., 2009; Joshi and Sharma, 2008; Sharma, 1998, 2003; Joshi, et al. 2012).

A certain degree of combination of the deterministic and probabilistic approaches is also conceivable (CWC guidelines, 2011). Both the approaches have their own advantages and disadvantages. Due to uncertainties involved and the assumptions made at almost every step of the approaches, it is advisable to use both the approaches conservatively to finalize the strong ground motion to be recommended for a site. For instance, in a zone with a well-documented history of damaging earthquakes one may choose the deterministic scenario, paying some consideration to the relative frequency of past events rather than to the MPE or MCE events. The 50 year reference term can be roughly regarded as the expected engineering lifetime of an ordinary building (this life time may increase to about 100 years for a bridge or tunnel and to 100–150 years for a dam). Since probabilistic seismic hazard studies cover all possible damaging events in a region and estimate the cumulative effect they can help in making an objective pre-selection of the basic features of the scenario earthquake. For instance, the latter could be the event that occurs on the seismogenic structure (fault or area source) that contributes most to the hazard at the chosen return period, and has the magnitude and location contributing most. Such exercise in PSHA is termed as deaggregation. Probabilistic ground-motion maps are normally developed on a regional basis for a single "reference" class of ground conditions, e.g., soft rock. The ground motion is then modified according to the site conditions given in form of soil types where frequency dependent factors are used as multiplying factors. Sometimes the attenuation relationships used contain the local site effect in form of shear wave velocities to be provided as the representation of the sites.

While the two approaches generally point to different requirements on the side of risk analysts and managers, the deterministic approach is more intuitive and may be more appealing to local administrators because it assumes the occurrence of a well-defined earthquake (in terms of magnitude and source-to-site distance) that is used as input for all the successive steps of the risk analysis.

The final results of DSHA and/or PSHA define the seismic hazard at the site of interest. Logic trees provides a convenient form for formal and quantitative treatment of uncertainties. Various models being used for estimation of intermediate results, logic trees are used to combine all these branches to give a recommended strong ground motion for the site. Probabilities that represent the relative likelihood or degree of belief that the branch represents the correct value or state of the input parameter are assigned to each branch of the logic tree. These probabilities are assessed conditionally on the assumption that all the branches leading to that node represent the true state of the preceding parameters. Because they are conditional probabilities for an assumed mutually exclusive and collectively exhaustive set of values, the sum of the conditional probabilities at each node is unity. The probabilities are usually based on subjective judgments because the available data are often too limited to allow for statistical analysis, and because scientific judgment is needed to weigh alternative interpretations of the available data. The logic tree approach simplifies these subjective assessments because the uncertainty in a single parameter is considered individually with all other parameters leading up to that parameter assessment assumed to be known with certainty. Thus, the nodes of the logic tree are sequenced to provide for the conditional aspects or dependencies among the parameters and to provide a logical progression of assumptions from the general to the specific in defining the input parameters for an evaluation (Abrahamson and Bommer, 2005; McGuire et al., 2005; Musson et al., 2005).

6 Local Site Characterization

The recent disastrous earthquakes world over have amply demonstrated large concentration of damage and consequent loss of life in specific pockets due to site dependent factors related to local site conditions. In addition to the ground motion characteristics being dependent on source, and path, the local site effects play important role and local geological conditions can strongly influence ground motion during earthquakes. This includes the effects of soil column below the site, topography, basin/basin-edge, surface and subsurface topography and strong lateral discontinuity present. Significant differences in structural damage in the basin as compared with the surrounding exposed rocks or even in the basin itself from place to place have been observed during the earthquakes (e.g., Michoacán, 1985; Spitak, 1988; Loma Prieta, 1989; Northridge, 1994 and Hyogoken-Nambu earthquake, 1995) (Narayan et al., 2002). Local site effects are generally taken into consideration in seismic provisions of building codes by means of a specified amplification factor applied to the spectral acceleration coefficient. It remains controversial whether in practice it is sufficient to consider 1D models or whether 2D or 3D models are necessary to evaluate realistically the site effects, since 1D models neglect completely the surface waves.

For ground response analysis to quantify the local site effects, the subsoil profile should be clearly understood along with its geometry in 2D and 3D. Further, densely populated area for seismic microzonation requires much more refined information of soil in terms of realistic three-dimensional (3D) seismic velocity model with a spatial scale appropriate for such application. Engineering seismology provides geophysical methods for such endeavor. Various methods are available for determination of subsoil profile. Cross-hole geophysical testing is generally conducted in the near surface (upper 100 m) for site-specific engineering applications. The dynamic elastic module of material can be determined from the knowledge of the insitu density, and P- and S-wave velocity. Since procedures to determine material densities are standardized, acquiring detailed seismic data yields the required information to analytically assess a site. Low-strain material damping and inelastic attenuation values can also be obtained from cross-hole surveys. However, the most robust application is the ability to define insitu shear wave velocity profiles for engineering investigations associated with earthquake engineering.

Seismic methods viz. cross-hole, down-hole, up-hole, i.e., borehole seismic, surface refraction, surface wave analysis methods and steady state vibration are commonly used for onshore investigation of insitu shear wave velocity.

Number of different methods such as ground penetrating radar, electrical resistivity imaging, seismic refraction and seismic reflection are available in engineering seismology. Each has its advantages and limitations and proper selection of the method is very important depending upon the site conditions and requirement of the survey. It is, therefore, important to understand basic principles and limitations of various methods. Sometimes use of more than one method is desirable to build confidence as geophysical methods harness the contrast in the different physical properties of the soil and rock, and the techniques are indirect. Geophysical surveys provide information about the magnitudes and variation of the physical properties of the subsurface. Some specific applications include; (1) mapping target horizons or units (e.g., bedrock depth/topography, water table, etc.), (2) providing information on subsurface architecture and its complexity (e.g., deltaic structures, buried valleys, existence of clay units within sands), and (3) identifying subsurface anomalies (e.g., contaminant plumes, salt water intrusion, cavities, faults).

There are several techniques available in the literature to estimate near surface effect from recorded time histories (Nakamura, 1989; Lermo and Chavez-Garcia, 1993; Suzuki et al., 1995; Huang and Teng, 1999; Theodulidis and Bard, 1995; Atkinson and Cassidy, 2000; Motazedian, 2006; Vladimir et al., 2007; Langston et al., 2010). Commonly, the site effect are estimated by dividing the spectrum obtained at the target site with that observed at a nearby reference site which preferably at bedrock. But the availability of such a reference site in nearby is not common. Lermo and Chavez-Garcia (1993) introduced a method to compute site effects at a particular site by dividing horizontal component of shear wave spectra with vertical component of spectra at that site. The amplification effect is observed more on the horizontal component than vertical component (Nakamura, 1989).

7 Strong Ground Motion

Engineering seismology works as a bridge between seismology and earthquake engineering. It is in this context it is important to provide an introduction to the strong ground motion characteristics vis-a-vis dynamics of earthquake response of buildings and other structures. When an earthquake moves the base of a structure, its masses experience forces of inertia as the structure attempts to follow the motion of the ground. The resultant motions of the structure depend on the amplitude and other features of the ground motion, dynamic properties of the structure, and characteristics of the materials of the structure and foundation. By Newton's laws of motion, the force on each mass is equal to mass into acceleration, which can be partitioned into the acceleration with respect to the ground, plus the acceleration of the ground. The resulting equations of motion typically reduce to the familiar form of those for a structure with an immovable base subject to external forces. The equivalent external forces are applied at each mass point and are equal to the mass times the acceleration of the ground. Hence, earthquake forces for most structures are characterized by the time history of the ground acceleration, rather than the ground velocity or displacement. This is not the case for extended structures such as large bridges and dam, nor for cases in which soil–structure interaction is important.

Since the acceleration time history characterizes the forces earthquakes exert on structures, it is desirable to know what forms potentially damaging accelerations can take. The strong-motion acceleration is generally characterized by amplitude, duration and frequency content. Acceleration highlights the high frequency content of the ground motion, while the velocity emphasizes mid-period motions in the range around 1- to 2-second periods. The displacement is typically dominated by periods of a few to several seconds. The peak values of ground motion in terms of Peak Ground Acceleration (PGA), Peak Ground Velocity (PGV) and Peak Ground Displacement (PGD) have been used to define the strong ground motion as input for earthquake engineering which has its own advantages and disadvantages. The duration of shaking is also one of the important parameters.

Frequency content is a term used to characterize the way the acceleration contains energy at different frequencies and is thereby more hazardous to some structures than to others. The frequency content of strong-motion acceleration is commonly measured by the response spectrum, which gives a spectrum of the maximum responses of simple structures with a range of natural frequencies and a specific damping, and by the well-known Fourier spectrum. From the viewpoint of function and safety, the amplitude of earthquake response of a structure is perhaps best measured against the capacity of the structure to resist motions of different amplitudes, first without damage, then with damage, but without serious threat to the integrity of the structure, and finally against the level of response at which collapse is imminent.

The response spectrum (Benioff, 1934; Housner, 1952) is perhaps the most basic tool of earthquake engineering. A response spectrum is a plot of the maximum response amplitude (displacement, velocity or acceleration) versus time period of many linear single degree of freedom oscillators to a given ground motion. The resulting plot can be used to select the response of any linear SDOF oscillator, given its natural frequency of oscillation. It directly gives the maximum response of single-degree-of-freedom structures to particular ground motions and through modal analysis, also is applicable to the earthquake response of multi-degree-of-freedom structures. In addition, it describes the frequency content of ground motions in a way that is more fundamental to earthquake response than traditional Fourier analysis, because of the inclusion of the effects of damping. The response spectrum also forms the basis for the construction of design spectra, such as those in building codes and in design criteria for major projects.

Another important characteristic of strong ground acceleration that has become more apparent in recent years as records have accumulated is the existence of large pulses within near field motions. Pulse-like motions have been observed in the Parkfield earthquake of 1966, the Imperial Valley earthquake of 1979, as well as the more recent Landers (1992), Northridge (1994), Kobe (1995), and Chi-Chi,

Taiwan (1999), earthquakes, among others. Some of the important historical accelerograms, including El- Centro (1934 and 1940), Taft (1952), and Tokachioki (1968) showed some pulses, but were more suggestive of broadband motion. Most of the records in this later group were obtained at moderate epicentral distances, rather than in the near field. Because the structural response to a pulse of acceleration is different in important ways from the response to random motion, understanding the source and effects of large pulses has become an important effort in engineering seismology.

Summarily, with the earthquake engineering perspective an earthquake accelerogram is a random function that may be thought of as being composed of a nonperiodic sequence of acceleration pulse. It is the area of a pulse that is a measure of its effectiveness in producing structural vibrations. The amplitude of the pulse, i.e., the maximum acceleration, is often used to indicate the severity of the ground motion. This is satisfactory if the pulse durations are similar in all earthquakes, but it is not a reliable measure if the pulses have different durations. The intensity of the strong phase of shaking is characterized by the size and shape of the pulses, i.e., the maximum accelerations and the number of zero crossings per second. The effect of the ground shaking also depends upon the number of pulses, i.e., the duration of strong shaking. The most significant engineering characteristics of the ground motion are exhibited by the response spectrum or by the Fourier spectrum of the ground acceleration (Housner, 1959; Hudson, 1956). In most earthquakes the vertical ground acceleration is from one-third to two-thirds as intense as the horizontal acceleration and has higher frequency components, i.e., it has approximately 50% more zero crossings per second. The amplitude of the ground accelerations decreases with distance from the causative fault. The higher frequency components attenuate more rapidly than the lower frequency components so that the spectrum at a distance will be relatively depressed in the high frequency end (Housner 1959).

8 Prediction of Strong Ground Motion

Earthquake engineering requires reliable estimation of the strong ground motion that could be experienced during the life-time of the considered engineering project or structure. The estimation of such ground shaking estimates, either in terms of strong-motion intensity parameters (e.g., peak ground acceleration and response spectral ordinates) or acceleration time histories, is one of the major goals of engineering seismology.

Recordings of ground motions in previous earthquakes are used to derive empirical equations that may be used to estimate values of particular ground-motion parameters for future earthquake scenarios. Such empirical equations have been derived for a variety of parameters but the most abundant are those for predicting PGA and ordinates of acceleration response spectra (Sharma, 1998). The two parameters, earthquake magnitude and source-to-site distance, are always included in ground-motion prediction equations. The equations are simple models for a very complex phenomenon and as such there is generally a large amount of scatter about the fitted curve. The residuals of the logarithmic values of the observed data points are generally found to follow a normal or Gaussian distribution about the mean and hence the scatter can be measured by the standard deviation. Predictions of PGA at the 84-percentile level (i.e., one standard deviation above the mean value) will generally be as much as 80% higher than the median predictions. Even though great progress has been made within engineering seismology during the past few decades, there are still a number of difficulties in providing accurate ground-motion predictions. These include the robust prediction of earthquake ground motions in regions where the strong-motion networks were only installed in the past couple of decades (or there are no networks) and/or seismic activity is moderate and hence little data is available (Douglas, 2006; Douglas et al., 2006; Sharma et al., 2009), prediction of long-period (>2 s) response spectral displacements, because of long-period noise and choices within the strong-motion data processing (Akkar and Bommer, 2007); and the reduction of the large aleatoric variabilities within empirical ground-motion prediction equations (Douglas, 2003).

9 Strong Ground Motion Synthetics

Recorded earthquake ground accelerations have properties similar to non stationary random functions, but, as yet, not enough ground motions have been recorded to determine precisely the statistical properties. Because of the incomplete sample of earthquake recordings, artificially generated accelerograms are used in some studies. These are generated by specified stochastic processes. They have the advantage that accelerograms of any duration can be generated and the statistical properties of the accelerograms are known.

The point source stochastic model is globally accepted as a reliable technique for ground motion simulation. The one-dimensional seismological model assumes that the point source is concentrated at a point source-to-site distance is much larger than the dimension of the source. Hence, the point source approximation for far field is valid. Time series generated at a site comprises of deterministic and random aspects of ground shaking which are contributed by derived theoretical amplitude spectra and random spectra form Gaussian white noise, respectively.

There are important factors that influence the ground motion from large earthquakes that are not included in the stochastic point source model, such as the effect of faulting geometry, distributed rupture and rupture homogeneity. To consider these finite faults–effects in the ground motion modeling, Hartzell (1978) proposed subdividing the fault surface of earthquake into a grid of sub-sources, each of which could be treated as a point source. The contribution to the ground motion can be summed at the observation site, over all of the sub-sources comprising the fault, considering proper delays of sub-sources due to rupture propagation. The basic ideas have been implemented in many articles (Irikura, 1983; Irikura and Kamae, 1994; Bour and Cara, 1997). There are a number of techniques which can be used for simulation of strong ground motion namely, the composite source modeling technique (e.g., Somerville et al., 1991; Saikia, 1993; Zeng et al., 1994; Yu et al., 1995); the empirical Green's function technique (e.g., Hartzell, 1978; Kanamori, 1979; Hadley and Helmberger, 1980; Irikura, 1986; Hutchings, 1985; Irikura and Kamae, 1994, Joshi et al., 2012) and the method of stochastic simulation of high-frequency ground motion (e.g., Housner and Jennings, 1964; Hanks and McGuire, 1981; Boore and Atkinson, 1987).

10 Conclusions

Engineering seismology is the bridge between science and engineering for earthquake engineering. The basic knowledge of observational seismology, theoretical seismology and strong motion seismology leads to understand the various aspects of engineering seismology. The main objective of engineering seismology is the specification of strong ground motion as a result of seismic hazard assessment in a form which can be directly used by earthquake engineers for earthquake engineering designs or retrofitting of existing structures. It is of paramount importance to understand the physical process of earthquake generation, propagation in the medium and its recording at the site. The inverse problem consists of knowing the source parameters, characteristics of medium and site characteristics. The prediction of strong ground motion is made considering those characteristics of strong ground motion which are important and useful in earthquake engineering.

References

Abrahamson, N. A., & Bommer, J. J. (2005). Probability and uncertainty in seismic hazard analysis. *Earthquake Spectra, 21*, 603–607. ISSN:8755-2930.

Akkar, S., & Bommer, J. J. (2007). New empirical prediction equations for peak ground velocity derived from strong-motion records from Europe and the Middle East. *Bulletin of the Seismological Society of America, 97*(2), 511–530.

Anagnos, T., & Kiremidjian, A. S. (1988). A review of earthquake occurrence models for seismic hazard analysis. *Journal of Probabilistic Engineering Mechanics, V3*, 3–11.

Anderson, J. G., & Trifunac, M. D. (1978). Uniform risk functionals for characterization of strong earthquake ground motion. *Bulletin of the Seismological Society of America, 68*(1), 205–218.

Atkinson, G. M., & Cassidy, J. (2000). Integrated use of seismograph and strong motion data to determine soil amplification in the Fraser delta: Results from Duvall and Georgia state earthquakes. *Bulletin of the Seismological Society of America, 90*, 1028–1040.

Benioff, H. (1934). The physical evaluation of seismic destructiveness. *Bulletin of Seismological Society of America, 24*, 398–403.

Boore, D. M., & Atkinson, G. M. (1987). Stochastic prediction of ground acceleration and spectral response parameters at hard rock sites in Eastern North America. *Bulletin of Seismological Society of America, 77*, 440–467.

Bour, M., & Cara, M. (1997). Test of a simple empirical Green's function method on moderate-sized earthquakes. *Bulletin of the Seismological Society of America, 87*, 668–683.

Cluff, L. S., Patwardhan, A. S., & Coppersmith, K. J. (1980). Estimating the probability of occurrences of surface faulting earthquakes on the Wasatch Fault zone, Utah. *Bulletin of the Seismological Society of America, 70*, 463–478.

Coppersmith, K. J. (1981). Probabilities of earthquake occurrences on San Andres fault based geologic risk. *Eos, 19*(17), 322.

Cornell, C. (1968). Engineering seismic risk analysis. *Bulletin of the Seismological Society of America, 58*, 1583–1606.

Cornell, C. A., & Winterstein, S. R. (1988). Temporal and magnitude dependence in earthquake recurrence models. *Bulletin of the Seismological Society of America, 78*(4), 1522–1537.

CWC Guide Lines. (2011). *Guidelines for preparation and submission of site specific seismic study report of river valley project to national committee on seismic design parameters*. Central Water Commission, Govt. of India. Retrieved from http://cwc.gov.in/main/downloads/Guidelines.pdf

Das, R., Wason, H. R., & Sharma, M. L. (2012a). Homoginisation of earthquake catalogue for North East India and adjoining region. *Journal of Pure and Applied Geophysics (PAGEOPH), 169*, 725–731.

Das, R., Wason, H. R., & Sharma, M. L. (2012b). Magnitude conversion to unified moment magnitude using orthogonal regression relation. *Journal of Asian Earth Sciences, 50*, 44–51.

Douglas, J. (2003). Earthquake ground motion estimation using strong-motion records: A review of equations for the estimation of peak ground acceleration and spectral ordinates. *Earth Science Reviews, 61*, 43–104.

Douglas, J. (2006). Difficulties in predicting earthquake ground motions in metropolitan France and possible ways forward. *Géosciences, 4*, 26–31.

Douglas, J., Bungum, H., & Scherbaum, F. (2006). Ground-motion prediction equations for southern Spain and Southern Norway obtained using the composite model perspective. *Journal of Earthquake Engineering, 10*(1), 33–72.

Esteva, L. (1970). Seismic risk and seismic design decisions. In R. J. Hansens (Ed.), *Seismic design of nuclear power plants*. Cambridge, MA: MIT Press.

Guagenti-Grandori, E., & Molina, D. (1984). Semi-Markov processes in seismic risk analysis. *Proceedings, International Symposium of Semi-Markov Processes and Their Applications, Brussels.*

Hadley, D. M., & Helmberger, D. V. (1980). Simulation of ground motions. *Bulletin of the Seismological Society of America, 70*, 617–610.

Hagiwara, Y. (1974). Probability of earthquake occurrence as obtained from a Weibull distribution analysis of crustal strain. *Tectonophysics, 23*(3), 313–318.

Hanks, T. C., & McGuire, R. K. (1981). Character of high frequency ground motion. *Bulletin of the Seismological Society of America, 71*, 2071–2095.

Hartzell, S. H. (1978). Earthquake aftershocks as Green's functions. *Geophysical Research Letters, 5*, 1–4.

Herbindu, A., Kamal, K., & Sharma, M. L. (2012a). Site amplification and frequency-dependent attenuation coefficient at rock sites of Himachal region in NW Himalaya, India. *Bulletin of the Seismological Society of America, 102*(4), 1497–1504.

Herbindu, A., Sharma, M. L., & Kamal, K. (2012b). Stochastic ground-motion simulation of two Himalayan earthquakes: Seismic hazard assessment perspectives. *Journal of Seismology, 16*, 345–369.

Housner, G.W. (1952) Spectrum intensities of strong motion earthquakes. *Proceedings of the Symposium of Earthquake and Blast Effects on Structures, Earthquake Engineering Research Institute.*

Housner, G. W. (1959). Behavior of structures during earthquakes. *Journal of the Engineering Mechanics Division, ASCE, 85*, 109–129.

Housner, G. W., & Jennings, P. C. (1964). Generation of artificial earthquakes. *Journal of the Engineering Mechanics Division, ASCE, 90*, 113–150.

Huang, C. H., & Teng, T. L. (1999). An evaluation of H/V ratio vs. spectral ratio for the site-response estimation using the 1994 Northridge earthquake sequences. *Pure and Applied Geophysics, 156*, 631–649.
Hudson, D. E. (1956). Response spectrum techniques in engineering seismology. *Proceedings of the World Conference on Earthquake Engineering. Earthquake Engineering Research Institute and the University of California, Berkeley.*
Hutchings, L. (1985). Modeling earthquakes with empirical Green's functions (abs.) *Earthquake Notes, 56*, 14.
Irikura, K. (1983). Semi-empirical estimation of strong ground motions during large earthquakes. *Bulletin of the Disaster Prevention Research Institute, 33*, 63–104.
Irikura, K. (1986). Prediction of strong ground acceleration motion using empirical Green's function. *Proceedings of 7th Japan Earthquake Engineering Symposium*, 151–156.
Irikura, K., & Kamae, K. (1994). Estimation of strong ground motion in broad-frequency band based on a seismic source scaling model and an Empirical Green's function technique. *Annali di Geofisica, 7*(6), 1721–1743.
Jordanovski, L.R., Todorovska, M.I. and Trifunac, M.D. (1991). A model for assessment of the total loss in a building exposed to earthquake hazard (Rep. No. CE 92-05). *Los Angeles, CA: Department of Civil Engineering, University of Southern California.*
Joshi, A., Kumari, P., Kumar, S., Sharma, M. L., Ghosh, A. K., Agrawal, M. K., & Ravikiran, R. (2012). Estimation of model parameter of Sumatra earthquake using empirical Green's function technique and generation of hypothetical earthquake scenario for Andaman Island, India. *Natural Hazards, 62*, 1081–1108.
Joshi, G. C., & Sharma, M. L. (2008). Uncertainties in estimation of Mmax. *Journal of Earth Sciences Systems, 117*(S2), 671–682.
Kanamori, H. (1979). A semi empirical approach to prediction of long period ground motions from great earthquakes. *Bulletin of the Seismological Society of America, 69*, 1645–1670.
Kiremidjian, A. S., & Anagnos, T. (1984). Stochastic slip-predictable model for earthquake occurrences. *Bulletin of the Seismological Society of America, 74*(2), 739–755.
Langston, C. A., Chi Chiu, S. C., & Lawrence, Z. (2010). Array observations of micro-seismic noise and the nature of H/V in the Mississippi embayment. *Bulletin of the Seismological Society of America, 99*(5), 2893–2911.
Lermo, J., & Chavez-Garcia, F. (1993). Site effect evaluation using spectral ratios with only one station. *Bulletin of the Seismological Society of America, 83*, 1574–1594.
McGuire, R. K., Cornell, C. A., & Toro, G. R. (2005). The case for using mean seismic hazard. *Earthquake Spectra, 21*(3), 879–886.
Motazedian, D. (2006). Region specific key seismic parameters of earthquakes in northern Iran. *Bulletin of the Seismological Society of America, 96*(4A), 1383–1395.
Musson, R. M. W., Toro, G. R., Coppersmith, K. J., Bommer, J. J., Deichmann, N., Bungum, H., Cotton, F., Scherbaum, F., Slejko, D., & Abrahamson, N. A. (2005). Discussion—Evaluating hazard results for Switzerland and how not to do it: A discussion of 'problems in the application of the SSHAC probability method for assessing earthquake hazards at Swiss nuclear power plants' by J-U Klügel. *Engineering Geology, 82*(1), 43–55.
Nakamura, Y. (1989). A method for dynamic characteristics estimation of subsurface using microtremor on the ground surface. *Quarterly Report of Railway Technical Research Institute (RTRI), 30*(1), 25–33.
Narayan, J. P., Sharma, M. L., & Kumar, A. (2002). A seismological report on the January 26, 2001 Bhuj, India earthquake. *Seismological Research Letters, 73*(3), 343–355.
Nishioka, T., & Shah, H. C. (1980). Application of the Markov chain on probability of earthquake occurrences. *Proceedings of the Japanese Society of Civil Engineering, 1*, 137–145.
Patwardhan, A. S., Kulkarni, R. B., & Tocher, D. (1980). A semi Markov model for characterizing recurrence of great earthquakes. *Bulletin of the Seismological Society of America, 70*, 323–347.
Saikia, C. K. (1993). Ground motion studies in great Los Angles due to Mw = 7.0 earthquake on the Elysian thrust fault. *Bulletin of the Seismological Society of America, 83*, 780–810.

Savy, H. B., Shah, H. C., & Boore, D. M. (1980). Nonstationary risk model with geophysical input. *Journal of the Structural Division, ASCE, 106*(ST1), 145–164.

Sharma, M. L. (1998). Attenuation relationship for estimation of peak ground horizontal acceleration using data from strong motion arrays in India. *Bulletin of the Seismological Society of America, 88*, 1063–1069.

Sharma, M. L. (2003). Seismic hazard in Northern India region. *Seismological Research Letters, 74* (2), 140–146.

Sharma, M. L., & Lindolhm, C. (2012). Earthquake hazard assessment for Dehradun, Uttarakhand, India, including a characteristic earthquake recurrence model for the Himalaya Frontal Fault (HFF). *Pure and Applied Geophysics (PAGEOPH), 169*, 1601–1617.

Sharma, M. L., Douglas, J., Bungum, H., & Kotadia, J. (2009). Ground motion predicting equations on data from the Himalayan and Zagros regions. *Journal of Earthquake Engineering, 13*(8), 1191–1210.

Somerville, P. G., Sen, M. K., & Cohece, B. P. (1991). Simulation of strong ground motions recorded during the 1985 Michoacan, Mexico and Valparaiso Chile earthquakes. *Bulletin of the Seismological Society of America, 81*, 1–27.

Suzuki, T., Adachi, Y., & Tanaka, M. (1995). Application of micro tremor measurement to the estimation of earthquake ground motion in Kushiro City during the Kushiro-Oki earthquake of 15 January 1993. *Earthquake Engineering and Structural Dynamics, 24*, 595–613.

Theodulidis, N. P., & Bard, P.-Y. (1995). Horizontal to vertical spectral ratio and geological conditions: An analysis of strong motion data from Greece and Taiwan (SMART-1). *Soil Dynamics and Earthquake Engineering, 14*, 177–197.

Tripathi, J. N., Singh, P., & Sharma, M. L. (2012). Variation of seismic coda-wave attenuation in the Garhwal region, north western Himalaya. *Journal of Pure and Applied Geophysics (PAGEOPH), 169*(1–2), 71–88.

Vagliente, V. H. (1973). Forecasting risk inherent in earthquake-resistant design (Tech. Rep. No. 174), Stanford, CA: *Department of Civil Engineering, Stanford University.*

Veneziano, D., & Cornell, C. A. (1974). Earthquake models with spatial and temporal memory for engineering seismic risk analysis (Research Rep. No. R 74–78). *Cambridge, MA: Department of Civil Engineering, MIT.*

Vere-Jones, D., & Davies, R. B. (1966). A statistical survey of earthquakes in the main seismic region of New Zealand, Part 2, Time series analysis. *New Zealand Journal of Geology and Geophysics, 9*, 251–284.

Vere-Jones, D., & Ozaki, T. (1982). Some examples of statistical estimation applied to earthquake data, I. Cyclic poisson and self-exciting models. *Annals of the Institute of Statistic and Mathematics, 34*(Part B), 189–207.

Vladimir, Y. S., Chin-Hsiung, L., & Wen-Yun, J. (2007). Application of horizontal to-vertical (H/V) Fourier spectral ratio for analysis of site effect on rock (NEHRP-class B) sites in Taiwan. *Soil Dynamics and Earthquake Engineering, 27*(4), 314–323.

Wason, H. R., Das, R., & Sharma, M. L. (2012). Magnitude conversion problem using general orthogonal regression. *Geophysical Journal International, 190*(2), 1091–1109.

Yu, G., Khattri, K. N., Anderson, J. G., Brune, J. N., & Zeng, Y. (1995). Strong ground motion from the Uttarkashi, Himalaya, India, earthquake: Comparison of observations with synthetics using the composite source model. *Bulletin of the Seismological Society of America, 85*, 31–50.

Zeng, Y., Anderson, J. G., & Su, F. (1994). A composite source model for computing realistic synthetic strong ground motions. *Geophysical Research Letters, 21*, 725–728.

Chapter 2
Instrumentation Seismology in India

Atindra Kumar Shukla, Ravi Kant Singh, and Rajesh Prakash

1 Introduction

Though, occurrence of an earthquake is not a frequent phenomenon but the intrinsic nature of earthquake to unleash devastation instantaneously in a large area without leaving much scope for prevention of hazard after the occurrence of the event and its unpredictability, place the earthquake hazard in a special category of natural disaster and require special disaster management framework. This requires both short-term and long-term, sustained efforts and a comprehensive planning and strategy for both post and pre-disaster management.

Two important scientific approaches are currently adopted worldwide to enhance our resilience to this natural hazard. The first one, aimed at providing long-term protection to life and property, involves estimation of the earthquake hazard in different areas of a region and its translation into engineering specifications for earthquake-resistant construction and land-use patterns in the area. The second, concomitantly adopted with the first, is to keep a constant vigil on the evolving character of ground motion signals emitted by earthquakes from a wide area surrounding the final site of rupture and in case of occurrence of significant earthquake provide real-time information of epicentral parameters of earthquake to the agencies involved in relief and rescue operations to manage post earthquake disaster. Both these steps use ground motion records, as raw data for detecting portents of an impending catastrophe. An earthquake ground motion monitoring network is thus the basic requirement to implement rationally designed earthquake disaster mitigation and management strategies.

Instrumental seismology deals with ability to measure true ground motion, its recording and transmission capabilities through seismic sensor, recorder and communication system respectively. The recent advancement particularly, digital

A. K. Shukla (✉) · R. K. Singh · R. Prakash
India Meteorological Department, New Delhi, India

M. L. Sharma et al. (eds.), *Advances in Indian Earthquake Engineering and Seismology*, https://doi.org/10.1007/978-3-319-76855-7_2

technology has ensured the realization of high levels of sensitivity and stability due to capability of digital recorders to operate over a wide dynamic range, and thus provide data for events of low as well as of high magnitude. The digital technology also ensures that the same dynamic range is achieved in data transmission via modern communication technology and permits high levels of data compaction with attendant advantages in transmission as well as storage and archival.

In India instrumental seismology started more than a century ago, with the first seismological observatory of the country set up at Alipore (Kolkata) by the India Meteorological Department (IMD) in 1898. Since then, a lot of development has been made in all the three sections of instrumental seismology, i.e., sensing the true ground motion, recording of ground motion with synchronized timing and real-time reliable communication, due to the advancement in development of digital technology.

2 Instrumental Seismology

Instrumental seismology comprises of *seismic sensor, recorder and communication system* and are the basic components of a seismological network. The great advancement has been taken place since initial stage of these instruments.

2.1 Seismic Sensors

Seismic sensors are the instruments that measure ground motion generated by earthquake, volcanic eruption and other seismic sources. Since the measurements of ground motion are achieved in a moving reference frame (the earth's surface), seismic sensors are mostly based on the inertia of a suspended mass, which will tend to remain stationary in response to external motion. The relative motion between the suspended mass and the ground will then be a function of the ground's motion. Earlier, sensors were made with sophisticated mechanical systems and were able to have a low resonance frequency down to about 1.0 Hz. Use of electronics has made it possible to increase the capability of measuring wider frequency band of ground motion and presently in use. According to capability of measuring frequency band of ground motion, seismic sensor can broadly be classified in Short Period, Broad Band and Very Broad Band.

The *Short-Period (SP) seismometers* measure signals from approximately 0.1 to 100 Hz, with a corner frequency at 1 Hz. They have a flat response to ground velocity for frequencies greater than this corner frequency. Typical examples are the Kinemetrics SS-1, the Geotech S13, and the Mark Products L-4C. The 4.5-Hz exploration-type geophone also belongs in this group. This sensor provides reasonably good signals down to about 0.3 Hz at a fraction of the cost of the 1.0-Hz sensor.

The *Broadband (BB) seismometers* have a flat response to ground velocity from approximately 0.01 to 50 Hz. Typical examples are the Guralp CMG40T seismometer with frequency range from 0.03 to 50 Hz and the Wieland–Streckeisen seismometer STS2 with a frequency range from 0.008 to 40 Hz. Several other sensors are available in the market.

The *Very Broadband (VBB) seismometers* measure frequencies from below 0.001 Hz to approximately 10 Hz and are also able to resolve earth tide. Typical examples are the Wieland–Streckeisen STS1 seismometer with frequency range from 0.0028 to 10 Hz and the STS-2 with frequency range from 0.0083 to 50 Hz, Nanometrics make Trillium-240 has a response flat to velocity from 240 s to 35 Hz, the 151 series broadband seismometer, "Observer" (Reftek make), a force-balance feedback sensor available with frequency bandwidth of 0.0083 Hz (120 s)–50 Hz, and Guralp make CMG-3T also has flat velocity response in the frequency bandwidth of 0.0083 Hz (120 s)–50 Hz.

In a conventional passive seismometer, the mass is allowed to move and inertial force produced by a seismic ground motion deflects the mass from its equilibrium position, and the displacement or velocity of the mass is then converted into an electric signal. This principle of measurement is now used for short-period seismometers only. Most of broadband seismometers are based on the force-balance principle (FBA), in which sensor mass is held nearly motionless relative to the frame by an electronic negative feedback loop. This type of sensors is called as active sensors. In this type of sensors, the motion of the mass relative to the frame is measured, and the feedback loop, applies a magnetic or electrostatic force to keep the mass nearly motionless. The voltage needed to produce this force is the output of the seismometer, which is recorded digitally. The FBA principle is now the heart of nearly all modern strong motion and broadband sensors used for recording in a large frequency band like 120 s to 50 Hz. Now the sensors based on molecular electronic transducers (MET) (http://www.mettechnology.com) are also in use. These are a class of inertial sensors (which include accelerometers, gyroscopes, tilt meters, seismometers and related devices) based on an electrochemical mechanism. METs capture the physical and chemical phenomena that occur at the surface of electrodes in electrochemical cells as the result of hydrodynamic motion. They are a specialized kind of electrolytic cell designed so that motion of the MET, which causes movement (convection) in the liquid electrolyte, can be converted to an electronic signal proportional to acceleration or velocity. MET sensors (Hurd and Lane, 1957; Fusca, 1957; Wittenborn, 1959; Larkam, 1965) have inherently low noise and high amplification of signal (on the order of 10^6).

2.2 Recorder

The next challenge is to be able to faithfully record the signals from the seismometer without losing the quality of ground motion. Two types of recorder are presently in used.

Analog Recorder: In an opto mechanical seismograph, a ray of light is made to fall on a mirror fixed to the mass of the seismometer and the reflected light is recorded at a distance on a photographic paper. An example of one of the earliest seismographs of this type is Milne–Shaw seismograph, which records the horizontal component of ground motion. Another important example of this type of seismograph is *Wood–Anderson seismograph* developed in the 1920s and also records horizontal component of ground motion. The pendulum consists of a cylindrical mass (generally made of copper), which is attached to a vertical suspension wire. During ground motion, the cylinder rotates around the axis of the wire and a mirror attached to the cylinder reflects a light beam on the photographic paper wrapped on a recorder drum. A horse-shoe magnet surrounding the copper cylinder acts as a damping (eddy current) device. Analog recording is still used since it gives a very fast and easy overview of the seismicity. However, the dynamic range is very limited (at most 200 db in contrast to 108 db).

Digital Recorders: The development of high dynamic range and broadband sensors would not have improved instrumental seismology a lot, unless the recording technique had followed the development. The last 30 years have seen a fast development in digital recording, which almost has been able to keep up with the improvement in sensor development. The digital recorders consist of digitizers and recording system.

Digitizers: In digitizers which are also known as analog-to-digital converters (ADC), continuously varying analog signal is converted to discrete samples of the analog signal. Digitizers are usually classified as 12 bit, 16 bit or 24 bit according to the number of discrete values the digitizer uses. The best digitizers can usually resolve 100 nV, which then ideally should correspond to the number 1. A 12-bit converter has 212 levels and a 24-bit converter 224 levels or ±211 and ±223, respectively. This corresponds to dynamic ranges of 2048 and 8.4×10^6, respectively. Now most digitizers are based on 24-bit analog-to-digital converters (ADC), however, a few have a true 24-bit performance, since shorting the input gives out counts significantly different from zero.

Recorders: Digitizer and recorder are often one unit although they physically are two separate parts. Recording is now mostly based on hard disks, although some units use solid-state memory. While many units on the market today use microcontrollers in order to save power, the trend is to use complete single-board PCs due to the low power consumption of these new-generation PCs. The advantage of using real PCs with "normal" operating systems is that all standard software, e.g., communication, can be used instead of the manufacturer special fix for his special recorder. Since large hard disks can be used for recording, recorder memory is no longer a problem and continuous recording is now easy to do. However, nearly all recorders also have a facility for additionally storing through Secure Digital (SD), Secure Digital Extended Capacity (SDXC), Secure Digital High Capacity (SDHC), miniSD, microSD, miniSDHC, microSDHC and compact flash (CF) cards. Now a days almost all the available digitizers (Nanometrics: TAURUS, Reftek: 131, Guralp: DM24, etc.) have CF cards as storage options along with RAM.

2.3 Communication

Communication plays a very important role in opeartion seismology to keep a constant vigil and in case of occurrence of a significant earthquake, make it possible to provide real-time information of epicentral parameters of earthquake to the agencies involved in relief and rescue operations to manage post earthquake disaster. In seismology there are several different kinds of physical data transmission links in use, from simple short wire lines to satellite links (e.g. dedicated leased lines, Dial-up communication, Internet and VSAT communication) at global distances. They differ significantly with respect to data throughput, reliability of operation, maximal applicable distances, robustness against damaging earthquakes, and in cost of establishment, cost of operation, and required maintenance. VSAT communication seems to be the most reliable and efficient medium for weak motion seismic telemetry network for real-time operation. For sounding alert strong motion may also need real-time communication. Presently taking advantage of availability of efficient two-way communication facility, field stations are being operated without any manpower and virtually operated remotely from central receiving stations.

3 Seismic Network

The seismic network is define as a group of stations working together jointly for data collection and analysis. Earlier in general stations were operating independently and data was being send to central place for analysis with a time lag of hours, depending upon the means of communications, such as telephone, telegrams etc. In real sense, the concept of seismic network started in 1960s, when ground motion recorded signals were transmitted in real time by wire or radio link to a central recording station, where all data was recorded with central timing and used for analysis (Havskov and Alguacil, 2002). Initially it was mainly for monitoring of microearthquake and the distances between stations were a few kilometers to a couple of 100 km. In India, this concept was implemented by deployment of seismic telemetry network in the North-East region around Tejpur (Assam) by the National Geophysical Research Institute (NGRI), Hyderabad, using one-way radio communication based on line of sight. Subsequently, by deployment of digital equipments under India Meteorological Department's National Seismological Network during 1995–1998, field stations were networked with telephone dial-up communication facility, in which data was being pulled up from a field station in regular interval through telephone dial-up modem at central receiving station, set up at IMD H.Q., New Delhi (Bhattacharya et al., 2001a, b; Bhattacharya and Dattatrayam, 2000; Srivastav et al., 2001, 2003; Shukla et al., 2001, 2002). With the evolution of communication capabilities, distances between the stations is no more problem and data can be transferred to thousands of kilometers in real time to cover the whole world, and seismic networks are now not limited to local networks. The real-time networking concept with increased spacing between field stations was implemented in the India Meteorological Department during 1998–2000. The India Meteorological Department (IMD) set up a more robust telemetry network around a populated area of the National Capital Territory of Delhi, using a more reliable VSAT communication for real-time data transmission from field station to central receiving station at IMD H.Q., New Delhi (Shukla et al., 2001, 2002). A few field stations of this network were as far as 300 km.

In the aftermath of the great Sumatra earthquake of December 26, 2004, which was followed by unprecedented tsunamis, a Real-Time Seismic Monitoring Network (RTSMN) system, comprising 17 stations of the national network, was set up by IMD, to monitor and automatically report, in least possible time, large magnitude under-sea earthquakes capable of generating tsunamis from the two prominent sources, viz., the Makran coast in the north Arabian sea region and the Andaman–Nicobar–Sumatra island arc region (Dattatrayam et al., 2009). Recently IMD has also set up again a VSAT based seismic telemetry network for near real-time monitoring of earthquake activity in the Northeast India region and automatically reporting, in least possible time (Shukla et al., 2011).

Now the use of global communication networks such as the internet or public phone system in seismic stations or at the nodal hub, the concept of network has changed. Such stations usually have a local seismic signal recording capability and sometimes there might not be any real-time data transmission to a given site. However, these stations still can be defined to be in a network since they are all connected to the global communication network. Any computer network can in principle be used to collect data from a number of stations in what functionally is a seismic network. By defining a seismic network in this way, the distinction between local, regional, and global networks does not exist anymore in terms of hardware, data transmission, and acquisition but is merely a question of how the data collection software is set up to handle communication, data collection, and processing. Thus the network can be defined as physical or virtual seismic network (Ottemoller and Havskov, 1999).

A physical seismic network (usually local) consists of closely linked remote seismic stations. The remote stations detect the ground motion and usually send data in real time to a central recording station for event detection and analysis. This type of network covers both the old analog systems and the current digital systems.

A *virtual seismic network* consists of stations selected among many stations connected to the global communication network or a public phone system. A station can also be a central recording station of a physical network. The remote stations must be capable of event detection and local recording and data is normally not sent to a central recording system in real time. The remote stations must have two-way communication capability (duplex or at least semi-duplex). The central recording station can manually or automatically connect to selected remote stations and download triggered and/or continuous data and make intelligent evaluation of possible events. The IRIS/Global Seismic Network (GSN) and public domain SEISNET system running on Unix are examples of virtual seismic network (Ottemoller and Havskov, 1999).

Both kinds of network (physical and virtual) require some medium of communication depending upon the network type, its use, etc. For example, strong-motion seismic networks generate far less data compared to weak-motion networks and therefore, their designs might differ significantly. Seismic data transmission links that are fully acceptable for strong-motion data may be inadequate for weak-motion data and data transmission links used in weak-motion field may be an absolute overkill and too expensive for strong-motion networks.

4 Physical Seismic Network in India

In India several physical seismic networks are being operated by different institutions such as NGRI Hyderabad, Regional Research Laboratory (RRL), Jorhat (now North East Institute of Science and Technology), Indian Institute of Technology (IIT) Roorkee, and IIT Kharagpur. The India Meteorological Department (IMD) being a nodal agency for earthquake monitoring operates two physical seismic networks: first one in and around Delhi established in 1999–2000 and second one in Northeast India commissioned and became operational in June 2011. Both of the physical seismic networks of IMD are briefly described in the following paras.

Delhi Seismic Telemetry Network: A 16-element VSAT-based digital seismic network in and around Delhi with central receiving station (CRS) at IMD, H.Q., New Delhi has been set up from September 2000 to April 2002 (Fig. 2.1). Of these, 13 field stations are within 200 km radius from CRS. Taking advantage of VSAT communication the remaining three field stations have been setup at Joshimath, Kalpa and Jaisalmer for monitoring regional seismicity and near real-time monitoring of seismicity of the northern Himalayan region. The field stations are unmanned and continue ground motion data in digital form is transmitted online to central receiving station in near real time. Each field station is consist of short-period seismic sensor Geotech make Teledyne S-13, Reftek 72 series, Reftek digital data acquisition system (DAS) powered from 12V 26 AH batteries charged through the solar panel, VSAT communication facility powered from the battery charged from UPS. Ten field stations have single component and six stations have three components (three S-13 sensors).

Northeast (NE) Telemetry Network: Recently IMD has setup a state-of-the-art seismic telemetry network in the Northeast India region with the following objectives (Shukla et al., 2011), similar to that of any state-of-the-art seismic network operating in the world (Havskov and Bungum, 1987; Hansen et al., 1989; Bungum et al., 1985; Kvamme and Hansen, 1989; Kvaerna and Mykkeltveit, 1985):

1. Near real-time monitoring of earthquake activity, autolocation and on occurrence of significant earthquake dissemination of earthquake epicentral parameters to the agencies involved in relief and rescue operation

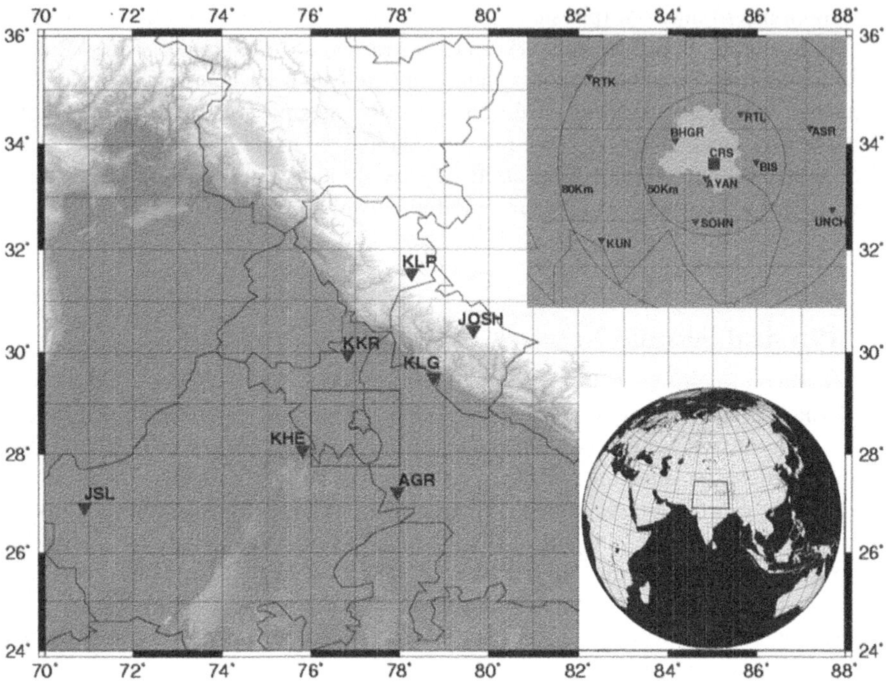

Fig. 2.1 VSAT-based 16-element Delhi telemetry network with close-up view of array in NCR domain. [*AGR*- Agra, *ASR*- Asauara, *AYN*- Ayanagar, *BHG*- Bhadurgarh, *BIS*- Bisrakh, *JOSH*- Joshimath, *JSL*- Jaisalmer, *KHE*- Khetri, *KKR*- Kurukshetra, *KLG*- Kalagarh, *KLP*- Kalpa, *KUN*- Kuldal, *RTK*- Rohtak, *RTL*- Rataul, *SOHN*- Sohana, *UNCH* Unchagaon, *CRS*- Central Receiving Station]

2. To improve detection capability in NE region of the country and to obtain better determinations of earthquake source parameters such as epicenter, depth, and sense of faulting
3. To improve the seismotectonic models for the area, including a better integration of the seismicity data with other geological and geophysical data
4. To obtain a better understanding of the wave propagation characteristics in NE India, including effects of lateral heterogeneities within the crust and upper mantle

This network consists of 21 seismological field observatories and a central receiving station (CRS) at IMD H.Q., New Delhi and also at CSO Shillong simultaneously (Fig. 2.2). State wise distribution of field stations and their locations are detailed in Table 2.1.

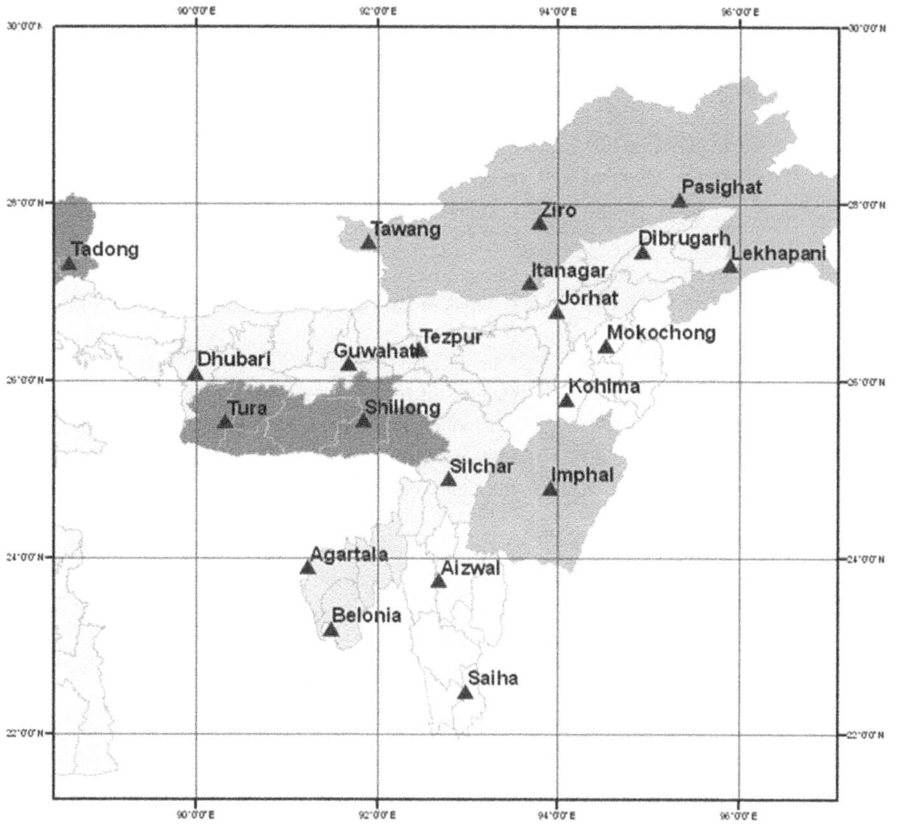

Fig. 2.2 Location map of field observatories of NE seismic telemetry network

4.1 Network Configuration

4.1.1 Field Stations

All the 21 field stations in Northeast India are equipped with:

Digital broadband (BB) seismometer [Reftek make; Type, Triaxial, Orthogonal; Frequency Response: 0.0083 Hz (120 s)–50 Hz; sensitivity: 2000 V/m/s; Power consumption: ~1.1 W]

Strong-motion accelerometer (SMA) [Reftek make, Full-scale range: > ±3 g; Full-scale output: ±10 V; Type, Force-balance accelerometer; Linearity: ±1% of full scale; Frequency response: DC- >500 Hz MA]

Data acquisition system (DAS) [Reftek make; channel: 6 (3 for BB; 3 for SMA); sampling rate: 40 sps for BB seismometer 200 sps for SMA; Storage: RAM 16 MB and CF card 16 GB; Recording format: PASSCAL; Power consumption: 2.1 W (6 channel, writing to disk, with communication)]

Table 2.1 State-wise distribution of the 21 NE telemetry stations and their locations

Station	Lat (°N)	Long (°E)	Location of observatory
Arunachal Pradesh (4)			
Pasighat	28.06	95.34	Campus of DC, Pasighat
Tawang	27.58	91.91	Campus of DC, Tawang
Yupia	27.12	93.70	Campus of DC, Yupia
Ziro	27.80	93.80	Campus of DRDA, Monipolyang, Lower Subansiri, Ziro
Assam (7)			
Dhubari	26.10	90.00	Campus of DC, Dhubari
Dibrugarh	27.48	94.93	Campus of District Disaster Management Authority (DDMA)
Guwahati	26.20	91.70	Campus of IIT Guwahati
Jorhat	26.80	94.00	Campus of Jorhat Engineering College
Lekhapani	27.32	95.90	Seismological Observatory, Lekhapani, Tinsukia
Silchar	24.90	92.80	Meteorological Office, Silchar, Cachhar
Tezpur	26.36	92.48	Residence campus of ADC Tezpur
Manipur (1)			
Imphal	24.80	93.93	Seismological Observatory, Chingmeirong, Lamlong
Meghalaya (2)			
Tura	25.55	90.33	Seismological Observatory, Ring Rey Forest Tila, West Garo Hills
Shillong	25.56	91.85	Central Seismological Observatory, Meath House, East Khasi Hills Opposite 3rd Mile, Upper Shillong
Mizoram (2)			
Aizwal	23.75	92.70	Campus of Department of Geology and Mines, Directorate of Industries
Saiha	22.50	93.00	Campus of Disaster Management & Rehabilation, O/o DC Saiha
Nagaland (2)			
Kohima	25.80	94.10	Campus of Nagaland Legislative Assembly Complex
Mokokchung	26.41	94.53	Campus of DC Mokokchung
Tripura (2)			
Agartala	23.90	91.25	Meteorological Centre Agartala, Agartala Airport
Belonia	23.20	91.50	Dak Bungalow premises
Sikkim (1)			
Tadong	27.34	88.61	Seismological Observatory, Tadong

VSAT communication [Hughes make HN 7000 series VSAT equipment operating on Ext-C band; 1.2-m antenna at 18 sites (including CRS); 1.8-m antenna at Tawang, Ziro, Pasighat, and Lekhapani; snow cover shield at Tadong, Tawang, Ziro, and Pasighat]

Solar-activated power system [for seismic equipment: 12 V 40 Watt (with SMF batteries 12 V 42 AH: 1 No. with charge controller 12 V 10 A): for communication equipment: 12 V 80 W (with SMF batteries 12 V 100 AH: 2 No with charge controller 12 V 30 A): Total solar panel: six per site].

The BB seismometers are covered with thermally insulated casing and SMA are placed on a pillar embedded to hard ground/rock (Fig. 2.3).

The time stamped ground motion data through local GPS is recorded at field stations by BB at 40 samples per second (sps) in continuous mode and SMA at 200 sps in event mode and stored in RAM of 16 MB and CF card of 16 GB of data acquisition system (Fig. 2.4). This data is transmitted automatically to CRS located at IMD H.Q., New Delhi, and also CSO Shillong in near real time through VSAT communication operating on extended C band using multicasting facility of communication for autolocation and dissemination at CRS of IMD H.Q., New Delhi.

Fig. 2.3 BB and SMA installed on seismic pillar embedded to hard ground/rock at field station

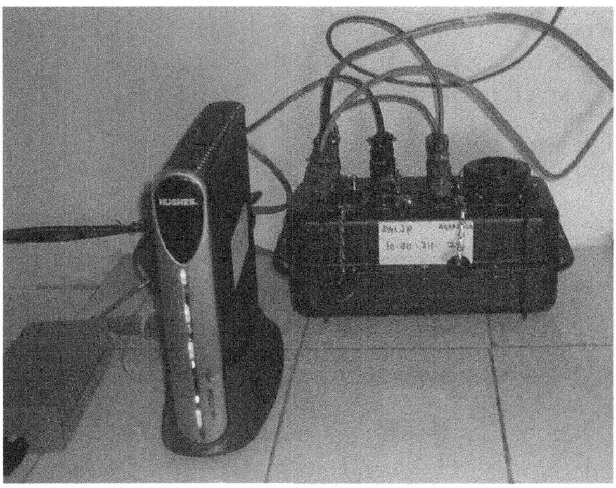

Fig. 2.4 VSAT communication indoor unit and DAS at field station

4.1.2 Central Receiving Station (CRS)

The data transmitted from the field stations is received by acquisition servers running in HOT STANDBY mode at CRS New Delhi and CSO Shillong. The real-time received data is also displayed in a large-screen LCD monitor.

The NE telemetry network makes use of a state-of-the-art autolocation software called SEISCOMP3 developed by Gempa GmbH, Potsdam, Germany; running on these servers at CRS New Delhi only; autolocate the earthquake events on receiving four to five field stations data in shortest possible time and issue an audio visible alert. The autolocated hypocentral parameters (latitude, longitude, depth, magnitude, origin time of the earthquake) are continuously updated on receiving more data (Fig. 2.5a, b).

The information of hypocentral parameters is being disseminated by an auto-dissemination system through SMS, email, and fax to predefined users/authorities. At CSO Shillong, data is also made available for off-line analysis, storage, etc. and displayed on a large-screen LCD monitor.

The processed data of NE telemetry network along with data of other observatories of the national network is archived at National Seismological Data Centre of the India Meteorological Department. The data of NE telemetry network has also been integrated with the national network for operational use. The network configuration of field station and CRS at New Delhi is shown in Fig. 2.6.

4.1.3 Autolocation Software (SeisComP3)

The first version of SeisComP was developed for the GEOFON network and further extended within the MEREDIAN project under the lead of GEOFON/GFZ Potsdam and ORFEUS. Originally SeisComP was designed as a high-standard fully automatic data acquisition and (near) real-time data processing tool for quality control, event detection and location as well as dissemination of event alerts. A first prototype of SeisComP3 developed by the GITEWS/GEOFON development group was released in May 2007. The latest Seiscomp3 released in Potsdam on Sep 2010 is in use in NE telemetry network. SeisComP3 provides the following features:

- Data acquisition
- Data quality control
- Data recording
- Real-time data exchange
- Network status monitoring
- Real-time data processing
- Issuing event alerts
- Waveform archiving
- Waveform data distribution

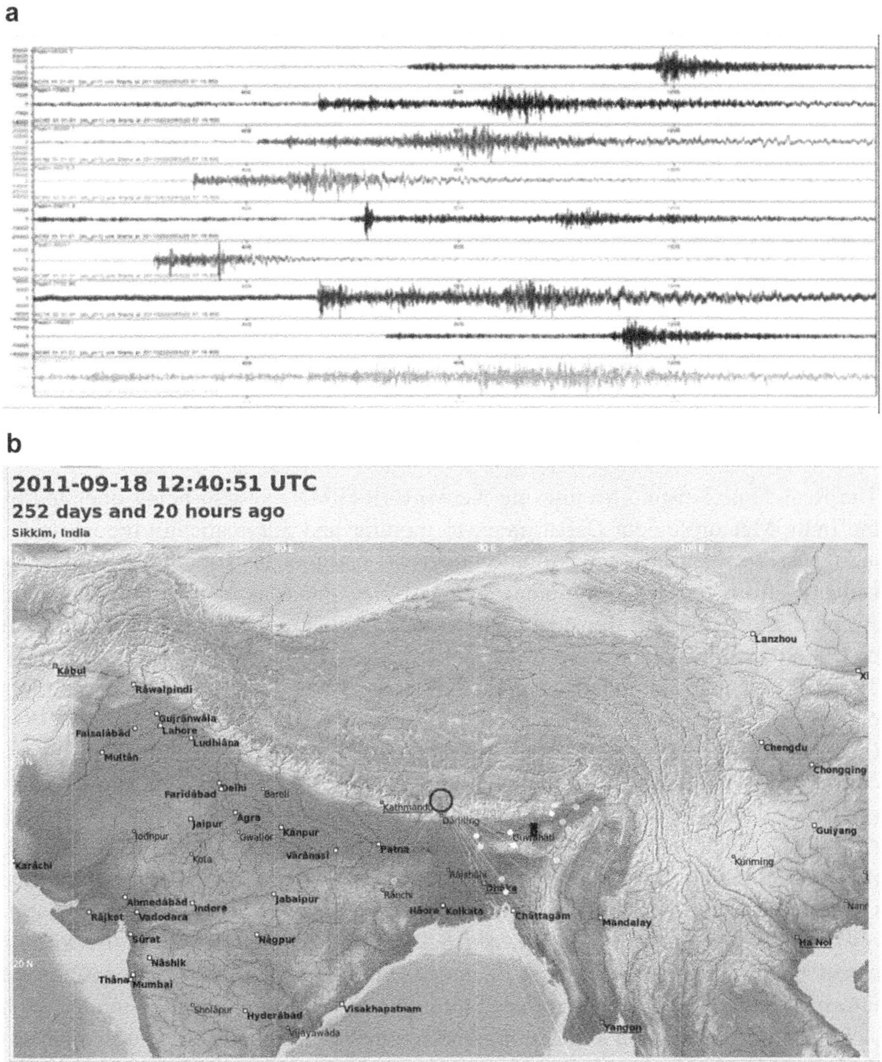

Fig. 2.5 (**a**) Real-time display of data received from field stations. (**b**) Real-time display of autolocated Sikkim earthquake

- Automatic event detection and location
- Interactive event detection and location
- Event parameter archiving
- Easy access to relevant information about stations, waveforms and recent earthquakes

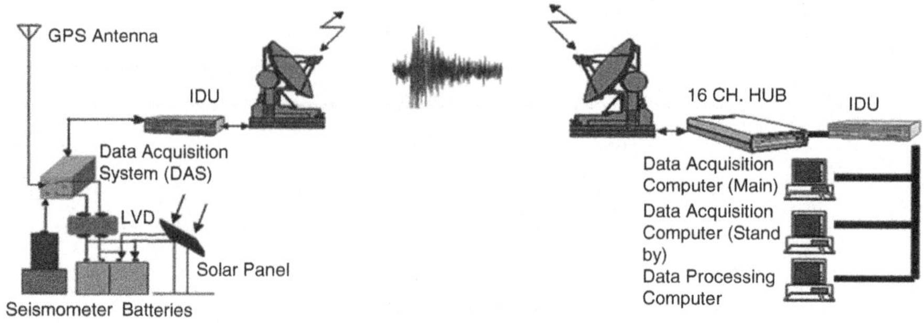

Fig. 2.6 Telemetry network configuration

5 Virtual Seismic Network

The Real-Time Seismic Monitoring Network (RTSMN) system being operated by the India Meteorological Department, to monitor and automatically report, in the least possible time, large-magnitude under-sea earthquakes capable of generating tsunamis from the two prominent sources, viz., the Makran coast in the north Arabian sea region and the Andaman–Nicobar–Sumatra island arc region consists of 17 field stations which are physically networked through VSAT communication. To provide better azimuthal coverage for precise location of earthquakes falling outside the coverage area of the RTSMN, more than 100 seismic stations of IRIS (Incorporated Research Institutions for Seismology) global network have also been configured in the RTSMN system making use of virtual network concept. RTSMN system makes use of a state-of-the-art autolocation software called "Response Hydra (V-1.47)," developed by the National Earthquake Information Center (NEIC) and the US Geological Survey (USGS). The Response Hydra also provides real-time Centroid Moment Tensor (CMT) and Moment Tensor (MT) solutions for earthquakes of magnitude ranging between 5.5–7.5 and 6.0–8.0, respectively, to help assess the tsunamigenic potential of under-sea earthquakes. The final earthquake information/products are disseminated through various communication modes, such as SMS, fax, and email in a fully automated mode, to all the concerned user agencies dealing with relief and rehabilitation-related work. The earthquake-related information is also posted on IMD's website (www.imd.gov.in) (Dattatrayam et al. 2009).

Acknowledgements Thanks to all officers and staff associated in setting up of Seismic Networks in India Meteorological Department, particularly, Delhi Seismic Telemetry and NE Seismic Telemetry Networks.

References

Bhattacharya, S. N., & Dattatrayam, R. S. (2000). Recent advances in seismic instrumentation and data interpretation in India. *Current Science, 79*(9), 1347–1358.

Bhattacharya, S. N., Shukla, A. K., Prakash, R., Gautam, J. L., & Verma R. K. (2001a). Seismicity of National Capital Territory Delhi. *Proceedings of International Conference on Earthquake Strategies with Particular Reference to India, New Delhi,* 1–10.

Bhattacharya, S. N., Srivastav, S. K., Dattatrayam, R. S., Shukla, A. K., Baidya, P. R. & Suresh, G. (2001b). Seismic monitoring and data processing system. *Proceedings of Seismic Hazards*, 231–238.

Bungum, H., Mykkeltveit, S., & Kvaerna, T. (1985). Seismic noise in Fennoscandia, with emphasis on high frequencies. *Bulletin of the Seismological Society of America, 75*, 1489–1513.

Dattatrayam, R. S., Bhatnagar, A. K., Suresh, G., Baidya, P. R. & Gautam, J. L. (2009). Real time earthquake monitoring for early warning of Tsunamis in Indian Ocean region. *Mausam*, Diamond Jubilee Volume, 253–264.

Fusca, I. (1957). Navy wants industry to share burden of solion development. *Aviation Week, 66* (26), 37.

Hansen, R. A., Bungum, H., & Alsaker, A. (1989). Three recent larger earthquakes offshore Norway. *Terra Nova, 1*, 284–295.

Havskov, J., & Alguacil, G. (2002). *Instrumentation in earthquake seismology*. Springer, 358.

Havskov, J., & Bungum, H. (1987). Source parameters for earthquakes in the northern North Sea. *Norsk Geologisk Tidsskrift, 67*, 51–58.

Hurd, R. M., & Lane, R. N. (1957). Principles of very low power electrochemical control devices. *Journal of the Electrochemical Society, 104*, 727–730.

Kvaerna, T., & Mykkeltveit, S. (1985). *Propagation characteristics of regional phases recorded at NORSAR*. NORSAR Semiannual Technical Summary, April–September 1985, NORSAR, Kjeller, Norway, 21–29.

Kvamme, L. B., & Hansen, R. A. (1989). The seismicity in the continental margin areas of Northern Norway. In S. Gregersen & P. W. Basham (Eds.), *Earthquakes at North Atlantic passive margins: Neotectonics and postglacial rebound,* Hingham, MA: Kluwer, 429–440.

Larkam, C. W. (1965). Theoretical analysis of the Solion polarized cathode acoustic linear transducer. *Journal of the Acoustical Society of America, 37*, 664–678.

Ottemoller, L., & Havskov, J. (1999). Seisnet: A general purpose virtual seismic network. *Seismological Research Letters, 70*(5), 522–528.

Shukla A. K., Prakash, R., Shukla H. P., & Bhattacharya, S. N. (2001). *Digital seismic VSAT based telemetry system for National Capital Region: A Perspective.* Proceedings of 32nd IETE Mid-Term Symposium, 10–17.

Shukla, A. K., Prakash, R., Shukla, H. P., Gupta, H. V., Bhattachaya, S. N., & Srivastava, S. K. (2002). Delhi Seismic Telemetry Network and analysis of digital waveform data collected through this network. *Proceedings of 12th Symposium on Earthquake Engineering, IIT Roorkee,* 105–113.

Shukla, A. K., Prakash, R., Pandey, A. P., Singh, R. K., Mandal, H. S., Singh, D., & Baidya, P.R. (2011). *Overview of seismic hazard and mitigation in North East India Region.* IMD Publication.

Srivastav, S. K., Bhattacharya, S. N., Shukla, A. K., Shukla, H. P., Baidya, P. R., & Dua, S. C. (2001). *Near real time seismic monitoring system in and around National Capital Region Delhi.* Proceedings of Seismic Hazards, 221–230.

Srivastava, S. K., Bhattacharya, S. N., Gupta, H. V., Dattatrayam, R. S., Shukla, A. K., Baidya, P. R., Prakash, R., Suresh, G., Shula, H.P., Ghose, A.K., Singh, H., Kondal, J.S., Gautam, J.L., Dua, S.C. & Ravindran, V.V. (2003). Recent developments in seismic monitoring and analysis capabilities in India. *Geological Society of India*, Memoir, 54, 231–251.

Wittenborn, A. F. (1959). Analysis of a logarithmic solion acoustic pressure detector. *Journal of the Acoustical Society of America, 31*, 474.

Chapter 3
Strong-Motion Instrumentation: Current Status and Future Scenario

Ashok Kumar and Himanshu Mittal

1 Introduction

Strong-motion seismographs are instruments designed to record the time history of strong ground motions where the traditional high-gain seismographs used to routinely locate earthquakes go off scale. They are often called accelerographs because they measure acceleration of the ground. Ground motion time- histories recorded during past earthquakes in a region provide valuable information about the expected characteristics of ground motion at a site during a future earthquake in that region. These records are essential for evaluating earthquake resistant design procedures, estimation of attenuation characteristics, assessment of seismic hazard and earthquake risk. Accordingly, analysis of strong ground motion data leads to a better understanding of the potential effect of strong shaking during earthquakes. The ground motion characteristics determined from strong-motion records are studied in various terrain and rock conditions and have been related to various earthquake parameters. The analysis of data has also helped in understanding the soil–structure interaction, effects of soil deposits, topography and other effects. More than half of the area of India is susceptible to strong ground motions from earthquakes; therefore, it is essential to know about the probable characteristics of strong ground motion of future earthquakes in this region.

The philosophy and recording in strong-motion instruments differ from conventional seismographs in very many ways. The conventional seismographs are very sensitive and as mentioned earlier, go out of scale if installed in the region of strong

A. Kumar (✉)
Professor (Retired), Department of Earthquake Engineering, Indian Institute of Technology Roorkee, Roorkee, Uttarakhand, India
e-mail: akmeqfeq@iitr.ac.in

H. Mittal
Department of Geosciences, National Taiwan University, Taipei, Taiwan

M. L. Sharma et al. (eds.), *Advances in Indian Earthquake Engineering and Seismology*, https://doi.org/10.1007/978-3-319-76855-7_3

earthquakes. The conventional seismographs run for 24 h a day, whereas, strong-motion instruments trigger when the ground motion exceeds a threshold values. Conventional seismographs are sensitive to ambient noise, and therefore, some of the low frequencies which match with that of ambient noise are required to be filtered. However, low frequencies are most crucial for engineering applications and therefore strong-motion instruments do not have such filters. But the present scenario in instrumentation has merged the two requirements. The presently available digital accelerographs have very wide dynamic range (108 dB or more) and therefore, are capable of recording very small as well as strong motions. Digital adaptive filters in the instrument can take care of ambient noise.

2 Strong-Motion Accelerographs

Rapid development in the technology of communication and computer-related instrumentation and its impact on seismological instrumentation have enabled scientists and engineers to address themselves to several earthquake-related studies which were difficult to be carried out until about a decade back. Wide dynamic range (more than 108 dB) of the presently available instruments can now serve for recording strong motion as well as weak motion. Capabilities to communicate between instruments at site with central recording station or head quarters through (direct or through satellite) and telephone have drastically reduced the time gap between the occurrence of event and analysis/interpretation of records. In fact this encouraged countries like Mexico, Taiwan and Japan to install early warning systems by detecting and analyzing events and sending warning signals to nearby populated areas giving a lead time of tens of seconds to about a minute to general public to react. Research in India on early warning system is also in progress and it is hoped that one such system in Northern India will be operational soon. Near real-time analysis and dissemination are being commonly achieved by networks in the United States, Japan and Taiwan. Efforts are being made to achieve similar capabilities in India also. Capability to record menu selectable pre event and availability of large memory in small PCMCIA memory cards with required robustness for field environment are commonly available in these instruments. These capabilities have been further strengthened by GPS signals which can give absolute time of trigger with less than few millisecond errors.

2.1 Sensor

The most commonly used sensors in digital strong-motion instruments are force-balanced or servo accelerometer (FBA). These sensors have high natural frequency (about 200 Hz) so that a flat frequency response from DC to about 100 Hz can be obtained. The damping of the accelerometer is about 70% of critical. The dynamic range of the presently available accelerometers is more than 120 dB. With the introduction of FBA, the philosophy of seismic instruments has undergone substantial changes. Among the conventional seismic sensors, an accelerometer can not be used to record the ground motion of microearthquakes for the simple reason that the system is too stiff to produce any measurable deflection of the mass. However, in FBA, the feedback current which is the measure of the ground acceleration, can be amplified and measured quite accurately. Thus the concepts of sensors shifted from mechanical measurement of the conventional sensors to electronic control and measurements employing force balance sensors. The disadvantages of FBA are its large size (about 150 gms for uniaxial and about 400 gms for triaxial) and its high cost (about US$ 700 for uniaxial and about US$ 2000 for triaxial). As the design of FBAs use mechanical coils for feedback, its size and cost are not likely to be reduced in the future.

Recent advancements made in micro electrical machined sensors (MEMS) have drastically changed the overall scenario of sensor specifications. MEMS technology has enabled to reduce the size of accelerometers to be the same as that of an IC chip. Its large-scale use in spaceships, aeroplanes, and automobiles has encouraged a very fast-track development in its technology. Although, MEMS sensors are available in the market since last more than 10 years but till few years back these were not used in seismological applications due to their poor dynamic range. However, since the last few years, we have seen good-quality MEMS sensors with dynamic range of more than 100 dB getting available in the market at a price which is much less than the conventional FBA. In the next few years, further improvements in sensor specifications and reduction in prices of such sensors are expected.

2.2 Recorder

Analog signals from sensors are amplified to match the full-scale voltage of sensor with that of AD converter and this analog signal is also low-pass filtered to remove effect of aliasing during digitization. Most of the recorders are now using delta–sigma-type AD converters which perform digitization at very high sampling rate and have inbuilt digital filters to perform decimation to get the record at the desired sampling rates. Most of the manufactures of accelerographs are using 24-bit delta–sigma AD converters which provide resolution of 18–19-bit accuracy at 200 SPS (the accuracy of delta–sigma AD converters is dependent on sampling rate). Strictly speaking such AD converters are 1-bit delta–sigma modulators followed by digital decimation filter. Most of the manufacturers of digital accelerographs use independent ADC chips for each channel of recording so that channel to channel skew is zero (synchronous sampling). Earlier, manufacturers used one ADC chip by multiplexing three channels and thus data from such instruments used to have channel to channel skew of few milliseconds. The synchronous sampling has enabled proper phase determination which may be important for structural response studies.

The recorders are controlled by microprocessor and each manufacturer of digital accelerograph uses microprocessor chips of their own choice. Although the requirements of basic functioning of digital accelerographs are very simple in comparison with the capabilities of the microprocessors but the processors currently used are not capable enough to perform real-time processing of the recorded data (transducer correction, removal of noise, calculation of response spectra, FFT, etc.).

The digital data is recorded on solid-state memory. In the earlier models of digital accelerographs, DRAM or SRAM chips were used with backup batteries. Nowadays, PCMCIA-type flash memory cards are used which have several advantages over conventional memory chips. These cards can be easily plugged in or taken out by users at the site and can have very large memory with very little additional cost and negligible extra current requirement. PCMCIA is an international standard and has fixed size, pin connections and protocol. Several types of PCs have PCMCIA slots where these cards can be used and read like an external drive (with some additional software). Typically, 4 MB memory card for a triaxial accelerograph is good enough for obtaining a few hours of recording at 200 SPS. It is apparent that in today's and the future technology, requirement of large memory will not be a problem.

All the parameters of the recorders like pre-event time, trigger threshold, mode of trigger, post-event time, post-event threshold, etc. are software selectable. All such instruments have built-in real-time clock and have provision to synchronize the clock with transmitted standard time signals like GPS, thus obtaining accuracy of less than 5 milliseconds.

2.3 Communication

Digital accelerographs, of all brands, have standard ports which can be connected to a computer. Till a few years back, the communication used to be done through RS232 (serial port) but nowadays most accelerographs are communicating through USB port and RJ45 (LAN port). Software for communication of computer with the instruments are supplied by the manufacturers generally in desired platforms like Windows, Unix, Linux, etc. The instruments are capable to be connected with modem or Internet for remote communication. The supplied software also has option for communicating directly or through modem.

The facility of remote communication through modem/Internet has substantially eased the maintenance of instrument as now the instruments at site can be connected easily from the headquarters. In some of the instruments there are facilities of inverse dialing in which, in the case of an event, the instrument dials the headquarters and dumps automatically the entire record or header part of the record onto the computer at the headquarters. Some of the instruments provide real-time data stream which can be telemetered to central recording stations. The rapid advancement in communication is sure to make an impact on capabilities of strong-motion information. Use of combination of satellite communication, Internet communication and real-time operating systems will substantially enhance capabilities of the system.

2.4 *Processing of Data*

As of now, the recorded data is processed off-line after retrieval on the computer. The processing includes transducer correction, filtering to remove high-frequency and low-frequency noise, calculation of response spectra, Fourier spectra, velocity history, displacement history etc. Several methods have been suggested by different authors to process the strong-motion data so that maximum information with desired accuracy can be extracted from strong-motion records.

The advancement in real-time digital signal processing (DSP) technology is likely to have its effect or strong-motion instruments also. The future strong-motion instruments should have DSP chips which should be able to perform the entire processing of the recorded time history as desired by user in real time. Another important function of DSP chips will be removal of ambient noise in real time through adaptive filter. Such filters will be extremely important as the strong-motion instruments will be required to record microearthquakes also.

3 Use of Strong-Motion Data

3.1 *Ground Motion Records*

Ground motion records obtained from strong-motion accelerographs are extremely useful for earth science and engineering communities. Earth scientists use these records for understanding the source mechanism and attenuation characteristics, whereas engineering community use these data for improvement in their skills for design and analysis of earthquake-resistant structural systems. Although the ultimate objective of both the communities is to mitigate disasters caused by earthquakes but the research for engineering application is likely to bring material benefit to mankind in a short span of time, whereas the work of earth scientists is much more difficult and can bring long-term benefits by way of understanding physics of earthquake.

Loss of lives during earthquakes is caused by strong ground shaking which results into collapse of buildings, bridges and other man-made structures. It is, therefore, imperative that for earthquake disaster mitigation, structures be built to resist earthquake-induced shaking levels during their lifetime. Strong ground motion records during earthquakes in the form of time history or spectra provide the basic information for earthquake engineering. These records are inputs for evaluation of seismic hazard of an area and in improving design practices and eventually provide the basics for earthquake engineering research. Analysis of strong ground motion data has accordingly lead to a better understanding of the potential effect of strong shaking during earthquakes. The ground motion characteristics determined from strong-motion records are studied in various terrain and rock conditions and have been related to various earthquake parameters. The analysis of data has also helped in the understanding of the soil–structure interaction, presence of soil deposits, topography and other effects, which are essential for microzonation of an area.

Strong ground motion data are also basic input to provide information needed to define adequately, seismic hazards at national and regional level. Preparation of seismic hazard maps will be an important requirement for engineering, insurance and planning purposes in the near future. All seismic hazard studies, whether deterministic or probabilistic require strong ground motion data to determine attenuation and seismic source characteristics. These studies will eventually lead to optimum reduction in uncertainties for seismic zonation maps and to promote integration of seismological engineering and disaster response activities.

Strong ground motion data are also very important for geotechnical engineers. Records from good rock–soil station pairs can be used to validate methods used by them to provide site-specific ground motion. An understanding of soil–structure interaction and its effect can substantially reduce the cost of construction and retrofitting. Although procedures for considering such effects in design have been recommended in some codes but these are based on theory, which needs to be supported with field data of strong-motion records. Strong-motion data has also been used to determine damping and stiffness of ground during earthquake. Other areas where geotechnical engineers can use these data are seismic safety evaluation of earth and rock-fill dams, study of ground motion of liquifiable sites during and after liquification, study of pile and other types of foundation during earthquakes.

3.2 *Records of Structural Response*

Structural engineers install strong-motion instruments in major structures like bridges, dams, multistoried buildings, nuclear power plants, and other lifeline systems to understand behavior of such systems during earthquakes and validate the analytical models used by them. The recorded response of such structures provides insight about performance of the structure as well as precise information on vibration characteristic and response of the system. The parameters, which are sought to be derived from such records, are natural frequencies, mode shapes, damping, relative displacement and acceleration distribution. For lower levels of excitation, these parameters can be determined through linear analysis. However, when the response is large and nonlinear, more rigorous analysis will be required by using system identification techniques to determine the parameters of linear or nonlinear model of the structure that best fit the recorded response.

Most structures are designed to remain elastic for code-level design forces and relative displacements, and some structural members are expected to go in the plastic range during large excitations. Thus, normally one can expect that there should be no damage to the structure if the records show that the building response is linear and earthquake forces and relative displacements are smaller than the design values. The shortening of natural frequency during the shaking, larger base shear than the design base shear or larger relative displacements is an indication of structural nonlinearity and the structural damage. From the structural drawings, it is possible to analyze the structures to determine the maximum relative displacement that the structure can withstand as well as maximum base shear that causes members to fail and also approximate reduction in the natural frequency which can be indicative of failure. These values can then be compared with the values derived from the records and failure pattern if any can be studied.

There can be several studies which can be taken up from the structural response records obtained from earthquakes. Some are listed below:

1. Evaluation of analytical models and methods that are presently available for seismic analysis of buildings
2. Estimating the response of the building and particularly the damage that the building can suffer from future strong ground motion
3. Development of a simplified method of analysis to estimate the expected overall and local deformations
4. Evaluation of codal provisions in regard to earthquake-resistant design of such buildings
5. Evaluation of the effects of site geology and local topography on seismic response of structures
6. Evaluation of present approach of system identification techniques of inferring the dynamic characteristics of buildings from its recorded response
7. Preparation of data bank about behavior of different types of structures

Strong-motion instrumentation of multistoried buildings and special structures have been done in several countries like the United States, Japan, Mexico, Korea, Switzerland, New Zealand, India, etc. In the state of California in the United States, the code of practice requires installation of at least three numbers of three component accelerographs in each multistoried building. In addition to the above, there are hundreds of extensively instrumented buildings in California which have more than 15 sensors installed at various locations in the building. Northridge earthquake of 1994 in the United States produced one of the most valuable database of ground and earthquake building responses in history. In this earthquake, 116 free field installations and 77 extensively instrumented structures maintained by California Strong-Motion Instrumentation Program produced records. In addition to the above several extensively instrumented buildings maintained by other agencies and several hundred code-required accelerographs also produced records. The analysis of the records obtained has resulted into a large number of publications in various journals, conferences and seminars as well as research reports.

4 Historical Background in Indian Context

India has a two-pronged earthquake problem. There is severe seismic hazard along the Himalayan belt and also at the western margin of the country in the state of Gujarat. The Indian plate pushes into the Asian plate at the high rate of 15–20 cm/

year (Bilham et al., 1998), and as a direct result of the collision between the Indian and the Asian plates the state of stress in the Indian plate is high, which in turn increases the earthquake hazard, particularly in northern India along the Himalayan collision zone. This process has given rise to three major thrust planes: (e.g., Gansser, 1964; Molnar and Chen, 1982) the Main Central Thrust (MCT), the Main Boundary Thrust (MBT), and the Main Frontal Thrust (MFT). The region has experienced several great earthquakes in the past 100 years or so (1897 Assam; 1905 Kangra; 1934 Bihar–Nepal; 1950 Assam). The Himalayan geodynamics and the occurrence of great earthquakes are well summarized by Seeber and Armbruster (1981), Khattri (1999), and Bilham and Gaur (2000). During the last episode of strain release, a 750-km-long segment, which lies between the eastern edge of the 1905 rupture zone and the western edge of the 1934 earthquake, remained unbroken (Fig. 3.1). This segment, called the central seismic gap, continues to be under high strain (Singh et al., 2002). Large earthquakes occurred in this seismic gap in 1803 and 1833, but the magnitudes of these earthquakes were less than 8, and hence, they were not gap-filling events (Khattri, 1999; Bilham, 1995). Based on these considerations and on a shortening rate of 20 mm/year across the Himalayas (Lyon-Caen and Molnar, 1985; Avouac and Tapponnier, 1993; Gahalaut and Chander, 1997; Bilham

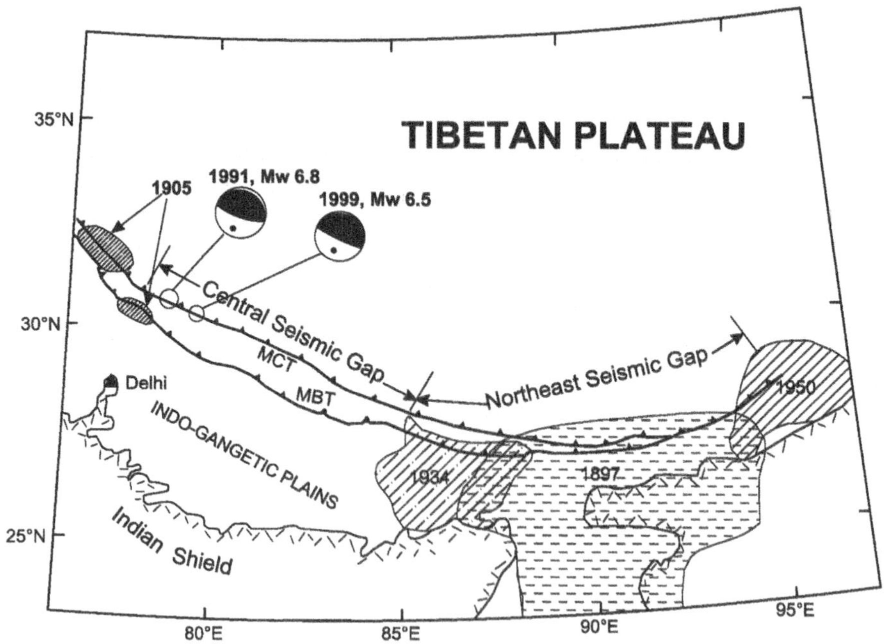

Fig. 3.1 Tectonic map of the Himalayas. The different two seismic gaps are shown. The segment between the rupture areas of the 1905 and 1934 earthquakes is known as the central seismic gap, and the segment between the rupture areas of the 1934 and 1950 earthquakes is known as the northeast seismic gap. MCT, main central thrust; MBT, main boundary thrust (modified after Seeber and Armbruster, 1981)

et al., 1998), Khattri (1999) has estimated the probability of occurrence of a great Mw 8.5 earthquake in the gap in the next 100 years to be 0.59.

The northeastern region of India is also regarded as one of the most seismically active regions worldwide. A seismic gap called northeast seismic gap is there between the 1950 Assam earthquake and the 1934 Bihar–Nepal earthquake (Fig. 3.1), where no major earthquake occurred in recent time. The past seismicity data from 1897 shows that the northeastern region has experienced two great earthquakes with magnitudes above 8.0 and about 20 large earthquakes with magnitudes varying between 8.0 and 7.0 (Kayal et al., 2006). These devastating earthquakes occurred in northeastern India when population was 10 times less than at present; if such earthquakes were to occur in the near future, they would be much more devastating, thus emphasizing the need for seismic hazard estimation in northeastern India as well. Studies related to hazard estimation depend on the availability of strong ground motion records from past earthquakes.

The strong-motion program in India was started in the mid-1960s when analog accelerographs named as RESA (Roorkee Earthquake School Accelerograph) and another low-cost strong-motion instrument known as the structural response recorder (SRR) were developed at the Department of Earthquake Engineering (DEQ), IIT Roorkee. Initially these instruments were installed in some river valley projects like Bhakra, Pong, Talwara, Tehri, etc. Later, in 1976, the research project INSMIN (Indian National Strong-Motion Instrumentation Network) was funded, on the recommendation of the Planning Commission, by the Department of Science and Technology (DST), Government of India, for fabrication, installation, maintenance and operation of RESA-V and SRRs.

Further, on the recommendation of the International Association of Earthquake Engineering (IAEE), funds were sanctioned by the National Science Foundation, USA, in 1982, to DEQ for installation of an array of 50 analog accelerographs in the Shillong region of northeastern India from which fifty analogue accelerographs (imported) were installed. In 1986, through funding from DST about 50 instruments in Himachal Pradesh, 40 in western Uttar Pradesh hills and 30 instruments in Arunachal Pradesh, Assam and Bihar were installed (Chandrasekaran and Das, 1992). All these instruments have now outlived their lives and most of them are nonfunctional and unrepairable. However, during their lifetime, this instrumentation has provided useful strong ground motion data which was disseminated and widely used for research.

Strong-motion instrumentation program in our country got big boost and was substantially strengthened in 2004 when DST sanctioned a project for installation of 300 strong-motion accelerographs in this region (Mittal et al., 2006). Under this project, state-of-the art instruments were procured and successfully installed in different parts of the country covering seismic Zone IV and V and some of the thickly populated areas of Zone III of seismic zoning map of India (BIS, 2002). This network was further strengthened in 2007 when another project entitled "Strong Motion Instrumentation Network in Delhi" was sanctioned by DST to IITR. Under this project 20 digital strong-motion accelerographs were installed in the Delhi region (Kumar et al., 2012).

5 Current Status of Strong-Motion Program in India

There are several organizations which are operating strong-motion instruments in India. The Institute of Seismological Research, Gandhinagar, is operating a network of about 40 strong-motion instruments in Gujarat, IIT Kharagpur is operating a network of about 12 stations in Sikkim; NGRI, Hyderabad, is operating about 10 stations in Kutchh and Indian Meteorological Department (IMD) is operating about 50 instruments installed in their observatories located all over India. In addition to the above, several river valley projects and some research institutes are also operating strong-motion network for specific studies. But the biggest network of about 300 strong-motion accelerographs is being operated by the Department of Earthquake Engineering, IIT Roorkee. This instrumentation will be described here.

5.1 Indian National Strong-Motion Instrumentation Network

The strong-motion instrumentation network of IITR covers the Indian Himalayan range from Jammu and Kashmir to Meghalaya. The accelerographs were procured in September 2005, and their installation started in November 2005. In total, 293 strong-motion stations have been installed in the states of Himachal Pradesh, Punjab, Haryana, Rajasthan, Uttarakhand, Uttar Pradesh, Bihar, Sikkim, West Bengal, Andaman and Nicobar, Meghalaya, Arunachal Pradesh, Mizoram, and Assam. A map showing the location of these instruments is given in Fig. 3.2.

The first phase of the project concerned the choice of the new sites for installations. Site selection has been generally a compromise between network geometry, logistics and safe installation. Based on these criteria, sites falling in seismic Zones IV and V of the country and in some densely populated cities of Zone III were selected and instrumented. Average station-to-station distance is kept to be 40–50 km, which ensures triggering of at least two or more accelerographs (set at trigger level of 5 gals) if a magnitude 5 or larger earthquake occurs anywhere in northern or northeast India. Typically the instruments are installed in a room on the ground floor of preferably a government-owned, one- or two-storied building where proper logistics are available, which means that the instrument is secure from tampering and 220-V AC power supply is available. In Delhi, however, free field instruments were installed inside a specially fabricated housing. Schematic diagram of a typical installation is shown in Fig. 3.3.

5.2 Networking

Modern telecommunications are utilized to allow networking of these accelerographs. An understanding with the National Informatics Centre (NIC), Government of India was achieved to permit the use of their NICNET facility at district headquarters. Telecommunication links to the entire instrumentation are

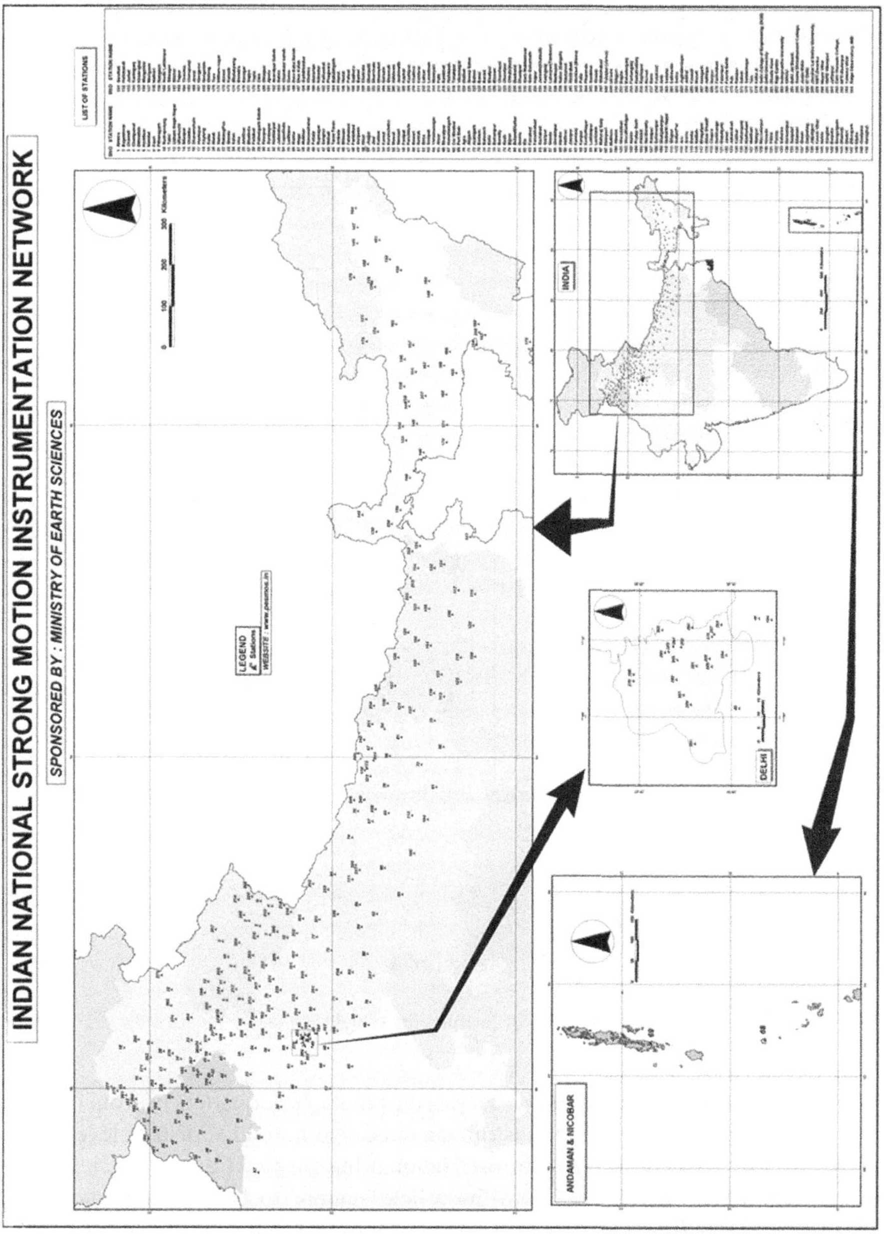

Fig. 3.2 Map showing the location of instruments along the Himalayan belt. Inset: map showing the location of instruments in Andaman Nicobar and Delhi

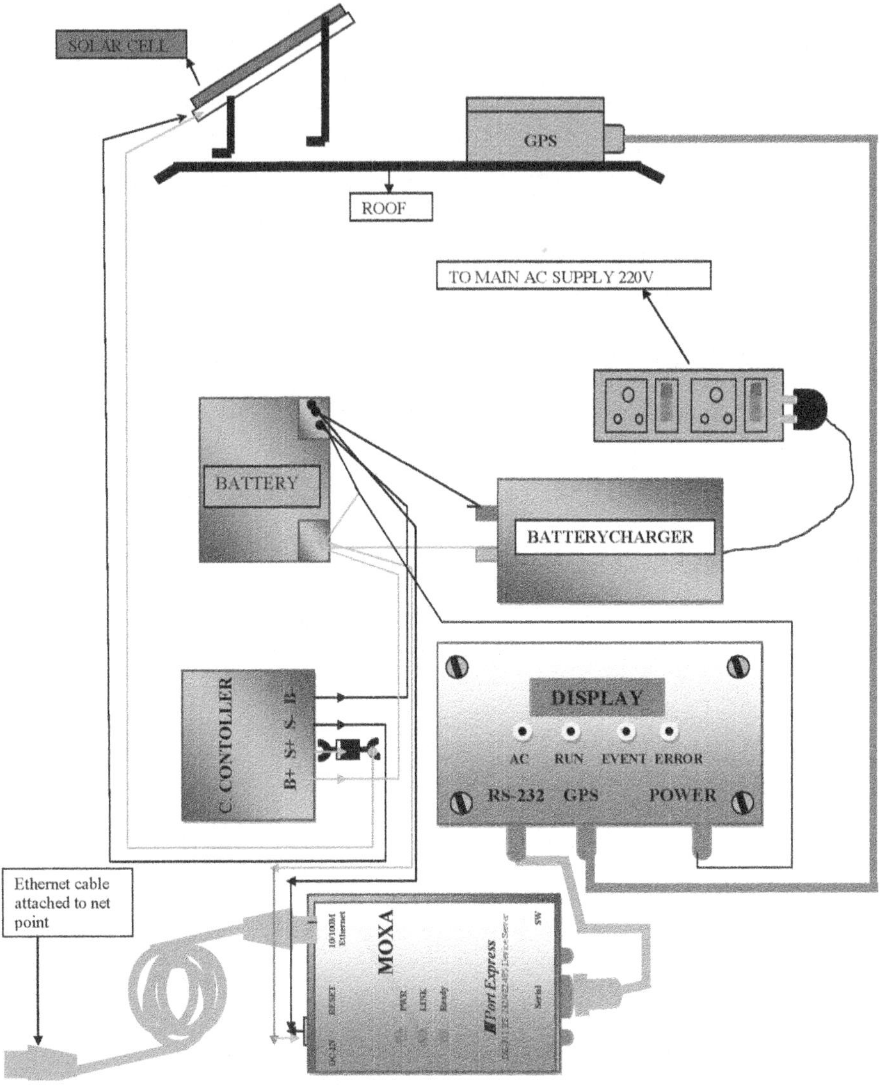

Fig. 3.3 A figure showing the installation of instruments in field

planned so that each instrument can be accessed from the headquarters or from other remote locations. Currently, all the instruments installed in field at district level are connected to the VSAT/leased line of the National Informatics Centre (NIC). Data flows through VSAT/leased line from these field stations to NIC headquarters in Delhi. From Delhi to Roorkee, data flows on a 2-Mbps leased line of Bharat Sanchar Nigam Limited (BSNL) (Fig. 3.4). The instruments situated in subdivision/towns are connected through state wide area network (SWAN). All the 20 installations of Delhi

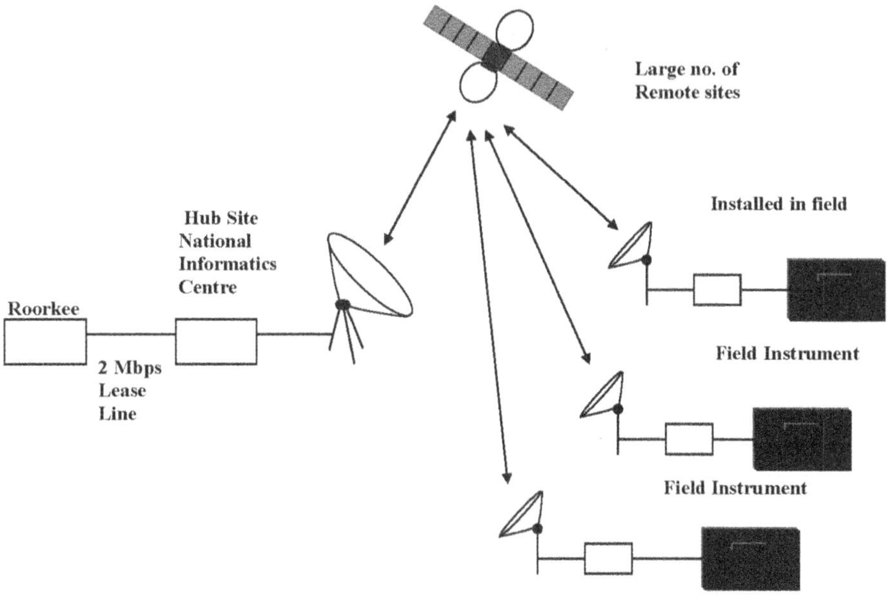

Fig. 3.4 A diagram showing the connectivity of instruments from field to Roorkee

are connected to Roorkee through the Mahanagar Telephone Nagar Limited (MTNL) network. With this connectivity, about 240 accelerograph stations can be accessed from headquarters at Roorkee. These instruments are now routinely checked remotely for their health and for data downloads when earthquakes occur.

5.3 *Instruments*

All 300 strong-motion accelerographs consist of internal AC-63 GeoSIG triaxial force-balanced accelerometers and GSR-18 GeoSIG 18-bit digitizers with external GPS. The 12 strong-motion accelerographs installed in Delhi are K-2 (Kinemetrics K-2's) with internal accelerometer (model EpiSensor) and 18-bit digitizer. The recording for all instruments is in trigger mode at a sampling frequency of 200 sps. The triggering threshold was initially set at 0.005 g for all the instruments. The recording is done on a 256-MB GeoSIG or 1-GB Kinemetrics compact flash card, respectively.

5.4 *Performance*

This instrumentation network has recorded around 170 earthquakes comprising of about 500 time histories (till August 2012) in a span of 5 years. Several prominent earthquakes in the northern Himalayas as well as in the Northeast Himalayas were recorded. Figure 3.5 shows example of a strong-motion accelerograph record from the *Mw* 4.1 Delhi Haryana border region earthquake that occurred on November 25, 2007, and Fig. 3.6 shows a strong-motion accelerograph record from the *Mw* 6.5

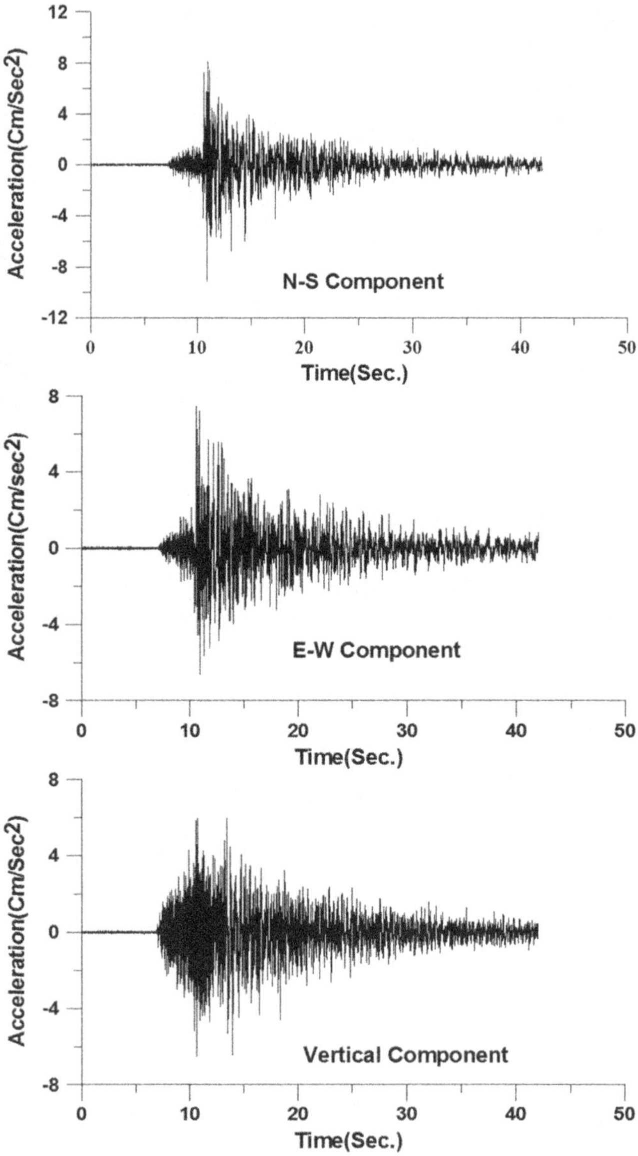

Fig. 3.5 A sample strong-motion (Kinemetrics) record of a Mw 4.1 earthquake from the Delhi Haryana border region recorded at the IMD station (epicentral distance 28 km) on November 25, 2007

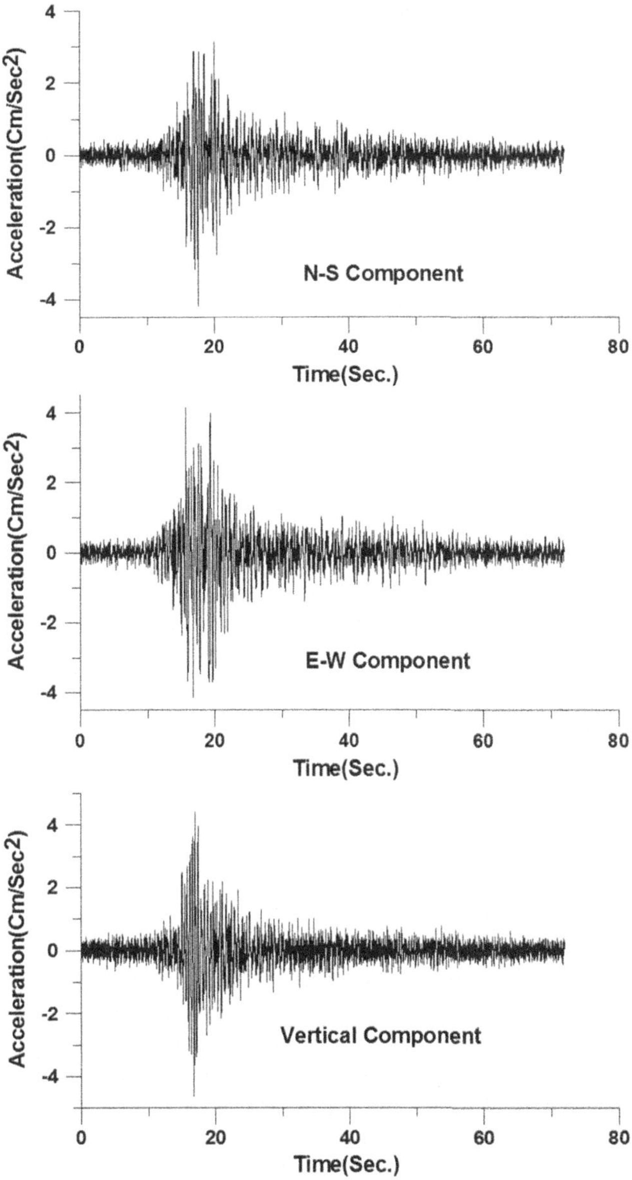

Fig. 3.6 A sample strong-motion (GeoSig) record of a Mw 6.5 earthquake from Hindukush region recorded at the Keylong station (epicentral distance 717 km) on September 17, 2010

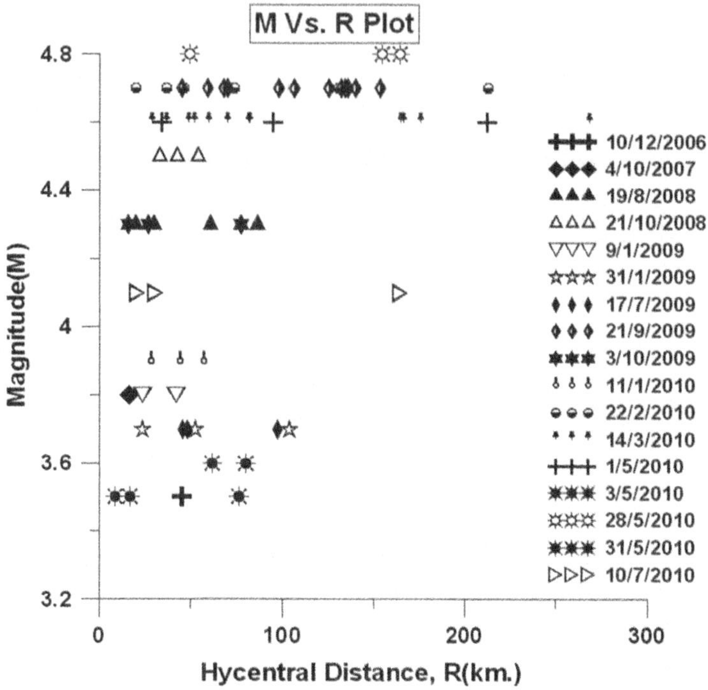

Fig. 3.7 Plot for <5 magnitude multi-station-recorded earthquakes in the northern Himalayas

Hindukush region earthquake that occurred on September 17, 2010. Records were obtained for near-source earthquakes at distances of 10 km and for a far-source earthquake at a distance of 1000 km. Based on experience with the performance of the instruments, the trigger thresholds of all the instruments placed in Himachal Pradesh, Uttarakhand and some of the northeast instruments were lowered to 0.002 g from 0.005 g. By doing this, we can now record lower-magnitude earthquakes. After due monitoring, we are planning to lower the triggering threshold of all the other instruments connected through NICNET. Figures 3.7 and 3.8 show plots of magnitude (M) vs. hypocentral distance (R) for <5 and $\geq$5 magnitude earthquakes in the northern Himalayas, while Fig. 3.9 shows a plot of M vs. R for multi-station-recorded earthquakes in the North-Eastern Himalayas.

5.5 *Processing of Records*

All downloaded accelerograms are processed before dissemination. Computer programs developed by the first author of this paper are used to process the records. As the headers of ASCII files are entirely different for records obtained from GeoSIG

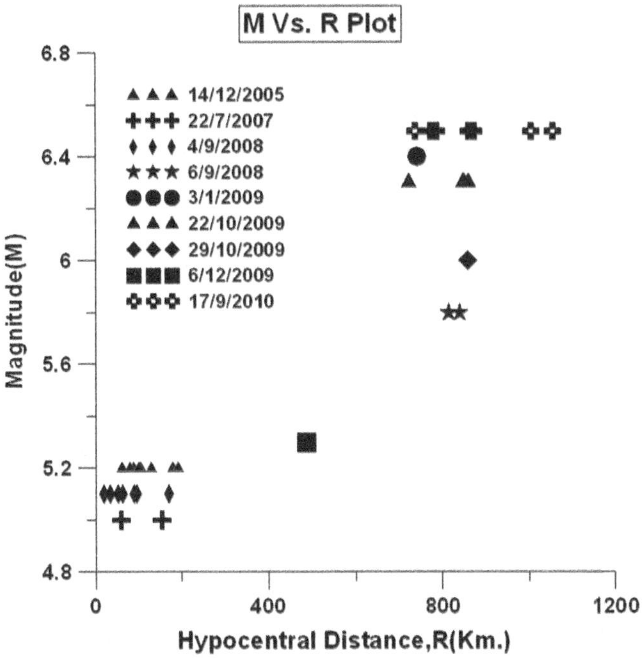

Fig. 3.8 Plot for ≥5 magnitude multi-station-recorded earthquake in the northern Himalayas

and Kinemetrics accelerographs, for ease of execution, it was essential to develop two different computer programs. In general both computer programs read headers of ASCII files of records, generates the output header as shown in Fig. 3.10, baseline corrects the record and rotates the horizontal components to get N–S and E–W component. Instrument correction of records is not done, since natural frequency of sensors is quite high (about 200 Hz) and thus it will not make any difference. Also no frequency filters are applied and it is left to users to apply filters according to their need. However, raw accelerograms are available to users on request. A typical data file for an earthquake record is given in Fig. 3.10. The *Mw* 5.1 Chamoli earthquake of December 14, 2005, is the first earthquake recorded by the digital instrumentation. In total, 8 instruments of the network were triggered by this earthquake and provided records. Table 3.1 gives the computed peak displacement for three components of various stations during this earthquake.

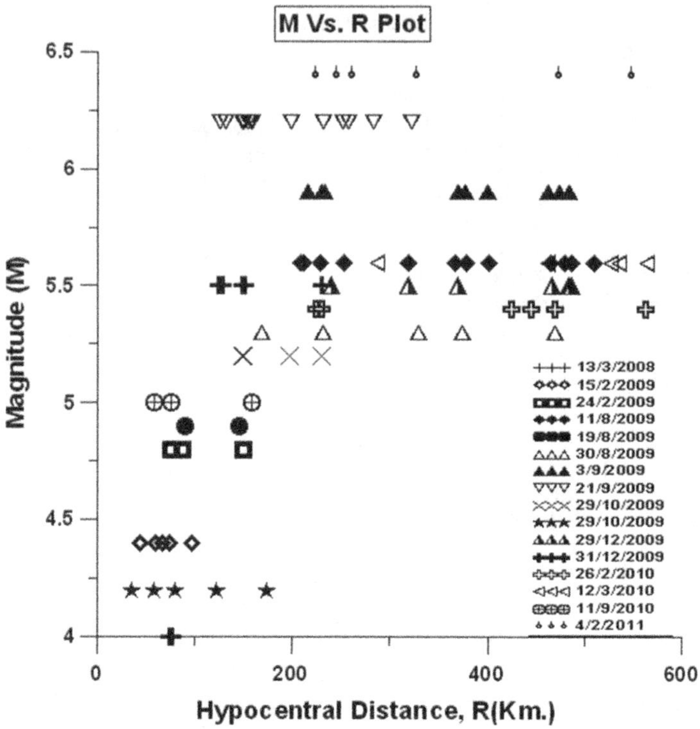

Fig. 3.9 Plot for multi-station-recorded earthquakes in NE Himalayas

6 Concluding Remarks

Research in the fields of earthquake engineering or engineering seismology requires strong-motion data for several applications, such as the development of ground motion prediction equations, the evaluation of shaking scenarios, the definition of probabilistic hazard maps, and the formulation of seismic codes.

India is one of the hazardous countries in the world and several moderate to strong earthquakes occurred since the early 1970s, when an early-stage strong-motion monitoring began. During the past 40 years, strong-motion data were mainly collected by the strong-motion network by IIT Roorkee. The data recorded by IIT network are disseminated through the website www.pesmos.in. At present the data recorded by IIT network are collected and processed to improve the data bank. The main restriction of the IIT network is represented by the data transmission system that does not allow a real-time acquisition of waveforms. The future developments will concern the data transmission system, through the upgrade of the communication system, so that data processing is done in real time.

Fig. 3.10 Example of header file of each component

```
Origin Time      17/9/2010 19:21:09 UTC
Lat.             36.5 N
Long.            70.8 E
Depth (Km)       167.0
Magnitude        6.5
 Region          Hindukush
 Above details taken from IMD

Station Code     KLG
Station Lat.     32.559 N
Station Long.    77.001 E
Station Height(m) 3129.0
Record Time      17.09.2010 19:23:45.109
Sampling Rate    200. Hz
 Record Duration  71.895 Sec.
 Direction       E-W (E positive)
Max. Acceleration -3.982 cm/sec**2
 Base Line Corrected Time History
 Acceleration data in cm/sec**2

  -0.075
  -0.117
   0.014
   0.044
  -0.027
..............................................
..............................................
```

Table 3.1 Computed longitudinal, transverse, and vertical peak ground displacement of different stations (modified after Mittal et al., 2006)

		Peak displacement (cm) values of LTV components		
Stations	Hypocentral distance	Longitudinal	Transverse	Vertical
Bageshwar	127.13	−0.0413	−0.0424	−0.0140
Chamoli	54.30	0.4162	0.4152	0.1212
Champawat	190.18	0.0179	0.0279	0.0279
Pauri	97.28	0.0280	−0.0182	−0.0182
Rudarprayag	74.61	0.0391	0.0649	−0.0362
Tehri	101.73	−0.0304	0.0528	−0.0293
Uttarkashi	83.81	−0.0514	0.0329	0.0225

Acknowledgements The authors wish to thank the Department of Science and Technology (DST), Government of India, and the Ministry of Earth Sciences (MoES) for providing funds to execute different projects. We are thankful to the faculty of the Department of Earthquake Engineering and to the Dean of Sponsored Research and Industrial Consultancy, Indian Institute of Technology, Roorkee, for their constant support and encouragement. We are also thankful to the technical and administrative staff of the Department of Earthquake Engineering, who have made important contributions in this project.

References

Avouac, J., & Tapponnier, P. (1993). Kinematic model of active deformation in central Asia. *Geophysical Research Letters, 20*, 895–898.

Bilham, R. (1995). Location and magnitude of the Nepal earthquake and its relation to the rupture zones of the contiguous great Himalayan earthquakes. *Current Science, 69*, 101–128.

Bilham, R., & Gaur, V. K. (2000). The geodetic contribution to Indian seismotectonics. *Current Science, 79*, 1259–1269.

Bilham, R., Blume, F., Bendick, R., & Gaur, V. K. (1998). The geodetic constraints on the translation and deformation of India: Implications for future great Himalayan earthquakes. *Current Science, 74*, 213–229.

BIS. (2002). *Criteria for earthquake resistant design of structures, part I—general provisions and buildings*. Bureau of Indian Standards, IS1893 [part I].

Chandrasekaran, A. R., & Das, J. D. (1992). Strong motion arrays in India and analysis of data from Shillong array. *Current Science, 62*, 233–250.

Gahalaut, V. K., & Chander, V. (1997). On interseismic elevation changes and strain accumulation for great thrust earthquakes in the Nepal Himalaya. *Geophysical Research Letters, 24*, 1011–1014.

Gansser, A. (1964). *Geology of the Himalayas*, New York: Interscience, 289.

GSI. (2000). *Seismotectonic Atlas of India and its Environs*. Kolkata: Geological Survey of India.

Kayal, J. R., Arefiev, S. S., Barua, S., Hazarika, V., Gogoi, N., Kumar, A., Chowdhury, S. N., & Kalita, S. (2006). Shillong plateau earthquakes in northeast India region: Complex tectonic model. *Current Science, 19*(1), 109–114.

Khattri, K. N. (1999). An evaluation of earthquakes hazard and risk in northern India. *Himalayan Geology, 20*, 1–46.

Kumar, A., Mittal, H., Sachdeva, R., & Kumar, A. (2012). Indian strong motion instrumentation network. *Seismological Research Letters, 83*(1), 59–66.

Lyon-Caen, H., & Molnar, V. (1985). Gravity anomalies, flexure of the Indian plate, and structure, support, and evolution of the Himalaya and Ganga basin. *Tectonics, 4*, 513–538.

Mittal, H., Gupta, S., Srivastava, A., Dubey, R. N., & Kumar, A. (2006). *National strong motion instrumentation project: An overview*. 13th Symposium on Earthquake Engineering, Indian Institute of Technology, Roorkee, December 18–20, 107–115.

Molnar, P., & Chen, W. P. (1982). Seismicity and mountain building. In K. Hsu (Ed.), *Mountain building processes*, New York, NY: Academic Press, 41–57.

Seeber, L., & Armbruster, J. G. (1981). Great detachment earthquakes along the Himalayan arc and long-term forecasting. In *Earthquake prediction: An international review, Maurice Ewing series 4*, Washington, DC: American Geophysical Union, 259–277.

Singh, S. K., Mohanty, W. K., Bansal, B. K., & Roonwal, G. S. (2002). Ground motion in Delhi from future large/great earthquakes in the central seismic gap of the Himalayan arc. *Bulletin of Seismological Society of America, 92*, 555–569.

Chapter 4
Regression Relations for Magnitude Conversion for the Indian Region

H. R. Wason, Ranjit Das, and M. L. Sharma

1 Introduction

Earthquake catalogs contain data about time of occurrence, hypocentral coordinates, earthquake magnitude, and other earthquake information. The data is derived from the observations of different seismic wave types, generally from heterogeneous networks of seismometers in space–time domain. Earthquake catalogs provide useful inputs for studies dealing with seismicity, seismotectonics, internal structure of the earth and seismic hazard assessment. A homogeneous earthquake catalog is of critical importance for the study of the earthquake occurrence patterns in space and time, seismic hazard estimates, land use planning, seismic microzonation and other seismological applications. These studies require well defined and consistently determined earthquake magnitudes. Historical seismic records usually do not fulfill this criterion.

Ever since the first magnitude scale (M_L) was proposed by Richter (1935) for earthquake occurrences in Southern California based on seismic body wave recordings from a network of standard Wood-Anderson torsion seismometers, a need had been felt by seismologists for extension of this magnitude scale to seismic wave types and data produced in different earthquake environments. These efforts culminated in formulation of additional magnitude scales, namely body wave magnitude (m_b/m_B), surface wave magnitude (M_s), and moment magnitude (M_w). These magnitude estimates utilize different information of the seismic source process and, hence, are manifested with variable measurement error.

H. R. Wason (✉) · M. L. Sharma
Department of Earthquake Engineering, Indian Institute of Technology Roorkee, Roorkee, Uttarakhand, India

R. Das
National Research Center for Integrated Natural Disaster Management, Santiago, Chile

M. L. Sharma et al. (eds.), *Advances in Indian Earthquake Engineering and Seismology*, https://doi.org/10.1007/978-3-319-76855-7_4

The concept of body wave magnitude was proposed initially by Gutenberg (1945a, b) and later reworked by Gutenberg and Richter (1956). Determination of body wave magnitude, m_B, is based on the ratio of maximum amplitude to period of P or S waves with periods up to about 10 sec recorded by intermediate to long period instruments. The body wave magnitude, m_b, reported in ISC and NEIC bulletins is based on the first 5 sec arrival of short period P waves recorded by short period instruments. The surface wave magnitude, M_s, is estimated using amplitude and corresponding period of Rayleigh waves with periods ranging between 10 and 60 sec based on Prague formula (Vanek et al., 1962) for shallow focal depths.

The local (M_L), body wave (m_b), and surface wave (M_s) magnitude scales suffer from a major shortcoming that they do not behave uniformly for all magnitude ranges and also exhibit saturation effects for large earthquakes at different magnitude thresholds. These limitations can result in under- or overestimation of earthquake magnitudes in certain magnitude ranges. In order to get rid of the saturation disadvantage present in the existing scales, Kanamori (1977) and Hanks and Kanamori (1979) proposed a new magnitude scale based on the seismic moment, called moment magnitude (M_w) defined by

$$M_w = \left(\frac{2}{3}\right) \log M_0 - 10.7 \tag{4.1}$$

where M_0 is the seismic moment in dyne-cm. The M_w scale as defined above does not saturate for large earthquakes as it is directly proportional to the logarithm of the seismic moment which in turn is related to earthquake source physics (average slip, fault plane area, rigidity modulus) depicting a uniform behavior for all magnitude ranges. Hence, moment magnitude is considered to be the most reliable magnitude to describe the size of earthquakes. Due to the explicit advantages of the M_w scale over other magnitude scales, it is desirable to compile earthquake catalogs in M_w for the purposes of seismic hazard assessment and other important seismicity studies. Different types of magnitudes when used in seismic hazard studies result in increase in uncertainties which affect the ultimate results of seismic risk assessment having direct bearing on economic and societal aspects. This leads to the requirement of homogenization of catalogues to a single scale of magnitude.

Various statistical regression procedures are in use for obtaining magnitude conversion relations required for preparation of homogenized earthquake catalogs, namely Standard Linear Regression (SR), Inverted Standard Regression (ISR), Orthogonal Regression (OR) and General Orthogonal Regression (GOR). In SR, it is assumed that the predictor (independent) variable is either error free or the order of its error is very small compared to the measurement error of the dependent variable. In this regression, vertical residuals are minimized. On the other hand, ISR is obtained by minimizing the horizontal offsets to the best fit line as the role of the dependent and independent variables gets reversed.

General orthogonal regression (GOR) takes into account the errors on both the variables (Madansky, 1959; Fuller, 1987; Gusev, 1991; Ristau, 2009) and is based on the minimization of the squares of the statistical Euclidean distances:

$$\sum_{i=1}^{n}\left\{\frac{(Y_i-\beta_0-\beta_1 x_i)^2}{\sigma_u^2}+\frac{(X_i-x_i)^2}{\sigma_e^2}\right\} \tag{4.2}$$

where (X_i, Y_i) and (x_i, y_i) are the observed and the true values, respectively, of the n data pairs of the independent and the dependent variables, β_0 and β_1 represent the intercept and the slope of the GOR line, and σ_e^2 and σ_u^2 denote the errors of X_i and Y_i, respectively. GOR relations between different magnitude types based on different data sets have been reported in several studies (e.g., Castellaro et al., 2006; Castellaro and Bormann, 2007; Ristau, 2009; Das et al., 2011, 2012, 2013; Wason et al., 2012). It has been observed by Das et al. (2017) that the Eq. (4.2) is incorrectly represented in some studies (e.g., Eq. (4.8) of Castellaro et al., 2006).

In this chapter, regression relations for conversion of m_b and M_s magnitudes to M_w applicable for the Indian region using SR, ISR and GOR procedures have been described. A unified homogeneous earthquake catalog prepared using the scaling relations presented here can serve as a reference catalog for reliable seismic hazard estimates and other seismicity studies for the Indian region, in particular, for Northeast India.

2 Earthquake Catalogs for the Indian Region

Authentic historical seismicity record for the Indian region extends to only about past 500 years or so. After a general compilation of some Indian earthquakes by Baird-Smith (1844), the first detailed catalog for earthquakes occurring in and near India from historical times to the year 1869 was prepared by Oldham (1883). Later on from 1937 onwards, some significant catalog compilations have been done by several researchers, although most cover short durations of seismicity record and/or earthquake magnitude range. For the peninsular region, catalogs by Gubin (1968), Kelkar (1968), Chandra (1977), and Guha and Basu (1993) have been prepared. The catalog by Chandra (1977) was prepared using event data from 33 different sources from 1594 to 1975 covering latitudes 5° N–28° N and longitudes 67.5° E–90° E. While updating the list of earthquakes, Chandra (1977) also pointed out some inconsistencies in the data in the earthquakes reported by Gubin (1968).

The Indian Society of Earthquake Technology published an earthquake catalog compiled by Bapat et al. (1983) for India and its neighborhood region bounded by latitudes 0° N–50° N and longitudes 50° E–100° E, from ancient times to 1979. This catalog contains significant historical earthquakes for India and adjoining region.

Gupta et al. (1986) prepared a catalog of 504 historical earthquakes for Northeast India and adjoining area for the period 1897–1962. In this catalog focal depths for around 89 earthquakes are also reported. This catalog is considered as one of the base catalogs for Northeast India for historical seismicity. Other significant catalogs for India and neighboring region include and Bhattacharya (1998). All these catalogs use different magnitude types and are heterogeneous in nature.

Recently, efforts have been made by some researchers to prepare unified homogeneous earthquake catalogs for the Indian region or specific areas like Northeast India region (e.g., Thingbaijam et al., 2008; Yadav et al., 2009; Das, 2013). Magnitude conversion relations used for conversion of different magnitude types to moment magnitude in different studies are, however, based on different regression procedures.

3 Regression Procedures

For homogenization of earthquake catalogs, it is important to know how different magnitude determinations compare with each other and their associated measurement errors. For applying SR, firstly, the independent variable should either be error free or have negligible error, and secondly, data and uncertainties should be normally distributed. These requirements are usually not fulfilled in the event magnitude data reported by different agencies. Standard least-squares regression minimizes the vertical offsets to the best fit line. ISR is similar to SR but instead minimizes the horizontal offsets to the best fit line. When both the magnitude types involved in the regression have measurement errors, it is appropriate to use GOR. The GOR procedure requires the knowledge of error variance ratio (η) between the two magnitude types. The GOR procedure is described in detail in the literature (e.g., Madansky, 1959; Kendall and Stuart, 1979; Fuller, 1987; Carroll and Ruppert, 1996; Castellaro et al., 2006; Das et al., 2011, 2014), and a brief description is included here.

Let us assume that two variables M_y and M_x are linearly related and that their measurement errors ε and δ are independent normal variates with variances σ_ε^2 and σ_δ^2, respectively. We can thus write

$$m_y = M_y + \varepsilon, \tag{4.3}$$

$$m_x = M_x + \delta, \tag{4.4}$$

and the regression-like model

$$M_y = \alpha + \beta M_x + \hat{\varepsilon}, \tag{4.5}$$

where

$$\hat{\varepsilon} = \varepsilon + \delta. \tag{4.6}$$

The error variance ratio is given by

$$\eta = \frac{\sigma_\varepsilon^2}{\sigma_\delta^2}, \tag{4.7}$$

where $\sigma_\varepsilon^2 = \sigma_{m_y}^2$ and $\sigma_\delta^2 = \sigma_{m_x}^2$, provided that σ_ε^2 and σ_δ^2 are constants.

If $s_{m_y}^2, s_{m_x}^2$ and $s_{m_x m_y}$ denote the sample variance of the M_y and M_x and the sample covariance, respectively, the general orthogonal estimator of slope is given by (Fuller 1987)

$$\hat{\beta} = \frac{s_{m_y}^2 - \eta s_{m_x}^2 + \sqrt{\left(s_{m_y}^2 - \eta s_{m_x}^2\right)^2 + 4\eta s_{m_x m_y}^2}}{2 s_{m_x m_y}}, \tag{4.8}$$

and the estimator of the intercept is

$$\hat{\alpha} = \overline{m}_y - \hat{\beta}\,\overline{m}_x, \tag{4.9}$$

where $\overline{m}_y$ and $\overline{m}_x$ denote the average values.

As demonstrated by Fuller (1987), the following formulas can be used to estimate the errors on the regression parameters.

$$\hat{\sigma}_\beta = \frac{\hat{\sigma}_{m_x}(n-1)(n+\hat{\beta}^2)\hat{\sigma}_\delta + (\sigma_\delta)^2(n-1)(\eta+\beta^2)^2 - (n-2)(-\hat{\beta}\,\hat{\sigma}_\delta)^2}{(n-2)(n-1)\sigma^2{}_{m_x}}, \tag{4.10}$$

and

$$\hat{\sigma}^2{}_\alpha = \frac{(n-1)(\eta+\hat{\beta}^2)\hat{\sigma}_\delta}{n(n-2)} + \overline{m}_x{}^2\hat{\sigma}_\beta^2, \tag{4.11}$$

where

$$\hat{\sigma}\, m_x = \frac{\sqrt{\left(s_{m_y}^2 - \eta s_{m_x}^2\right)^2 + 4\eta s_{m_x m_y}^2} - \left(s_{m_y}^2 - \eta s_{m_x}^2\right)}{2\eta}, \tag{4.12}$$

and

$$\hat{\sigma}_\delta = \frac{\left(s_{m_y}^2 - \eta s_{m_x}^2\right) - \sqrt{\left(s_{m_y}^2 - \eta s_{m_x}^2\right)^2 + 4\eta s_{m_x m_y}^2}}{2\eta}. \tag{4.13}$$

4 Measurement Errors in Different Magnitude Types

International Seismological Center (ISC), UK (http://www.isc.ac.uk/iscbulletin/search/catalogue/); National Earthquake Information Center (NEIC), USGS, USA (http://neic.usgs.gov/neis/epic/epic-global.htm) and HRVD (since 1976 and now operated as Global Centroid-Moment-Tensor project at Lamont Doherty Earth Observatory (LDEO) http://www.globalcmt.org /CMTsearch.html) are three earthquake databases from where the data can easily be accessed by the users. Event data for the Indian region is also provided by the India Meteorological Department (IMD), New Delhi. The different magnitude data contain measurement errors whose estimates are generally not provided by corresponding databases.

In order to examine the measurement errors associated with different magnitude determinations, Das et al. (2011) compiled body wave magnitudes for 348,423 events from ISC and 238,525 events from NEIC and surface wave magnitude values for 81,974 events from ISC and 16,019 events from NEIC. In addition, moment magnitude estimates by HRVD (CMT solutions) for 27,229 events from 1976 to May 2007 and for 7634 events from NEIC are also considered. The measurement errors for different magnitude types as estimated by Das et al. (2011) and other investigators are summarized in Table 4.1.

5 Magnitude Conversion Relations

For deriving the scaling relations between different magnitude types and the moment magnitude, events magnitude data from ISC, NEIC, GCMT and other region-specific databases (e.g., IMD for the Indian region) are generally considered. Magnitude conversion relations for the Indian region using different regression techniques as reported in different studies are discussed in the following subsections. SR

Table 4.1 Error estimates of different magnitude types (m_b, M_s, and M_w)

		Error Estimates		
Study	Catalog/Events Data	m_b	M_s	M_w
Kagan (2003)	Global catalog (1980–2000)	$\sigma = 0.25$ (10,496 events) $4.5 \leq m_b \leq 7.0$	$\sigma = 0.2$ (1746 events)	$\sigma = 0.12$ (992 events) $6 \leq M_w \leq 8$
Scordilis (2006)	Global catalog (1964–2003)	$\sigma = 0.2$ (215,163 events)	$\sigma = 0.17$ (25,960 events)	$\sigma = 0.11$ (3756 events) $5 \leq M_w \leq 8$
Castellaro et al. (2006)	Italy catalog (1981–1996)	$\sigma = 0.37$	$\sigma = 0.28$	$\sigma = 0.18$
Thingbaijam et al. (2008)	Northeast India catalog (1976–2006)	$\sigma = 0.25$	$\sigma = 0.20$	$\sigma = 0.15$
Das et al. (2011)	Global catalog (1964–2007)	$\sigma = 0.2$	$\sigma = 0.12$	$\sigma = 0.09$

and ISR approaches are the most commonly used techniques for magnitude conversion assuming one of the variables (magnitudes) to be fixed and free from error. If the measurement errors present in both the magnitudes represented by the dependent and the independent variables are taken into consideration, then GOR is the preferred regression as SR and ISR approaches in this case may yield incorrect estimates of the conversion magnitude.

5.1 *Surface Wave Magnitude to Moment Magnitude* (M_s/M_w)

The surface wave magnitudes estimated by ISC and NEIC have been compared in various studies and are found to be equivalent (Utsu, 2002; Scordilis, 2006; Das and Wason, 2010), since both the M_s estimates are based on the same technique. Scordilis (2006) verified the equivalence of $M_{s,ISC}$ and $M_{s,NEIC}$ magnitudes by deriving a SR relation between them, whereas a GOR relationship was derived by Das et al. (2011) in this regard.

Scordilis (2006) derived the following SR relations between M_s and M_w using a global data set for shallow focus events for the period 1978–2007:

$$M_w = 0.67\ (\pm 0.005)\ M_s + 2.07\ (\pm 0.03), 3.1 \leq M_s \leq 6.1 \quad (4.14)$$

$$M_w = 0.99\ (\pm 0.02)\ M_s + 0.08\ (\pm 0.13), 6.2 \leq M_s \leq 8.7 \quad (4.15)$$

Das et al. (2011) obtained GOR relationships between M_s and M_w using 24,807 events for the Indian region as given below:

$$M_w = 0.67\ (\pm 0.0005)\ M_s + 2.12\ (\pm 0.0001), 3.1 \leq M_s \leq 6.1 \quad (4.16)$$

$$M_w = 1.06\ (\pm 0.0002)\ M_s - 0.38\ (\pm 0.006), 6.2 \leq M_s \leq 8.7 \quad (4.17)$$

5.2 *Body Wave Magnitude to Moment Magnitude* (m_b/M_w)

Some studies have been devoted to compare the differences in the estimates of body wave magnitudes reported by ISC and NEIC. Scordilis (2006) observed some bias between the two magnitudes for the magnitude range $2.5 \leq m_{b,NEIC} \leq 7.3$, by fitting a SR line using global data set. Das et al. (2011) considered 23,281 and 22,960 m_b event data from ISC and NEIC, respectively, for the period of January 1, 1976 to May 31, 2007, which have also $M_{w,GCMT}$ values assigned, and observed that the average difference between $m_{b,ISC}$ and $m_{b,NEIC}$ is of the order of ±0.04 m.u., but the absolute average difference is found to be ±0.2 m.u. Utsu (2002) also observed a slight bias between the two m_b determinations. Therefore, Das et al. (2011) considered the two sets of data separately for derivation of regression relations between m_b and M_w.

5.2.1 SR Relations

Scordilis (2006) derived a global SR relationship between $m_{b,\mathrm{ISC}}$ and M_w for the magnitude range $3.5 \leq m_{b,\mathrm{ISC}} \leq 6.2$, as

$$M_{w,\mathrm{GCMT}} = 0.85\ (\pm 0.04) m_{b,\mathrm{ISC}} + 1.03(\pm 0.23) \tag{4.18}$$

A similar relation for the magnitude range $2.9 \leq m_{b,\mathrm{ISC}} \leq 6.2$ was obtained by Das et al. (2011) in the form

$$M_{w,\mathrm{GCMT}} = 0.65\ (\pm 0.003)\ m_{b,\mathrm{ISC}} + 1.65\ (\pm 0.02) \tag{4.19}$$

5.2.2 GOR Relations

Recently, Wason et al. (2012) derived a GOR relationship between m_b and M_w in the form

$$M_{w,\mathrm{GCMT}} = 1.13(\pm 0.04)\ m_{b,\mathrm{ISC}} - 0.416\ (\pm 0.23), \tag{4.20}$$

using an improved GOR procedure taking $\eta = 0.2$.

5.2.3 Regression Relations for Northeast India Region

For conversion of body wave magnitudes to moment magnitudes, several regional regression relations for Northeast India region have been developed (e.g., Thingbaijam et al., 2008; Yadav et al., 2009; Das et al., 2012, 2013).

Yadav et al. (2009) derived SR relationships for conversion of m_b to M_w in the magnitude range 4.7–6.8 and 4.6–6.4 using 90 $m_{b,\mathrm{NEIC}}$ and 157 $m_{b,\mathrm{ISC}}$ events, respectively, as given below:

$$M_{w,\mathrm{GCMT}} = 0.80\ (\pm 0.6)\ m_{b,\mathrm{NEIC}} + 1.03\ (\pm 0.31), \text{with } R^2 = 0.68 \text{ and } \sigma = 0.19 \tag{4.21}$$

and

$$M_{w,\mathrm{GCMT}} = 1.08\ (\pm 0.5)\ m_{b,\mathrm{ISC}} - 0.24\ (\pm 0.29), \text{with } R^2 = 0.72 \text{ and } \sigma = 0.26 \tag{4.22}$$

Thingbaijam et al. (2008) derived the following GOR relation based on 30 event data and assuming $\eta = 1$:

$$M_w = 1.3691\ (\pm 0.211)\ m_{b,\mathrm{ISC}} - 1.7742\ (\pm 1.139), \text{for } 4.4 \leq m_{b,\mathrm{ISC}} \leq 6.7 \tag{4.23}$$

Das et al. (2012) derived SR and GOR relationships for the magnitude range $4.7 \leq m_{b,ISC} \leq 6.6$, as given below:

$$M_w = 1.060\ (\pm 0.05)\ m_{b,ISC} + 0.151\ (\pm 0.263), \tag{4.24}$$

$$M_w = 1.4\ (\pm 0.0043)\ m_{b,ISC} + 1.98\ (\pm 0.122), \quad \eta = 0.36. \tag{4.25}$$

Similarly, Das et al. (2012) derived SR and GOR relationships for $m_{b,NEIC}$ to $M_{w,GCMT}$ in the magnitude range $4.6 \leq m_{b,NEIC} \leq 6.8$, using 146 events, as follows:

$$M_w = 0.983\ (\pm 0.054)\ m_{b,NEIC} + 0.216\ (\pm 0.286) \tag{4.26}$$

$$M_w = 1.37(\pm 0.006)\ m_{b,NEIC} - 1.77\ (\pm 0.1567), \quad \eta = 0.36. \tag{4.27}$$

In the absence of region-specific relations, corresponding global relations can be used for conversion of m_b to M_w, particularly for lower magnitudes.

5.3 *Local Magnitude to Moment Magnitude* (M_L/M_w)

The equivalence of M_L and M_w magnitudes has been discussed by many investigators (e.g., Thingbaijam et al., 2008; Ristau, 2009; Yadav et al., 2009). Yadav et al. (2009), however, emphasized that M_L and M_w are not equivalent for large magnitude events. In a study specifically for Northeast India region, Baruah et al. (2012) reported that M_w is equivalent to M_L with an uncertainty of about 0.13 m.u. on average.

5.4 *Intensity to Moment Magnitude* (I_{max}/M_w)

Gutenberg and Richter (1956) derived the empirical relationship:

$$M_w = 2/3\ I_{max} + 1 \tag{4.28}$$

where I_{max} is the maximum intensity (MMI). As the above relation is based on the California seismicity, it may not yield correct estimates of M_w for other regions including the Indian region. Using 126 global historical earthquakes MMI intensities, Das et al. (2012) derived a GOR-based relation between M_w and intensity (I_{max}), greater than or equal to VI, in the following form:

$$M_{w,NEIC} = 0.762\ (\pm 0.001)\ I_{max} + 0.865\ (\pm 0.014) \tag{4.29}$$

The M_w estimates yielded by the above empirical relationship are found to largely match with the range of values associated with different seismic zones in the seismic zoning map of India as per the code IS 1893 (Part 1):2002 (BIS 2002). The above

GOR relation (Eq. 4.29) can be used for conversion of maximum intensity (MMI) observed to M_w for historical earthquakes, where no reliable magnitude information is available.

6 Conclusions

A homogeneous earthquake catalog is of critical importance for the study of earthquake occurrence patterns in space and time, seismic hazard estimates, land-use regulation and other seismological applications. Different types of regression relationships are in use to convert different magnitude types into a preferred magnitude depending on the order of measurement error present or assumed in the two magnitudes involved. However, GOR is considered to be the most appropriate when both the variables (magnitude types) contain measurement errors provided their error variance ratio is known.

Region-specific regression relationships for the Indian region for magnitude conversion as reported in different studies based on SR, ISR, and GOR approaches (Thingbaijam et al., 2008; Yadav et al., 2009; Das et al., 2011, 2012; Wason et al., 2012) are discussed in this chapter. In addition, some SR regression relations derived by Scordilis (2006) based on global data are also included to cover the entire magnitude range. Choice of appropriate regression relationships is an important aspect of homogenization of earthquake catalogs, since error propagation during magnitude conversion process can introduce serious bias in "*b*" value and consequently on the seismic hazard assessment for a seismic region.

References

Baird-Smith, R. (1844). Memoirs of Indian earthquakes. *Journal of Asiatic Society of Bengal, 13*(2), 964–983.

Bapat, A., Kulkarni, R., & Guha, S. (1983). *Catalogue of earthquakes in India and neighbourhood from historical period up to 1979*. Roorkee: Indian Society of Earthquake Technology.

Baruah, S., Bora, P. K., Duarah, R., Kalita, A., Biswas, R., Gogoi, N., & Kayal, J. R. (2012). Moment magnitude (M_w) and local magnitude (M_L) relationship for earthquakes in Northeast India. *Pure and Applied Geophysics*, https://doi.org/10.1007/s00024-012-0465-9. *169*.

Bhattacharya, S. N. (1998). A perspective of historical earthquakes in India and its neighbored up to 1900. *Mausam, 49*, 375–382.

BIS. (2002). *IS 1893(Part 1):2002, Indian Standard Criteria for Earthquake Resistant Design of Structures, 5th revision*. New Delhi: Bureau of Indian Standards.

Carroll, R. I., & Ruppert, D. (1996). The use and misuse of orthogonal regression in linear errors-in-variables models. *The American Statistician, 50*(1), 1–6.

Castellaro, S., & Bormann, P. (2007). Performance of different regression procedures on the magnitude conversion problem. *Bulletin of the Seismological Society of America, 97*, 1167–1175.

Castellaro, S., Mulargia, F., & Kagan, Y. Y. (2006). Regression problems for magnitudes. *Geophysical Journal International, 165*, 913–930.

Chandra, U. (1977). Earthquakes of peninsular India: A seismotectonic study. *Bulletin of the Seismological Society of America, 67*, 1387–1413.

Das, R. (2013). Probabilistic seismic hazard assessment for Northeast India Region. Ph.D. thesis, Earthquake Engineering Department, Indian Institute of Technology Roorkee, India.

Das, R., & Wason, H. R. (2010). Comment on "a homogeneous and complete earthquake catalog for Northeast India and the adjoining region". *Seismological Research Letters, 81*, 232–234.

Das, R., Wason, H. R., & Sharma, M. L. (2011). Global regression relations for conversion of surface wave and body wave magnitudes to moment magnitude. *Natural Hazards, 59*, 801–810.

Das, R., Wason, H. R., & Sharma, M. L. (2012). Magnitude conversion to unified moment magnitude using orthogonal regression relation. *Journal of Asian Earth Sciences, 50*, 44–51.

Das, R., Wason, H. R., & Sharma, M. L. (2013). General orthogonal regression relations between body wave and moment magnitudes. *Seismological Research Letters, 84*, 219–224.

Das, R., Wason, H. R., & Sharma, M. L. (2017). Reply to "comments on 'Unbiased estimation of moment magnitude from body-and surface-wave magnitudes' by Ranjit Das, H.R. Wason and M.L. Sharma and 'Comparative analysis of regression methods used for seismic magnitude conversions' by P. Gasperini, B. Lolli, and S. Castellaro" by Pujol. *Bulletin of the Seismological Society of America, 108*(1), 540–547.

Fuller, W. A. (1987). *Measurement error models*. New York: Wiley. 458.

Gubin, I. E. (1968). Seismic zoning of the Indian peninsula. *Bulletin of the International Institute of Seismology and Earthquake Engineering, 5*, 109–139.

Guha, S. K., & Basu, P. C. (1993), *Catalog of earthquakes ($M \geq 3.0$) in peninsular India*. Atomic Energy Regulatory Board, Tech. Document No. TD/CSE-1, 1–70.

Gupta, H. K., Rajendran, K., & Singh, H. N. (1986). Seismicity of the north–East India region, part I: The database. *Journal Geological Society of India, 28*, 345–365.

Gusev, A. A. (1991). Intermagnitude relationships and asperity statistics. *Pure and Applied Geophysics, 136*, 515–527.

Gutenberg, B. (1945a). Amplitudes of surface waves and the magnitudes of shallow earthquakes. *Bulletin of the Seismological Society of America, 35*, 57–79.

Gutenberg, B. (1945b). Magnitude determination for deep-focus earthquakes. *Bulletin of the Seismological Society of America, 35*, 117–130.

Gutenberg, B., & Richter, C. F. (1956). Earthquake magnitude, intensity, energy, and acceleration. *Bulletin of the Seismological Society of America, 46*, 105–145.

Hanks, T., & Kanamori, H. (1979). A moment magnitude scale. *Journal of Geophysical Research, 84*, 2348–2350.

Kagan, Y. Y. (2003). Accuracy of modern global earthquake catalogs. *Physics of the Earth and Planetary Interiors, 135*(2–3), 173–209.

Kanamori, H. (1977). The energy release in great earthquakes. *Journal of Geophysical Research, 82*, 2981–2987.

Kelkar, Y. N. (1968). Earthquakes in Maharashtra in last 300 years. *Kesari Daily* (Marathi newspaper), Pune, January 7, 1968.

Kendall, M. G., & Stuart, A. (1979). *The advanced theory of statistics* London: Griffin, 2(4), 748.

Madansky, A. (1959). The fitting of straight lines when both variables are subject to error. *Journal of the American Statistical Association, 54*, 173–205.

Oldham, T. (1883). A catalogue of Indian earthquakes from the earliest times to the end of AD 1869. *Memoirs - Geological Survey of India, 19*(3), 1–53.

Richter, C. F. (1935). An instrumental earthquake magnitude scale. *Bulletin of the Seismological Society of America, 25*, 1–31.

Ristau, J. (2009). Comparison of magnitude estimates for New Zealand earthquakes: Moment magnitude, local magnitude, and teleseismic body-wave magnitude. *Bulletin of the Seismological Society of America, 99*, 1841–1852.

Scordilis, E. M. (2006). Empirical global relations converting M_S and m_b to moment magnitude. *Journal of Seismology, 10*, 225–236.

Thingbaijam, K. K. S., Nath, S. K., Yadav, A., Raj, A., Walling, M. Y., & Mohanty, W. K. (2008). Recent seismicity in Northeast India and its adjoining region. *Journal of Seismology, 12*, 107–123.

Utsu, T. (2002). Relationships between magnitude scales. In W. H. K. Lee, H. Kanamori, P. C. Jennings, & C. Kisslinger (Eds.), *International handbook of earthquake and engineering seismology part,* Amsterdam: Academic Press, 733–746

Vaněk, J., Zatopek, A., Karnik, V., Riznichenko, Y. V., Saverensky, E. F., Solov'ev, S. L., & Shebalin, N. V. (1962). Standardization of magnitude scales. *Bulletin of the Academy of Sciences of the USSR Geophysics Series, 2*, 108. (in English).

Wason, H. R., Das, R., & Sharma, M. L. (2012). Magnitude conversion problem using general orthogonal regression. *Geophysical Journal International, 190*(2), 1091–1096.

Yadav, R. B. S., Bormann, P., Rastogi, B. K., Das, M. C., & Chopra, S. (2009). A homogeneous and complete earthquake catalog for Northeast India and the adjoining region. *Seismological Research Letters, 80*, 609–627.

Chapter 5
Tsunami Hazards and Aspects on Design Loads

S. A. Sannasiraj

1 Introduction

Tsunamis, a name derived from Japan, meaning *harbor waves* are generated due to sudden displacement of sea floor due to underwater seismic activities, landslides, or volcanic eruption. The impact of a meteorite on the ocean might also displace large volume of water, resulting tsunamis. These are generally characterized as shallow-water waves with period ranging from few seconds to few hours and of speed greater than 600 km/h near the generation point. As they reach shallow-water depths, their speed and length reduce. Furthermore, as the period of the wave remains the same, more water is forced between the wave crests causing the height of the wave to increase. Because of this *shoaling* effect, a tsunami that was unnoticeable in deep water might grow to wave heights of several meters. If the trough of the tsunami wave reaches the coast first, a phenomenon called *drawdown* occurs, where it appears that the sea level has dropped considerably. Drawdown is followed immediately by the crest of the wave and can grasp people observing the drawdown curiously. When the crest of the wave hits, sea level rises (called *run-up* at the landfall location).

Since tsunami waves result in severe destruction to livelihood and damage to coastal structures, scientists and engineers grew interest in assessing the potential of the damages caused by tsunamis. The knowledge on characteristics of the tsunami, their generation, propagation and how they differ from the waves observed on the beach are of great importance in prediction and early warning. This is an important aspect to minimize the damages in the future. Estimating the flooding area of the coastal zone caused by the tsunami waves is essential for tsunami hazard mitigation.

S. A. Sannasiraj (✉)
Department of Ocean Engineering, Indian Institute of Technology Madras, Chennai, Tamilnadu, India
e-mail: sasraj@iitm.ac.in

M. L. Sharma et al. (eds.), *Advances in Indian Earthquake Engineering and Seismology*, https://doi.org/10.1007/978-3-319-76855-7_5

In order to determine run-up of long waves, different theoretical and experimental studies have been performed. The Federal Emergency Management Agency's Coastal Construction Manual (2005) and the City and County of Honolulu Building Code (2000) provide design guidelines for buildings exposed to tsunamis. Okada et al. (2005) provided design guidelines for tsunami refuge buildings based on the estimation of tsunami loads from the laboratory study (Asakura et al., 2000). This chapter mainly covers the extent of coastal devastation caused due to the tsunami.

2 Characteristics of Tsunami

In this section, let us consider the perspectives of tsunami only in terms of its impact on the coasts. The tsunami wave propagates from its source of generation in all possible directions. The major wave propagation direction is dictated by the false plane (in the case of earthquake), i.e., the generated wave crest is parallel to the false plane. For example, 2004 Indian Ocean Tsunami, the fault plane is about 1000 km in north–south direction and about 200 km in east–west direction. Hence, a large wave propagates along east–west direction.

Similar to the definition of wind waves, the basis tsunami wave characteristics are wavelength and wave period. However, these two characteristics are only related to water depth in the case of tsunami, i.e., simply follows shallow-water wave propagation case due to the relatively larger wavelength compared to water depth. In other words, the speed of tsunami and the dispersive nature of the tsunami are mainly constrained by the water depth. On the landfall point, the two main characteristics related to tsunami coastal impact are run-up and inundation.

2.1 *Wave Characteristics*

The wavelength (L) of the tsunami is decided by the displaced water mass during its origination. A large displacement of water mass (proportional to the displaced underwater land mass) makes a tsunami wave since the energy propagates as an oscillating wave in a medium such as water. The wavelength is set as very large relative to wave amplitude since the horizontal distribution of energy (from the source such as fault plane) is larger than the vertical transfer of momentum against gravity. Having said this, the restoring force for a tsunami wave is gravity. Now following long wave theory for a gravity wave, the wavelength can be expressed using the dispersion relation as,

$$L = T\sqrt{gd} \tag{5.1}$$

in which, T is the tsunami wave period, g is the acceleration due to gravity and d is the water depth. Following the above relation, one can understand that in a given

Table 5.1 Speed of long waves in various water depths

Water depth (d) (m)	Speed (C) (m/s)	Speed (C) (km/h)
2000	140	500
1000	99	350
200	44.3	160
50	22.1	80
20	14	50
10	10	36

water depth, if the wavelength is longer, the wave period should also be higher. These two parameters (L and T) have strong interdependency for the constant water depth. In general, the period is of the order of 10 min to few hours.

From Eq. (5.1), the speed of the wave (C) can be deduced:

$$C = \frac{L}{T} = \sqrt{gd} \tag{5.2}$$

It is interesting to note that the speed of tsunami is only dependent on water depth. Table 5.1 presents typical magnitudes of tsunami speed in various water depths.

That is, while a tsunami propagates in deeper water more than 1 km deep, the average speed of the wave is more than 400 km/hr. The wave reaches the continental shelf, say in the case of Bay of Bengal with the origin along off the coast of Sumatra, in about few hours. However, the wave speed substantially reduced in the continental shelf of the order of 100 km/hr and takes an average 2 hours to make a landfall along the east coast of India. Hence, even if the tsunami is realized after it reaches the continental shelf, one has sufficient time to take preventive measures to avoid human loss and to reduce economic loss. In this aspect, our tsunami warning systems provided as ample breathing time to take further preventive measures.

The tsunami wave speed in a very shallow depth is further accelerated by its elevation (η) following the relation:

$$C = \sqrt{g(d + \eta)} \tag{5.3}$$

The above relation is used to estimate wave speed if the tsunami elevation is comparable to the water depth.

Another important aspect on long waves like tsunami propagation is its propagation over longer distances without any loss of energy. A long gravity wave propagates without loss of energy for larger distances which make a tsunami wave easily sweeps the regional seas. The 2004 tsunami had made landfall at east coast of Africa and the 1960 Chilean tsunami propagated across the Pacific Ocean to reach the Japanese coasts.

2.2 *Wave Transformation*

2.2.1 Diffraction, Refraction and Reflection

Similar to wind- wave, the tsunami wave is subjected to phenomenon such as diffraction, refraction, and reflection. The wave diffracts while encounters an island or headland like projection, for example, the southernmost Indian coast act as a headland for the tsunami encounter. The variation in bathymetry refracts the wave, in which the change in wave direction happens according to Snell's law. The direction change results in the reduction of the width between adjacent wave rays and then the wave height between the adjacent wave rays increases to conserve wave energy between the rays. A similar phenomenon appears around a shoal under the sea. A steeper coast further reflects substantial part of wave energy compared to a flatter slope, in which the wave spends its energy for its propagation and deformation.

By considering the above wave phenomena, our Indian continents' southernmost tip is a precarious region since the tsunami energy gets concentrated due to refraction. However, in the presence of SriLankan island, the wave also diffracts and the net energy concentration occurs from the southern tip of Kanyakumari to about 50–60 kms on its west. This is the combined effect of diffraction and refraction. Figure 5.1 shows the energy concentration due to such a combined effect.

The tsunami energy is also focused in a converging ocean basin such as say, V-shaped. A tsunami intruding into the converging basin is reflected from the shores of both sides of the basin and is concentrated in the innermost part of the basin.

2.2.2 Shoaling

While the wave propagates from a deeper water to shallower region, the wavelength decreases as defined in Eq. (5.1). The conservation of energy leads to the increase in

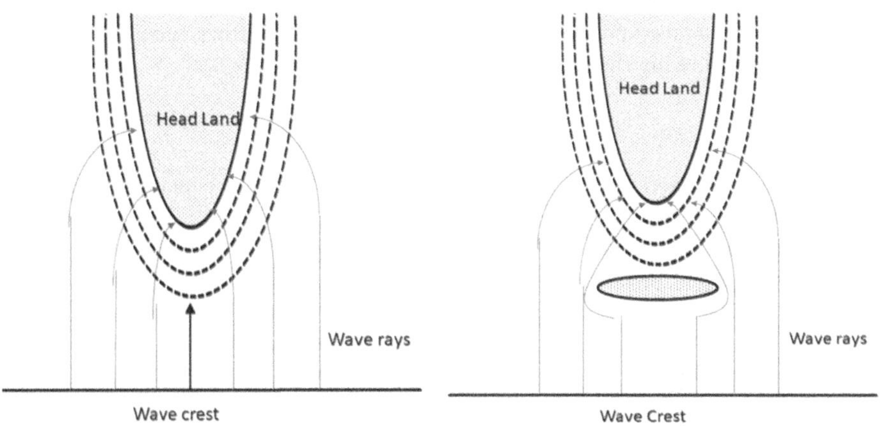

Fig. 5.1 Concentration of wave energy due to refraction and combined diffraction and refraction

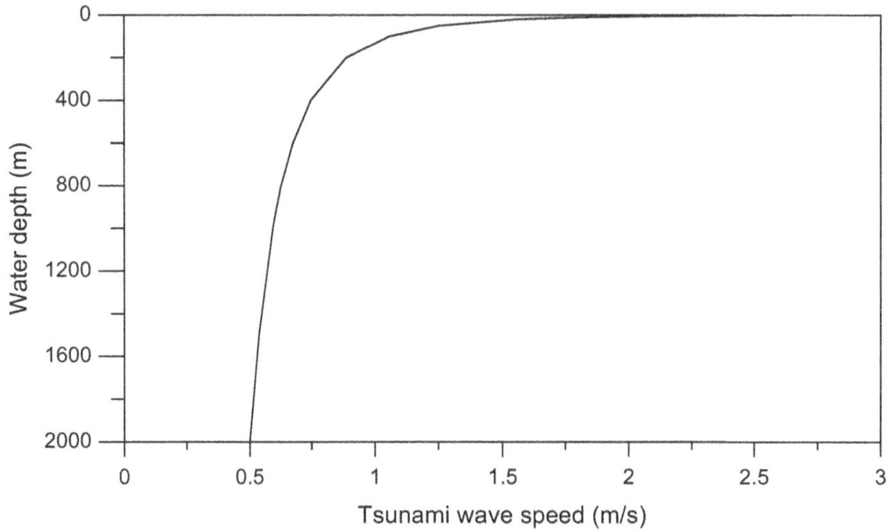

Fig. 5.2 Long wave evolution from deep water to near shore

wave amplitude. The increase in wave amplitude is called shoaling. In general, the tsunami is unnoticeable in deep waters, i.e., the amplitude is of the order of few centimeters to less than a meter. Due to the decrease of wavelength of the order of ten folds, the wave amplitude increases more than 1 m up to 10–20 m. The tsunami amplitude (η_2) in a relatively shallower location (d_2) can be calculated, while a tsunami amplitude (η_1) in deeper water, d_1 is known:

$$\eta_2 = \eta_1 \left(\frac{d_1}{d_2}\right)^{1/4} \tag{5.4}$$

From the above equation, it can be inferred that a tsunami wave amplitude of 0.5 m while at the entrance of the continental shelf with a water depth of 200 m amplifies to 1 m in a water depth of 10 m. In most of the context, the tsunami height is represented at 5–10 m depth for most of the design calculations. In a water depth of 5 m, the wave amplifies to about two and half times. Figure 5.2 presents the wave evolution from an origin water depth, say 0.5 m new born wave in a water depth of 2000 m towards the coast. It is assumed here that no loss of energy occurs during the wave propagation. However, if the wave height could not be sustained, it breaks and the wave approaches the shore as a bore. This effect is considered later to estimate the wave impact force.

2.2.3 Nearshore Tsunami Deformation

While tsunami approaches the nearshore region, it changes its profile by the way of transforming its energy. The potential head transforms to kinetic energy, in which

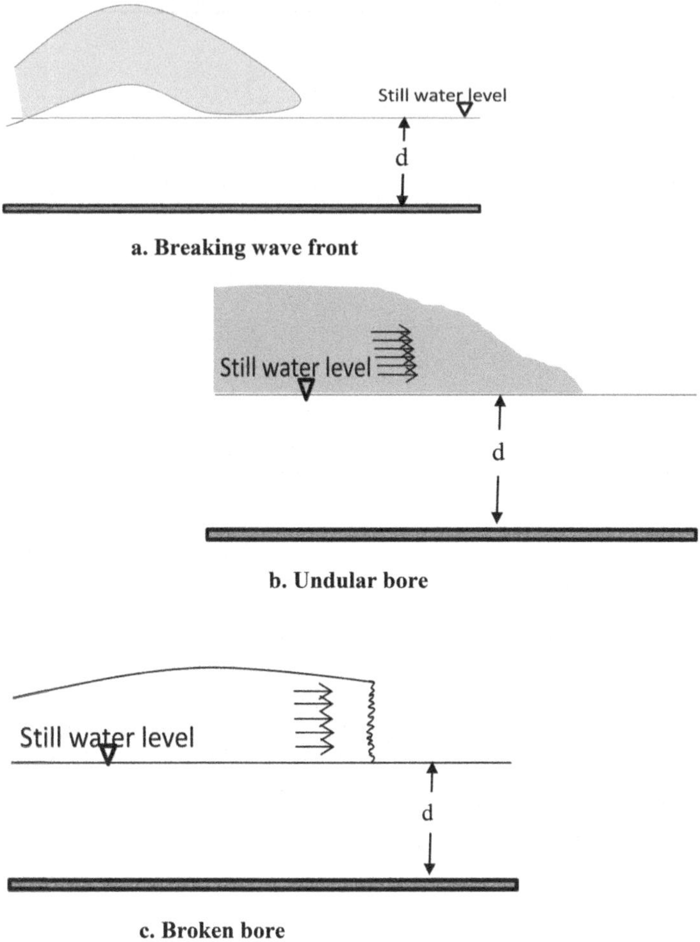

Fig. 5.3 Tsunami wave deformation near the coast

the wave either breaks or moves like a broken bore or undular bore. Figure 5.3 depicts the above deformation process. The factors affecting the deformation process are water depth, beach slope, wave height, and length modulated by the bottom bathymetry. A mild slope beach produces an undular bore while a relatively shorter tsunami approaches the coast. In this way, the tsunami does not lose its energy while making a landfall and might cause huge impact. In Nagapattinam coast, the mild slope in the nearshore region allows the entire tsunami energy to enter the human habitat region. The undular bore formed by a shorter wave breaks if the wave height reaches about 0.83 times the water depth in front of it. The bore-like formation can induce impulsive impact forces on the port and coastal structures and might cause large destruction of the structure. The bore formation might also occur at the tsunami landfall and the propagation speed can be approximately estimated by

$$C = \beta\sqrt{gd} \tag{5.5}$$

where β is the ground roughness parameter and takes the value of 0.7 for a very rough surface and 2.0 for a very smooth surface. To make a conservative design, the above value of wave speed can be taken as particle velocity. However, this would lead to uneconomical design, if the tsunami is very infrequent.

2.2.4 Tsunami Wave Dispersion

Unlike ocean surface waves, tsunami waves are non-dispersive. This can be clearly understood from Eq. (5.2) that the wave speed is independent of the wave period, i.e., either a short or a long wave, the tsunami speed is a function of only water depth. However, the tsunami wave disperses similar to a solitary wave dispersion in a very shallow waters. It is essentially due to the development of wave nonlinearity expressed in terms of *H/d* ratio. In this aspect, the wave period also plays a role. The wave speed in this case can be derived from the second-order nonlinear wave theory.

$$C = \sqrt{gd}\left\{1 - \frac{(kd)^2}{6}\right\} \tag{5.6}$$

where, k is the wave number ($k = 2\pi/L$). The wave dispersion happens in a relatively shallow water and in particular, when a tsunami propagates from a deeper water to the continental shelf. This makes splitting of tsunami waves into multiple waves and hence, the coast sees multiple landfalls even from a single generation source.

2.3 Tsunami Landfall Characteristics

Three basic landfall characteristics of tsunami are run-up, inundation and time of landfall. The last one becomes important only when multiple waves sweep the coast. And in other cases, it is of importance only for hydrodynamicist to understand different wave propagation patterns.

2.3.1 Run-Up

The vertical height of maximum water surface elevation at the tsunami landfall point is called as run-up at that particular location of the coast. The datum in general is mean sea level or chart datum. It varies in a wider range even within say 10 km coastal stretch. This is due to the fact of tsunami intensity at the origin, nearshore shallow-water bottom topography, the tsunami approach angle and any obstruction such as island in front of the coast under consideration. In most of the calculations, it is simple to assume that tsunami is a shallow-water wave. Analytical solution of the shallow-water equations for the prediction of long wave run-up over smooth plane beaches can be derived. The run-up of non-breaking long waves theoretically and obtained the following simple power law for the prediction of solitary wave run-up on a smooth plane beach:

$$\frac{R}{d} = 2.831\sqrt{\cot\beta}\left(\frac{H}{d}\right)^{5/4} \tag{5.7}$$

where R is the run-up, H is the wave height, d is the water depth, and β is the inclination angle of the plane beach. The run-up is defined at the landfall point. However, the water depth in the calculation of run-up needs a definite value. It is of in general to adopt a water depth of 10 m to estimate the run-up. After the tsunami makes a landfall, the run-up is estimated from the local maxima along the horizontal inundation stretch.

2.3.2 Inundation

The other important terminology in tsunami impact on coastal zone is inundation. It is the horizontal distance over which the sea spreads its tongue during the tsunami landfall. It is the function of run-up and the onshore land topography. The coastal morphological features of the particular shoreline play a major role in dictating the inundation levels. However, if there is a creek, river mouth or backwaters at the coast, the impact due to inundation will be more severe due to the rush up of water through the weak openings at the coast. It causes inundation from both sides of land sandwiched between ocean and backwaters.

3 Tsunami Hazards

The tsunami imposes so-called hazards only on the coastal zone and nearshore waters. In deep waters, its influence is generally unnoticed by the human community. The major water circulation in deep water might affect the atmospheric processes and, then, impact in the climatology such as change in monsoon pattern and cyclonic systems. The tsunami-induced hazards can be generally categorized into impacts on coastal zone, influence on the nearshore sailing vessels and impact inside the ports.

3.1 Coastal Zone Impacts

The effect of tsunami on the coasts can be broadly classified under four major categories, namely: flotation, impact, receding and submergence. During the initial tsunami run-up toward the coast, the movable and loose objects such as cars, boats and people, can be dragged by the water mass by the way of flotation. Fishing boats and other small vessels are usually found to be dragged a few kilometers upstream (along backwaters/river course), and many of them will be left on bridges, jetties, and elevated bunds. Considerable damage will also be caused by the floating debris, including boats and cars that became dangerous projectiles that crashed into the buildings and broke power lines. The flotation of navigational vessels also damage the berthing structures, particularly the top coping beam and pile heads are dented.

Secondly, the mass movement of water towards the coast resulted in an impact force on the shore-based structures. Even heavy objects such as overhead cranes on the berth or a vertical wall either a masonry wall as a component of residential building or as shore protection structure are subjected to damages heavily. The impact force also destabilizes the floating objects resulting to mass destruction along with the effect of flotation.

The acceleration of the moving water mass onto the land decelerated quickly due to the elevation of the land. The quick deceleration of the moving water induced the movement of the water in the opposite direction towards the sea. The reversal in the flow of water towards the sea has also been accelerated by the natural slope of the land. Hence, the wave had the tremendous power to drag away any loose materials on its way toward the sea, which forms the third category of the devastation. One of the severe impacts during the receding of water is the scour around foundations and trees. Strong, tsunami-induced currents lead to the erosion of foundations leading to the collapse of bridges and seawalls. The houses are mostly damaged by foundation failures as a result of their complete exposure.

Finally, the fourth category is the damages aggravated due to the inundated water which could not drain out to sea immediately. The inundated seawater in the lake and ponds affected the groundwater, and most of the cultivable lands near the coast are polluted due to seawater intrusion. The long duration of the inundation caused most deaths due to the spread of diseases, power outages and machinery malfunction.

3.2 Port Impacts

Once a long wave such as tsunami enters into the partially closed basin such as harbors with narrow entrances, the wave persistently oscillate for many hours/days without losing its energy. It is due to the nature of the long wave phenomenon discussed in the Sect. 2. Figure 5.4 shows the tidal record measured after the incidence of 2004 Indian Ocean tsunami at Chennai port. The time of landfall occurrence of tsunami is marked. It can be seen that the smooth variation in the

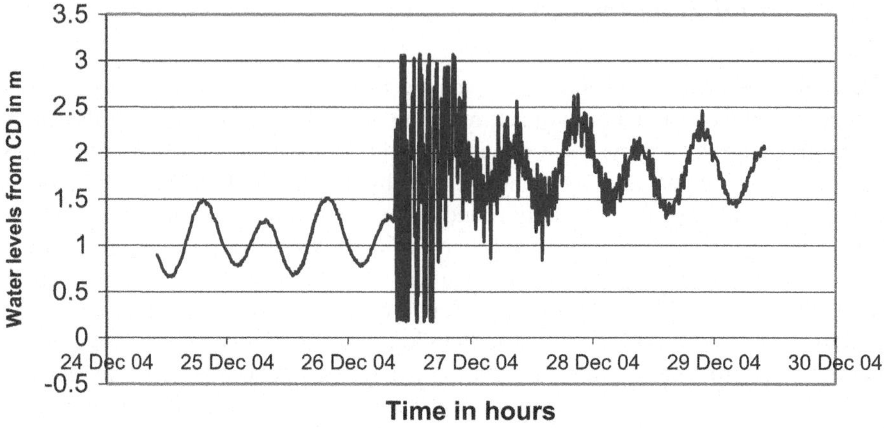

Fig. 5.4 Tidal record at Chennai port after the incidence of 2004 tsunami

water surface elevation is perturbed for the following couple of days. This meant that the waves of few minutes up to 20 min got superposed over the tidal wave. The resulting oscillation affects the regular loading and unloading operations, if the amplitude exceeds a threshold limit. More importantly, the maneuvering of the vessels inside the harbor might be greatly affected.

In addition, if the wavelength of one such long wave has been of the harmonic order of basin's characteristic length, then so-called resonance would occur, known as harbor resonance. The chance of occurrence of resonance is high since once trapped, the long wave forms into many wave components in the range of 10 min to an hour. The corresponding wavelength is of the order of few kilometers to several tens of kilometers depending on the water depth in the basin. Hence, the wavelength might form in terms of multiples of basin length/width. Once resonance is set, the wave amplitude increases to multifold unless there is a strong dampening mechanism. Most of the artificial harbors protected with rubble mound breakwaters have natural dampening system in terms of absorbing face of the breakwaters.

The harbor oscillation inside the basin thus depends on:

1. The tsunami wave period that enters into the basin;
2. The characteristic dimensions of the harbor basin;
3. Absorption and reflection characteristics of basin boundaries dictated by the type of breakwater boundaries and berth face.

The resonant period of oscillation (T) inside a harbor basin with characteristic length (L) and breadth (B) is given by:

$$T = \frac{2}{\sqrt{gd}}\left[\left(\frac{n}{L}\right)^2 + \left(\frac{m}{B}\right)^2\right]^{\left(-\frac{1}{2}\right)} \tag{5.8}$$

$$n = 0, 1, 2, 3\ldots\ldots \quad \text{and} \quad m = 0, 1, 2, 3$$

Here, n and m are integer numbers representing each mode of oscillation following the definition of Raichlen (1996). For the characteristic dimension of the Chennai harbor basin with $d = 15$ m and $L = 1000$ m, a resonant tsunami period for $n = 1$ and $m = 0$ is of the order of 3 min. The first mode oscillation is severer than higher modes ($n > 1$). The above equation is valid in a closed basin. Depending on the width opening at the harbor entrance, the above equation should be modified.

4 Indian Ocean Tsunami

It is of important to note the facts of Indian Ocean Tsunami which occurred due to a massive underwater earthquake of the West Coast of northern Sumatra, Indonesia on 26th December 2004, 7:58:53 am local time. Over 170,000 people lost their lives in this disaster. Areas near to the epicenter in Indonesia, especially Aceh, were devastated by the earthquake and tsunamis. The tsunamis also affected Phuket and surrounding areas in Thailand, Penang in Malaysia, Sri Lanka, India, and places as far as Somalia in Africa. Let us examine only the coastal devastation. Since, there is no proper evidence of tsunami along Indian coasts, the available measurements from the known severe event form the basis for design. Hence, one can adopt the measurements from 2004 tsunami as design basis tsunamis for design till more refined data are available.

Tables 5.2 and 5.3 present the measured run-up level along Tamil Nadu coast and Andaman and Nicobar Islands, respectively. Figure 5.5 presents the run-up level and Fig. 5.6 presents the inundation level along the Tamil Nadu coast.

The coastal region of Gujarat is vulnerable to tsunamis from great earthquakes in Makran coast. Earthquake of magnitude 7 or more may be dangerous. For the Indian region, two potential sources have been identified, namely, Makran coast and Andaman to Sumatra region.

The numerical model predicted run-up heights of December 26th Tsunami indicate that the run-up heights along Indian coast will be of the order of 2–4 m.

Table 5.2 Measured tsunami run-up height along Tamil Nadu coast

Location	Tsunami run-up height above MSL
From north Tamil Nadu to South Chennai (upto Muttukadu)	2–5 m
Kovalam to Pondicherry	4–7 m
Cuddalore to Adirampattinam	5–11 m
From north of Tuticorin to Kanyakumari	2–5 m
West coast of Tamil Nadu	4–7 m

Table 5.3 Measured tsunami run-up height along Andaman Islands

Location	Run-up height above MSL	Water level difference after the December 26 earthquake
Mayabandhar	4 m	−0.4 m
Havelock and Neil Island	3–4.5 m	0.0
Port Blair	3–6 m	+1 m
Little Andaman Islands (hut bay)	7–8 m	+1.5 m

Note: Positive water level difference indicates the submergence of land, and hence water level with reference to land level comes up

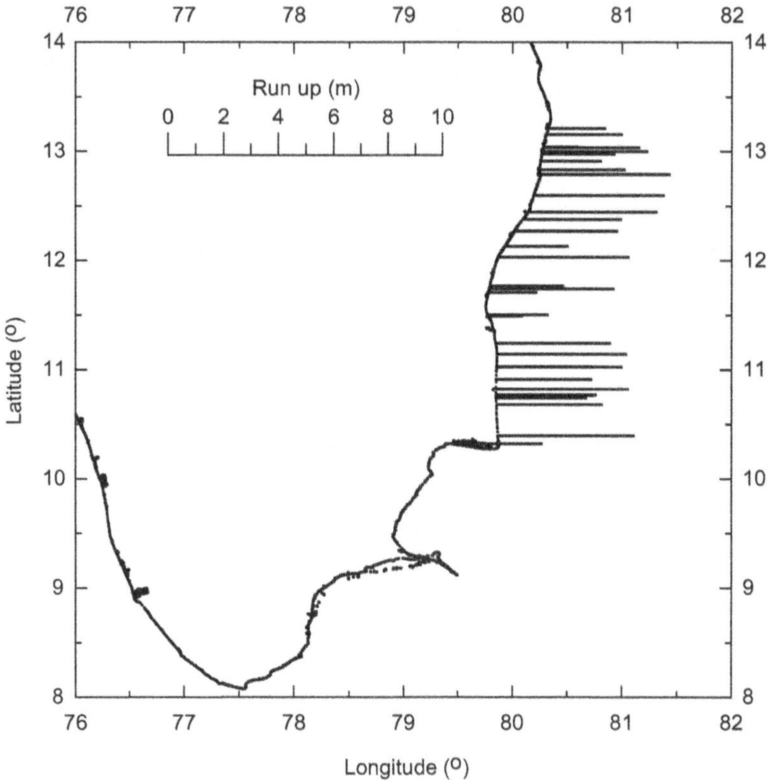

Fig. 5.5 Tsunami run-up levels along northern Tamil Nadu coast (Sundar et al., 2007)

4.1 Observed Impacts Along Indian Coasts

During the evacuation, people making an attempt to move out from shore by cars were caught by the wave, and those who ran towards the upper floors escaped from the tsunami. This had happened in Andaman Island and near Cuddalore Port.

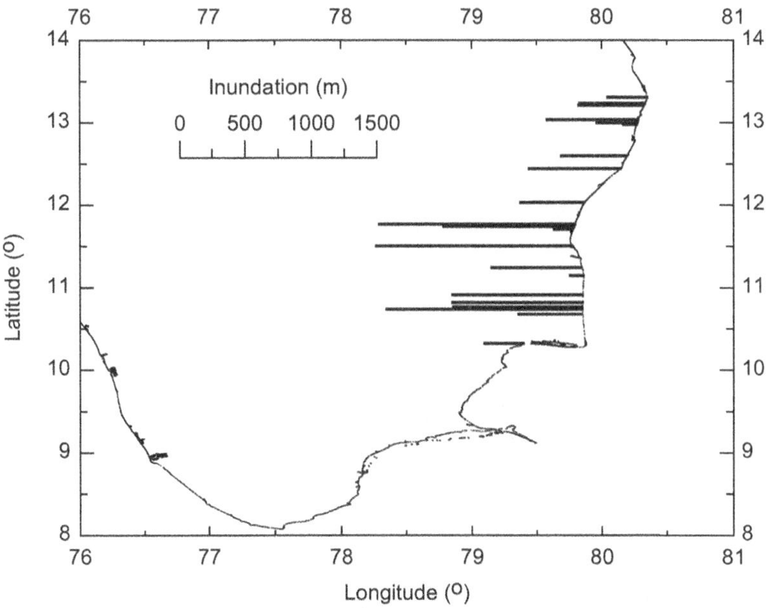

Fig. 5.6 Tsunami inundation along northern Tamil Nadu coast (Sundar et al., 2007)

Photo 5.1 Total displacement of bridge decks at Melamanakudithurai

The displacement of simply supported bridge decks from its support piers due to the impact force and the flotation effects in Melamanakudithurai is shown in Photo 5.1. The completely damaged walls of a church located at about 50 m from the shoreline in the Kezhaamanakudithurai village exhibits the might of the tsunami

Photo 5.2 Damaged Seawall and Church at Kezhaamanakudithurai

impact. In fact, this shoreline was protected against erosion by seawalls, the armor units of which weighted more than 1ton and were displaced by the moving mass of the water in a few pockets (Photo 5.2). Stones weighing 100 kg were lifted and thrown over the thatched roofs, about 4 m above the mean sea level. The immediate property damages in this village were mainly due to the impact force of the moving water.

Many of the buildings with reinforced concrete columns close to the shore were found to be in good condition, except that their foundations had experienced severe scour. The thatch houses were completely washed away and the brick masonry walls had collapsed in Colachel as shown in Photo 5.3. However, the transparent structures like open piled support jetty in Colachel (Photo 5.4) which is just offshore the brick masonry house referred in the above photo has been withstood without any damages. Photo 5.5a, b depicts the damages to the masonry building and complete collapse of thatch houses in Pillathopputhurai.

The scour during tsunami retreat had opened up the foundation of many shore buildings (Photo 5.6). The entire basins of the fishing and commercial harbors were found to be deepened to an extent of 0.5–2 m. Even though it is advantageous in terms of dredging requirements, the unsupported height of the berthing and other associated structures increased leading to concerns on their stability. The only consolation due to the tsunami was the opening of sandbars near the mouths of several rivers (Ramesh et al., 2005). Along the east coast of India, the net littoral sediment drift of about 1.0 million m^3 per year is towards the north (Sundar, 2005), and hence, the river mouths are usually found to be closed during most of the year. The great Indian Ocean tsunami resulted in the removal of sandbars, and a few of the rivers, where the mouth was almost permanently closed like Cooum River in Chennai city, south of Chennai port, have been opened.

Photo 5.3 Damaged brick masonry building at Colachel

Photo 5.4 Undamaged jetty at Colachel

Photo 5.5 (**a**, **b**) Damage to residential buildings at Pillathopputhurai

The inundation in a dry dock of a port at Port Blair (Photo 5.7) resulted in the collapse of two boats. A coastal wall of length 18 km had been constructed at about 100 m inland along the coast of Karaikal to prevent penetration of seawater into the cultivable land during high tide. This wall collapsed during the tsunami resulting in large stagnation of water leading to infertility of the agriculture land for

Photo 5.6 Damaged residential buildings at Kootapuli

approximately a 16 km stretch along the coast. Ramachandran et al. (2005) have reported the ecological impacts due to tsunami for Nicobar Islands, whereas Natesan and Kalaivani (2006) have reported the impact to the coast of Tamil Nadu.

Photo 5.7 Flotation in the dry dock

5 Tsunami Forces on Coastal Structures

Large bodies such as vertical faced walls needs to be designed considering hydrostatic force, hydrodynamic force and impact force due to waves and floating object. The buoyant force and static pressure head on the sea and lee sides of the structure form under hydrostatic pressure head. For small water piercing structures such as pile supported jetty, wharves etc., only the hydrodynamic force and impact force are to be considered.

5.1 Hydrostatic Force

The hydrostatic force can be estimated from

$$\begin{aligned} F_h &= \frac{1}{2}\rho g h(2d + 6\eta - h) + \frac{u_t^2 h_1}{2g} \quad \text{for}\, h_1 = h \\ F_h &= \frac{1}{2}\rho g h_1^2 + \frac{u_t^2 h_1}{2g} \quad \text{for}\, h_1 < h \\ h_1 &= \min(h, d + 3\eta) \end{aligned} \tag{5.9}$$

where ρ is the seawater density, g is the gravitational acceleration, h is the height of the structure, d is the water depth at the toe of the structure, η is the tsunami wave

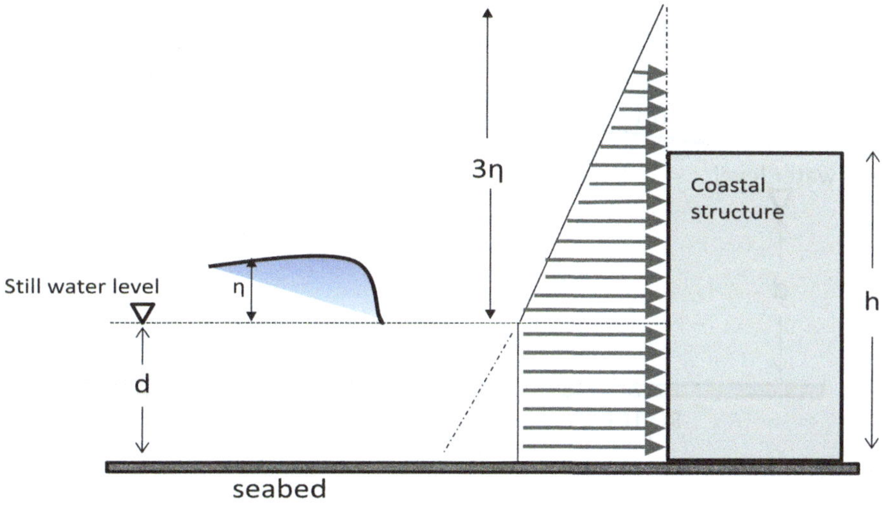

Fig. 5.7 Tsunami incidence on the coastal structure

height at 10 m water depth offshore, and u_t is the water particle velocity which can be assumed to be equal to the maximum tsunami flow velocity which is equal to $2\sqrt{g(d+\eta)}$. The tsunami run-up at the structure is adopted as three times the tsunami wave height at 10 m water depth offshore as shown in Fig. 5.7. The static pressure force due to flow velocity head is adopted to be uniform throughout the water depth. However, for transparent structures such as open-pile type structures, the hydrostatic force can be neglected.

Nakano and Paku (2005) have conducted extensive surveys in the areas of Sri Lanka and Thailand that were damaged by the tsunami generated by Sumatra earthquake. Data was collected on damage to several tens of coastal structures that had little influenced by the presence of obstructions, including buildings, RC pillars, brick walls and elevated water reservoirs. Using these data, the validity of the design equation for tsunami load is examined. The value of α, that is the dividing line between damage and no damage, is calculated separately for wall members and column members:

$$p_{\mathrm{m}} = \rho g(\alpha\eta_{\max} - z) \tag{5.10}$$

where,

p_{m} Maximum tsunami wave pressure ($0 \leq z/\eta_{\max} \leq 3$)
z Height of the relevant portion from ground level
$\eta_{\max}$ Maximum inundation depth
α the impact coefficient

Calculating the value of α in several tens of examples of damage revealed that wall members were undamaged roughly when $\alpha > 2.5$; therefore, the value $\alpha = 3.0$ for the coefficient in the tsunami wave pressure equation is considered appropriate.

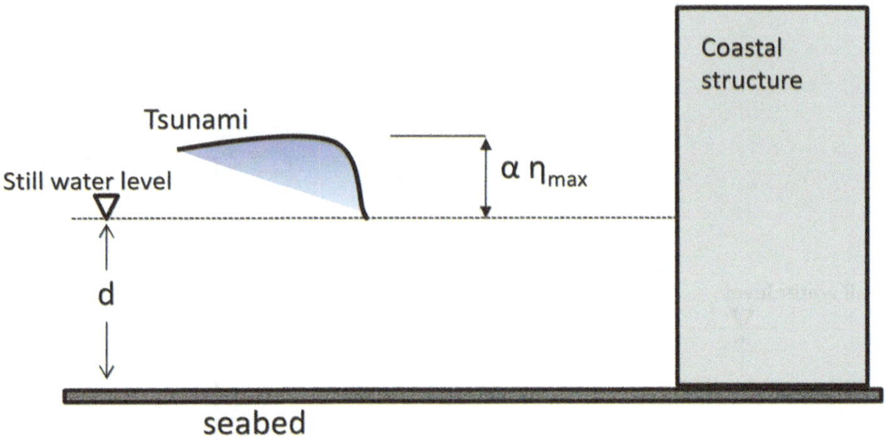

Fig. 5.8 Experiment setup

$\alpha = 2$ is found to be the dividing line between damaged and undamaged conditions of column members.

Further, the tsunami-induced wave pressure is formulated based on experiment as shown in Fig. 5.8 by Ikeno et al. (2001). The water pressure is measured by bore wave level of the flush tank divided at the gate:

$$p_{\mathrm{m}}(z) = 2.2\alpha\rho g\,(\eta_{\max} - z/3) \tag{5.11}$$

where p_{m} is the maximum tsunami wave pressure ($0 \leq z/\eta_{\max} \leq 3$: above the surface of still water), $\eta_{\max}$ tsunami wave amplitude and α, the impact coefficient (=1.36).

5.1.1 Leeside Water Pressure While Overtopping

During the tsunami inundation on the lee side of seawalls, the pore water pressure increases due to the saturation of soil. This pore water pressure might exist for longer period which led to destabilization of the structure. Hence, provision of pore pressure release system in wall-type structures and drains in seawalls is recommended.

5.1.2 Buoyant Force

The submergence of structure due to the rise in water level reduces the effective weight of the structural elements which in other words, reduces the stiffness of the structure. This can be adopted in terms of upward force, i.e., buoyant force, F_{B}:

$$F_{\mathrm{B}} = \rho g V \tag{5.12}$$

where V is the submerged volume of structural elements.

5.2 Hydrodynamic Wave Force

Due to the nature of the long tsunami wave, drag force is induced on relatively small characteristic dimension of the structural elements. The hydrodynamic drag force, F_d can be estimated from

$$F_d = \frac{1}{2}\rho C_d A u_t^2 \tag{5.13}$$

where C_d is the drag coefficient = 1.2 for circular structures, 2.0 for square buildings, and 1.5 for wall sections where the length of the wall is less than the wavelength. A is the projected area of the elements on the plane normal to the flow direction. The above equation is valid if the structure dimension (B) is small relative to tsunami wave length (L), i.e., $B << L$.

On the other hand, the tsunami wave force (F_H) is represented as the combination of drag, inertia, impulse and hydraulic gradient forces by Omori et al. (2000):

$$F_H = \frac{1}{2}\rho C_D u|u|B\eta + \rho C_M \dot{u} BL\eta + \frac{1}{2}\rho C_S(\theta)u|u|B\eta + \rho g BL\eta\frac{d\eta}{dx} \tag{5.14}$$

where

B width of structure
L length of structure
C_D drag coefficient (= 2.05)
C_M mass coefficient (= 2.19)
$C_S(\theta)$ impact coefficient (=3.6 tanθ)
θ angle of wave
u velocity of tsunami wave in the direction of advance
$\dot{u}$ acceleration of tsunami wave in the direction of advance
η inundation depth

5.2.1 Tsunami Force on Inland Structure

Iilzuka and Matsutomi (2000) proposed the equation to estimate the hydrodynamic tsunami force (F_{HD}) based on the damages of houses when tsunami hit in Japan. The equation is formulated in order to make the relationship between drag forces and damage degrees of houses:

$$F_{HD} = \frac{1}{2}\rho C_D u^2 h_f B_h \tag{5.15}$$

where,

C_D drag coefficient (=1.1 $\approx$ 2.0)
u velocity in inland
h_f inundation depth in front of the structure
B_h width of structure

5.3 Impact Force

The impact force needs to be considered for any structure elements facing the water front. The impact is caused not only due to impulsive wave but also due to the floating objects. Tsunamis often approach the shore in the formation of bores depending on the bottom bathymetry. The impulsive force can be reduced if the structure is elevated. The pressure impulse is an integrated quantity as a function of maximum wave impact and the duration of its attack. Figure 5.9 presents a typical variation of pressure impulse on an elevated building. Here, d_o is the bottom clearance of the building, d_1 is the tsunami front height reaching the building and d is the height of the building from its bottom elevated position up to maximum tsunami run-up level. $z/d = 0$ indicates the free surface and $z/d = -1$ indicates the bottom wall level. Hence, for $d_o = 0$, i.e., the building is located at the natural ground level, the maximum pressure is induced. The effect of reduction in pressure impulse is significant even when the bottom clearance is small: for the case of $d_b/d_1 = 0.05$, 24% reduction of the maximum pressure impulse from the case of no clearance. This estimate has been made using pressure impulse theory and validated from experimental measurements.

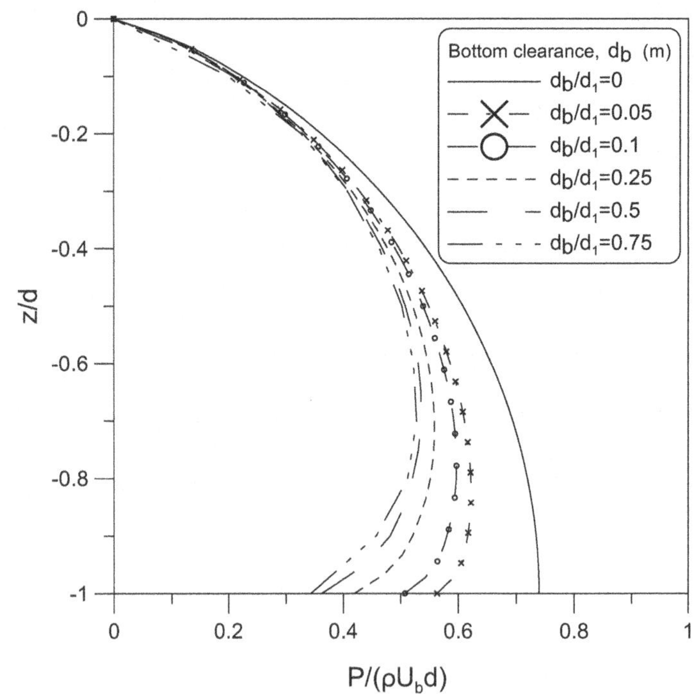

Fig. 5.9 Variation of pressure impulse on the vertical face of the building with different bottom clearances

The impact force due to a floating object is equal to the mass of the probable floating objects times the acceleration induced into the floating objects by the tsunami flow velocity. In a harbor, the severe impact may be due to the flotation of boats and ships.

The impact force, F_I can be assumed to act at still water level, i.e., at $(h + 3\eta)$ above the seabed level where tsunami landfall occurs:

$$F_{\mathrm{I}} = m\frac{du_{\mathrm{b}}}{dt} = m\frac{u_{\mathrm{t}}}{\Delta t} \tag{5.16}$$

where, u_{b} is the velocity of probably floating object and during the tsunami wave propagation, the floating body is assumed to activate from its rest to the flow velocity, u_{t} ($du_{\mathrm{b}} = u_{\mathrm{t}}$). Δt is the impact duration which is the time between initial contact of the floating body with the building and the maximum impact force. This impact duration depends on the natural frequency of the structure. This value can be 1.0 s for wooden structure, 0.5 s for steel structure and 0.1 s for reinforced concrete structure.

5.4 Scour Depth

One of the severest devastation effects of the tsunami wave is the scour under the foot of the structure while the tsunami wave receded with a speed proportional to the slope of the land topography and the tsunami height. The effect of scour exposes the free length of the coastal structures which makes say, the embedment length of the pile become less than the design requirement. The caisson-type block structural elements can be destabilized with the scour on the seaside and the excess water pressure on the land side of the block. The scour depth for coast-based structures such as berthing structures should be adopted to fix the founding depth of the structure. Table 5.4 presents an estimate of maximum scour for various soil conditions. The scour depth may be suitably reduced if there is a first-level defense in front of the structure such as vegetation, sand dune, breakwater etc. It should be noted that the Chennai harbor basin got deepened on an average 2 m during the 2004 tsunami. Such measurement for the loose silty sand is less than given in the table due to the presence of breakwaters.

Table 5.4 Recommended scour depth in front of the obstruction

	Distance from shoreline	
	100 m	>100 m
Loose sand	0.8 d	0.6 d
Dense sand	0.5 d	0.35 d
Soft silt	0.5 d	0.25 d
Stiff silt	0.25 d	0.15 d
Soft clay	0.25 d	0.15 d
Stiff clay	0.1 d	0.05 d

d water depth level in front of the pile/structure

6 Summary

The tsunami impact on coastal zone has been analyzed based on two major concerns: saving lives and reducing human suffering and limiting economic losses and promoting sustainable development. It is well understood that the tsunami disaster cannot be prevented. It is our task to reduce its effect. An integrated multi-hazard approach with an emphasis on cyclone and tsunami is an essential task before us. The infrequent low probable tsunami hazard is coupled with frequent cyclonic hazard to justify the investments. The specific engineering measures range from the construction of elevated shelters for the safety of people to the plantation of mangroves/coastal forests for reducing the impact on the coastal zone. Wherever possible/required, the hard measures of coastal protection such as seawalls or coral reefs and harbor like protection using breakwaters can be planned. A detailed environmental impact study is an essential task before implementing any hard measures. This is because looking at an once in blue- moon event, the day-to-day operational activities such as fishing tourism should not be affected. The bio-shield can be considered with due respect on coastal morphology. A revised planning strategy may be adopted for identified vulnerable structures. This is an essential task for critical infrastructures such as power stations, warehouses, oil and other storages tanks etc. Wherever possible, retrofitting of such structures can be implemented.

References

Asakura, R., Iwase, K., Ikeya, T., Takao, M., Fujii, N., & Omori, M. (2000). An experimental study on wave force acting on on-shore structures due to overflowing tsunamis. *Proceedings of Coastal Engineering of Japan Society of Civil Engineering, 47*, 911–915.

City and County of Honolulu Building Code (CCH). (2000). Department of Planning and Permitting of Honolulu, Chapter 16, Article 11, Honolulu, HI.

FEMA. (2005). *Coastal construction manual* (Rep. No. 55, 3rd ed.). Washington, DC: Federal Emergency Management Agency.

Iilzuka, H., & Matsutomi, H. (2000). Damage prediction of a tsunami flood. *Proceedings of the Coastal Engineerings of Japan Society of Civil Engineering, 47*, 381–385.

Ikeno, M., Mori, N., & Tanaka, H. (2001). Experimental research about wave force of a bore and breaking wave and the action and the impulsive force of drifting timber. *Proceedings of the Coastal Engineerings of Japan Society of Civil Engineering, 48*, 846–850.

Nakano, Y., & Paku, C. (2005). *Studies on lateral resistance of structures and tsunami load caused by the* 2004 Sumatra earthquake: Part-1 outline of survey. Summaries of technical papers of Annual Meeting Architectural Institute of Japan (Kinki).

Okada, T., Sugano, T., Ishikawa, T., Ohgi, T., Takai, S. and Kamabe, C. (2005), Structure design method of buildings for tsunami resistance (proposed). English translation of Japanese report. *The Building Letter*. November 2004. The Building Center of Japan.

Omori, M., Fujii, N., Kyotani, O. (2000). The numerical computation of the water level the flow velocity and the wave force of the tsunami which over flow the perpendicular revetments. *Proceedings of the Coastal Engineerings of Japan Society of Civil Engineering, 47*, 376–380.

Raichlen, F. (1996). "Harbor resonance", interaction of structures and waves. In A. T. Ippen (Ed.), *Coastline and estuarine hydrodynamics*, McGraw Hill: New York, 281–315.

Ramachandran, S., Anitha, S., Balamurugan, V., Dharanirajan, K., Irene Preeti Divien, M., Senthilvel, A., Hussain, I. S., & Udayaraj, A. (2005). Ecological impact on tsunami on Nicobar islands (Camorta, Katchal, Nancowry and Trinkat). *Current Science, 89*(1), 195–200.

Ramesh, R., Rajkumar, A. N., & Ramachandran, P. (2005). Initial assessment of the tsunami inundation and sediment characteristics in Chennai, South India. In S. M. Ramasamy & C. J. Kumanan (Eds.), *Tsunami: The Indian context,* New Delhi: Allied Publishers, 183–199.

Sundar, V. (2005). Behaviour of shoreline between groin field and its effect on the tsunami propagation, Proc. Fifth International Symposium on Ocean Wave Measurement and Analysis, WAVES2005, Madrid, Spain, 323.

Sundar, V., Sannasiraj, S. A., Murali, K., & Sundaravadivelu, R. (2007). Run-up and inundation along Indian coasts due to 2004 Indian Ocean Tsunami. *Journal of Waterway, Port, Coastal and Ocean Engineering, ASCE, 133*(6), 401–413. Special issue.

Natesan, U., & Kalaivani, S. (2006). Tsunami induced water quality changes in Buckingham canal, Chennai region, India. In S. M. Ramasamy, C. J. Kumanan, R. Sivakumar, & B. Singh (Eds.), *Geomatics in Tsunami,* New Delhi: New India Publishing Agency, 191–198.

Part II
Geotechnical Earthquake Engineering

Chapter 6
Developments in Geotechnical Earthquake Engineering in Recent Years: 2012

Shamsher Prakash and Vijay K. Puri

1 Introduction

Dynamic loads on soil may be imposed by earthquakes, machines, blasting, traffic, and constructions operations. The problems associated with the dynamic loading of soils include but are not limited to dynamic soil properties, liquefaction of soils, design of shallow and deep foundations in non-liquefying and liquefying soils, design of retaining structures, stability of slopes, design of foundations for machines and vibratory equipment, and man-made vibrations. Significant developments have taken place in the last two to three decades in almost all areas of soil dynamics following the 1964 earthquakes in Niigata (Japan) and Alaska (USA) which has lead to better understanding of the problems and safe design procedures. In this chapter developments in the following areas which are of immense interest to geotechnical profession are presented.

1. Liquefaction of sand, silt, and clay mixtures
2. Stability of rigid retaining structures
3. Prediction and performance of piles under dynamic load
4. Piles in liquefiable soils
5. Seismic bearing capacity and settlement of shallow foundations

S. Prakash (✉)
Department of Civil Engineering, Missouri University of Science and Technology, Rolla, MO, USA
e-mail: Prakash@mst.edu

V. K. Puri
Department of Civil and Environmental Engineering, Southern Illinois University, Carbondale, IL, USA
e-mail: puri@engr.siu.edu

M. L. Sharma et al. (eds.), *Advances in Indian Earthquake Engineering and Seismology*, https://doi.org/10.1007/978-3-319-76855-7_6

2 Liquefaction of Sand Silt, and Clay Mixtures

Liquefaction of sands has been studied extensively in the past. The state-of-the-art on liquefaction behavior of saturated cohesionless soils has progressed to a stage that reasonable estimates of liquefaction potential can be made based on laboratory investigations or on simple in situ test data such as standard penetration values (N_1) or $(N1)_{60}$ or cone penetration data, and the experience during the past earthquakes, (Youd and Idriss, 2001; Youd et al., 2001). Fine-grained soils such as silts, clayey silts, and sands with fines and silty soils were considered non-liquefiable. However, the observations following several recent earthquakes indicate that many cohesive soils had liquefied. These cohesive soils had clay fraction less than 20%, liquid limit between 21%–35%, plasticity index between 4 and 14%, and water content more than 90% of their liquid limit. Kishida (1969) reported liquefaction of soils withup to 70% fines and 10% clay fraction during Mino-Owar, Tohankai, and Fukui earthquakes. Observations during several other earthquakes show evidence of liquefaction in silty and clayey soils. This led to study of liquefaction and cyclic mobility of fine-grained soils. It is now known that all soils including sands, silts, clays, and gravels and their mixtures may liquefy depending upon the seismic and environmental factors.

Fine-grained soils susceptible to liquefaction (before 2000) appear to have the following characteristics (based on Chinese criteria):

Percent finer than 0.005 mm (5 μ microns) < 15%
Liquid limit < 35%
Water content > 90% of liquid limit

Seed et al. (2001) observed that there is significant controversy and confusion regarding the liquefaction potential of silty soils (and silty/clayey soils) and also coarser, gravelly soils and rockfills. Finn et al. (1994), Perlea et al. (1999) and Andrews and Martin (2000) have provided general criteria about liquefaction susceptibility of soils with fines.

Studies undertaken at MST (formerly UMR) in the early 1980s have identified the effect of plasticity of soil on the liquefaction of silts. Dynamic triaxial tests were conducted on 73.65-mm (diameter) and 147.3-mm (high) samples of two different silts A and B (Table 6.1) to determine the effect of soil plasticity on susceptibility to

Table 6.1 Properties of silty soils used in MST investigation

	Soil A	Soil B
Percent finer than 75 μ (0.075 mm)	93–98	96–98
Natural water content (%)	18–26	–
Liquid limit	32.0–36.0	24.2–26.6
Plasticity index	9–14 (mostly ~10)	1.6–1.8
Clay content (<2 μm)	2.0–7.2%	
Specific gravity of soil particles	2.71	2.725
Particle size D_{50} mm	0.06	0.022

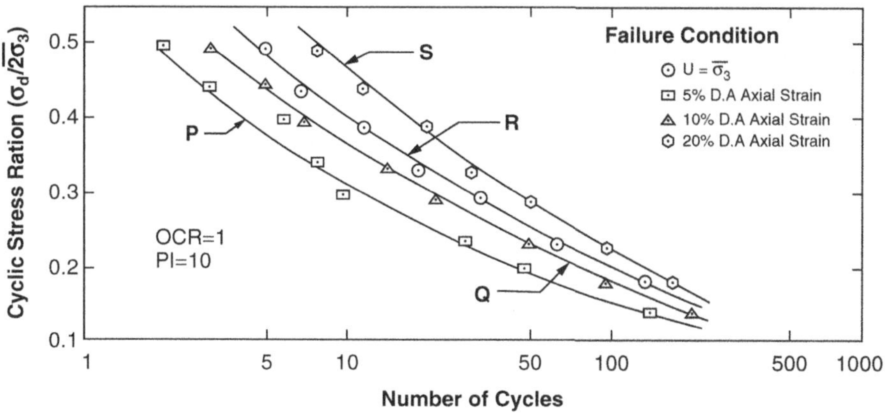

Fig. 6.1 Cyclic stress ratio versus number of cycles for reconstituted saturated samples, silt A, for $\overline{\sigma}_3 = 15$ psi

liquefaction. The index properties of these silts are given in Table 6.1. Soil A is naturally occurring silt. The PI of this silt was altered by adding the clay fraction obtained from this soil itself (Puri 1984). The tests on silt A were conducted at PI = 10, 15, and 20. The PI of silt B was varied in the low plasticity range by adding kaolinite. The tests on silt B were conducted at PI = 1.7, 2.6, and 3.4 (Sandoval, 1989).

A typical data for the tests on silt A is shown in Fig. 6.1. It is seen from this figure that for the case of silt samples tested as a part of this investigation, the failure defined by 5 or 10% double amplitude axial strains occurs before the condition of initial liquefaction defined by $\mu = \overline{\sigma}_3$ occurs. Figure 6.2 shows the effect of plasticity index on cyclic stress ratio inducing 5% DA strain in a given number of load cycles. Increase in PI value is seen to increase the cyclic stress ratio. The trend of the data from other tests was similar with the exception that for the case of PI = 20, the condition $\mu = \overline{\sigma}_3$ did not develop within the range of cyclic load applications used in this study.

Typical results of the investigation on samples of silt B showing the effect of plasticity index (PI = 1.7%, 2.6% and 3.4%) on the cyclic stress ratio causing initial liquefaction in any given number of cycles are shown in Fig. 6.3. It is seen from this figure that the cyclic stress ratio causing liquefaction in a given number of cycles decreases with the increase in plasticity index. It was observed during the testing phase that the cyclic loading of plastic silts results in pore pressure build up which becomes equal to the initial effective confining pressure resulting in development of stage initial liquefaction. This is just opposite the case when PI is 10% or greater.

Combining results for silts A and B with cyclic stress ratio (CSR), normalized at void ratio of 0.74, (Prakash and Guo, 1998) leads to results as shown in Fig. 6.4. It is observed from this figure that for PI values of less than about 4, the cyclic stress ratio causing liquefaction in any given number of cycles decreases with an increase in PI values. For PI values beyond about 4, the cyclic stress ratio causing initial liquefaction in any given number of cycles increases with an increase in the PI values.

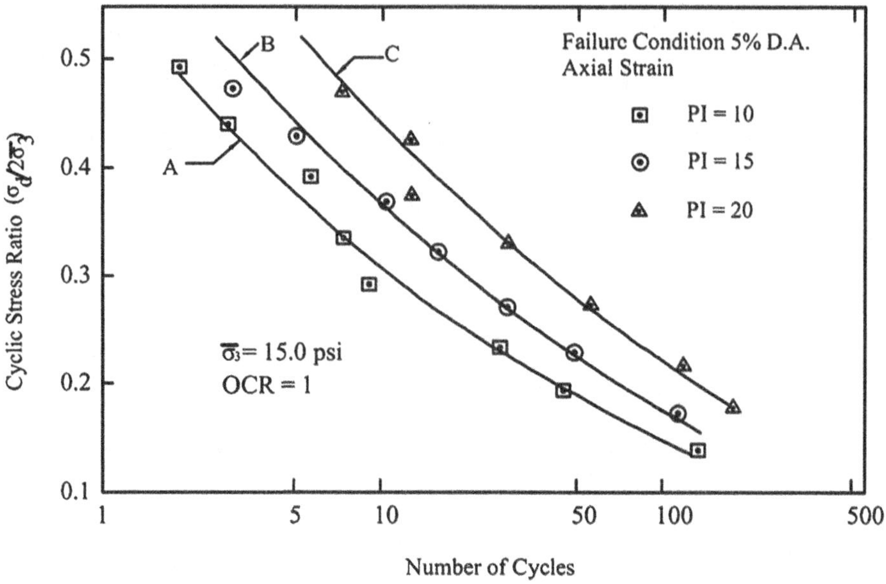

Fig. 6.2 Cyclic stress ratio versus number of cycles for reconstituted saturated samples, silt A, for $\overline{\sigma} = 10$ psi (Puri, 1984)

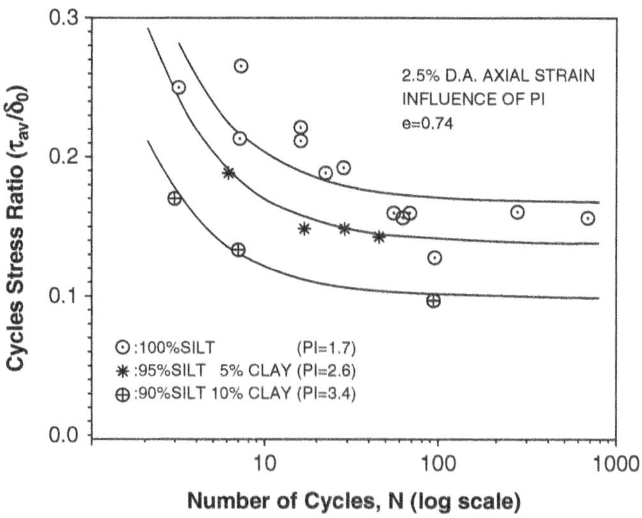

Fig. 6.3 Cyclic stress ratio versus number of cycles for low plasticity silts for inducing initial liquefaction condition at 15 psi effective confining pressure: PI = 1.7, 2.6, and 3.4; for density 97.2–99.8 pcf; and $w = 8\%$ (Sandoval, 1989; Prakash and Sandoval, 1992)

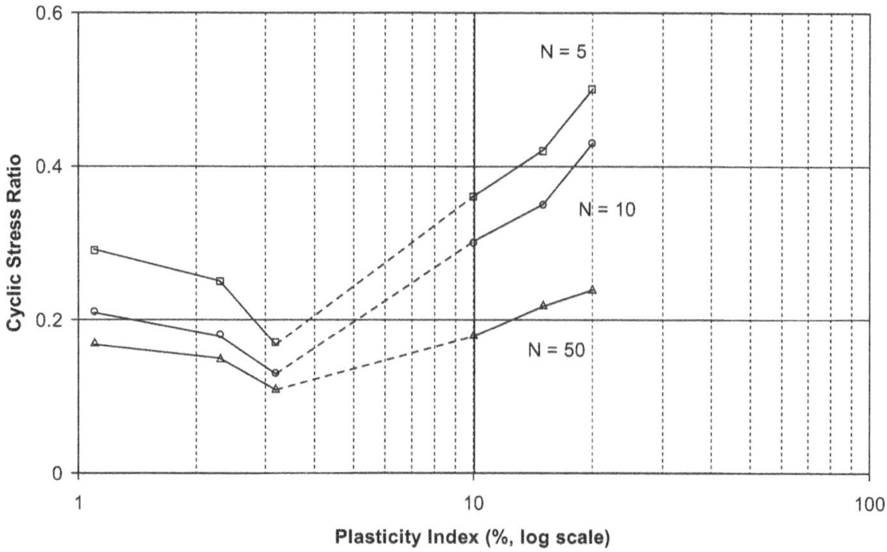

Fig. 6.4 Cyclic stress ratio versus plasticity index for silt–clay mixtures (CSR normalized to initial void ratio $e_0 = 0.74$) (Prakash and Guo, 1998)

Based on these results, it may be observed that there is a critical value of plasticity index (PI) at which saturated samples of silt–clay mixtures have a minimum resistance to cyclic loading or highest susceptibility to liquefaction. It is worth mentioning here that data of El Hosri et al. (1984) on undisturbed sample (Fig. 6.5) also suggests a similar effect of PI on cyclic stress ratio causing liquefaction as observed during the present investigation.

2.1 Recent Studies 2001–2012

Andrews and Martin (2000) suggested percentage of clay fraction and liquid limit as a criteria to identify fine-grained soils which may or may not be prone to liquefaction. They recommended the matrix in Table 6.2 for evaluating liquefaction susceptibility of these soils.

Seed et al. (2001) proposed guidelines to estimate liquefaction of soils with significant fines content. These guidelines are shown in graphical form in Fig. 6.6. The chart in Fig. 6.6 is divided into three zones. Zone A soils are considered potentially susceptible to "classic cyclically induced liquefaction, if the water content is greater than 80% of the LL." Zone B soils are considered potentially liquefiable with detailed laboratory testing recommended, if the water content is greater than 85% of the LL. Zone C soils (outside Zones A and B) are considered generally not susceptible to classic cyclic liquefaction, although they should be checked for potential loss in strength due to cyclic loading. Figure 6.6 also shows

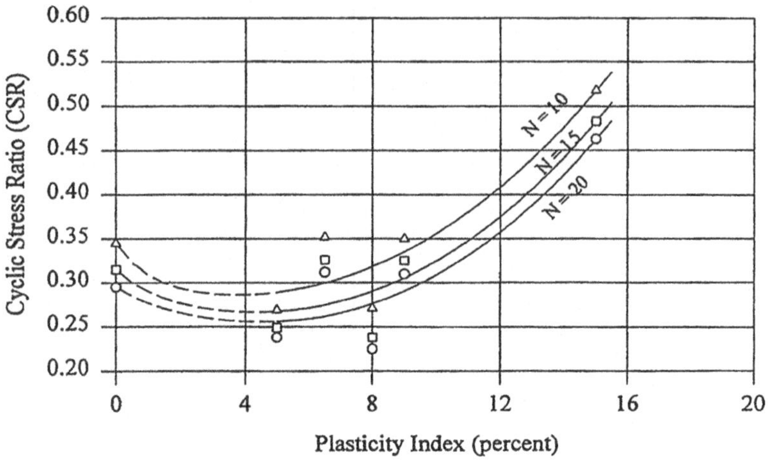

Fig. 6.5 Normalized cyclic stress ratio versus plasticity index on undisturbed samples (after El Hosri et al., 1984 and Prakash and Guo, 1998)

Table 6.2 Criteria recommended by Andrews and Martin (2000) for evaluating the liquefactions of fine-grained soils

	LL[a] < 32	LL[a] ≥ 32
Minus 2 μm fraction <10%	Susceptible to liquefaction	Further studies required (*consider plastic non-clay sized grain*)
Minus 2 μm fraction ≥10%	Further studies required (*consider non-plastic clay sized grains*)	Not susceptible to liquefaction

[a]LL (liquid limit) determined by Casagrande-type percussion apparatus

that soils with PI < 12 and water contents greater than 85% of the LL were susceptible to liquefaction (note that PI = 12 is the upper boundary on Seed et al. (2001)) for Zone A in Fig. 6.6. Soils with 12 < PI < 20 and water contents greater than 80% of the LL were "systematically more resistant to liquefaction but still susceptible to cyclic 'mobility'" (Bray et al., 2004).

Boulanger and Idriss (2004, 2008) reviewed available experimental literature on the monotonic and cyclic undrained shear loading behavior of fine-grained soils (>50% passing#200 sieve), and concluded that these soils could reasonably be grouped into *soils that exhibit either sand-like or clay-like stress-strain characteristics.* They have used liquid limit and plasticity index as the criteria to distinguish between sand-like or clay-like behavior as shown in Fig. 6.7. Figure 6.7 shows the range of Atterberg limits for the transition between sand-like and clay-like behavior. Boulanger and Idriss (2008) have remarked that the sand-like behavior was observed only for soils with PI ≤ 3.5. They have also suggested procedures that are best used to estimate potential strains and strength loss during earthquake loading. These procedures are different for soils that exhibit sand-like behavior versus those that

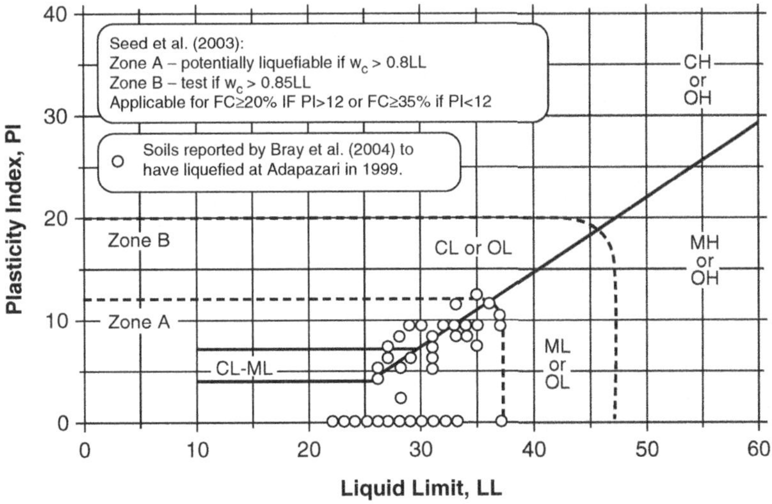

Fig. 6.6 Plasticity chart showing the recommendations by Seed et al. (2001) and the Atterberg limits of fine-grained soil reported by Bray et al. (2004) to have "liquefied" at 12 building sites during the 1999 Kocaeli earthquake

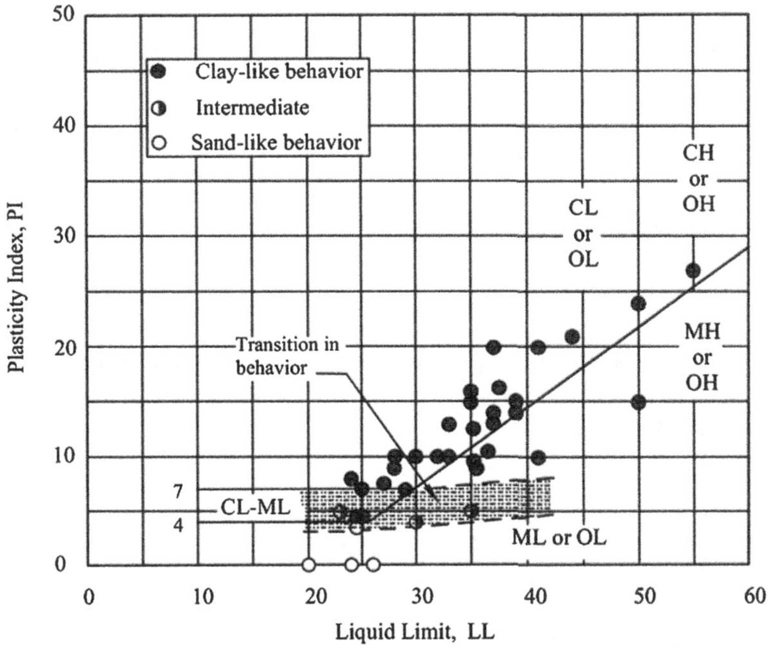

Fig. 6.7 Plasticity chart for identifying clay-like, sand-like, and intermediate behavior (Boulanger and Idriss, 2008)

exhibit clay-like behavior during monotonic and cyclic undrained shear loading (Boulanger and Idriss 2004). It is recommended that *soils exhibiting sand-like behavior* be evaluated using SPT- and CPT-based methodologies and that the term "*liquefaction*" be reserved for these types of soils. It is further recommended that *soils exhibiting clay-like behavior* be evaluated using procedures appropriate for clays and that term "*cyclic failure*" be used to describe failure in these soils. The key features for clay-like behavior are that the soil's undrained shear strength is intimately related to its consolidation stress history, and the soil's cyclic shear resistance is closely related to its monotonic undrained shear strength (Boulanger and Idriss, 2004).

2.2 *Overview of Liquefaction of Fine-Grained Soils*

It is observed from the discussion in the previous paragraph that there is no well-defined criteria to estimate the liquefaction susceptibility of fine-grained soil with plasticity. This potential source of confusion can be partly alleviated by adopting different terms to describe the onset of significant strains in *sand-like* versus *clay-like* soils, and thus it is recommended that the term "liquefaction" be reserved for describing sand-like soil behaviors, and the term "cyclic failure" be used for describing clay-like soil behaviors (Boulanger and Idriss, 2004). Fine-grained soils that liquefied during the 1994 Northridge, 1999 Kocaeli, and Chi-Chi earthquakes often did not meet the clay-size criterion of the Chinese recommendation. Cyclic testing of a wide range of soils found to liquefy in Adapazari during the Kocaeli earthquake confirmed that these fine-grained soils were susceptible to liquefaction. It is not the amount of "clay-size" particles in the soil; rather, it is the amount and type of clay minerals in the soil that best indicate liquefaction susceptibility (Bray et al., 2004).

It was determined that

1. Plasticity index appears to be a better indicator of liquefaction susceptibility.
2. Loose soils with $PI < 12$ and $wc/LL > 0.85$ were susceptible to liquefaction.
3. Loose soils with $12 < PI < 18$ and $wc/LL > 0.8$ were systematically more resistant to liquefaction.
4. Soils with $PI > 18$ tested at low effective confining stresses were not susceptible to liquefaction.

Other factors which control liquefaction and/or cyclic mobility are:

5. Confining pressure.
6. Initial static shear stress.
7. Stress path.
8. The location of a soil on the Casagrande plasticity chart and, or in combination with, the use of the "C" descriptor, (USCS) (e.g., CH, CL, SC, and GC) are considered as non-liquefiable.

There are some other considerations as follows:

9. Liquefiable fine-grained soils should have LL < 35 and plot below the A-line or have PI < 7.
10. Seed et al. (2001) state that soils with LL < 37 and PI < 12 are potentially liquefiable, and those with 37 < LL < 47 and 12 < PI < 20 require laboratory testing.
11. Plito (2001) found that soils with LL < 25 and PI < 7 are liquefiable, and soils with 25 < LL < 35 and 7 < PI < 10 are potentially liquefiable, and soils with 35 < LL <50 and 10 < PI < 15 are susceptible to cyclic mobility.

It, therefore, appears at this time (2012) that:

1. Chinese criterion for liquefaction of silty–clayey soils does not work.
2. Liquefaction or cyclic mobility of fine-grained soils depends on their sand-like or clay-like behavior in static undrained tests.
3. Cyclic mobility of clay may depend upon the factors listed below:
   (a) Plasticity index
   (b) wc/LL ratio
   (c) Initial static shear stress
   (d) Confining pressure
   (e) Stress path

2.2.1 An Observation

First cyclic tests on sands were performed in 1960s (Seed and Lee 1966). It took almost 40 years to more or less fully understand liquefactions of sands. First cyclic tests in silts were performed in 1984 (El-Hosri and Puri). It may take us some more time to understand the cyclic mobility of clays and clayey silts. So, we are away from the final understanding of liquefaction behavior of clays and clayey silts.

3 Stability of Rigid Retaining Structures

The rigid retaining walls that are generally designed for lateral earth pressure of Coulomb or Rankine, investigations on the performance of retaining wall during earthquakes have shown that in some cases the retaining walls have failed either by sliding away from the backfill or due to combined action of sliding and rocking displacements (Prakash and Puri, 2003; Iai, 1998; Wu, 1999; Wu and Prakash, 2001). A realistic design should account for the displacements of the retaining wall during an earthquake. Displacement of the retaining wall is also an important consideration in performance-based design.

Rafnsson and Prakash (1991) developed a model for simulating the response of rigid retaining walls subjected to seismic loading. This model consisted of a rigid

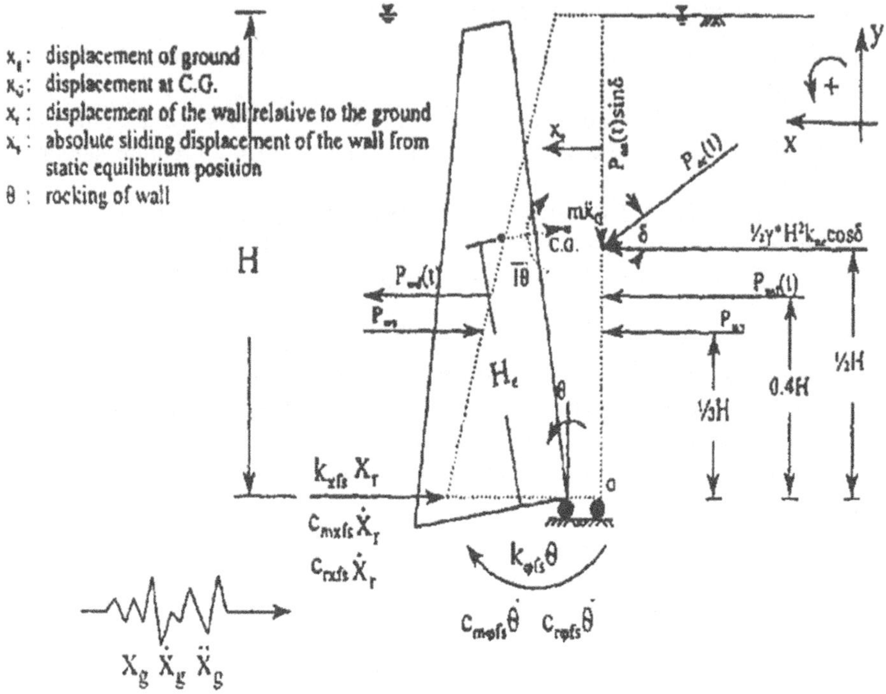

Fig. 6.8 Force diagram of forced vibration of rigid retaining wall with submerged pervious backfill (Wu, 1999)

wall resting on the foundation soil and subjected to a horizontal ground motion and analyzed the problem as a case of combined sliding and rocking vibrations including the effect of various important parameters such as soil stiffness in sliding, soil stiffness in rocking, geometrical damping in sliding, geometrical damping in rocking, material damping in sliding, and material damping in rocking. Only dry backfill was considered, and seismic ground motion was represented by an equivalent sinusoidal motion. This model was further modified to accommodate both (Fig. 6.8) the dry and submerged backfills (Wu 1999; Wu and Prakash 2001; Wu et al. 2010). This model realistically estimates the dynamic displacement due to coupled sliding and rocking and accounts for nonlinear soil stiffness and material and geometrical damping.

3.1 Typical Results Using Wu (1999) Model

A wall 4 m high (Fig. 6.9a) with granular backfill and foundation soil is used for illustration of typical results obtained by using Wu (1999) model. The retaining wall was subjected to Northridge earthquake of January 17, 1994 (Fig.6.9b). The

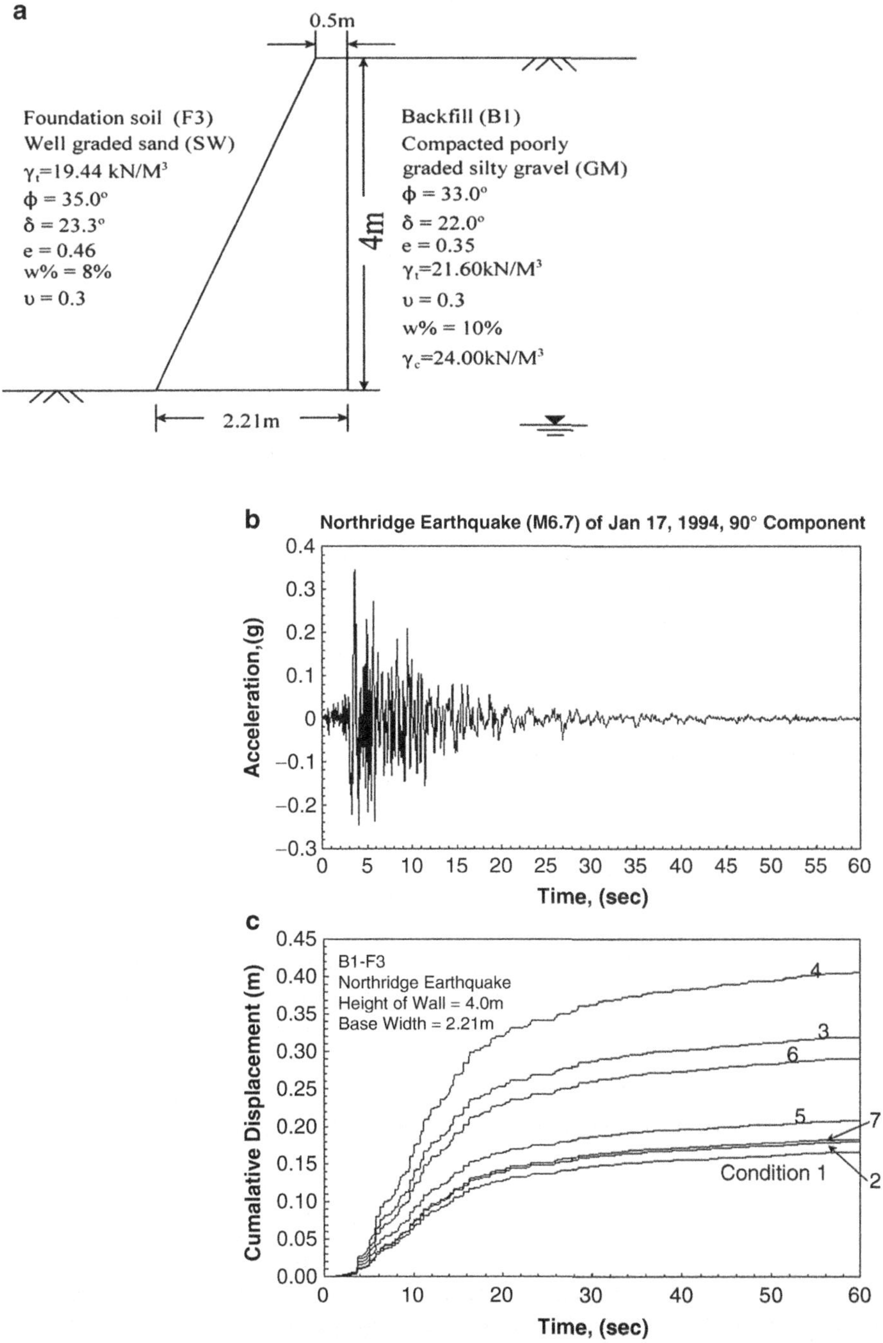

Fig. 6.9 (**a**) Geometry of retaining wall and properties of backfill and foundation soil; (**b**) accelerogram of Northridge earthquake of Jan. 17, 1994; and (**c**) computed displacements for various field conditions

Table 6.3 Engineering properties for foundation soil and backfill (Wu, 1999)

	Soil type	γ_d (kN/m^3)	φ (deg)	δ (deg)	Void ratio	ν	c (kN/m^2)	PI	w%
Foundation soil (F)									
F-1	GW	21.07	37.5	25.0	0.25	0.3	–	–	6
F-2	GP	19.18	36.0	24.0	0.36	0.3	–	–	6
F-3	SW	18.00	35.0	23.3	0.46	0.3	–	–	8
F-4	SP	16.82	34.0	22.7	0.56	0.3	–	–	10
F-5	SM	15.70	33.0	22.0	0.68	0.3	–	4	15
F-6	SC	14.00	30.0	20.0	0.88	0.3	–	13	25
F-7	ML	14.15	32.0	21.3	0.85	0.3	9.57	4	14
Backfill (B)									
B-1	GM	19.6	33.0	22.0	0.35	0.3	–	–	10
B-2	GP	18.9	34.0	22.7	0.40	0.3	–	–	8
B-3	SP	15.6	34.0	22.7	0.69	0.3	–	–	8

All properties of backfill are for the condition of 90% of the "Standard Proctor"

engineering properties of different backfill and foundation soils used in the analysis are given in Table 6.3, and the various field conditions to which a retaining wall may be subjected are given in Table 6.4. The displacements were computed on the assumption that the base width has been designed as for field condition 1 (Table 6.4) and displacements computed for Northridge earthquake for field conditions 1 through 7. Nonlinear soil modulus and strain-dependent damping values are used in this solution. The magnitude of this earthquake is M 6.7, and peak ground acceleration is 0.344 g. Figure 6.9c shows displacements of the 4-m-high wall under seven field conditions. These displacements are also listed in Table 6.5.

An examination of Table 6.5 shows that sliding displacements (column 2) are close to 30–40% of the total displacement (column 5). Maximum total displacements occur in field condition 4, i.e., submerged wall with pervious backfill.

According to Eurocode, the permissible displacement is 10.32 cm (300 × α_{max}, where α_{max} is 0.344 in Northridge earthquake). Sliding displacement in conditions 3 and 4 exceeds this value. For practical design field conditions 1–7 are typical.

3.2 Soil and Wall Heights Used to Develop Design Charts

Wu (1999) has studied seven soil conditions for foundation soil F1–F7 and three soils for backfill B1-B3 (Table 6.3). Thus 21 combinations for foundation and Tables 3, 4 and 5 to appear after this and before (Fig. 6.9) backfills soils were investigated. Rigid walls heights investigated are 4m, 5m, 6m, 7m, 8m, 9m, and 10 m. Table 6.6 lists cumulative displacements for B1–F1.

Table 6.4 Loading conditions and corresponding parameters for dynamic displacements

	Field condition	Parameters for Static condition	Parameters for Dynamic condition
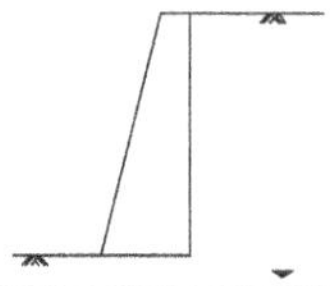	*Condition 1* Moist backfill moist foundation soil	$\gamma^* = \gamma_t$ $P_{ws} = 0$	$\gamma^* = \gamma_t$ $\Psi = \tan^{-1}\left(\frac{\alpha_h}{1 \mp \alpha_v}\right)$ $P_{wd}(t) = 0$
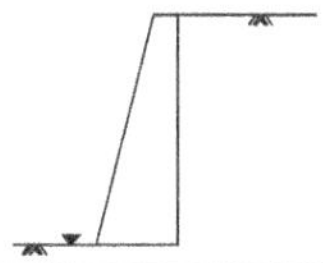	*Condition 2* Moist backfill saturated foundation soil	$\gamma^* = \gamma_t$ $P_{ws} = 0$	$\gamma^* = \gamma_t$ $\Psi = \tan^{-1}\left(\frac{\alpha_h}{1 \mp \alpha_v}\right)$ $P_{wd}(t) = 0$
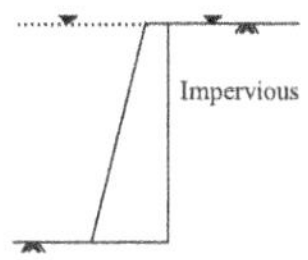	*Condition 3* Submerged with impervious backfill	$\gamma^* = \gamma_{sat} - \gamma_w$ $P_{ws} = 0$	$\gamma^* = \gamma_{sat} - \gamma_w$ $\Psi = \tan^{-1}\left(\frac{\gamma_{sat}}{\gamma_{sat} - \gamma_w} \frac{\alpha_h}{1 \mp \alpha_v}\right)$ $P_{wd}(t) = 7/12 \times \alpha_h \times \gamma_w \times H'$
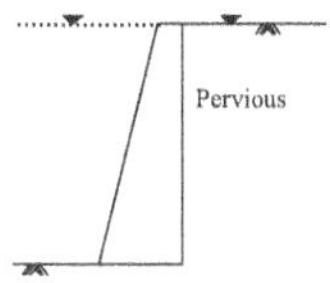	*Condition 4* Submerged with pervious backfill	$\gamma^* = \gamma_{sat} - \gamma_w$ $P_{ws} = 0$	$\gamma^* = \gamma_{sat} - \gamma_w$ $\Psi = \tan^{-1}\left(\frac{\gamma_d}{\gamma_{sat} - \gamma_w} \frac{\alpha_h}{1 \mp \alpha_v}\right)$ $P_{wd}(t) = 2 \times 7/12 \times \alpha_h \times \gamma_w \times H'$
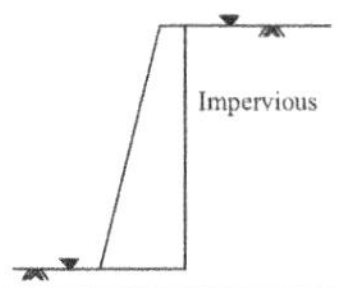	*Condition 5* Perched with impervious backfill	$\gamma^* = \gamma_{sat} - \gamma_w$ $P_{ws} = ½ \times \gamma_w \times H^2$	$\gamma^* = \gamma_{sat} - \gamma_w$ $\Psi = \tan^{-1}\left(\frac{\gamma_{sat}}{\gamma_{sat} - \gamma_w} \frac{\alpha_h}{1 \mp \alpha_v}\right)$ $P_{wd}(t) = 0$
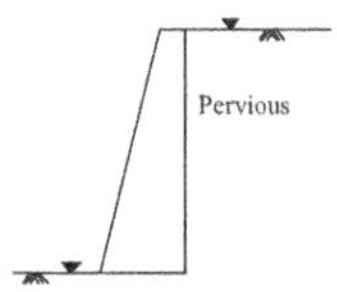	*Condition 6* Perched with pervious backfill	$\gamma^* = \gamma_{sat} - \gamma_w$ $P_{ws} = ½ \times \gamma_w \times H^2$	$\gamma^* = \gamma_{sat} - \gamma_w$ $\Psi = \tan^{-1}\left(\frac{\gamma_d}{\gamma_{sat} - \gamma_w} \frac{\alpha_h}{1 \mp \alpha_v}\right)$ $P_{wd}(t) = 7/12 \times \alpha_h \times \gamma_w \times H'$
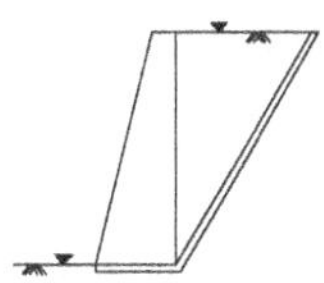	*Condition 7* Perched with sloping drain	$\gamma^* = \gamma_{sat}$ $P_{ws} = 0$	$\gamma^* = \gamma_{sat}$ $\Psi = \tan^{-1}\left(\frac{\alpha_h}{1 \mp \alpha_v}\right)$ $P_{wd}(t) = 0$

3.3 *Vertical vs Inclined Walls*

In order to economize on design of walls, several cases of 6.0-m-high retaining walls were analyzed for typical cases of foundation soil condition varying from well-graded gravel (GW) to silt (ML) and the backfill soil varying from silty gravel (GM) to poorly graded sand (SP). Ground motions corresponding to El Centro,

Table 6.5 Displacement of 4-m-high wall for field condition 1–7

Field condition	Displacement				
	Sliding	Rocking		Total	
	m	degree	m	m	% of height
Column 1	Column 2	Column 3	Column 4	Column 5	Column 6
1	0.0622	1.48	0.1034	0.1656	4.1
2	0.0667	1.61	0.1126	0.1793	4.5
3	0.1168	2.90	0.2023	0.3191	8.0
4	0.1492	3.67	0.2564	0.4055	10.1
5	0.0759	1.89	0.1319	0.2078	5.2
6	0.1076	2.62	0.1830	0.2905	7.3
7	0.0682	1.64	0.1148	0.1830	4.6

Table 6.6 Cumulative displacement for several angles of inclination of the back of the wall subjected to Northridge earthquake condition ($B = 3.57$ m)

Inclination-on angle (degree)	Base width-h (m)	Cumulative displacement by fixed base width (3.57 m)			
		Sliding (m)	Rocking (degree)	Rocking (m)	Total (m)
+5.00°	3.81	0.0820	1.31	0.1374	0.2194
+3.75°	3.76	0.0820	1.30	0.1366	0.2186
+2.50°	3.70	0.0815	1.30	0.1361	0.2176
+1.25°	3.63	0.0808	1.29	0.1355	0.2163
0.00°	3.57	0.0808	1.29	0.1347	0.2155
−1.25°	3.50	0.0806	1.28	0.1338	0.2144
−2.50°	3.43	0.0805	1.27	0.1329	0.2134
−3.75°	3.35	0.0803	1.26	0.1320	0.2123
−5.00°	3.38	0.0801	1.25	0.1311	0.2112

Loma Prieta, and Northridge earthquakes were used in the analysis. Typical case of a reference retaining wall, 6.0 m high, with nine different inclination angles of the wall face in contact with the backfill "α" (0°, 1.25°, 2.5°, 3.75°, +5°, −1.25°, −2.5°, −3.75°, and −5°) subjected to Northridge earthquake is used for illustration. The negative angle at the back of the wall is the case of the wall resting on the backfill. Figure 6.10 shows cumulative displacement of the retaining wall away from the backfill due to combined sliding and rocking effects for $\alpha = -5^\circ$, 0° and $+5^\circ$ for a base width of 3.57 m. The foundation soil for this case was well-graded sand (SW), and the backfill consisted of submerged silt gravel (GM). It can be observed from Fig. 6.10 that the negative values of "α" result in somewhat smaller cumulative displacements compared to the case of vertical wall face ($\alpha = 0$) or for positive value of α within the range considered.

Another typical plot of cumulative displacement of a rigid retaining wall of 6.0 m height and having a base width of 4.61 m and subjected to Northridge earthquake motion is shown in Fig. 6.11. The foundation and backfill soils in this case were silt of low compressibility and silty gravel, respectively. The trend of the results in Fig. 6.11 is similar that in Fig. 6.10. Similar results were observed for other cases.

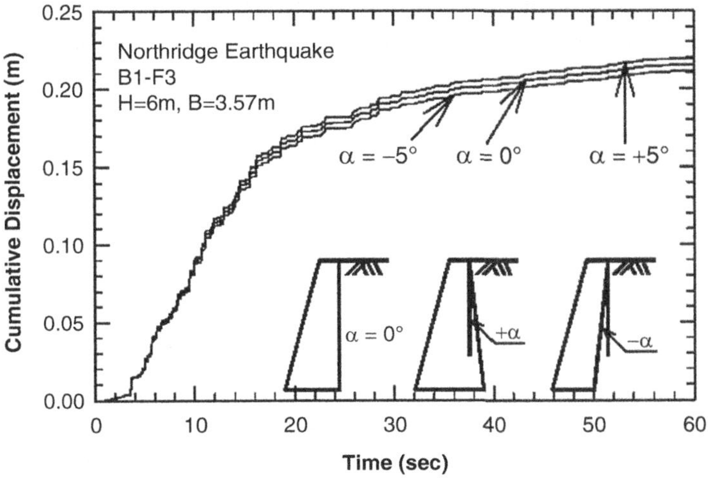

Fig. 6.10 Cumulative displacements of walls (B1–F3) with different inclinations with the vertical

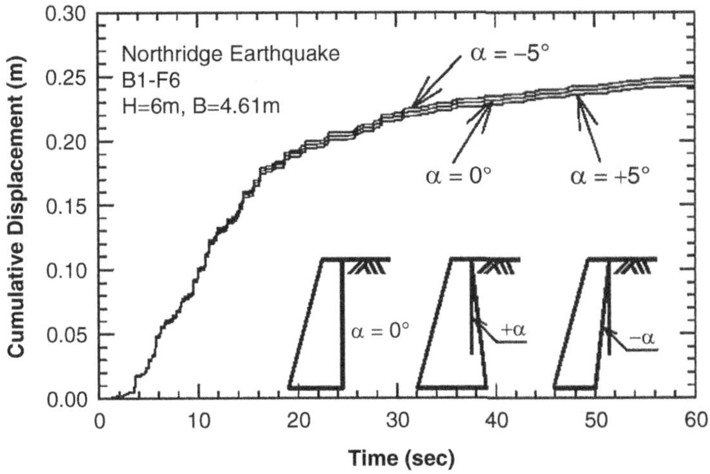

Fig. 6.11 Cumulative displacements of walls (B1–F6) with different inclinations with the vertical

It, therefore, appears that retaining walls be designed for permissible displacement for sliding only and then be built resting by a few degrees on the backfill. In this case this tilt is about 4° (3.67° maximum).

Table 6.6 shows a summary of new base widths and computed displacement for various inclinations. The computed cumulative sliding, rocking, and total displacements are also shown in this table. The base widths decreased from 3.57 to 3.38 m as the inclination changed from 0° to −5°, since the active earth forces decrease with negative inclination. Therefore, the base width was somewhat smaller for a wall with a negative inclination.

The angular rotation in rocking (Table 6.6) decreased from 1.29° ($\alpha = 0$) to 1.25° ($\alpha = -5°$), and the total displacements decreased slightly from 0.2155 to 0.2112 m. The cumulative displacements for these walls will not be significantly altered by changing the inclination at the back of the wall.

For the wall built as a leaning-type rigid retaining wall with $\alpha = -5°$ lying on the backfill, the wall experienced a rocking movement of 1.25° during the Northridge earthquake. Therefore, when the wall was subjected to the same earthquake event up to three or four times, the wall experienced a total rocking close to 5°. At this time, the wall may become vertical.

Further analysis was conducted for 21 backfill and foundations soil combinations for a typical reference wall 6 m high, subjected to three earthquakes. The backfill soil was varied from silty gravel to poorly graded sand, and the foundation soil varied from well-graded gravel to silt of low compressibility.

The results generally indicated that the design widths of foundations for 21 cases of backfill – foundation soil combinations used in the analysis, generally reduced as values of α varied from 0° to −5°. This may result in saving of 8–10% in the material cost. It is, therefore, recommended that rigid walls be constructed with a negative batter with the walls resting on the backfill. In this situation, these may be designed only for sliding displacements.

3.4 Recommended Design Procedure

1. Determine the section for static loading condition with FOS = 2.5 in bearing and FOS = 1.5 for sliding and tilting as a rigid body and no tension on the head.
2. Estimate the sliding displacement from Wu (1999). Use design charts for comparable, backfill and foundation soils, and comparable ground motion.
3. Compare these displacements with permissible displacements as per Eurocode ($300 \times \alpha_{max}$ mm).
4. If displacement in (2) is less than or equal that in (3), then design is OK, or else revise the sections of the wall for lower FOS in (1).

3.5 Typical Design Charts

Table 6.7 lists the sliding and total displacement and rotation of walls 4 m–10 m high and subjected to three ground motions of (1) El Centro (1940), (2) Northridge (1994), and (3) Loma Prieta (1989). Wu (1999) has presented 21 similar charts.

Table 6.7 Cumulative displacements for walls 4–10 m high with B1–F1 and field conditions 1, 2, and 7 (Table 6.4) subjected to El Centro, Northridge, and Loma Prieta earthquakes

		Cumulative displacement								
		El Centro[2]			Northridge[2]			Loma Prieta[2]		
H and B[1] (m)	Field con.	Sliding (m)	Rocking degree (m)	Total (m)	Sliding (m)	Rocking degree (m)	Total (m)	Sliding (m)	Rocking degree (m)	Total (m)
4	1	0.0895	2.52 (0.1760)	0.2655	0.0604	1.62 (0.1128)	0.1732	0.0052	0.11 (0.0074)	0.0126
(2.08)	2	0.0940	2.71 (0.1889)	0.2829	0.0642	1.75 (0.1221)	0.1862	0.0057	0.12 (0.0083)	0.0140
	7	0.0960	2.76 (0.1928)	0.2887	0.0652	1.79 (0.1247)	0.1899	0.0058	0.12 (0.0085)	0.0143
5	1	0.1058	2.54 (0.2515)	0.3273	0.0722	1.65 (0.1439)	0.2160	0.0068	0.12 (0.0104)	0.0172
(2.60)	2	0.1118	2.71 (0.2365)	0.3483	0.0766	1.77 (0.1546)	0.2311	0.0074	0.13 (0.0113)	0.0187
	7	0.1136	2.76 (0.2412)	0.3548	0.0779	1.81 (0.1578)	0.2357	0.0075	0.13 (0.0116)	0.191
6	1	0.1184	2.37 (0.2483)	0.3667	0.0809	1.54 (0.1615)	0.2424	0.0082	0.12 (0.0124)	0.0206
(3.22)	2	0.1235	2.53 (0.2654)	0.3889	0.0849	1.66 (0.1740)	0.2589	0.0087	0.13 (0.0138)	0.0225
	7	0.1225	2.58 (0.2736)	0.3961	0.0863	1.70 (0.1776)	0.2639	0.0089	0.13 (0.0140)	0.0229
7	1	0.1281	2.25 (0.2745)	0.4026	0.0880	1.47 (0.1794)	0.2674	0.0094	0.12 (0.0147)	0.0241
(3.84)	2	0.1335	2.39 (0.2923)	0.4258	0.0922	1.58 (0.1924)	0.2846	0.0101	0.13 (0.0163)	0.0264
	7	0.1357	2.44 (0.2979)	0.4336	0.0937	1.61 (0.1964)	0.2901	0.0103	0.14 (0.0167)	0.0270
8	1	0.1353	2.05 (0.2863)	0.4216	0.0931	1.34 (0.1871)	0.2802	0.0104	0.12 (0.0161)	0.0265
(4.56)	2	0.1407	2.18 (0.3048)	0.4455	0.0970	1.44 (0.2011)	0.2981	0.0112	0.13 (0.0178)	0.0289
	7	0.1428	2.22 (0.3106)	0.4535	0.0985	1.47 (0.2052)	0.3037	0.0114	0.13 (0.0181)	0.0295
9	1	0.1442	2.05 (0.3213)	0.4655	0.0998	1.35 (0.2122)	0.3120	0.0117	0.12 (0.0192)	0.0309
(5.08)	2	0.1498	2.17 (0.3405)	0.4903	0.1035	1.44 (0.2267)	0.3303	0.0127	0.13 (0.0211)	0.0339
	7	0.1521	2.21 (0.3470)	0.4991	0.1051	1.47 (0.2312)	0.3364	0.0130	0.14 (0.0216)	0.0345
10	1	0.1499	1.91 (0.3373)	0.4816	0.1034	1.26 (0.2195)	0.3229	0.0128	0.12 (0.0205)	0.0334
(5.80)	2	0.1558	2.01 (0.3515)	0.5073	0.1073	1.34 (0.2342)	0.3415	0.0138	0.13 (0.0227)	0.0365
	7	0.1581	2.05 (0.3581)	0.5162	0.1089	1.37 (0.2388)	0.3477	0.0141	0.13 (0.0232)	0.0372

[1]*H* height of wall, *B* base width
[2]Permissible displacements for three earthquakes according to Eurocode = $300 \times \alpha_{max}$
El Centro = 0.349 × 300 (mm) = 0.1047 m

3.6 Overview of Stability of Rigid Retaining Structures

The method proposed by Wu and Prakash (2001) for determination of seismic displacement of rigid retaining walls is simple and useful. The determination of displacement of retaining walls is an essential requirement in performance-based design. However the amount of seismic displacement that may be considered as acceptable, has no definite answer. Some guidelines based on experience or judgment are given below (Huang, 2005).

Eurocode (1994)
Permissible horizontal displacement = $300.a_{max}$ (mm), a_{max} = maximum horizontal design acceleration.
AASHTO (2002)
Permissible horizontal displacement = $250.a_{max}$ (mm)
Wu and Prakash (2001)
Permissible horizontal displacement = 0.02 H, H = height of retaining wall
Failure horizontal displacement = 0.1 H
Permissible differential settlement = 0.1–0.2 m (damage needing minor retrofit measures)
Severe differential settlement = > 0.2 m (damage needing long-term retrofit measures)

It may be noted from the above that Eurocode 8 (1994) and Wu and Prakash (2001) recommend using specified horizontal displacements of the retaining wall for evaluating its seismic performance. The Japanese Railway Technical Research Institute (1999) suggests the use of vertical differential settlement as the performance criterion which seems reasonable for traffic accessibility and retrofit purposes after the earthquake.

4 Comparison of Computed and Predicted Pile Response

The present methods for design of pile foundations subjected to dynamic loads are generally based on the models developed by Novak (1974) and Novak and El Sharnouby (1984) and also presented by Prakash and Puri (1988) and Prakash and Sharma (1990). The pile response under dynamic loads is generally determined by making simplified spring-mass models. The soil springs are obtained from the shear modulus of the soil or from the modulus of subgrade reaction. The seismic loading induces large displacements/strains in the soil. The shear modulus of the soil degrades and damping (material) increases with increasing strain. The stiffness of piles should be determined for these strain effects. Several researchers have attempted to make a comparison of the observed and predicted pile response. Prediction of pile behavior under dynamic loads depends upon pile dimensions and soil properties, which include; soil shear modulus, material damping, and also on geometrical damping, and frequency of operation and more importantly on strain level. Many investigators have attempted to bridge the gap between prediction and performance, by applying arbitrary correction factors to either soil stiffness, or damping or to both, based on linear as well as nonlinear solutions (Prakash and Puri, 2008).

Jadi (1999) and Prakash and Jadi (2001) reanalyzed the reported pile test data of Gle (1981) for the lateral dynamic loads and proposed reduction factors for the stiffness and radiation damping obtained by using the approach of Novak and El Sharnouby (1983). Gle (1981) tested four different single steel pipe piles at two different sites in Southeastern Michigan. The soil profiles at these sites were predominantly composed of clayey soils. Each pile was tested at several vibrator-operating speeds. A total of 18 dynamic lateral tests were conducted in clayey and silty sand media. The method of analysis used by Jadi (1999) is as follows.

4.1 Method of Analysis (Jadi, 1999; Prakash and Jadi, 2001)

1. Field data obtained from lateral dynamic tests performed by Gle (1981) on full-scale single piles embedded in clayey soils were collected.
2. Theoretical dynamic response was computed for the test piles, using Novak and El- Sharnouby (1983) analytical solution for stiffness and damping constants, with no corrections.
3. The soil's shear modulus and radiation damping used for the response calculations were arbitrarily reduced, such that measured and predicted natural frequencies and resonant amplitude matched.
4. The reduction factors obtained from step 3 were plotted versus shear strain at resonance without corrected G and "c." Two quadratic equations given below were developed to determine the shear modulus reduction factors (λ_G) versus shear strain (γ) and the radiation damping reduction factor (λ_C) versus shear strain (γ).

$$\lambda_G = -353500\,\gamma^2 - 0.00775\,\gamma + 0.3244 \tag{6.1}$$

$$\lambda_c = 217600\,\gamma^2 - 1905.56\,\gamma + 0.6 \tag{6.2}$$

   where, λ_G and λ_c are the reduction factors for shear modulus and damping and γ is shear strain at computed peak amplitude, without any correction.
5. For all the pile tests considered in this study, the empirical equations determined in 4 above were used to calculate shear modulus and radiation damping reduction factors. Predicted responses before and after applying the proposed reduction factors were then compared with the measured response.
6. To validate this approach, the proposed equations were used to calculate shear modulus and radiation damping reduction factors for different sets of field pile tests. The new predicted response was then compared to the measured response, both for Gle (1981) tests and two other cases.

4.2 Comparison of Observed and Predicted Pile Response (Jadi, 1999; Prakash and Jadi, 2001)

Using the reduction factors (Eqs. 6.1 and 6.2), Jadi (1999) and Prakash and Jadi (2001) calculated the pile response for the same site and made a comparison of observed and predicted pile response. Figure 6.12 shows prediction and performance of Gle's pile. In Fig. 6.12, it may be noted that the reduction factors for shear modulus and damping had been developed from tests by Gle. Therefore, this match is obvious.

4.2.1 Comparison with a Different Data Sets

In order to confirm the validity of the proposed method, the calculated dynamic response for different sets of experimental data from other sites were also checked. Two series of experimental data were analyzed. Blaney (1983) carried out two lateral dynamic tests on the single pile, embedded in the clayey soils. The first test was performed with a "WES" (Waterways Express Station) vibrator. For the second test an "FHWA" (Federal Highway Administration) vibrator was used. Figure 6.13 represents the predicted response computed by applying suggested shear modulus and radiation damping reduction factors and measured lateral dynamic response of the same pile. Resonant amplitudes matched, but computed natural frequency differs by about 40%. Figure 6.14 also confirms similar observation. However, these figures show that the predicted response with proposed reduction factors compares much

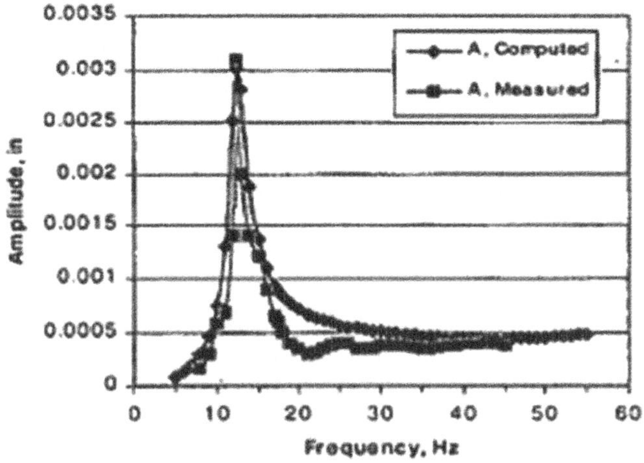

Fig. 6.12 Measured and reduced predicted lateral dynamic response for pile for lateral dynamic load test for pile K 16–7 $\theta = 5°$, Belle River Site (Jadi, 1999)

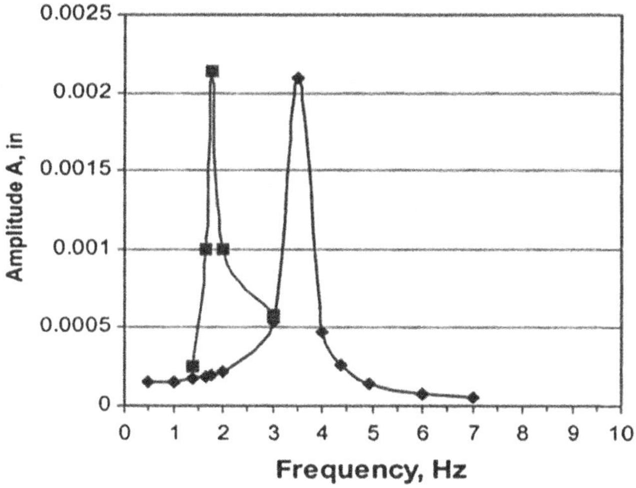

Fig. 6.13 Measured versus reduced predicted lateral dynamic response for the 2.4″ test pile using proposed reduction factors $\lambda_G = 0.044$ and $\lambda_c = 0.34$ (Jadi, 1999)

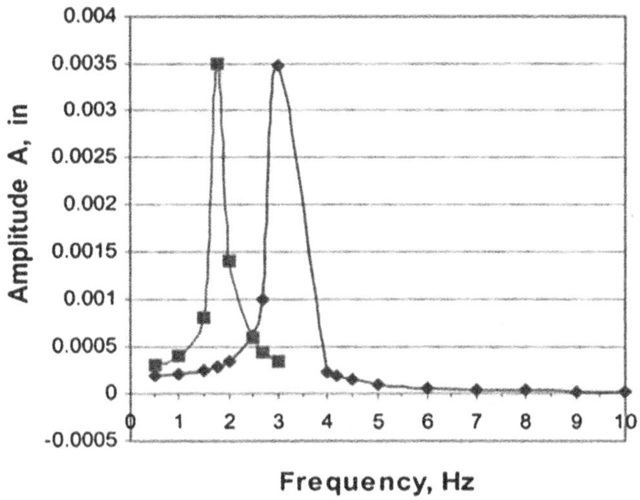

Fig. 6.14 Measured versus reduced predicted lateral dynamic response for the 2.4″ test pile using proposed reduction factors $\lambda_G = 0.044$ and $\lambda_c = 0.34$ (Jadi, 1999)

better with the measured response as compared to the predictions by Blaney (Jadi 1999). Jadi (1999) model for prediction of dynamic pile response is essentially a linear model with shear modulus and damping matched at maximum amplitude. Cambio (2012) improved upon Jadi's model as follows.

4.3 Cambio (2012) Model

Cambio (2012) analyzed the existing available pile test using DYNA5 and observed that the calculated natural frequencies are overestimated, and the resonant amplitudes are generally underestimated by DYNA5 which was attributed to overestimated shear modulus and damping in calculations. Cambio (2012) proposed an equivalent linear model to predict the response of piles in clay or fine silty soil and subjected to lateral vibrations. The model incorporates frequency-dependent parameters and the effects of soil nonlinearity by using strain-dependent values of shear modulus. To improve upon the computed response a set of reduction factors on soil shear modulus and total damping were determined through regression analysis until the predicted and the measured amplitudes and frequencies match. Empirical equations relating the reduction factors with soil shear strain, elastic properties of soils and piles, and pile geometry were developed and are given below:

$$\lambda_G = 0.912385 + 0.00165.\frac{L}{r_o} - 0.0001334.\frac{E_p}{G_{max}} - 1.407 \times 10^{-9}.E_{max} + 43.246\gamma_s \tag{6.3}$$

$$\lambda_c = 0.573217 - 119.542\gamma_s - 0.01182f_{max} \tag{6.4}$$

where
λ_G = Reduction factor for shear modulus of soil.
λ_c = Reduction factor for total damping in soil.
L = Pile length.
r_o = Radius of pile or equivalent radius for a noncircular pile.
E_p = Young's modulus of pile material.
G_{max} = Maximum shear modulus of soil.
E_{max} = Maximum value of Young's modulus of soil.
f_{max} = Maximum value of natural frequency.
γ_s = Shear strain in soil.

4.3.1 Comparison of Observed and Predicted Pile Response (Cambio, 2012)

Cambio (2012) checked her proposed model for predicting dynamic pile response against different set of reported pile test data. Figure 6.15 shows a comparison of observed and predicted pile response for lateral vibrations based on field tests of Gle (1981) and Cambio (2012) equivalent linear model. Another similar comparison with reported pile test data of El-Marsafawi et al. (1990) is shown in Fig. 6.16. The proposed model seems to make good predictions of pile response. Cambio (2012) also made a comparison of pile response predictions made by her proposed model with those made by Jadi (1999). One such typical comparison is shown in Fig. 6.17. Cambio's model seems to make better prediction in this case.

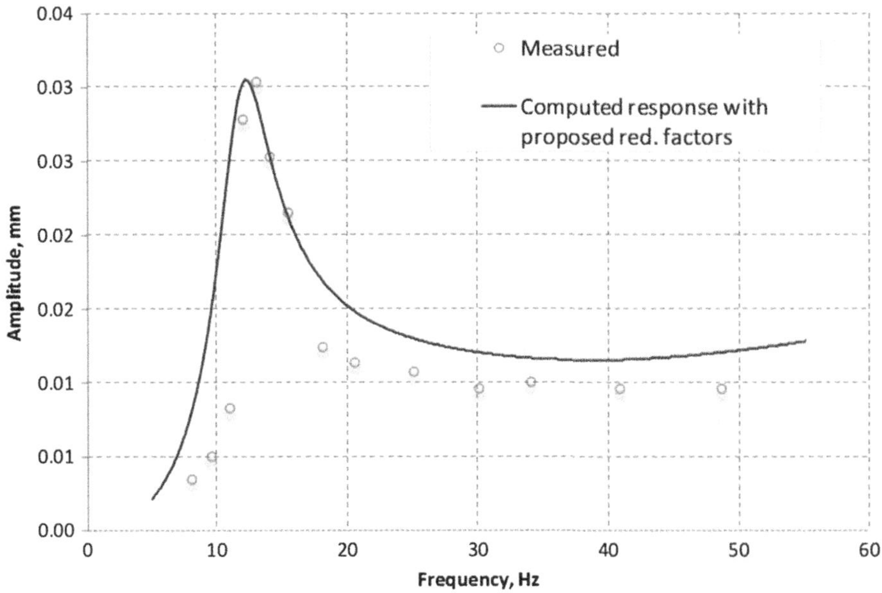

Fig. 6.15 Comparison of observed and predicted pile response using prosed reduction factors on pile L1810, $\theta = \mathring{5}$, (Gle 1981; Cambio, 2012)

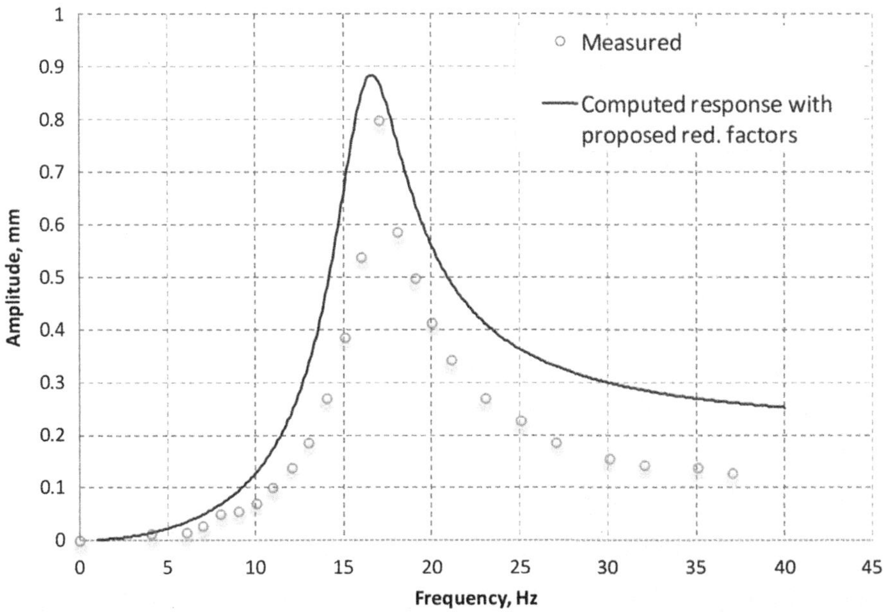

Fig. 6.16 Comparison of observed and predicted pile response (pile 2, El-Marsafawi et al., 1990; Cambio, 2012)

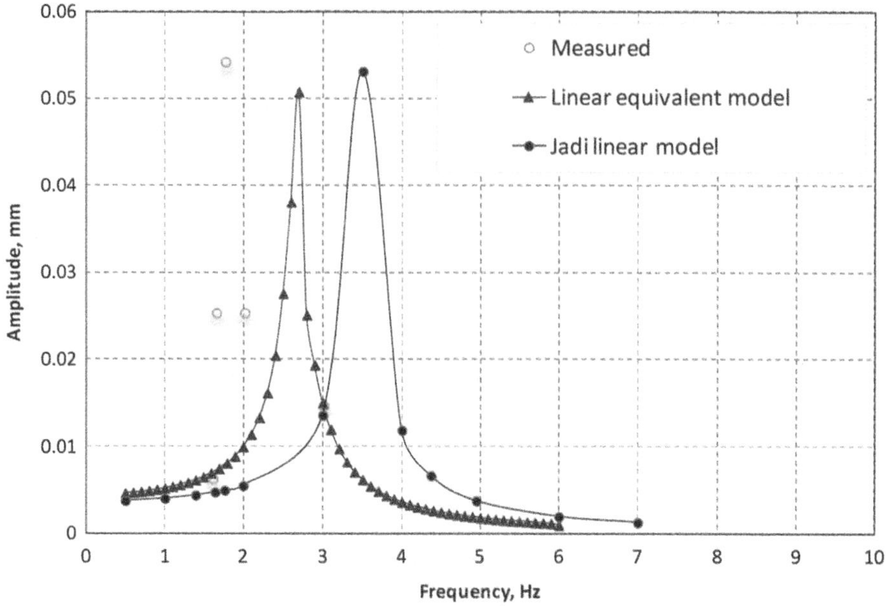

Fig. 6.17 Measured and predicted pile response (FHWA test data, Blaney (1983)) by Jadi's (1999) and Cambio (2012) models (Cambio, 2012)

4.4 Overview of Predicted and Observed Pile Response

Effort has been devoted in recent years to develop simple but realistic solutions for obtaining pile response under dynamic loads which shows reasonable match with observed data. Cambio (2012) makes good predictions for *d*ynamically loaded piles in clays and silty soils. The model has not been tried for piles in sands.

5 Piles in Soils Susceptible to Liquefaction

Excess pore pressures during seismic motion may cause lateral spreading resulting in large moments in the piles and settlements and tilt of the pile cap and the superstructure. Excessive lateral pressure may lead to failure of the piles which was experienced in the 1964 Niigata and the 1995 Kobe earthquakes (Finn and Fujita, 2004). Damage to a pile under a building in Niigata caused by about 1 m of ground displacement is shown Fig. 6.18. Displacement of quay wall and damage to piles supporting tank TA72 (Figs. 6.19 and 6.20) during 1995 Kobe earthquake has been reported by Ishihara and Cubrinovsky (2004). The seaward movement of the quay wall was accompanied by piles lateral spreading of the backfill soils resulting in a

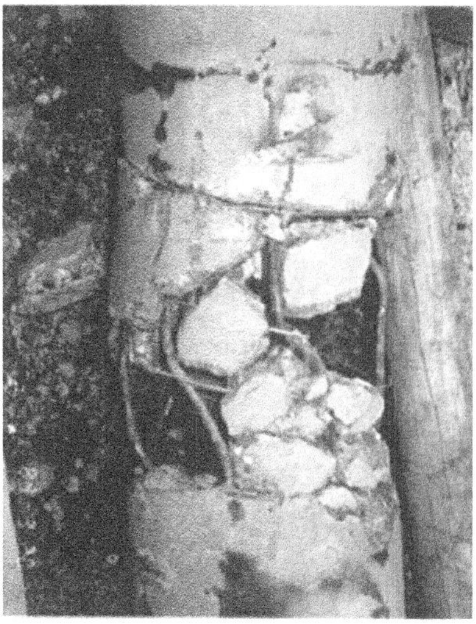

Fig. 6.18 A pile damaged by lateral ground displacement during 1964 Niigata earthquake (Yasuda et al., 1999)

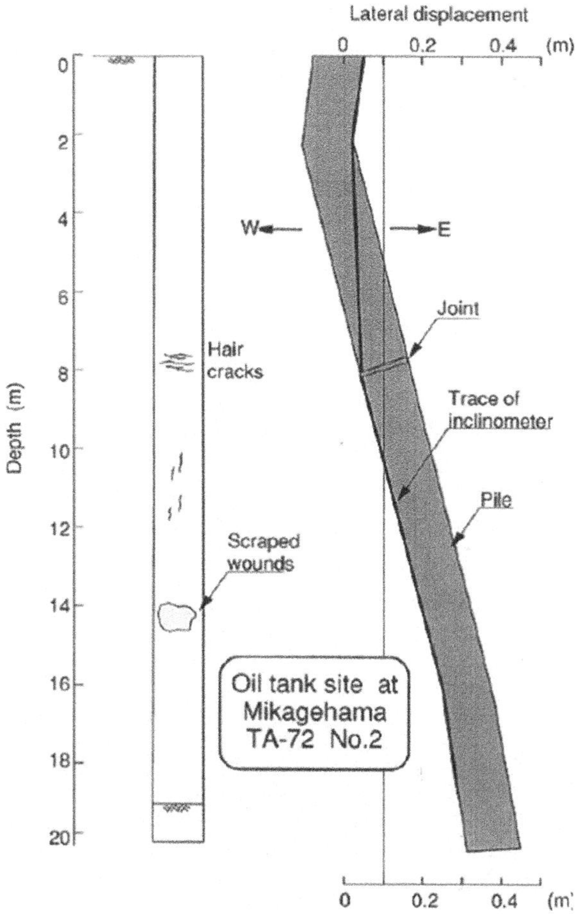

Fig. 6.19 Lateral displacement and cracking of pile no. 2 (Ishihara and Cubrinovsky, 2004)

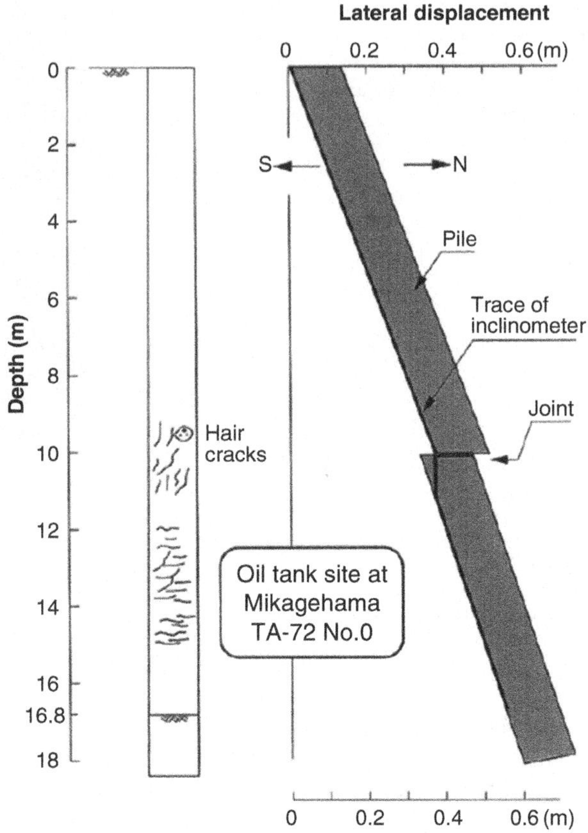

Fig. 6.20 Lateral displacement and cracking of pile no. 2 (Ishihara and Cubrinovsky, 2004)

number of cracks on the ground inland from the waterfront. The lateral ground displacement was plotted as a function of the distance from the waterfront.

The permanent lateral ground displacement corresponding to the location of Tank TA72 was somewhere between 35 and 55 cm (Ishihara and Cubrinovsky, 2004).

This observation indicates that liquefaction and resulting lateral spreading of the backfill soil seriously affected the pile performance.

The methods currently in use for design of piles in liquefying soil are:

1. The force or limit equilibrium analysis
2. The displacement or p–y analysis

5.1 *The Force or Limit Equilibrium Analysis*

The method of analysis is recommended in several Japanese design codes for analysis of pile foundations in liquefied soils undergoing lateral spreading (JWWA, 1997; JRA, 1996). The method involves estimation of lateral soil pressures

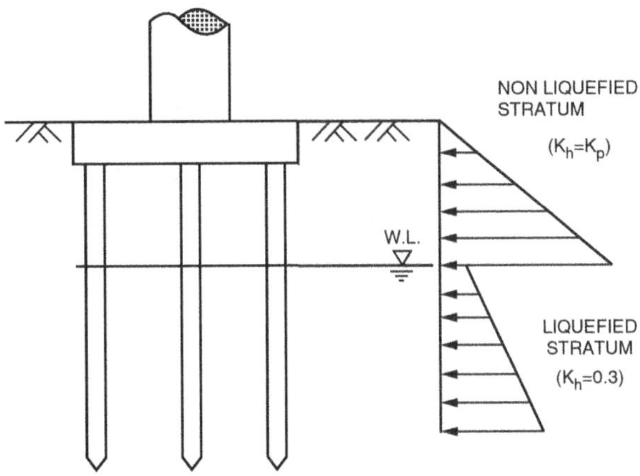

Fig. 6.21 Schematic sketch showing pressure distribution against the piles due to lateral soil flow associated with liquefaction (JWWA, 1997; Ashford and Juirnarongrit, 2004; Finn and Fujita, 2004)

on pile and then evaluating the pile response. A schematic sketch showing lateral pressures due to non-liquefied and liquefied soil layers is shown in Fig. 6.21. The non-liquefied top layer is assumed to exert passive pressure on the pile. The liquefied layer is assumed to apply a pressure which is about 30% of the total overburden pressure. This estimation of pressure is based on back calculation of case histories of performance of pile foundations during the Kobe earthquake. The maximum bending moment is assumed to occur at interface between the liquefied and non-liquefied soil layer.

5.2 *Displacement or P–Y Analysis*

This method involves making Winkler-type spring-mass model shown schematically in Fig. 6.22. The empirically estimated post-liquefaction free field displacements are calculated. These displacements are assumed to vary linearly and applied to the springs of the soil–pile system as shown in Fig. 6.22 (Finn and Thavaraj, 2001). Degraded p–y curves may be used for this kind of analysis. In the Japanese practice, the springs are assumed to be linearly elastic–plastic and can be determined from the elastic modulus of soil using semiempirical formulas (Finn and Fujita, 2004). The soil modulus can be evaluated from plate load tests or standard penetration tests. Reduction in spring stiffness is recommended by JRA (1996) to account for the effect of liquefaction. Such reduction is based on F_L (factor of safety against liquefaction). These reduction factors are shown in Table 6.8.

The US practice is to multiply the p–y curves by a uniform degradation factor "p," which is commonly referred to as the p-multiplier. The typical values of "p" range from 0.3 to 0.1. The values of "p" seem to decrease with pore water pressure increase (Dobry et al., 1995) and become 0.1 when the excess pore water pressure is 100%. Wilson et al. (2000) suggested that the value of "p" for a fully liquefied soil also

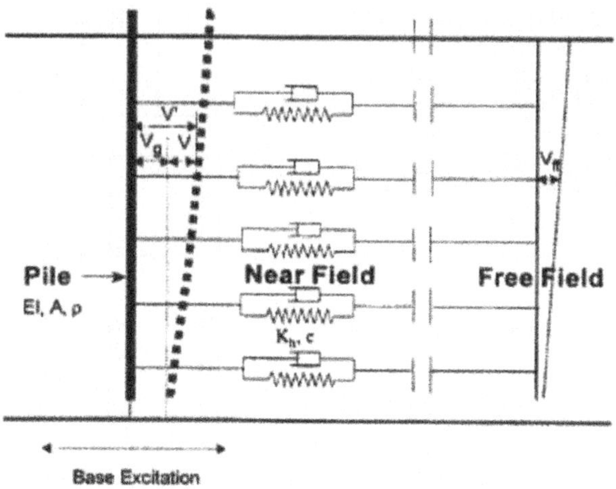

Fig. 6.22 A schematic sketch for Winkler spring model for pile foundation analysis (Finn and Thavaraj, 2001)

Table 6.8 Reduction coefficients for soil constants due to liquefaction (JRA, 1996)

Range of F_L	Depth from the present ground surface × (in)	Dynamic shear strength ratio R	
		$R \leqq 0.3$	$0.3 \leqq Ra$
$F_L \leqq 1/3$	$0 \leqq \times \leqq 10$	0	1/3
	$10 < \times \leqq 20$	1/3	1/3
$1/3 < F_L \leqq 2/3$	$0 \leqq \times \leqq 10$	1/3	2/3
	$10 < \times \leqq 20$	2/3	2/3
$2/3 < F_L \leqq 1$	$0 \leqq \times \leqq 10$	1/3	1
	$10 < \times \leqq 20$	1	1

depends on the initial relative density D_R. The values of "p" range from 0.1 to 0.2 for sand at about 35% relative density and from 0.25 to 0.35 for a relative density of 55%. It was found that the resistance of the loose sand did not pick up even at substantial strains, but the denser sand, after an initial strain range in which it showed little strength, picked up strength with increasing strain (Fig. 6.23). This finding suggests that the good performance of the degraded p–y curves, which did not include an initial range of low or zero strength, must be test specific and the p-multiplier may be expected to vary from one design situation to another.

Dilatancy effects may reduce the initial p–y response of the dense sands (Yasuda et al. 1999). Ashford and Juirnarongrit (2004) compared the force-based analysis and the displacement-based analysis for the case of single piles subjected to lateral spreading problems. They observed that the force-based analysis reasonably estimated the pile moments but underestimated pile displacements. The displacement analysis was found to make better prediction about the pile moment and the pile displacement.

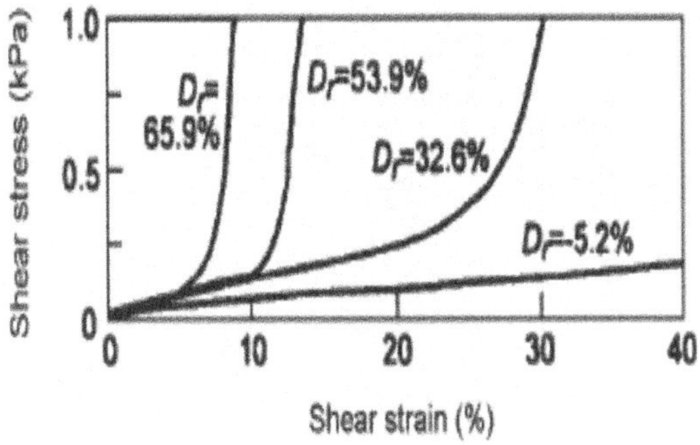

Fig. 6.23 Post-liquefaction undrained stress-strain behavior of sand (Yasuda et al., 1999)

5.3 *Further Comments on Piles in Liquefying Soil*

The force-based and displacement-based design procedure are based on limited number of observations. Liyanapathirana and Poulos (2005) developed a numerical model for simulating the pile performance in liquefying soil. They also studied the effect of earthquake characteristics on pile performance and observed that the "Arias intensity" and the natural frequency of the earthquake strongly influence performance of the pile in liquefying soil. Bhattacharya (2006) re-examined the damage to piles during 1964 Niigata and 1995 Kobe earthquakes and noted that pile failure in liquefying soil can be better explained as buckling-type failures. Recently, Madabhushi et al. (2010) have provided a detailed view of how the loss of strength due to liquefaction can influence pile capacity to axial loads. Failure may be induced by loss of bearing capacity at the pile tip or due to buckling induced by loss of resistance along the shaft or lateral spreading, all induced by soil liquefaction.

6 Shallow Foundations Under Seismic Loads

Shallow foundations may experience a reduction in bearing capacity and increase in settlement and tilt due to seismic loading as has been observed during several earthquakes. The foundation must be safe both for the static as well as for the dynamic loads imposed by the earthquakes. The earthquake associated ground shaking can affect the shallow foundation in a variety of ways:

1. Cyclic degradation of soil strength may lead to bearing capacity failure during the earthquake.
2. Large horizontal inertial force due to earthquake may cause the foundation to fail in sliding or overturning.
3. Soil liquefaction beneath and around the foundation may lead to large settlement and tilting of the foundation.
4. Softening or failure of the ground due to redistribution of pore water pressure after an earthquake which may adversely affect the stability of the foundation post-earthquake.

(a) (b)

Fig. 6.24 Examples of bearing capacity failures of shallow foundations in Adapazari (Yilmaz et al., 2004)

Bearing capacity failures of shallow foundations have been observed in Mexico City during Michoacan earthquake of 1985 (Mendoza and Avunit, 1988; Zeevart, 1991) and in city of Adapazari due to 1999 Kocaeli earthquake (Karaca, 2001; Bakir et al., 2002; Yilmaz et al., 2004). Typical examples of bearing capacity failure in Adapazari are shown in Fig. 6.24. The surface soils at the site of foundation damage belong to CL/ML group which are generally considered non-liquefiable. Settlements as much as 0.5–0.7 m have been observed in loose sands in Hachinohe during the 1968 Tokachi-Oki earthquake of magnitude 7.9. Settlements of 0.5–1.0 m were observed at Port and Roko Island in Kobe due to the Hyogoken-Nanbu ($M = 6.9$) earthquake.

Several research investigations, mostly analytical, have been conducted in the area of dynamic bearing capacity of foundations in the recent years. The more significant of these studies are briefly presented here.

6.1 Developments in Dynamic Bearing Capacity

The response of a footing to dynamic loads is affected by the (1) nature and magnitude of dynamic loads, (2) number of pulses, and (3) the strain rate response of soil. Shallow foundations for seismic loads are usually designed by the equivalent static approach. The foundations are considered as eccentrically loaded with inclined load (combination of vertical + horizontal load), and the ultimate bearing capacity is accordingly estimated. To account for the effect of dynamic nature of the load, the bearing capacity factors are determined by using dynamic angle of internal friction which is taken as 2- degrees less than its static value (Das, 1992). Building codes generally permit an increase of 33% in allowable bearing capacity when earthquake loads in addition to static loads are used in design of the foundation. This recommendation may be considered reasonable for dense granular soils, stiff to very stiff clays, or hard bedrocks but is not applicable for friable rock, loose soils susceptible to liquefaction or pore water pressure increase, sensitive clays, or clays likely to undergo plastic flow (Day, 2006).

Behavior of small footing resting on dense sands and subjected to static and impulse loads was experimentally investigated by Selig and McKee (1961). It was observed that the footing failed in general shear in static case, and local shear failure occurred in the dynamic case. Large settlements at failure were observed for the dynamic case. These experimental results indicate that for given value of settlement, the dynamic bearing capacity is lower than the static bearing capacity. This observation is further supported by results of experimental studies on small footings on surface of sand (Vesic et al., 1965) wherein dynamic bearing capacity was found to be about 30% lower than static bearing capacity. Therefore, the increase in bearing capacity permitted by codes should be taken with a caution. Recently several analytical studies on seismic bearing capacity of shallow footing have been reported. These studies used limit equilibrium approach with various assumptions on the failure surface. Plane failure surface shown in Fig. 6.25 was assumed by Richards et al. (1993), and equations and charts were developed to estimate seismic bearing capacity and settlement using foundation width, depth, soil properties, and horizontal and vertical acceleration components. This approach is used for its simplicity although the assumption of a plane failure surface may not be realistic.

Logarithmic failure surfaces shown in Fig. 6.26 were assumed by Budhu and Al-Karni (1993) to determine the seismic bearing capacity of soils. They suggested modifications to the equations commonly used for static bearing capacity to obtain the dynamic bearing capacity as follows:

$$q_{\mathrm{ud}} = \mathrm{c}\, N_{\mathrm{c}}\, S_{\mathrm{c}}\, d_{\mathrm{c}}\, e_{\mathrm{c}} + \mathrm{q}\, N_{\mathrm{q}}\, S_{\mathrm{q}}\, d_{\mathrm{q}}\, e_{\mathrm{q}} + 0.5\, \gamma\, B\, N_{\gamma}\, S_{\gamma}\, d_{\gamma}\, e_{\gamma} \tag{6.5}$$

where

N_c, N_q, and N_γ are the static bearing capacity factors.

S_c, S_q, and S_γ are static shape factors.

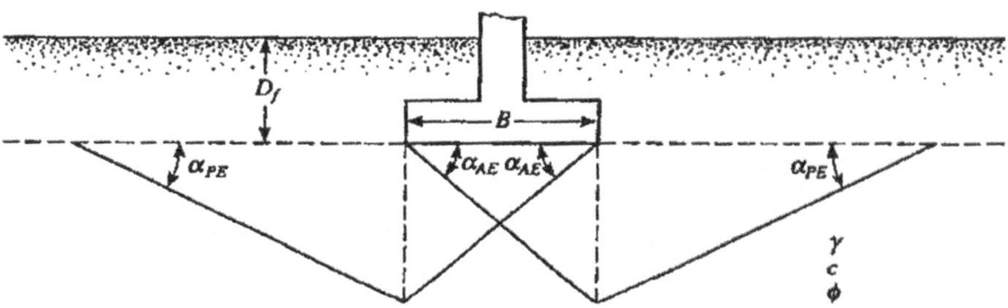

Fig. 6.25 Failure surface in soil for seismic bearing capacity assumed by Richards et al. (1993)

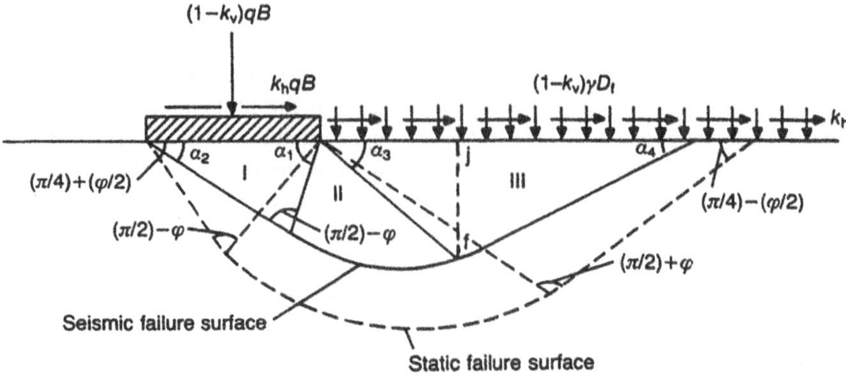

Fig. 6.26 Failure surfaces used by Budhu and Al-Karni (1993) for static and dynamic case

d_c, d_q, and d_γ are static depth factors.
e_c, e_q, and e_γ are the seismic factors estimated using the following equations:

$$e_c = \exp\left(-4.3k_h^{l+D}\right) \tag{6.6}$$

$$e_q = (1 - k_v)\exp\left[-\left(\frac{5.3k_h^{1.2}}{1 - k_v}\right)\right] \tag{6.7}$$

$$e_\gamma = \left(1 - \frac{2}{3}k_v\right)\exp\left[-\left(\frac{9k_h^{1.2}}{1 - k_v}\right)\right] \tag{6.8}$$

where
K_h and K_v are the horizontal and vertical acceleration coefficients, respectively.
H = depth of the failure zone from the ground surface and

$$D = \frac{c}{\gamma H}$$

$$H = \frac{0.5B}{\cos\left(\frac{\pi}{4} + \frac{\phi}{2}\right)}\exp\left(\frac{\pi}{2}\tan\phi\right) + D_f \tag{6.9}$$

D_f = depth of the footing.
φ = angle of internal friction.
c = cohesion of soil.

An experimental study was also conducted by Al-Karni and Budhu (2001) on model footing to study the response under horizontal acceleration and compared the results with the approach suggested in Budhu and Al-Karni (1993).

A study of the seismic bearing capacity of shallow strip footing was conducted by Chaudhury and Subba Rao (2005). The failure surfaces for the static and dynamic case are shown in Fig. 6.27. They used the limiting equilibrium approach and the

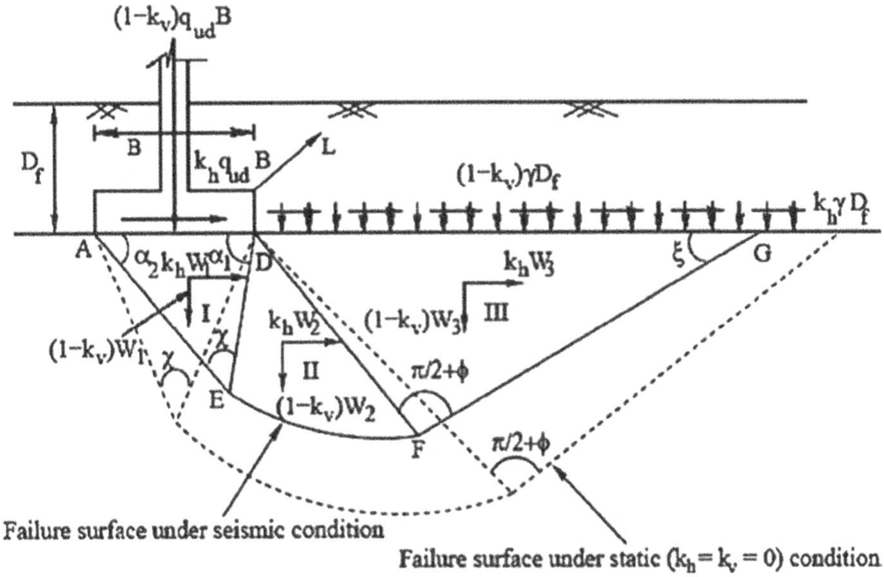

Fig. 6.27 Failure surfaces under static and seismic loading assumed by Chaudhury and Subba Rao (2005)

equivalent static method to represent the seismic forces and obtained the seismic bearing capacity factors.

The dynamic bearing capacity "q_{ud}" is obtained as

$$q_{\text{ud}} = c\,N_{\text{cd}} + q\,N_{\text{qd}} + 0.5\,\gamma\,B\,N_{\gamma d} \tag{6.10}$$

where N_{cd}, N_{qd}, and $N_{\gamma\text{d}}$ are seismic bearing capacity factors.

Values of N_{cd}, N_{qd}, and $N_{\gamma\text{d}}$ are shown in Fig. 6.28 for various combinations of k_{h}, k_{v}, and Φ.

As there is general lack of experimental data, it is difficult to verify which one of the above analytical approaches may be expected to provide reasonable estimates of seismic bearing capacity.

Yilmaz and Bakir (2009) used the finite element method (Fig. 6.29) to analyze the behavior of surface foundations resting on saturated soil under seismic loading and assessed the impact of the conventional approach of using undrained shear strength in estimating the foundation bearing capacity. They observed that assumption of undrained condition may lead to significant over estimation of moment carrying capacity of the foundation.

Gajan and Kutter (2009) provided the concept of contact interface model to estimate the load capacities, stiffness degradation, energy dissipation, and deformation of shallow foundations under combined cyclic loading. The "contact interface model" provides a nonlinear relation between cyclic loads and displacements of the footing–soil system during combined cyclic loading (vertical, shear, and moment).

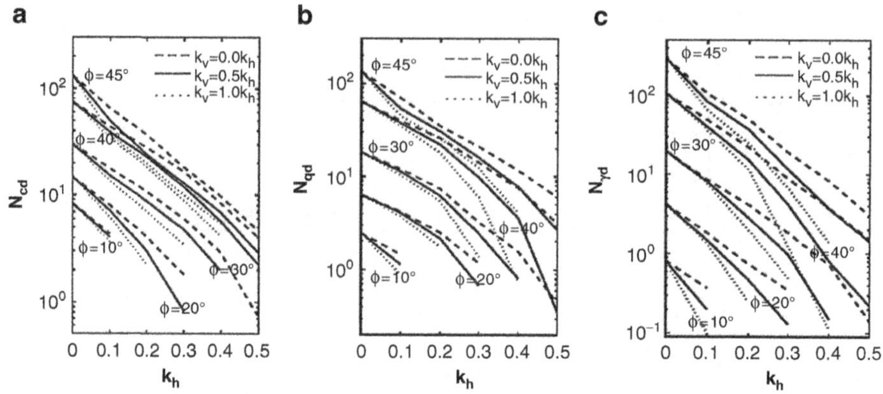

Fig. 6.28 Values of bearing capacity factors (**a**) N_{cd}, (**b**) N_{qd}, and (**c**) $N_{\gamma d}$ (Chaudhury and Subba Rao, 2005)

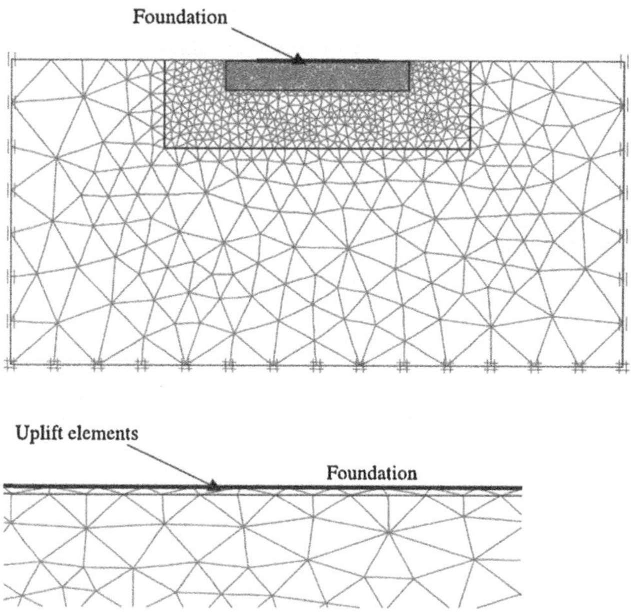

Fig. 6.29 Finite elements used by Yilmaz and Bakir (2009)

6.2 Settlement of Shallow Foundations Subjected Seismic Loading

A shallow foundation subjected to earthquake may undergo settlement and tilt due to the effect of loads and moments imposed on the foundation. The settlement tilt may be estimated using the procedure suggested by Prakash and Saran (1977) for the static case using the following equations:

$$\frac{S_e}{S_o} = 1.0 - 1.63\frac{e}{B} - 2.63\left(\frac{e}{B}\right)^2 + 5.83\left(\frac{e}{B}\right)^3 \tag{6.11}$$

$$\frac{S_m}{S_o} = 1.0 - 2.31\frac{e}{B} - 22.61\left(\frac{e}{B}\right)^2 + 31.54\left(\frac{e}{B}\right)^3 \tag{6.12}$$

where

S_o = settlement at the center of the foundation for vertical load only.

S_e = settlement at the center of the eccentrically loaded foundation (combined action of vertical load and moment).

S_m = maximum settlement of the eccentrically loaded foundation.

B = width of the foundation.

e = eccentricity given by $e = \frac{M}{Q}$.

Q = vertical load and M = moment.

The tilt of the foundation "t" may then be obtained from the following equation:

$$S_m = S_e + \left(\frac{B}{2} - e\right)\sin t \tag{6.13}$$

S_e, S_m, and "t" can thus be obtained if S_o can be determined. Prakash and Saran (1977) have suggested the use of plate load test to determine S_o.

Whitman and Richart (1967) and Georgiadis and Butterfield (1988) have suggested procedures for determining the settlement and tilt of the foundation subjected to static vertical loads and moments.

Richards et al. (1993) suggested the use of the following equation to estimate the seismic settlement of a strip footing.

$$S_{Eq}(m) = 0.174\frac{V^2}{Ag}\left|\frac{k_{h*}}{A}\right|^{-4} \tan\alpha_{AE} \tag{6.14}$$

where, S_{Eq} = seismic settlement (in meters), V = peak velocity for the design earthquake (m/s), A = acceleration coefficient for the design earthquake, and g = acceleration due to gravity (9.81 m/s^2). The value of tan α_{AE} in Eq. (6.14) depends on φ and k_h*. Figure 6.30 shows the variation of tan α_{AE} with k_h* for φ values from 15° to 40°.

6.3 Shallow Foundations in Liquefiable Soils

6.3.1 Design Aspects

Complete liquefaction of soil can lead to loss of strength due to increase in pore water pressure leading to bearing capacity failure and very large settlements. Therefore, it is not feasible to design shallow foundation resting directly on this type of soil. However, if the expected increase in pore water pressure is not likely to cause

Fig. 6.30 Variation of tan α_{AE} with k_h* and φ (After Richards et al., 1993)

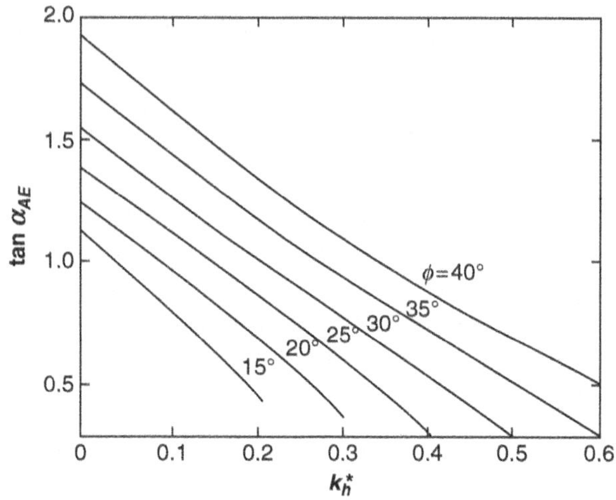

complete liquefaction, it may be possible to design shallow foundation on the basis reduced bearing capacity. The ultimate bearing capacity of a strip footing resting on the surface of a sand deposit may be obtained from Terzaghi's theory (Eq. 6.15).

$$q_{\text{ult}} = \frac{1}{2}\gamma B N_\gamma \tag{6.15}$$

where, q_{ult} = ultimate bearing capacity.

γ = unit weight of soil.

N_γ = Terzaghi's bearing capacity factor.

The reduced ultimate bearing capacity "$q_{\text{ult}R}$" may be obtained as

$$q_{\text{ult}R} = \frac{1}{2}(1 - r_u)\gamma_b B N_\gamma \tag{6.16}$$

where, γ_b = submerged unit weight of soil.

r_u = excess pore water pressure ratio $= \frac{u_e}{\overline{\sigma}}$.

u_e = excess pore water pressure developed due to ground shaking.

$\overline{\sigma}$ = effective overburden pressure.

To determine the value r_u, the factor of safety "FS_L" against liquefaction failure should be determined. The value r_u may then be obtained from Fig. 6.31. The foundation may then be designed using the reduced bearing capacity.

In case a foundation is to be designed resting on a non-liquefying soil layer lying over a liquefiable layer, then punching resistance of the footing in the upper layer must be determined and must be adequate to ensure safe performance of the foundation.

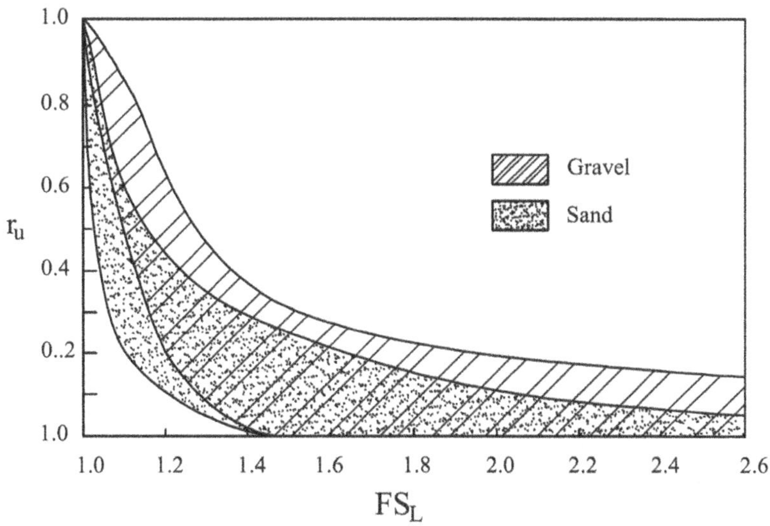

Fig. 6.31 Pore pressure ratio versus factor of safety against liquefaction failure (Marcuson and Haynes, 1990)

6.3.2 Settlement of Foundation in Liquefying Soil

Ishihara and Yoshimine (1992) have provided a chart to estimate the post-liquefaction volumetric strain of clean sand as function of factor of safety against liquefaction. This chart is shown in Fig. 6.32. This chart can be easily used if any of the corrected SPT values, cone resistance at the site, or maximum cyclic shear strain induced by the earthquake are known. The settlement of the layer is obtained by multiplying the volumetric strain with the thickness of the layer.

6.4 Foundation Performance on Liquefied Soil

Gazetas et al. (2004) studied tilting of buildings in 1999 Turkey earthquake. Detailed scrutiny of the "Adapazari failures" showed that significant tilting and toppling were observed only in relatively slender buildings (with aspect ratio: $H/B > 2$), provided they were laterally free from other buildings on one of their sides. Wider and/or contiguous buildings suffered small if any rotation. For the prevailing soil conditions and type of seismic shaking; most buildings with $H/B > 1.8$ overturned, whereas building with $H/B < 0.8$ essentially only settled vertically, with no visible tilting. Figure 6.33 shows a plot of H/B to tilt angle of the building. Soil profiles based on three SPT and three CPT tests, performed in front of each building of interest, reveal the presence of a number of alternating sandy- silt and silty- sand layers, from the

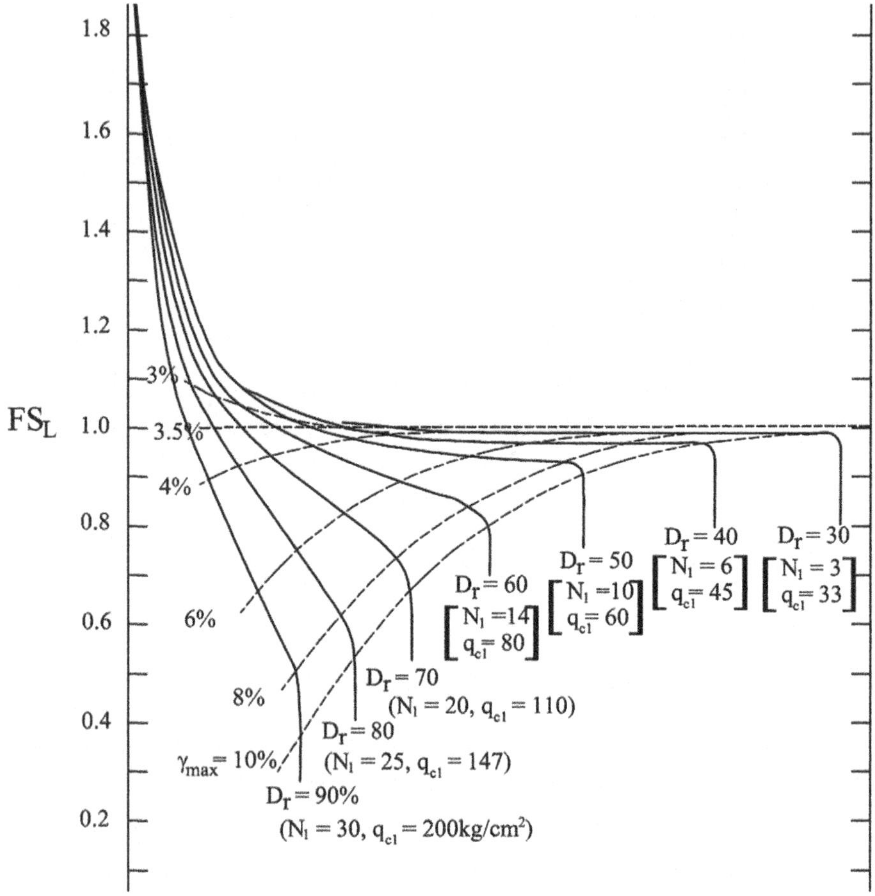

Fig. 6.32 Chart for post-liquefaction volumetric strain (Ishihara and Yoshimine, 1992)

Fig. 6.33 The angle of permanent tilting as a unique function of the slenderness ratio H/B (Gazetas et al., 2004)

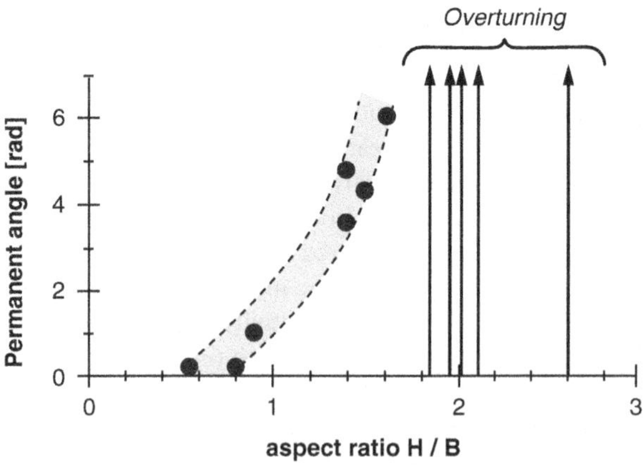

surface down to a depth of at least 15 m with values of point resistance $qc \approx (0.4\text{–}5.0)$ MPa. Seismo–cone measurements revealed wave velocities *Vs* less than 60 m/s for depths down to 15 m, indicative of extremely soft soil layers. Ground acceleration was not recorded in Tigcilar. Using in 1-D wave propagation analysis, the EW component of the Sakarya accelerogram (recorded on soft rock outcrop, in the hilly outskirts of the city) leads to acceleration values between 0.20 and 0.30 g, with several significant cycles of motion, with dominant period in excess of 2 sec. Even such relatively small levels of acceleration would have liquefied at least the upper-most loose sandy silt layers of a total thickness 1–2 m and would have produced excess pore water pressures in the lower layers Gazetas et al. 2004.

6.5 Overview on Seismic Design of Shallow Foundations

Considerable research effort has been devoted to define the failure surfaces below shallow foundations subjected to seismic loads as well as their settlements. However, the equivalent static approach is still commonly used for their design.

It may be emphasized here that for the case soils susceptible to liquefaction, (a) the foundation should not rest directly on soil layers that may liquefy as even lightly loaded foundations can sink into the soil, and (b) adequate thickness of non-liquefiable soil should be there to prevent damage to the foundation due to sand boils and surface fissuring. If these conditions are not met, then the ground improvement may be needed or the deep foundation should be provided.

References

Al-Karni, A. A., & Budhu, M. (2001). An experimental study of seismic bearing capacity of shallow footings. *Proceedings of the 4th International Conference on Recent Advances in Geotechnical Earthquake Engineering and Soil Dynamics and Symposium in Honor of Professor W.D. Liam Finn, CD-ROM, San-Diego, CA.*

American Association of State Highway and Transportation Officials (2002). *Standard specifications for highway bridges.* Sections 3 and 7.

Andrews, D. C. A., & Martin, G. R. (2000). Criteria for liquefaction of silty soils. *Proceedings of the 12th WCEE, Auckland, New Zealand.*

Ashford, S., & Juirnarongrit, T. (2004). Evaluation of force based and displacement based analysis for response of single piles to lateral spreading. *Proceedings of 11th International Conference on Soil Dynamics and Earthquake Engineering and 3rd International Conference Earthquake Geotechnical Engineering, University of California, Berkeley,* 1, 752–759.

Bakir, B. S., Sucuoglu, H., & Yilmaz, T. (2002). An overview of local site effects and the associated building damage during 17 August 1999 Izmit earthquake. *Bulletin of Seismological Society of America, 92*(1), 509–526.

Bhattacharya, S. (2006). A review of methods for pile design in seismically liquefiable soil. In *Design of foundations in seismic areas: Principles and applications.* Kanpur: NICEE, IIT.

Blaney, G. W. (1983). *Lateral response of a single pile in over consolidated clay to relatively low frequency harmonic pile –head loads and harmonic ground surface loads*. Ann Arbor, MI: University of Michigan.

Boulanger, R. W., & Idriss, I. M. (2004). New criteria for distinguishing between silts and clays that are susceptible to liquefaction versus cyclic failure. *25th Annual USSD Conference, Salt Lake City, Utah, June 6–10,* 357–366.

Boulanger, R. W., & Idriss, I. M. (2008). *Soil liquefaction during earthquakes*. Oakland, CA: EERI. MNO-12.

Bray, J. D., Sancio, R. B., Reimer, M. F., & Durgunoglu, T. (2004). Liquefaction susceptibility of fine-grained soils. *Proceedings of the 11th International Conference on Soil Dynamics and Earthquake Engineering and 3rd International Conference on Earthquake Geotechnical Engineering, Berkeley, CA, Jan. 7–9,* 1, 655–662).

Budhu, M., & Al-Karni, A. A. (1993). Seismic bearing capacity of soils. *Geotechnique, 43*(1), 181–187.

Cambio, D. (2012). *Prediction and performance of single piles in clay soil*. Ph.D. Dissertation, MST Rolla.

Chaudhury, D., & Subba Rao, K. S. (2005). Seismic bearing capacity of shallow strip footings. *Geotechnical and Geological Engineering, 23*(4), 403–418.

Das, B. M. (1992). *Principles of soil dynamics*. PWS Kent.

Day, R. W. (2006). *Foundation engineering handbook*. New York: McGraw Hill.

Dobry, R., Taboda, V. & Liu, L. (1995). Centrifuge modeling of liquefaction effects during earthquakes. *Proceedings of the 1st International Conference on Earthquake Engineering, Tokyo,* 3, 1291–1324).

El Hosri, M. S., Biarez, J., & Hicher, P. Y. (1984). Liquefaction characteristics of silty clay. *8th World Conference on Earthquake Engineering, Prentice-Hall Eaglewood Cliffs, NJ,* 3, 277–284.

El-Marsafawi, H., Han, Y., & Novak, M. (1990, August). *Dynamic experiments on two pile groups*. Research Report GEOT-20-90, Department of Civil Engineering, Western Ontario.

Eurocode 8 (EUROPEAN PRE-STANDARD). (1994). *Design provisions for earthquake resistance of structures-part 5: Foundations, retaining structures and geotechnical aspects*. The Commission of the European Communities.

Finn, W. D. L., & Fujita, N. (2004). Behavior of piles during earthquakes: Analysis and design. *Proceedings, Fifth International Conference on Case Histories in Geotechnical Engineering, New York, NY, April 13–17.*

Finn, W. D. L., Ledbetter, R. H., Fleming, R.L. Jr., Templeton, A.E., Forrest, T.W., & Stacy, S.T. (1994). Dam on liquefiable foundation: safety assessment and remediation. *Proc. 17th International Congress on Large Dams, Vienna*, 531–555.

Finn, W. D. L., & Thavaraj, T. (2001). Deep foundations in liquefiable soil. *Proceedings of the 4th International Conference on Recent Advance in Geotechnical Earthquake Engineering and Soil Dynamics, San Diego.*

Gajan, S., & Kutter, B. L. (2009). Contact interface model for shallow foundations subjected to combined cyclic loading. *Journal of Geotechnical and Geoenvironmental Engineering, ASCE, 135*(3), 407–419.

Gazetas, G., Apostou, M., & Anasta-Sopoular, J. (2004). Seismic bearing capacity failure and overturning of Terveler building in Adapazari 1999. *Proceedings of Fifth International Conference on Case Histories in Geotechnical Engineering. New York.* CD ROM –SOAP11 (1–51).

Georgiadis, M., & Butterfield, R. (1988). Displacements of Footings on Sand under Eccentric and Inclined Loads. *Canadian Geotechnical Journal, 25*(2), 199–212.

Gle, D. R. (1981). *The dynamic lateral response of deep foundations*. Ph.D. dissertation, University of Michigan, Ann Arbor.

Huang, C.-C. (2005). Seismic displacement of soil retaining walls situated on slope. *Journal of Geotechnical and Geo-environmental Engineering, ASCE, 131*(9), 1108–1117.

Iai, S. (1998). Rigid and flexible retaining walls during Kobe earthquake. *Proceedings of the Fourth International Conference on Case Histories in Geotechnical Engineering, St. Louis, MO, March 8–12*. CD-ROM, SOA-4, 108–127.

Ishihara, K., & Cubrinovsky, M., (2004). Case studies of pile foundations undergoing lateral spreading in liquefied deposits. *Proceedings of the Fifth international Conference on Case Histories in Geotechnical Engineering, New York, NY, April 13–17.*

Ishihara, K., & Yoshimine, M. (1992). Evaluation of settlements in sand in sand deposits following liquefaction. *Soils and Foundations, 32*(1), 173–188.

Jadi, H. (1999). *Predictions of lateral dynamic response of single piles embedded in clay*. MS thesis, University of Missouri, Rolla.

Japan Railway Technical Research Institute. (1999). *Design guidelines for railway structures-aseismic design*. Maruzen Co. Ltd.

JRA. (1996). *Seismic design specifications of highway bridges*. Tokyo: Japan Road Association.

JWWA. (1997). *Seismic design and construction guidelines for water supply facilities*. Tokyo: Japan Water Works Association.

Karaca, G. (2001). *An investigation into large vertical displacement experienced by the structures in Adapazari during 17 August 1999 earthquake*. MS thesis, Middle East Technical University, Ankara, Turkey.

Kishida, H. (1969). Characteristics of liquefied sands during Mino-Owari, Tohnankai, and Fukui earthquakes. *Soils and Foundations, 9*(1), 75–92.

Liyanapathirana, D. S., & Poulos, H. G. (2005). Seismic lateral response of piles in liquefying soil. *Journal of Geotechnical and Geo-Environmental Engineering, ASCE, 131*(12), 1466–1479.

Madabhushi, G., Knappett, J., & Haigh, S. (2010). *Design of pile foundations in liquefiable soil*. London: Imperial college Press.

Marcuson, W. F., & Haynes, M. E. (1990). Stability of slopes and embankments during earthquakes. *Proceedings of the ASCE/Pennsylvania Department of Transportation Geotechnical Seminar, Hershey, Pennsylvania.*

Mendoza, M. J., & Avunit, G. (1988). The Mexico earthquake of September 19, 1985-Behavior of building foundations in Mexico City. *Earthquake Spectra, 4*(4), 835–853.

Novak, M. (1974). Dynamic stiffness and damping of piles. *Canadian Geotechnical Journal, 11*(4), 574–598.

Novak, M., & El Sharnouby, E. B. (1983). Stiffness and damping constants for single piles. *Journal of Geotechnical Engineering, ASCE, 109*(7), 961–974.

Novak, Milos & El Sharnouby, E. B. (1984) Evaluation of Dynamic Experiments on Pile Group. *Journal of Geotechnical Engineering, 109*, 961–964.

Perlea, V. G., Koester, J. P., & Prakash, S. (1999). How liquefiable are cohesive soils? *Proceedings of the Second International Conference on Earthquake Geotechnical Engineering, Lisbon, Portugal,* 2, 611–618.

Plito, C. (2001). Plasticity based liquefaction criteria. *Proceedings of the 4th International Conference on Recent Advance in Geotechnical Earthquake Engineering and Soil Dynamics, San Diego.*

Prakash, S., & Guo, T. (1998). *Liquefaction of silts with clay content, Soil dynamics and earthquake engineering,Seattle, WA: ACSE,* I, 337–348.

Prakash, S., & Jadi, H. (2001). Predictions of lateral dynamic response of single piles embedded in fine soil. *Proceedings of the Fourth International Conference on Recent Advance in Geotechnical Earthquake Engineering and Soil Dynamics, San Diego, CA.*

Prakash, S., & Puri, V. K. (1988). *Foundation for machines analysis and design*. New York: John Wiley &Sons.

Prakash, S., & Puri, V. K. (2003). Geotechnical earthquake engineering and infrastructure development. *Proc. IGC-2003, Roorkee, India.*

Prakash, S., & Puri, V. K. (2008). Piles under earthquake loads. In *Geotechnical Earthquake Engineering and Soil Dynamics IV*, CD-ROM, May 18–22. Sacramento, CA

Prakash, S., & Sandoval, J. A. (1992). Liquefaction of low plasticity silts. *Journal of Soil Dynamics and Earthquake Engineering, 71*(7), 373–397.

Prakash, S., & Saran, S. (1977). Settlement and tilt of eccentrically loaded footings. *Journal Structural Engineering, Roorkee, 4*(4), 166–176.

Prakash, S., & Sharma, H. D. (1990). *Pile foundations in engineering practice*. New York: John Wiley and Sons.

Puri, V. K. (1984). *Liquefaction behavior and dynamic properties of loessial (silty) soils*. Ph.D. thesis, University of Missouri – Rolla, Missouri

Rafnsson, E. A., & Prakash, S. (1991). Stiffness and damping parameters for dynamic analysis of retaining wall. *Proceedings of the Second International Conference on Recent Advances in Geotechnical Earthquake Engineering and Soil Dynamics, St. Louis, MO,* III, 1943–1952.

Richards, R., Elms, D. G., & Budhu, M. (1993). Seismic bearing capacity and settlement of foundations. *Journal of Geotechnical Engineering Division, ASCE, 119*(4), 662–674.

Sandoval, J. A. (1989). *Liquefaction and settlement characteristics of silt soils*. Ph.D. thesis, University of Missouri – Rolla, MO.

Seed, H. B., & Lee, K. L. (1966). Liquefaction of saturated sands during cyclic loading. *Journal Soil Mechanics and Foundations Division, ASCE, 92*(SM6), 105–134.

Seed, R. B., Cetin, K. O., Moss, R. E. S., Kammerer, A. M., Wu, J., Pestana, J. M. & Riemer, M. F. (2001). Recent advances in soil liquefaction engineering and seismic site response evaluation. *Proceedings of the 4th International Conference on Recent Advance in Geotechnical Earthquake Engineering and Soil Dynamics, San Diego.*

Selig, E. T., & McKee, K. E. (1961). Static and dynamic behavior of small footings. *Journal Soil Mechanics and Foundation Division, ASCE, 87*(SM6), 29–47.

Vesic, A. S., Banks, D. C., & Woodward, J. M. (1965). An experimental study of dynamic bearing capacity of footing on sand. *Proceedings of the 6th INCSMFE, Montreal, Canada,* 2, 209–213.

Wilson, D. O., Boulanger, R. W., & Kutter, B. L. (2000). Observed seismic lateral resistance of liquefying sand. *Journal of Geotechnical and Geo-Environmental Engineering, ASCE, 126*(10), 898–906.

Wu, Y. (1999). *Displacement-based analysis and design of rigid retaining walls during earthquakes*. Ph.D. Dissertation, University of Missouri – Rolla.

Wu, Y., & Prakash, S. (2001). Seismic displacements of rigid retaining walls: State of the art. *Proceedings of the Fourth International Conference on Recent Advances in Geotechnical Earthquake Engineering and Soil Dynamics. Paper No. 7.05, San Diego, CA.*

Wu, Y., & Prakash, S. & Puri, V. K. (2010). On seismic design of retaining walls. *Earth Retention Conference, Geo-Institute, ASCE, Seattle, August 1–4.*

Yasuda, Y. N., Kiku, H., & Adachi, K. (1999). A simplified practical method for evaluating liquefaction induced flow. *Proceedings 7th U.S.- Japan Workshop on Earthquake Resistant Design of Lifeline Facilities and Countermeasures Against Liquefaction, November,* 311–320.

Yilmaz, M., & Bakir, B. S. (2009). Capacity of shallow foundations on saturated cohesionless soils under combined loading. *Canadian Geotechnical Journal, 46.*

Yilmaz, M., Pekcan, O., & Bakir, B. S. (2004). Undrained cyclic shear and deformation behavior of silt-clay mixtures of Adapazari, Turkey. *Soil Dynamics and Earthquake Engineering, 14*(7), 497–507.

Youd, T. L., & Idriss, I. M. (2001). Liquefaction resistance of soils: Summary report from the 1996 NCEER and 1998 NCEER/NSF workshops on evaluation of liquefaction resistance of soils. *Journal of Geotechnical and Geo-environmental Engineering, ASCE, 127*(10), 297–313.

Youd, T. L., Idriss, I. M., Andurus, R. D., Arango, I., Castro, G., Christian, J. T., Dobry, R., Finn, W. D. L., Harder, L. F., Haymes, M. E., Ishihara, K., Koester, J. P., Liao, S. S. C., Marcusson, W. F., Martin, G. R., Mitchell, J. K., Moriwaki, Y., Power, M. C., Robertson, P. K., Seed, R. B., & Stokoe, K. H. (2001). Liquefaction resistance of soils: Summary report from the 1996 NCEER and 1998 NCEER/NSF workshops on evaluation of liquefaction resistance of soils. *Journal of Geotechnical & Geo-environmental Engineering, ASCE, 127*(10), 817–833.

Zeevart, L. (1991). Seismo-soil dynamics of foundations in Mexico city earthquake, September 1985. *Journal of Geotechnical Engineering, 117*(3), 376–427.

Chapter 7
Seismic Analysis and Design of Retaining Walls and Shallow Foundations

Swami Saran, Hasan Rangwala, and S. Mukerjee

1 Introduction

This chapter presents a state-of-art review of two important problems in geotechnical earthquake engineering. Seismic analysis and design of (i) Retaining Walls and (ii) Shallow Foundations. Different methods available, their merits and demerits have been discussed critically. Finally recommendations have been made for the design of retaining walls and shallow foundations.

2 Rigid Retaining Wall

Analysis and design of a rigid retaining wall is an important problem in geotechnical earthquake engineering.

The seismic response of a rigid retaining wall needs to be treated as a soil-structure interaction problem. Wall movements and seismic earth pressures depend on the response of the soil underlying the wall, the response of the backfill, the inertial response of the wall itself and the nature of input motions. The methods available for the design of a rigid retaining wall can be grouped under three categories, namely (i) Force equilibrium based pseudo-static analysis, (ii) Pseudo-dynamic analysis, and (iii) Displacement-based analysis.

S. Saran (✉) · H. Rangwala · S. Mukerjee
Department of Earthquake Engineering, IIT Roorkee, Roorkee, India
e-mail: saranfce@iitr.ac.in

M. L. Sharma et al. (eds.), *Advances in Indian Earthquake Engineering and Seismology*, https://doi.org/10.1007/978-3-319-76855-7_7

2.1 Force Equilibrium Based Pseudo-static Analysis

A pseudo-static analysis is relatively simple and straight forward. In this method the complex dynamic effects of earthquake shaking are represented by equivalent constant horizontal and seismic accelerations.

Seismic active and passive earth pressures and their points of application, from the base of the wall, are required for carrying out a stability analysis of a wall located in an earthquake prone area. Initial studies in this direction were carried out in Japan, following the great Kanto Earthquake (1923, $M_w = 7.9$), by Okabe (1924) and Mononobe and Matsuo (1929). The method proposed by them, generally known as the Mononobe–Okabe (MO) method, is an extension of classical work of Coulomb (1776). Coulomb developed a method for obtaining static active and passive earth pressures for analyzing gravity walls retaining dry/moist cohesionless backfills.

Mononobe–Okabe modified Coulomb's theory for evaluating earth pressure by incorporating the inertia effects in terms of seismic coefficients A_h and A_v. Figure 7.1 shows a section of a wall of height H, inclined at an angle α with the vertical. It retains dry/moist cohesionless backfill having unit weight γ and angle of shearing resistance φ. Angle of wall friction is δ. The backfill is inclined at an angle i with the horizontal. Dynamic active and passive earth pressures are obtained using the following expressions:

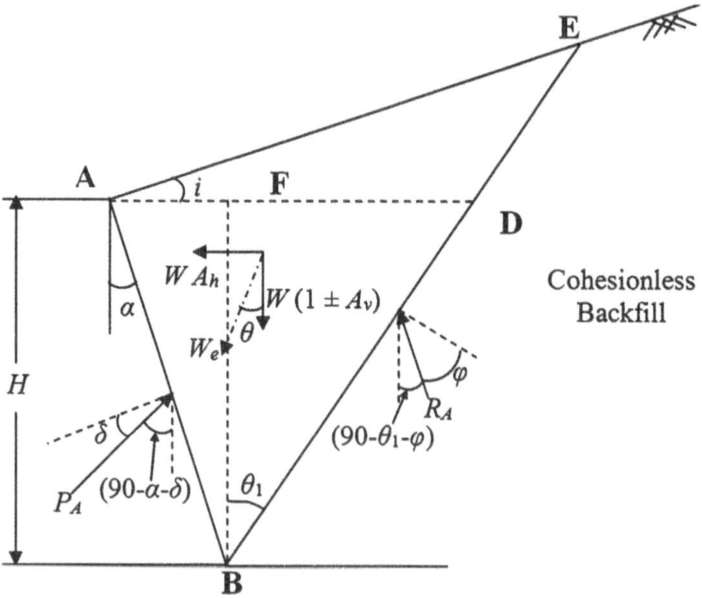

Fig. 7.1 Forces acting on the failure wedge (active state)

$$(P_{A\gamma})_{dyn} = 1/2X\gamma H^2 C_a, \text{and} \tag{7.1}$$

$$(P_{P\gamma})_{dyn} = 1/2X\gamma H^2 C_p \tag{7.2}$$

C_a and C_p are dynamic active and passive earth pressure coefficients, respectively, and are given by;

$$C_a = \frac{(1 \pm A_v)\cos^2(\phi - \theta - \alpha)}{\cos\theta\cos^2\alpha\cos(\delta + \alpha + \theta)}\left[1 + \sqrt{\frac{\sin(\phi + \delta)\sin(\phi - i - \theta)}{\cos(\alpha - i)\cos(\delta + \alpha + \theta)}}\right]^{-2}. \tag{7.3}$$

$$C_p = \frac{(1 \pm A_v)\cos^2(\phi + \alpha - \theta)}{\cos\theta\cos^2\alpha\cos(\delta - \alpha + \theta)}\left[1 - \sqrt{\frac{\sin(\phi + \delta)\sin(\phi + i - \theta)}{\cos(\alpha - i)\cos(\delta - \alpha + \theta)}}\right]^{-2} \tag{7.4}$$

$$\text{and } \theta = \tan^{-1}\frac{A_h}{1 \pm A_v}. \tag{7.5}$$

The expressions for C_a and C_p give two values of each, depending upon the direction of A_v. For design purposes the higher value of C_a and the lesser value of C_p shall be considered.

According to Zarrabi (1979), the critical angle of the planar rupture surface with the vertical, θ_{Acr}, in the active state, is as given below:

$$\theta_{Acr} = 90^\circ - \phi + \theta - \tan^{-1}\left[\frac{C_{1A} - \tan(\phi - \theta - i)}{C_{2A}}\right] \tag{7.6a}$$

where,

$$C_{1A} = \sqrt{\begin{array}{l}\tan(\phi - \theta - i)\{\tan(\phi - \theta - i) + \cot(\phi - \theta - \alpha)\} \\ \times\{1 + \tan(\delta + \theta + \alpha)\cot(\phi - \theta - i)\}\end{array}} \tag{7.6b}$$

$$C_{2A} = 1 + [\tan(\delta + \theta + \alpha)\{\tan(\phi - \theta - i) + \cot(\phi - \theta - \alpha)\}] \tag{7.6c}$$

Similarly, for the passive state, Ebelling and Morrison (1992) gave the expression for θ_{Pcr}, the critical angle of the planar rupture surface with the vertical as,

$$\theta_{Pcr} = 90^\circ + \phi - \theta - \tan^{-1}\left[\frac{C_{1P} + \tan(\phi - \theta + i)}{C_{2P}}\right] \tag{7.7a}$$

where,

$$C_{1P} = \sqrt{\begin{array}{l}\tan(\phi - \theta + i)\{\tan(\phi - \theta + i) + \cot(\phi - \theta + \alpha)\} \\ \times\{1 + \tan(\delta + \theta - \alpha)\cot(\phi - \theta + i)\}\end{array}} \tag{7.7b}$$

$$C_{2P} = 1 + [\tan(\delta + \theta - \alpha)\{\tan(\phi - \theta + i) + \cot(\phi - \theta + \alpha)\}]. \quad (7.7c)$$

For the static case, values of θ_{Acr} and θ_{Pcr} may be obtained from Eqs. 7.6 and 7.7 respectively, by substituting $\lambda = 0$. For a smooth vertical wall, retaining a horizontal backfill, ($\delta = \alpha = i = 0$).

Hence,

$$C_{1A} = \sqrt{\tan(\phi)\{\tan(\phi) + \cot(\phi)\}} = \sec\phi$$

and

$$C_{2A} = 1.$$

Therefore,

$$\theta_{Acr} = 90^\circ - \phi - \tan^{-1}[\sec\phi - \tan\phi] = 90^\circ - \phi - \left(45^\circ - \frac{\phi}{2}\right) = 45^\circ - \frac{\phi}{2}.$$

Similarly, $C_{1P} = \sec\phi$ and $C_{2P} = 1$ and $\theta_{Pcr} = 90^\circ + \phi - \tan^{-1}\$\$[\sec\phi + \tan\phi] = 45^\circ + \frac{\phi}{2}$.

Additional active and passive dynamic earth pressure, $(P_{Aq})_{dyn}$ and $(P_{Pq})_{dyn}$, on this wall due to a uniform surcharge of intensity q per unit area, on the inclined backfill surface, shall be taken as:

$$(P_{Aq})_{dyn} = \frac{qH\cos\alpha}{\cos(\alpha - i)} C_a \quad (7.8)$$

$$(P_{Pq})_{dyn} = \frac{qH\cos\alpha}{\cos(\alpha - i)} C_p \quad (7.9)$$

In the Mononobe–Okabe method, only two equilibrium conditions, namely $\Sigma H = 0$ and $\Sigma V = 0$ are satisfied and an uncertainty exists about the point of application. Mononobe and Matsuo (1932) suggested that the point of application of seismic active and passive earth pressures may be taken as 1/3 the height of the wall above its base. Saran and Prakash (1971) developed an analytical approach for obtaining seismic active earth pressure distribution along the height of a wall, the total active earth pressure and its point of application satisfying all the three conditions of static equilibrium, namely $\Sigma H = 0$, $\Sigma V = 0$ and $\Sigma M = 0$. Total earth pressure obtained from this approach is the same as that obtained from Mononobe–Okabe method. The point of application of the total seismic active earth pressure was found to be dependent upon ϕ, δ, α, i and the seismic coefficient A_h. This was in contradiction to the adopted practice. They also developed an expression for the critical wedge angle, which was taken corroborated by Zarrabi (1979).

Mononobe–Okabe's expressions, for obtaining total active and passive earth pressure, have been included in IS: 1893(Part III)—2012. To locate the point of application of dynamic earth pressures; (i) substitute $A_h = A_v = \lambda = 0$ in Eqs. (7.3) and (7.4), to obtain static active and passive earth pressure coefficients K_a and K_p,

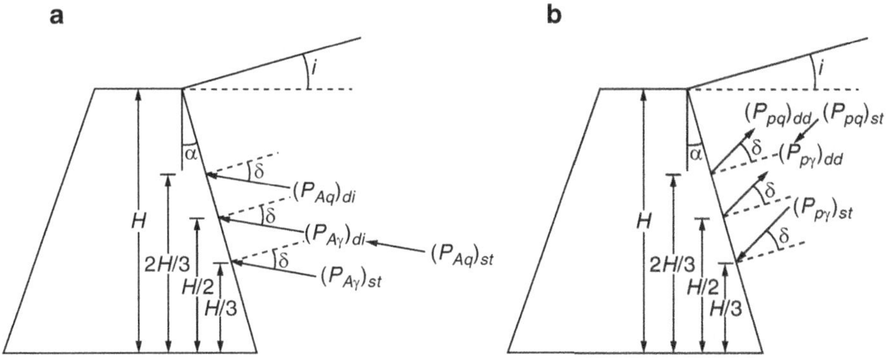

Fig. 7.2 Points of application for dynamics earth pressures. (**a**) Active case. (**b**) Passive case (Saran, 2012)

(ii) replacing C_a and C_p by K_a and K_p in Eqs. (7.1), (7.2), (7.8) and (7.9) static earth pressures $(P_{A\gamma})_{st}$, $(P_{P\gamma})_{st}$, $(P_{Aq})_{st}$ and $(P_{Pq})_{st}$ are obtained, and (iii) dynamic increment in the case of active earth pressure and dynamic decrement in the case of passive earth pressure are obtained by subtracting static earth pressure from dynamic earth pressure. Hence,

$$(P_{A\gamma})_{di} = (P_{A\gamma})_{dyn} - (P_{A\gamma})_{st} \tag{7.10}$$

$$(P_{P\gamma})_{dd} = (P_{P\gamma})_{st} - (P_{P\gamma})_{dyn} \tag{7.11}$$

$$(P_{Aq})_{di} = (P_{Aq})_{dyn} - (P_{Aq})_{st} \tag{7.12}$$

$$(P_{Pq})_{dd} = (P_{Pq})_{st} - (P_{Pq})_{dyn} \tag{7.13}$$

In the analysis and design of retaining walls, the earth pressures are considered to act as shown in Fig. 7.2a, b.

Kapila (1962) modified Culmann's (1866) graphical constructions for obtaining seismic active and passive earth pressures. These graphical constructions are applicable for backfills of any shape and different types of surcharge. They also permit the location of critical rupture surfaces. Procedures given by Kapila (1962) have been incorporated in IS: 1893(Part III)—2012.

In the current retaining wall design practice, total passive earth pressure is conservatively estimated using the analytical expression given by Richards and Elms (1979) (Davies et al., 1986; Das and Ramana, 2010).

For saturated backfill, the saturated unit weight of the soil shall be adopted in Eqs. (7.1) and (7.2). The dynamic increment (or decrement) in active and passive earth pressure for earthquake condition shall be computed from the expressions given in Eqs.(7.1), (7.2), (7.3) and (7.4) with the following modifications:

(a) The value of δ shall be taken as 1/2 the value of δ for dry/moist backfill.
(b) The value of λ shall be taken as:

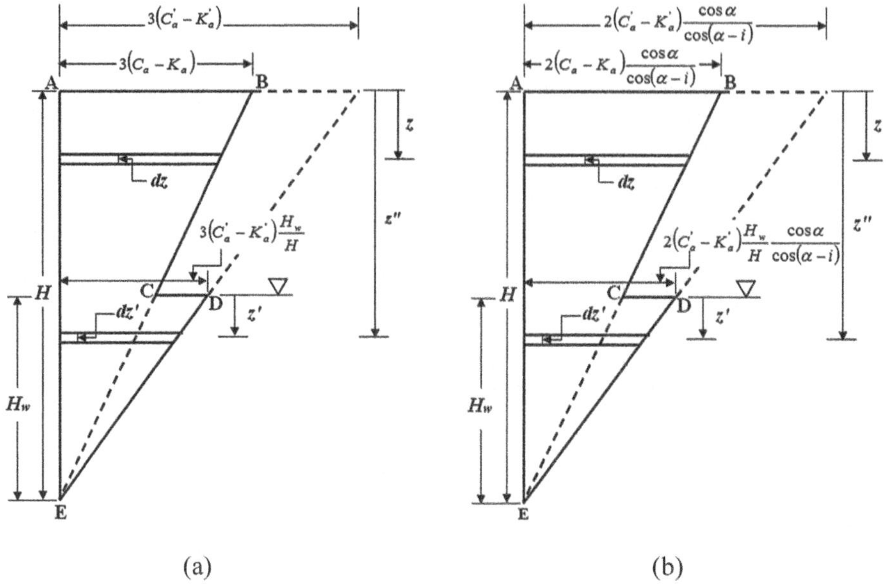

Fig. 7.3 Distribution of the ratio of lateral dynamic increment to the vertical effective pressure with the height of wall (Saran and Rangwala, 2012). (**a**) Due to weight of the backfill. (**b**) Due to surcharge on the backfill

$$\lambda = \tan^{-1}\left(\frac{\gamma_s}{\gamma_s - 10} \times \frac{A_h}{1 \pm A_v}\right) \tag{7.14}$$

where, γ_s is saturated unit weight of soil in kN/m^3,

(c) Submerged unit weight shall be adopted in Eqs. (7.1) and (7.2).
(d) From the value of earth pressure determined as above, subtract the value of earth pressure determined by putting $A_h = A_v = \lambda = 0$. The remainder shall be dynamic increment in the active case and dynamic decrement in the passive case.

Hydrodynamic pressure on account of water contained in earth fill shall not be considered separately as the effect of acceleration of water has been considered indirectly.

For analyzing a retaining wall having partially submerged backfill, Saran and Rangwala (2012) have proposed the use of Fig. 7.3a, b, in the active case, to obtain (i) lateral dynamic increment due to weight of the backfill, and (ii) lateral dynamic increment due to surcharge on the backfill. Similarly Fig 7.4a, b, in the passive case, is suggested to obtain (i) lateral dynamic decrement due to weight of the backfill and (ii) lateral dynamic decrement due to surcharge on the backfill.

They have also developed expressions for total dynamic increment/decrement and the location of their points of application measured from the base of the wall in

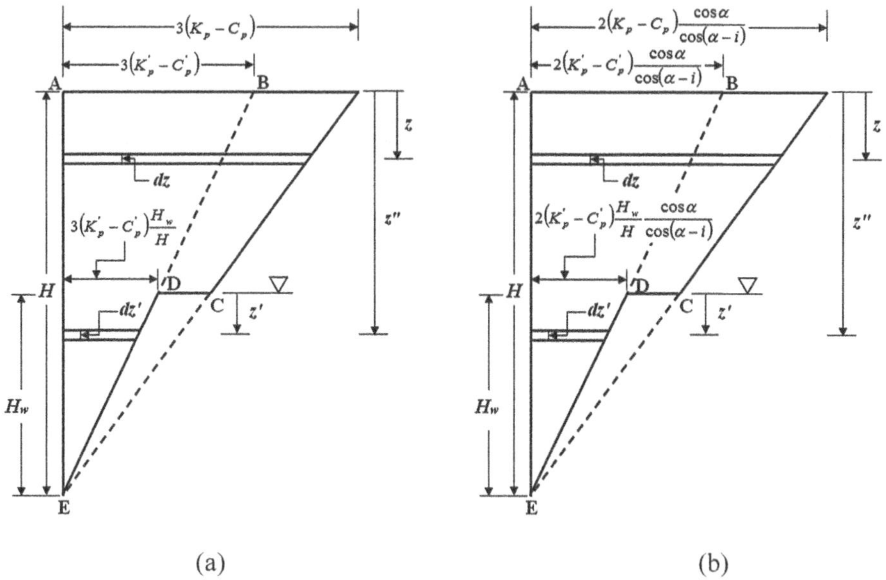

Fig. 7.4 Distribution of the ratio of lateral dynamic decrement due to the vertical effective pressure with the height of wall (Saran and Rangwala, 2012). (**a**) Due to weight of the backfill. (**b**) Due to surcharge on backfill

closed form. Expressions for location of point of application of dynamic increment/decrement in active/passive cases indicated that when backfill was dry or fully submerged, the dynamic component of earth pressure due to weight acts at mid height of the wall (Figs. 7.3a and 7.4a), while the dynamic component of earth pressure for surcharge acts at 2/3 height of wall from its base (Figs. 7.3b and 7.4b). Use of these figures have been incorporated in IS: 1893(Part III)- 2012.

After reviewing the results of experimental work based on small 1g shake table experiments, Seed and Whitman (1970) suggested that the point of applications of the dynamic increment thrust should be between one half to two-third the wall height above its base. However, the experimental and numerical (FEM) results obtained by Al-Atik and Sitar (2010) for cantilever retaining structures with a dry medium dense sand backfill indicates that the point of application should be at one third wall height as originally suggested by Mononobe and Matsuo (1932).

Seed and Whitman (1970) observed that the peak ground acceleration occurs only for one instant of time and does not have sufficient duration to cause significant wall movement. Thus, Seed and Whitman (1970) concluded that *"many walls adequately designed for static earth pressures will automatically have the capacity to withstand earthquake ground motions of substantial magnitudes and in many cases, special seismic earth pressure provisions may not be needed."* In fact, some well-documented case histories show that retaining structures designed only for static loading perform reasonably well under seismic loading (Clough and Fragaszy, 1977; Gazetas et al., 2004).

The solutions so far discussed considered the soil to be cohesionless. In real life situations, c–ϕ soils are used as backfills, especially when cohesionless soils are not available economically at construction sites. Attempts have been made in the past to develop analytical expressions for total dynamic active earth pressure considering c–ϕ backfill (Prakash and Saran, 1966; Saran and Prakash, 1968; Richards and Shi, 1994; Das and Puri, 1996; Saran and Gupta, 2003; Shukla et al., 2009; Shukla and Zahid, 2011). Prakash and Saran (1966) obtained dimensionless parameters for active earth pressure for retaining walls having horizontal backfill with uniformly distributed surcharge using the assumption of a plane failure surface. They have considered only horizontal seismic inertia force. Effect of tension crack has also been incorporated in the analysis. They optimized the pressure due to weight of the soil wedge, surcharge, and cohesion separately. Saran and Prakash (1968) extended this analysis for retaining walls inclined toward the fill, these are generally used in hydraulic structures, e.g., barrages and falls. Das and Puri (1996) extended the analysis proposed by Prakash and Saran (1966) to investigate the effects of vertical seismic inertial forces acting in the upward direction and considering an inclined backfill surface. Richards and Shi (1994) developed an analytical expression for the total dynamic active force from c-φ soil backfills by presenting a generalization of the elastoplastic solution for the free field with uniform acceleration for granular materials and considering both the horizontal and the vertical seismic forces. Saran and Gupta (2003) have carried out an analysis for determining seismic earth pressures behind an inclined wall, retaining an inclined cohesive frictional backfill, considering the effect of tension cracks. They have presented the results in the form of non-dimensional charts convenient for use in design. This method has also been incorporated in IS: 1893(Part III)—2012.

Shukla et al. (2009) developed an expression for the total dynamic active pressure against a smooth vertical retaining wall, based on the Coulomb sliding wedge mechanism concept, considering both horizontal and vertical seismic accelerations and maximizing the total active force to define a single wedge. They have considered a horizontal backfill. Shukla and Zahid (2011) extended this analysis considering a uniformly distributed surcharge on the backfill. Effect of tension crack has not been incorporated in their analysis; however they suggested replacing the cohesion value, in the expression for active earth pressure, by an average cohesion value as suggested by Lambe and Whitman (2008) and Shukla et al. (2009), the effect of tension cracks can be considered. Shukla and Habibi (2011) gave an analytical expression for the total passive earth pressure against a smooth vertical wall, having a horizontal c–φ soil backfill, considering both horizontal and vertical seismic inertia forces acting on the backfill. They have not considered surcharge. Shukla et al. (2011) extended this analysis for c–φ soil backfills having a uniformly distributed surcharge.

2.2 Pseudo-Dynamic Analysis

Pseudo-dynamic methods are more appropriate when compared to pseudo-static methods as they consider parameters like loading and site condition though the seismic force is represented by an equivalent inertial force. Pseudo-dynamic methods include the effects of loading, i.e., duration, frequency and acceleration as well as the effects of static and dynamic properties of the soil such as angle of internal friction, primary and shear wave velocities and amplification.

Steedman and Zeng (1990) developed an analysis, for a fixed-base vertical cantilever wall retaining a horizontal backfill, based on a series of centrifuge modeling tests. The results showed the phase difference in lateral acceleration in the backfill due to the propagation of shear waves from the base of the model toward ground surface. The effect of seismic excitation in the vertical direction has not been considered in this analysis.

For a sinusoidal base shaking of a retaining wall of height H (Fig. 7.5), the acceleration at depth z at time t, including the effect of amplification of the waves in the soil media, is given by:

$$A(z,t) = \left\{1 + \frac{H - z}{H}(f_a - 1)\right\} A_h g \sin \omega \left(t - \frac{H - z}{V_s}\right) \tag{7.15}$$

V_s is the shear wave velocity of the backfill soil, ω is frequency of shaking and f_a is amplification factor.

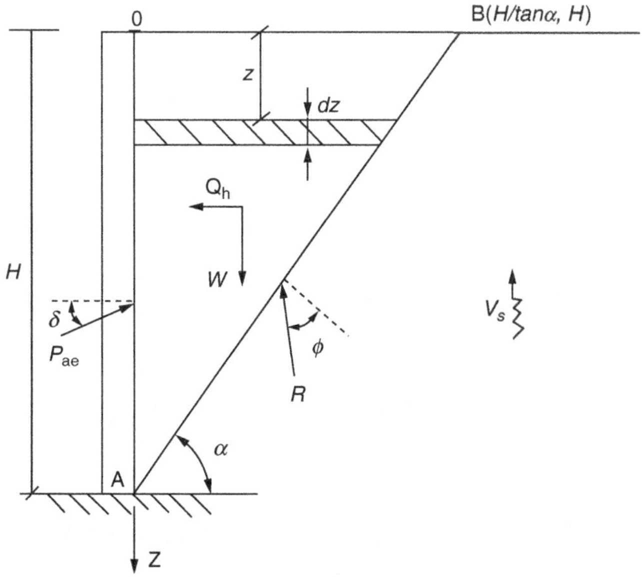

Fig. 7.5 A rough vertical wall subjected to seismic excitation (Steedman and Zeng, 1990)

The total active thrust acting on the wall is given by:

$$P_{ae} = \frac{1}{2}\gamma H^2 K_{ae} = \frac{Q_h \cos(\alpha - \varphi) + W \sin(\alpha - \varphi)}{\cos(\delta - \alpha + \varphi)} \tag{7.16a}$$

The horizontal inertia force Q_h, as a function of time, is given by:

$$Q_h = \int_0^H A(z,t)\rho\left(\frac{H - z}{\tan\alpha}\right)dz$$
$$= \frac{V_s \gamma H^2}{4\pi^2 f \tan\alpha} \times \left[\begin{Bmatrix} 2\pi H \cos\omega\zeta \\ +V_s/f(\sin\omega\zeta - \sin\omega t) \end{Bmatrix} + \frac{(f_a - 1)}{\pi}\begin{Bmatrix} 2\pi H(\pi H \cos\omega\zeta + V_s \sin\omega\zeta/f) \\ +V_s^2/f^2 \times (\cos\omega t - \cos\omega\zeta) \end{Bmatrix}\right] \tag{7.16b}$$

Variation of coefficient of seismic active earth pressure K_{ae} as a function of non-dimensional velocity, H/TV_s and soil amplification factor f_a is shown in Fig. 7.6. The distribution of dynamic earth pressure was given by:

$$p_{ae} = P_{ad} + P_{as}$$
$$= \frac{\cos(\alpha - \varphi)\gamma H^2}{\cos(\delta - \alpha + \varphi)\tan\alpha} z \sin(t - z/V_s) + \frac{\gamma z \sin(\alpha - \varphi)}{\sin(\delta - \alpha + \varphi)\tan\alpha}$$
$$+ \frac{\gamma A_h V_s (f_a - 1)\cos(\alpha - \varphi)}{2\pi f \cos(\delta - \alpha + \varphi)\tan\alpha} \times \left[\begin{matrix} -\cos\omega\left(t - \frac{z}{V_s}\right) - \frac{V_s}{\pi z f}\sin\omega\left(t - \frac{z}{V_s}\right) \\ + \frac{V_s^2}{2(\pi z f)^2}\left\{\cos\omega\left(t - \frac{z}{V_s}\right) - \cos\omega t\right\} \end{matrix}\right] \tag{7.16c}$$

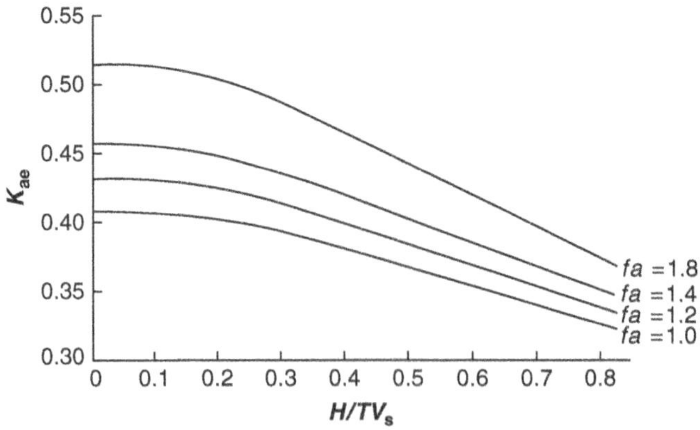

Fig. 7.6 Variation of earth pressure coefficient for different amplification factors, for $\varphi = 33°$, $\delta = 16$, $k = 0.20$ (Steedman and Zeng, 1990)

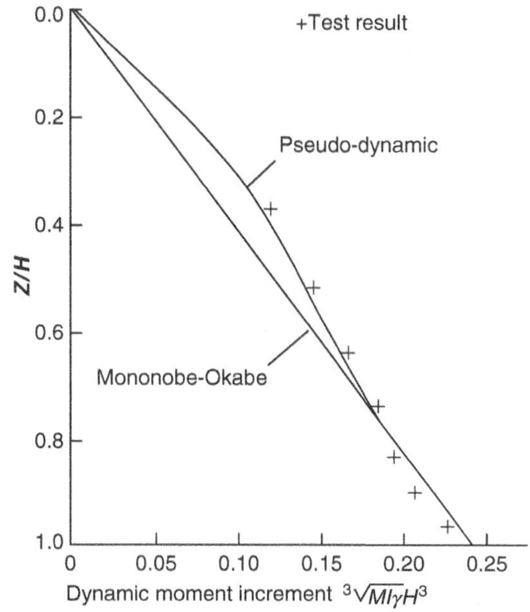

Fig. 7.7 Variation of dynamic moment increment with depth for $\varphi = 47°$, $\delta = 20°$, $A_h = 0.184, f_a = 2$, $G = 57$ MPa and $T = 1$ s (Steedman and Zeng, 1990)

A comparison of the variation of dynamic moment increment obtained from pseudo-dynamic analysis, pseudo-static analysis (Mononobe–Okabe theory) and centrifuge tests is shown in Fig. 7.7. The results of the pseudo-dynamic approach are in better agreement with the experimental results when compared to those obtained from the Mononobe–Okabe theory.

2.3 Displacement Based Analysis

Design of a retaining wall based on allowable displacement under dynamic conditions has gained importance in recent years. Methods available to compute displacements of a rigid retaining wall under earthquake loading are; (i) Pure translation, (ii) Analysis based on pure rotation, (iii) Richard–Elms model based on Newmark's approach, (iv) Nadim–Whitman model, and (v) Analysis based on Reddy–Saran–Viladkar model.

Nandakumaran (1973) proposed a mathematical model, based on soil mass for a SDOF mass-spring-dashpot system, considering pure translational movement. Results have been presented in Fig. 7.8.

Nandakumaran suggested the following procedure for obtaining the displacement of the wall:

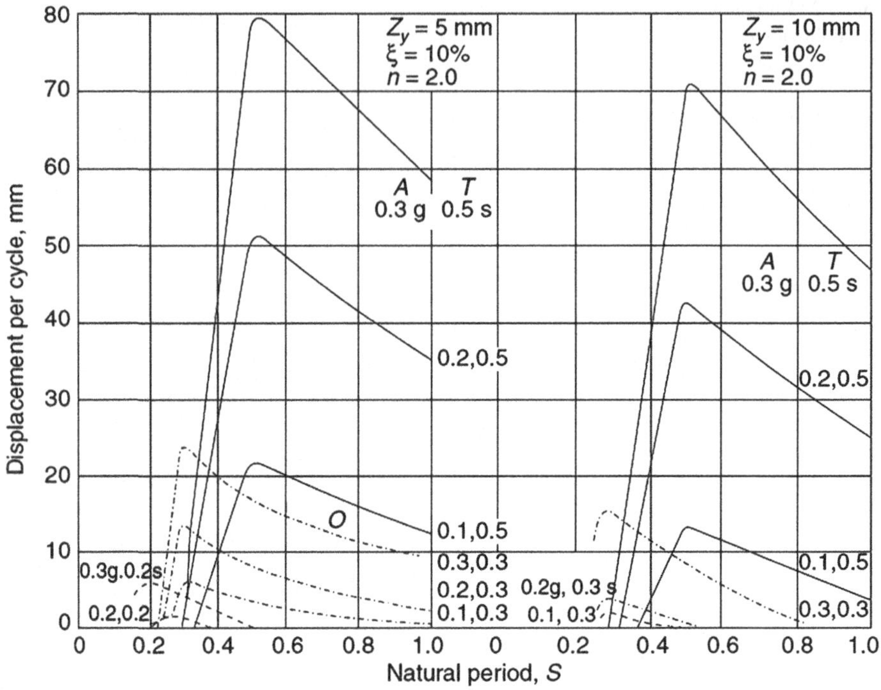

Fig. 7.8 Natural period versus slip per cycle (Nandakumaran, 1973)

1. Determine the natural period of the wall using the equation:

$$T = 2\pi\sqrt{\frac{m}{K}} \tag{7.17}$$

2. Select suitable values of yield displacement.
3. Determine the slip per cycle from Fig. 7.10 or other similar plots corresponding to a given yield displacement, natural period of wall and ground motion.
4. Compute the total slip for the ground motion.

Later Prakash et al. (1981) extended this solution for the case of pure rotation.

Newmark (1965) proposed a basic methodology for evaluating the potential deformation that would occur in an embankment dam subjected to earthquake loading by considering the equilibrium of a sliding block resting on a plane as shown in Fig. 7.9a. The sliding block was considered to be analogous to the sliding mass in an embankment. It was envisaged that block movement would be initiated and displacements would begin to occur once the inertial forces on the potential sliding mass exceed the base resistance. The acceleration at which the inertial forces become just sufficient to cause displacement to begin, was defined as yield

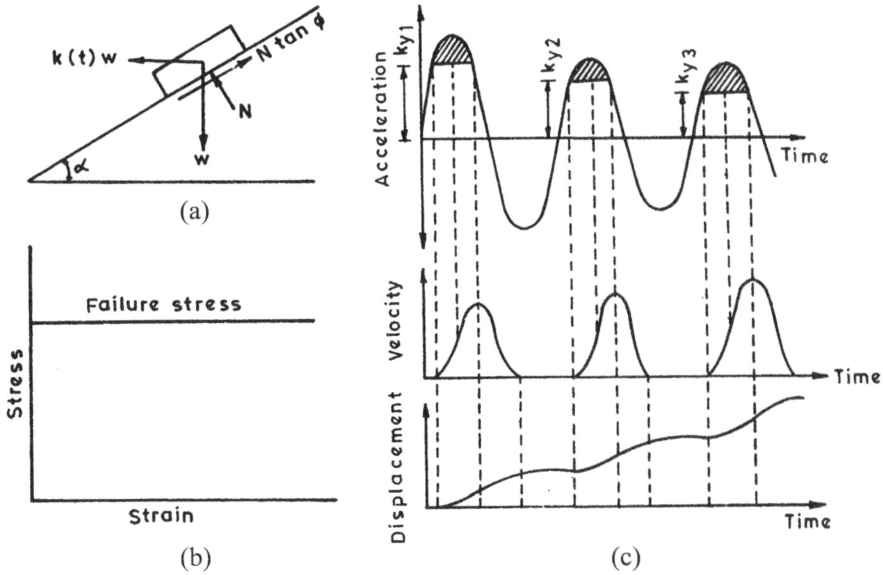

Fig. 7.9 (**a**) Forces on the sliding block, (**b**) Rigid plastic stress strain behavior of a material, (**c**) Integration of effective acceleration/velocity time history to determine velocity/displacement (Newmark, 1965)

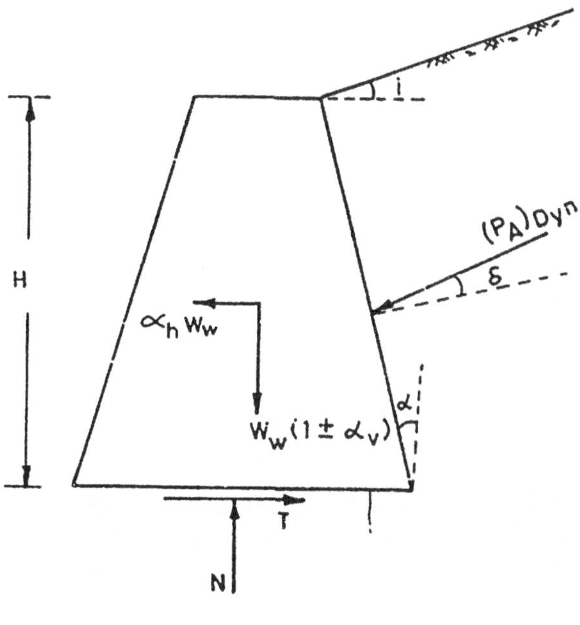

Fig. 7.10 Forces on a gravity wall (Richards and Elms, 1979)

acceleration (K_{y1}, K_{y2}, K_{y3} in Fig. 7.9c). Time histories of velocities and ultimately displacements of the sliding mass could be evaluated by successive integration of the shaded area as illustrated in Fig. 7.9c.

This analysis is essentially based on the rigid plastic behavior of the constitutive materials (Fig. 7.9b). Though this method was developed for a stability analysis of an earth dam, it has been used by Richards and Elms (1979) to compute the displacements of retaining walls. They have proposed a method for design of gravity retaining walls based on limiting displacement considering the wall inertia effect. The procedure developed by them is described below. A gravity retaining wall along with forces acting on it during earthquake is shown in Fig. 7.10.

Summing the forces in the vertical and horizontal directions, we get

$$N = W_w \pm A_v W_w + (P_A)_{dyn} \sin(\alpha + \delta) \tag{7.18}$$

$$T = A_h W_w + (P_A)_{dyn} \cos(\alpha + \delta) \tag{7.19}$$

For sliding,

$$T = N \tan \phi_b \tag{7.20}$$

ϕ_b = angle of internal friction of base soil

Solving Eqs. (7.18), (7.19) and (7.20), we get

$$W_w = \frac{(P_A)_{dyn}[\cos(\alpha + \delta) - \sin(\alpha + \delta)\tan \phi_b]}{(1 \pm A_v)\tan \phi_b - A_h} \tag{7.21}$$

Richards and Elms (1979) have given a design procedure based on a limited allowable wall movement, rather than on the assumption that the wall will not move at all. The suggested procedure is as follows:

1. Decide upon an acceptable maximum displacement, d.
2. Determine the design value of A_{hd} from Eq. (7.22) (Franklind and Chang, 1977).

$$A_{hd} = A_h \left(\frac{5A_h}{d}\right)^{\frac{1}{4}} \tag{7.22}$$

where,
A_h = Horizontal acceleration coefficient
d = Maximum displacement in mm

3. Using A_{hd}, determine the required wall weight, W_w, from Eq. (7.21). The value of A_v may be taken as 2/3 A_{hd}.
4. Apply a suitable factor of safety, say 1.5, to W_w.

The Richard–Elms model assumes a constant value of horizontal acceleration ($A_h g$) during slippage. But once the backfill begins to slip, compatibility of movement requires that the backfill has a vertical acceleration, thus causing a change in the horizontal acceleration.

Zarrabi (1979) considered the equilibrium of the wall and the backfill wedge separately and satisfied the continuity requirements for the failure surface. An iterative procedure was developed for computing the instantaneous values for the inclination of the failure plane, the dynamic active earth pressure and the acceleration of the wall, given the horizontal and vertical ground accelerations. The horizontal acceleration of the wall and the inclination of failure plane in the backfill are not constant in Zarrabi's model.

Generally, displacements computed using Zarrabi's model are slightly lower than those computed using the Richard–Elms model. Dynamic tests on model retaining walls performed by Lai (1979) indicated that Zarrabi's model predicts the movement of the wall more accurately than Richard–Elms model. Lai, also, observed the existence of a single rupture plane in the backfill in disagreement to Zarrabi's prediction. Later, Zarrabi's model was modified to consider a constant inclination of the failure plane in the backfill.

Both Richard–Elms and Zarrabi's models assume a rigid–plastic behavior of the backfill material. Hence the input ground acceleration is constant throughout the backfill. But due to more or less elastic behavior of the soil, at stress levels below failure, the input acceleration is not constant. Amplification of ground motion cannot be taken into account in the above models.

Nadim and Whitman (1983) used a two-dimensional plane – strain finite element model for computing permanent displacements, taking into account ground motion amplification. The slip element at the base of the wall has been assigned a very large value of normal stiffness, thus restraining the wall from vertical and rotational movements relative to its base. Thus, the wall undergoes only translational movements. Forces acting on the wall and soil wedge, in the model, are shown in Fig. 7.11a. The finite element (FE) mesh used by them is as shown in Fig. 7.11b. To understand the effect of ground motion amplification, typical results are shown in Fig. 7.11c. In this figure, R is the ratio of permanent displacement from the FE model to the permanent displacement from the rigid–plastic (Richard–Elms or Zarrabi's) model. f_1 is the fundamental frequency of wall and f is the frequency of ground motion. It can be seen that the effect of amplification of motion on displacement is greater when f/f_1 is greater than 0.3. The FE model predicts zero permanent displacement at high frequencies. This is because in the analysis only three cycles of base motion are considered during which steady state conditions cannot be achieved. However, it can be said that large values of $f/f_1 > 2$ are not of practical interest because displacements are very small. Nadim and Whitman (1983) suggested the following simple procedure for taking into account the effects of ground motion amplification in seismic design:

(i) Evaluate the fundamental frequency f_1 of the backfill for the design earthquake using one-dimensional amplification theory, using the following equation,

$$f_1 = V_1/4H \tag{7.23}$$

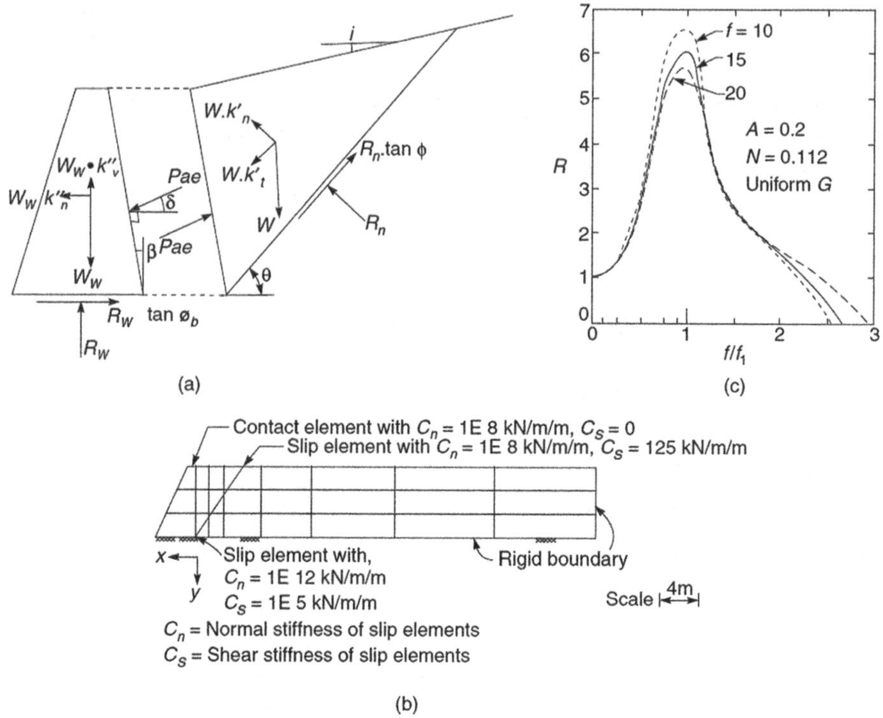

Fig. 7.11 (**a**) Forces acting on the wall and soil wedge, (**b**) Retaining wall and its finite element idealization, (**c**) Effect of ground motion amplification on permanent wall displacement (Nadim and Whitman, 1983)

where,
H = height of retaining wall in m, and
V_1 = peak velocity of earthquake in m/s
Also estimate the ground motion frequency, f.

(ii) If f/f_1 is less than 0.25, neglect, amplification of ground motion. If f/f_1 is in the vicinity of 0.5, increase the peak acceleration, A, and the peak velocity, V, of the design earthquake by 25–30%. If f/f_1 lies between 0.7 and 1, increase A and V by 50%. Obtain A_h as A/g.
(iii) Use the value of A_h, from the previous step, in Eq. (7.22) given by the Richard–Elms model for getting A_{hd} for the known value of displacement.
(iv) The value of A_{hd} estimated in step (iii) is used as the value of horizontal seismic coefficient in the Mononobe–Okabe analysis to calculate the lateral thrust for which the wall is designed. The value of vertical seismic coefficient may be taken as 2/3 A_{hd}.

Reddy et al. (1985) have chosen a mathematical model such that it represents translation and rotation simultaneously and therefore it has two degrees of freedom. In practice, the cross-section of the rigid retaining wall varies to a large extent. A

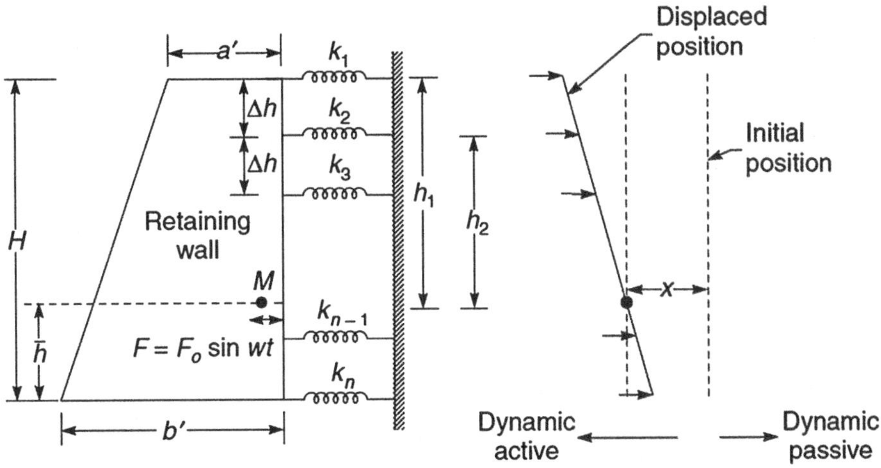

Fig. 7.12 Mathematical model for displacement analysis under dynamic condition (Reddy et al., 1985)

reasonable approximation is, therefore, made by lumping the mass of the rigid retaining wall at its center of gravity. The backfill soil is replaced by closely spaced independent elastic springs as shown in Fig. 7.12.

This method is based on the following assumptions:

1. The earthquake motion may be considered as an equivalent sinusoidal motion with uniform peak acceleration and the total displacement is equal to residual displacement per cycle multiplied by the number of cycles.
2. Soil stiffnesses (or spring constants) for displacement of wall toward the backfill and away from the backfill are different.
3. Soil participating in vibration, damping of soil and base friction are neglected.

It is difficult to determine analytically the soil mass that would participate in the vibration together with the wall when it undergoes translational and rotational motions simultaneously. Neglecting this soil mass, this method gives higher displacements and the solution is conservative. However, the mass of vibrating soil can be determined from experimental studies. In the case of pure translation, Nandakumaran (1973) had conducted experiments to determine the vibrating soil mass and concluded that it can be taken as equal to 0.8 times the mass of the Rankine's wedge. By adopting a similar technique, the soil mass vibrating along with the rigid retaining wall under combined rotational and translational motions can be obtained. Then it is added to the mass of the wall and lumped at the center of gravity and the analysis can be carried out without any change.

In soils, it is customary to consider values of damping such as 15 or 20% of critical damping, in view of larger energy absorption when compared to other engineering/structural material. In the present analysis however, energy absorption in the form of plastic displacement of the wall has been considered. Therefore, smaller damping values would be appropriate. Neglecting even this smaller damping, the displacement of the wall, by this method, will be more than the actual displacement.

The displacement of the retaining wall is greatly influenced by base friction. In case of walls in alluvial deposits and at the waterfront, translational motion is predominant. In some other cases, the walls may have predominantly rotational motion. But in general for any type of foundation soil, retaining walls undergo translational and rotational motions simultaneously. For rigid retaining walls, the stability is mainly due to its gravity, hence neglecting base friction; the analysis will lead to an overestimation of displacement.

To study the response characteristics of the system, two cases were considered, (i) in which plastic deformations do not occur (elastic system), and (ii) in which plastic deformations do occur (plastic system).

However, refinement of the model by including the vibrating soil mass, damping of the soil and base friction is needed so that the analysis can predict displacement values closer to the actual displacements.

3 Shallow Foundation

A seismic design of shallow foundations is a vital area of research in geotechnical earthquake engineering.

The strength of the soil reduces significantly when subjected to seismic excitation because of the large inertial stresses induced during shaking. This decrease in strength of soil leads to a significant reduction in the bearing capacity of the footing and increases the induced settlement; this reduced bearing capacity is termed as *seismic bearing capacity* of the footing and the increased settlement is termed as *seismic settlement.* Estimation of seismic bearing capacity and seismic settlement of footings resting on flat ground has received considerable attention of researchers in the last two decades. Although, in comparison to the extensive studies available on the static bearing capacity of shallow foundations only a limited amount of information is available on seismic bearing capacity of footings.

The methods available for design of shallow foundations may be divided into three categories, namely (i) Seismic bearing capacity from Pseudo-Static approach, (ii) Seismic bearing capacity from Pseudo-Dynamic approach, and (iii) Seismic settlement analysis.

3.1 Seismic Bearing Capacity from Pseudo-Static Approach

Sarma and Iossifelis (1990) determined the seismic bearing capacity factors (N_{cE}, N_{qE} and $N_{\gamma E}$) using the method of inclined slices used in slope stability analysis. The

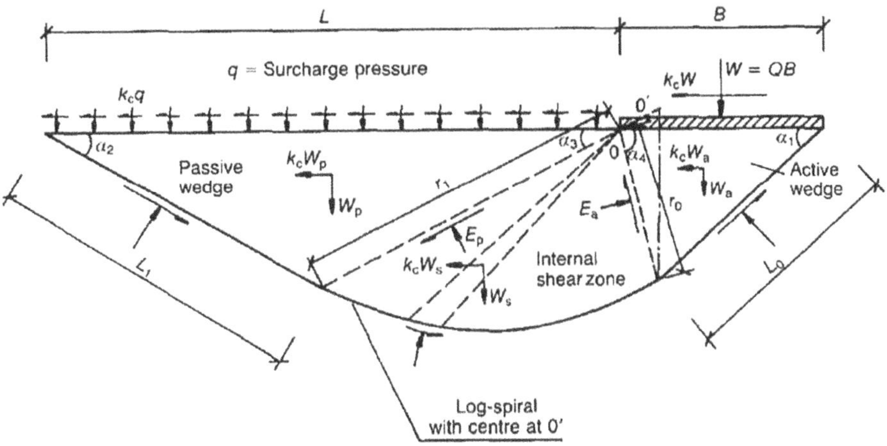

Fig. 7.13 Geometry of critical slip surface and inter-slice slip surfaces for general case (Sarma and Iossifelis, 1990)

method uses the limit equilibrium principle for soils exhibiting rigid–plastic properties and following the Mohr–Coulomb failure criterion. This method also allows internal shearing within the soil mass. The failure surface is assumed to be of the form shown in Fig. 7.13.

The failure surface contains an active wedge and a passive wedge and an internal shear zone sandwiched between the two wedges. The shape of the failure surface defining the shear zone is assumed to be a log-spiral.

The angles forming these wedges are unknown at the beginning and are found by iteration to produce the minimum bearing capacity factors. In Fig. 7.13, W_a is weight of active wedge, W_p is weight of passive wedge and W_s is weight of internal shear zone; inter-slice slip surfaces pass through O while centre of log-spiral is at O', while Q is bearing capacity of the footing.

Bearing capacity is expressed, in the classical form, with seismic bearing capacity factors;

$$Q_E = cN_{cE} + qN_{qE} + 0.5\,B\gamma N_{\gamma E} \quad (7.24)$$

The minimum value of Q_E gives the bearing capacity of the foundation. However, in practice, the minimum values of the three bearing capacity factors are determined independently of each other and therefore their use errs on the safe side. Results of the analysis are given in a graphical form as functions of the horizontal acceleration factor and the angle of internal friction of the soil. The analysis does not include the effect of vertical acceleration generated by the earthquake shaking.

Richards et al. (1993) presented a pseudo-static analysis using a Coulomb-type mechanism including inertial forces in the soil and on the footing. The classic 2D slip-line field (Fig. 7.14) obtained by Prandtl (1921) is used for the determination of static bearing capacity. Seismic bearing capacity factors were derived considering a

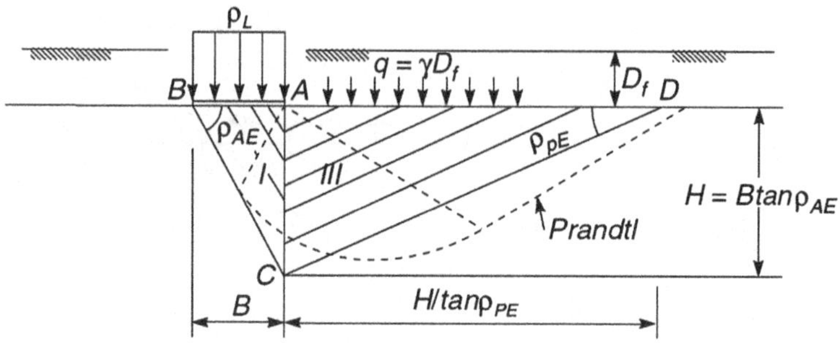

Fig. 7.14 Simplified static slip surface with Coulomb wedges ($\delta \neq 0$)

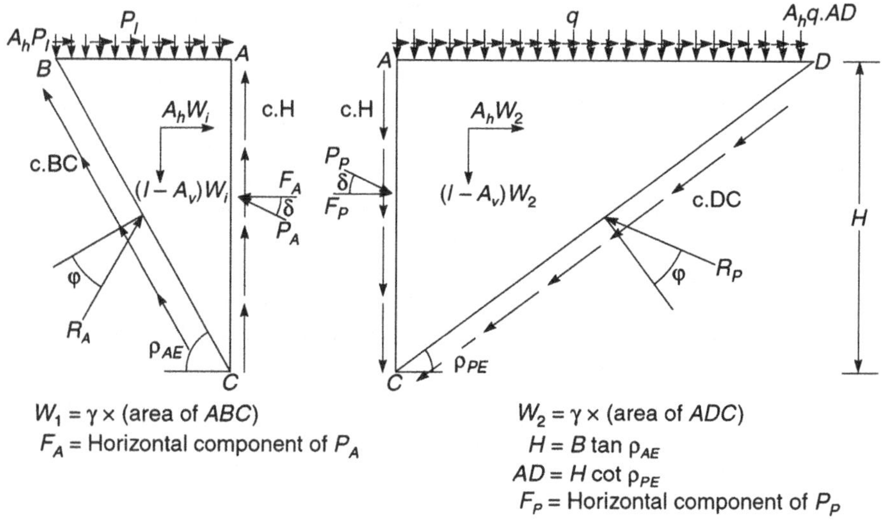

Fig. 7.15 Coulomb mechanism (with wall friction, δ)

planar rupture surface as proposed by Coulomb (1776) in developing classical earth pressure theories (Fig. 7.15) and are given as,

$$N_{\mathrm{qE}} = \frac{K_{\mathrm{PE}}}{K_{\mathrm{AE}}} \tag{7.25}$$

$$N_{\gamma E} = \tan\rho_{\mathrm{AE}}\left(\frac{K_{\mathrm{PE}}}{K_{\mathrm{AE}}} - 1\right) \tag{7.26}$$

$$N_{\mathrm{cE}} = \left(N_{\mathrm{qE}} - 1\right)\cot\phi \tag{7.27}$$

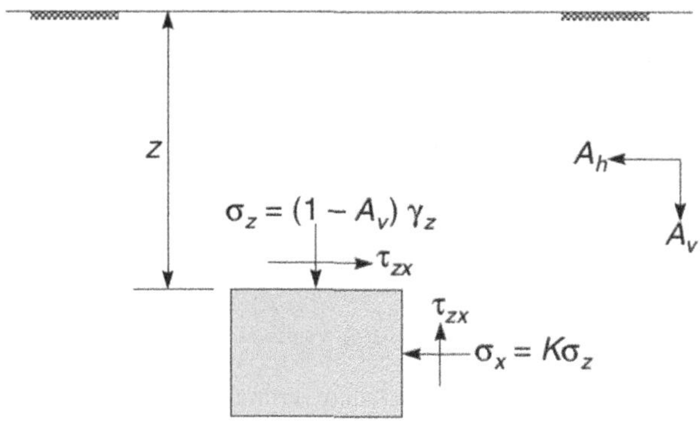

Fig. 7.16 Stresses in soil element (Budhu and Al-Karni, 1993)

where K_{AE} and K_{PE} are coefficients of dynamic earth pressure in the active and passive cases, respectively and can be obtained by Mononobe-Okabe equation and,

$$\rho_{AE} = \alpha + \tan^{-1}\left\{\frac{\sqrt{(1+\tan^2\alpha)(1+\tan(\delta+\theta)\cot\alpha)} - \tan\alpha}{1+\tan(\delta+\theta)(\tan\alpha+\cot\alpha)}\right\} \tag{7.28}$$

where $\alpha = \phi - \theta$, and

$$\theta = \tan^{-1}[A_h/(1 - A_v t)] \tag{7.29}$$

Budhu and Al-Karni (1993) considered a soil element at depth z, under free field conditions, and subjected to horizontal and vertical accelerations due to an earthquake, as shown in Fig. 7.16. The soil element is subjected to stresses as given below.

$$\sigma_x = K\,(1 - A_v)\gamma z \tag{7.30a}$$

$$\sigma_z = (1 - A_v)\gamma z \tag{7.30b}$$

$$\tau_{xz} = -A_h\gamma z \tag{7.30c}$$

where, σ_x and σ_z are the normal stresses, τ_{xz} is the shear stress, A_v and A_h are the vertical and horizontal acceleration coefficients, K is the lateral earth pressure coefficient and γ is the unit weight of the soil.

From the geometry of Mohr circle drawn for these stresses coefficient of earth pressure was obtained as expressed in Eq. (7.31);

$$K_{PE,AE} = \frac{\sin^2\phi + 1}{\cos^2\phi} + K_D\tan\phi \pm \sqrt{K_{sqrt}} \tag{7.31}$$

where,

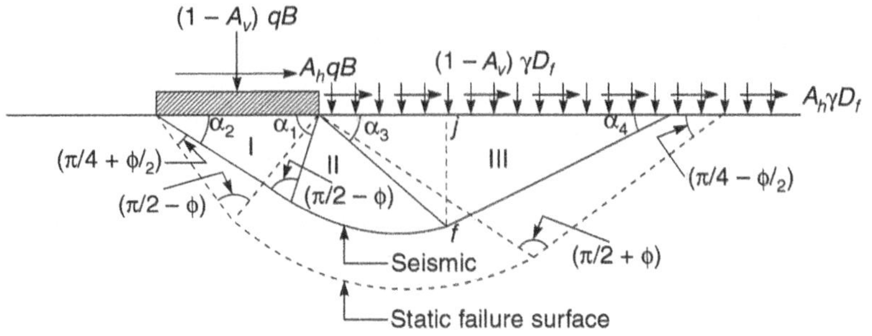

Fig. 7.17 Static and seismic failure surfaces (Budhu and Al-Karni, 1993)

$$K_{\text{sqrt}} = \left[K_{\text{D}}\tan\phi + \frac{\sin^2\phi + 1}{\cos^2\phi}\right]^2 + K_{\text{D}}^2 + 2K_{\text{D}}\tan\phi - \frac{4\tan^2\theta}{\cos^2\phi} - 1,$$

and

$$K_{\text{D}} = 2D/(1 - A_{\text{v}})$$

Seismic condition results in a non symmetric failure surface (Fig. 7.17) comprising of zones similar to the static case. However, the failure surface is shallower and subtends at an angle α_1, which is smaller than the static angle ($\pi/4 + \phi/2$). Expressions for the angles α_1, α_2, α_3 and α_4 have been given (Budhu and Al-Karni, 1993).

By considering vertical and horizontal equilibrium of forces, two equations for each bearing capacity factor were obtained. The value for each bearing capacity factor was found by iteration using different values of mobilization factor *m*, to satisfy the equilibrium conditions. A parametric study was performed using a computer program to calculate the bearing capacity factors for different values of seismic accelerations A_{h} and A_{v}, stability factor *D* and angle of internal friction ϕ. Seismic bearing capacity factors $N_{\gamma\text{E}}$, N_{qE} and N_{cE} are given as in Eqs. (7.32), (7.33) and (7.34), respectively.

$$N_{\gamma E} = \left(1 - \tfrac{2}{3}A_{\text{v}}\right)\exp\left(\frac{-9A_{\text{h}}^{1.1}}{1 - A_{\text{v}}}\right)N_{\gamma S} \tag{7.32}$$

$$\text{N}_{qE} = (1 - A_{\text{v}})\exp\left(\frac{-5.3A_{\text{h}}^{1.2}}{1 - A_{\text{v}}}\right)N_{\text{qS}} \tag{7.33}$$

$$N_{\text{cE}} = \exp\left(-4.3A_{\text{h}}^{(1+D)}\right)N_{\text{cS}} \tag{7.34a}$$

$$\text{where, } D = \frac{c}{\gamma(0.5Be^{\pi/2^{\tan\varphi}} + D_{\text{f}})} \tag{7.34b}$$

Dormieux and Pecker (1995) considered Prandlt's failure surface (Fig. 7.18) for the analysis and showed the effect of the inertial forces. The analysis was done with and without considering the seismic acceleration in the soil mass. The results were

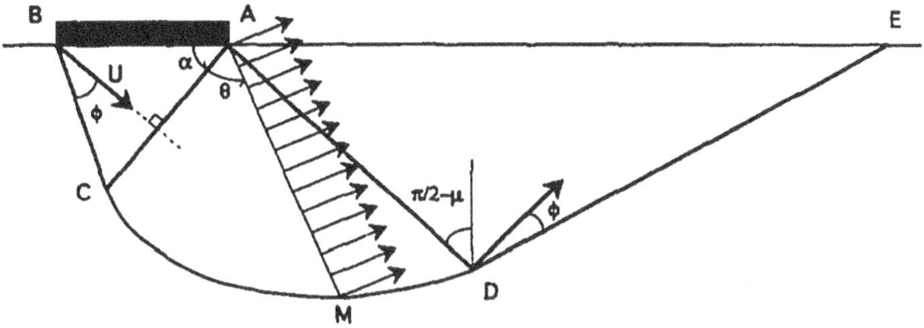

Fig. 7.18 Failure surface (Dormieux and Pecker, 1995)

found in good agreement with Richards et al. (1993) for smaller values of seismic coefficient.

Soubra (1999) solved the seismic bearing capacity problem using the upper-bound method of limit analysis using a pseudo-static approach. A soil-foundation system subjected to a horizontal seismic acceleration, undergoing translational movement, was assumed be fail in either of two distinct translational failure mechanism named herein as M1 and M2 (Figs. 7.19a, b). Failure mechanism M1, applicable for static cases only, was symmetrical about a vertical axis passing through the center of the footing. Failure mechanism M2, applicable for both static as well in seismic case, was non-symmetrical in nature.

Seismic bearing capacity factors were observed to be decreasing with an increase in the horizontal acceleration. The weight factor ($N_{\gamma E}$) was observed to be in good agreement with Richards et al. (1993) while the surcharge factor (N_{qE}) and cohesion factor (N_{cE}) were observed to be higher than those of Richards et al. (1993).

Choudhury and Rao (2005) presented a pseudo-static analysis assuming a non-symmetrical one sided failure surface (Fig. 7.20). The angle of failure for the passive wedge was considered as given below;

$$\xi = \frac{\pi}{4} - \frac{\varphi}{2} + \frac{\theta}{2} - \frac{1}{2}\sin^{-1}\left(\frac{\sin\theta}{\sin\varphi}\right) \tag{7.35}$$

Seismic bearing capacity factors were obtained by calculating passive pressure for all three zones for the three different cases; (i) weight only, (ii) surcharge only, and (iii) cohesion only. Wedge angles α_1 and α_2 were obtained by considering equilibrium of forces in the horizontal and vertical directions. Effect of vertical acceleration has also been studied by considering the three cases, namely (i) $A_v = 0$, (ii) $A_v = 0.5A_h$, and (iii) $A_v = A_h$. Results have been presented in the form of non-dimensional charts, varying the angle of internal friction from 10° to 40°, at intervals of 10° and horizontal acceleration coefficient A_h from 0 to 0.5, at intervals of 0.1.

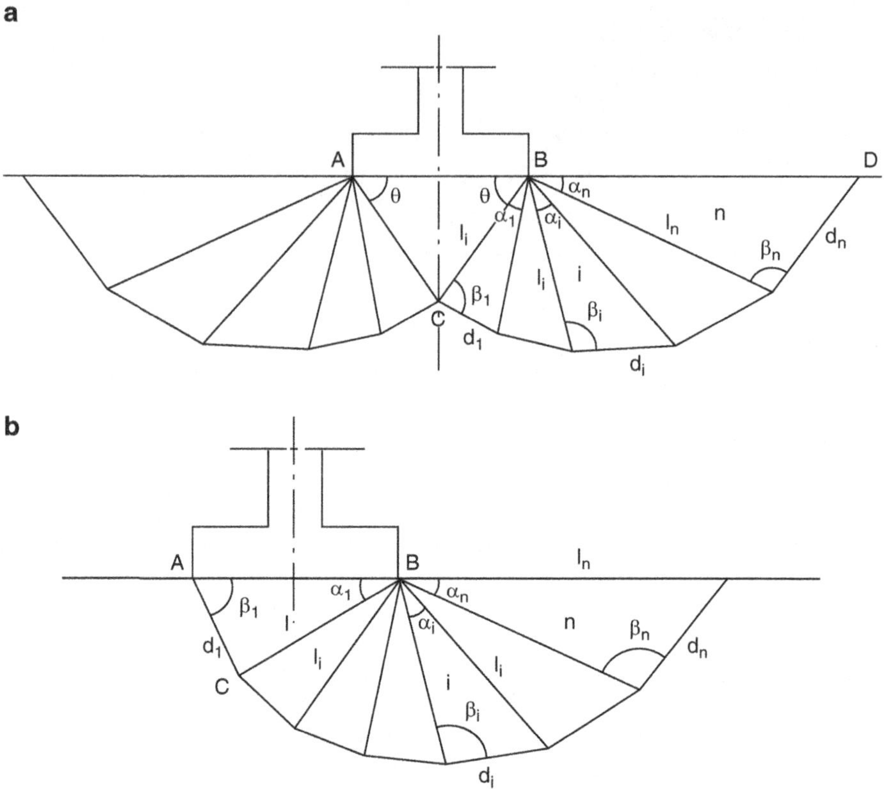

Fig. 7.19 Failure mechanisms M1 and M2 (Soubra, 1999). (**a**) Failure mechanism M1 for static bearing capacity analysis (**b**) Failure mechanism M2 for seismic bearing capacity analysis

Saran and Rangwala (2011) developed a pseudo-static analysis considering the effect of vertical and horizontal seismic acceleration, both on the foundation as well as within the soil mass beneath the foundation. Analysis was done assuming a failure surface comprising of three zones: a one sided unsymmetrical triangular zone, a log-spiral transition zone and a triangular passive zone (Fig. 7.21). The soil on the other side was considered having partially mobilized shear strength. The problem was solved for the three cases, namely, (i) $c = q = 0$, (ii) $c = \gamma = 0$ and (iii) $\gamma = q = 0$.

Variation of the point of application of passive pressure, acting on the central elastic wedge beneath the footing, with the coefficient of horizontal seismic acceleration was assumed to be as shown in Fig. 7.22. x_1 is multiplying factor, applied to the length DE, to obtain the location of the point of application of passive pressure, measured from the point E in case (i); x_2 is the same for cases (ii) and (iii).

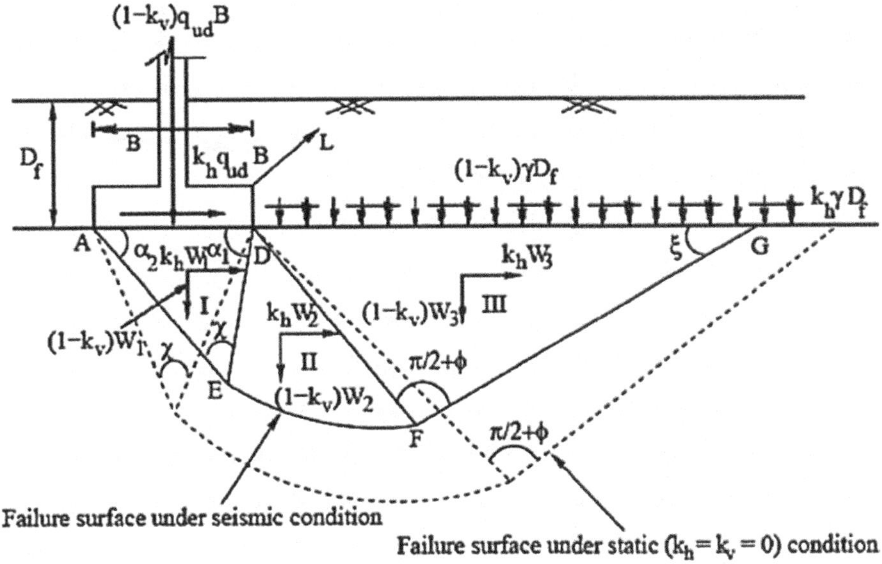

Fig. 7.20 Failure surface (Choudhury and Rao, 2005)

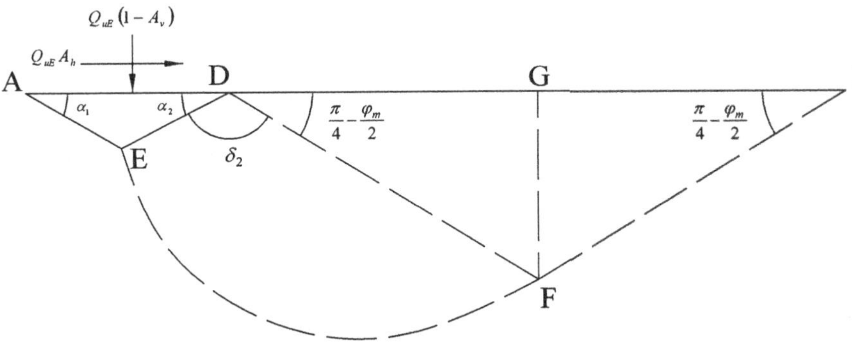

Fig. 7.21 Failure surface (Saran and Rangwala, 2011)

Seismic bearing capacity factors were obtained for the maximum value of m, which is obtained by iteration and satisfies all equilibrium conditions. Values of the ratios of seismic to static bearing capacity factors were given in tabular as well as graphical form.

Comparison of the bearing capacity factors obtained by different investigators, discussed above is presented in Figs. 7.23, 7.24 and 7.25.

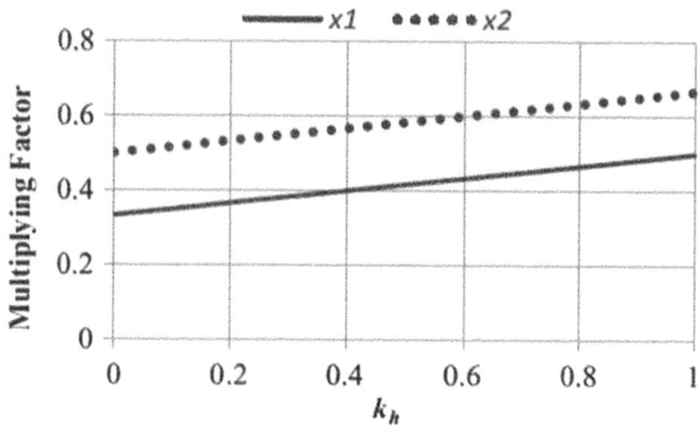

Fig. 7.22 Variation of x_1 and x_2 with the coefficient of horizontal seismic acceleration (Saran and Rangwala, 2011)

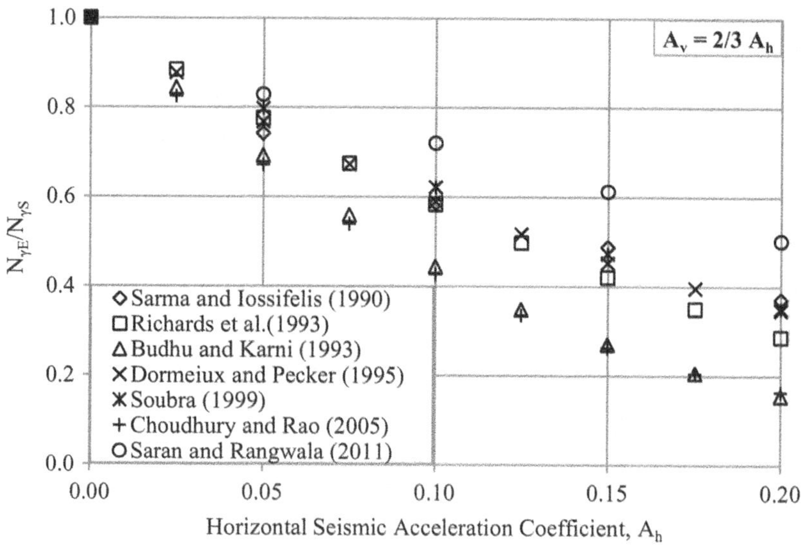

Fig. 7.23 Seismic to static bearing capacity factor, $N_{\gamma E}/N_{\gamma S}$, for $\varphi = 30°$

3.2 Seismic Bearing Capacity from Pseudo-dynamic Approach

Ghosh (2008) used a pseudo-dynamic approach to analyze, the bearing capacity problem, considering Coulomb's mechanism (Fig. 7.26). A parametric analysis was performed considering various parameters. The results were presented in the form of charts.

Ghosh and Choudhury (2011) proposed a pseudo-dynamic analysis for determining seismic bearing capacity of a footing. The pseudo-dynamic method include the effects of loading, i.e., duration, frequency and acceleration of shaking, as well as

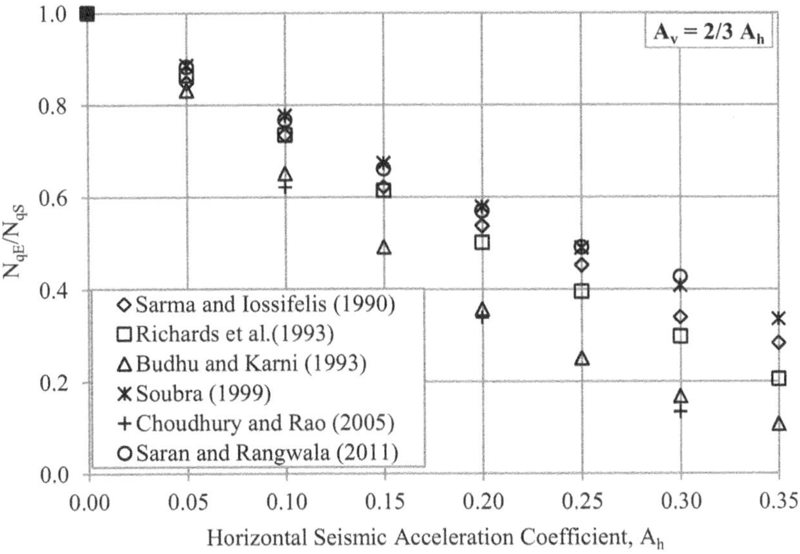

Fig. 7.24 Seismic to static bearing capacity factor, N_{qE}/N_{qS}, for $\varphi = 30°$

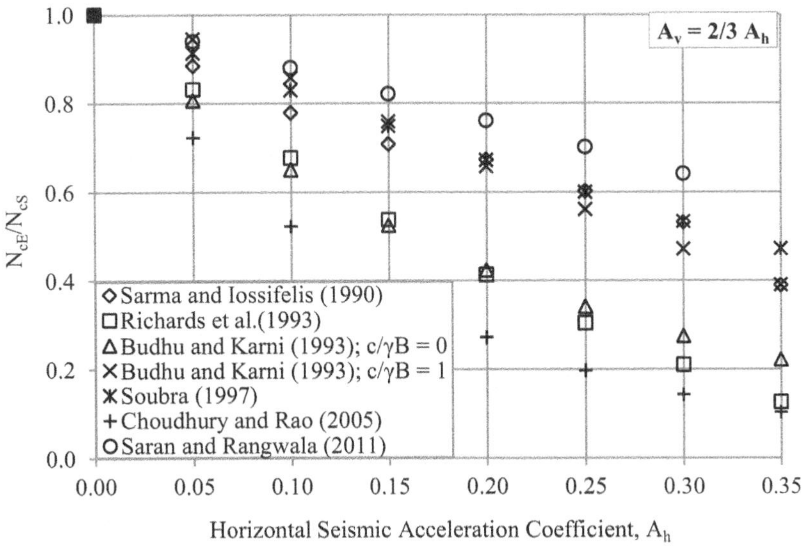

Fig. 7.25 Seismic to static bearing capacity factor, N_{cE}/N_{cS}, for $\varphi = 30°$

effects soil properties such as angle of internal friction, primary and shear wave velocities and soil amplification. The soil mass was assumed to fail as conceived by Richards et al. (1993). Bearing capacity factors were considered as given by the

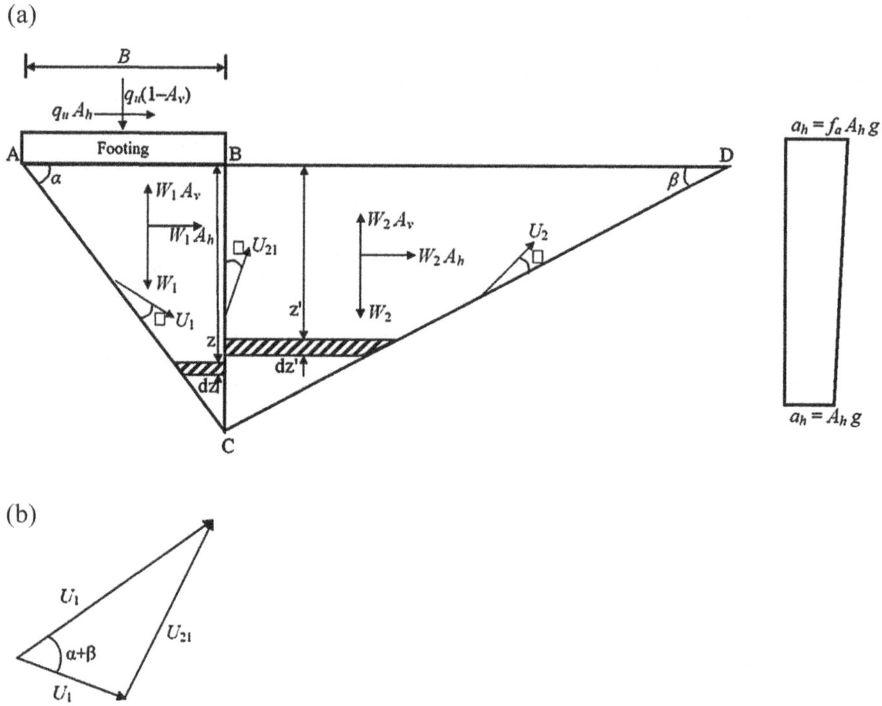

Fig. 7.26 (**a**) Collapse mechanism, (**b**) Velocity hodograph (Ghosh, 2008)

Eqs. (7.25), (7.26) and (7.27). However, the coefficients of dynamic earth pressure are as in Eq. (7.36),

$$K_{\mathrm{ae,pe}} = \frac{1}{\tan\alpha_{\mathrm{a,p}}}\frac{\sin(\alpha_{\mathrm{a,p}} \mp \varphi)}{\cos(\delta + \alpha_{\mathrm{a,p}} \mp \varphi)} \pm A_{\mathrm{h}}\left(\frac{TV_{\mathrm{s}}}{H}\right) \frac{\left[\begin{array}{l} m_1\cos(\alpha_{\mathrm{a,p}} \mp \varphi) + m_2\sin(\alpha_{\mathrm{a,p}} \mp \varphi) \\ +(f_{\mathrm{a}} - 1)\{m_3\cos(\alpha_{\mathrm{a,p}} \mp \varphi) + m_4\sin(\alpha_{\mathrm{a,p}} \mp \varphi)\} \end{array}\right]}{2\pi^2\tan\alpha_{\mathrm{a,p}}\cos(\delta + \alpha_{\mathrm{a,p}} \mp \varphi)} \tag{7.36}$$

where,

$$m_1 = \left[2\pi\cos 2\pi\left(\frac{t}{T} - \frac{H}{TV_{\mathrm{s}}}\right) + \left(\frac{TV_{\mathrm{s}}}{H}\right)\left(\sin 2\pi\left(\frac{t}{T} - \frac{H}{TV_{\mathrm{s}}}\right) - \sin 2\pi\left(\frac{t}{T}\right)\right)\right] \tag{7.37a}$$

$$m_2 = \left[2\pi\cos 2\pi\left(\frac{t}{T} - \frac{H}{TV_{\mathrm{p}}}\right) + \left(\frac{TV_{\mathrm{p}}}{H}\right)\left(\sin 2\pi\left(\frac{t}{T} - \frac{H}{TV_{\mathrm{p}}}\right) - \sin 2\pi\left(\frac{t}{T}\right)\right)\right] \tag{7.37b}$$

$$m_3 = \begin{bmatrix} 2\pi\left(\pi\cos 2\pi\left(\frac{t}{T}-\frac{H}{TV_s}\right)+\left(\frac{TV_s}{H}\right)\sin 2\pi\left(\frac{t}{T}-\frac{H}{TV_s}\right)\right) \\ +\left(\frac{TV_s}{H}\right)^2\left(\cos 2\pi\left(\frac{t}{T}\right)-\cos 2\pi\left(\frac{t}{T}-\frac{H}{TV_s}\right)\right) \end{bmatrix} \quad (7.37c)$$

$$m_4 = \begin{bmatrix} 2\pi\left(\pi\cos 2\pi\left(\frac{t}{T}-\frac{H}{TV_p}\right)+\left(\frac{TV_p}{H}\right)\sin 2\pi\left(\frac{t}{T}-\frac{H}{TV_p}\right)\right) \\ +\left(\frac{TV_p}{H}\right)^2\left(\cos 2\pi\left(\frac{t}{T}\right)-\cos 2\pi\left(\frac{t}{T}-\frac{H}{TV_p}\right)\right) \end{bmatrix} \quad (7.37d)$$

The failure angles for both the active and passive wedges, in the failure surface, were obtained by satisfying the equilibrium conditions. The minimum values of the bearing capacity factors with respect to time were obtained from a parametric study. The values of seismic bearing capacity obtained by this method were observed to be higher than those obtained from pseudo-static analysis.

3.3 Seismic Settlement Analysis

Al-Karni (1993) developed an analysis to obtain the seismic settlement of a shallow foundation. A footing will slide when applied seismic acceleration, A_h is more than critical acceleration, A_c. Figure 7.27 shows the proposed sliding block model, assuming plane strain condition, with all the participating driving and resisting forces.

Critical acceleration, A_c, is the acceleration at which the footing just begins to slide. It was assumed that at critical acceleration, the ultimate seismic bearing capacity, q_{uE}, was reduced to the allowable static bearing capacity, i.e.,

$$q_{aS} = q_{uE} \text{ or } q_{uS}/F = q_{uE} \quad (7.38)$$

$$\text{where, } q_{uS} = cN_{cS} + qN_{qS} + 1/2\gamma BN_{\gamma S} \quad (7.39)$$

$$\text{and, } q_{uE} = cN_{cE} + qN_{qE} + 1/2\gamma BN_{\gamma E} \quad (7.40)$$

Critical acceleration can be obtained by solving Eq. (7.38) numerically incorporating Eqs. (7.39) and (7.40). Critical acceleration, A_c, is dependent on F, A_v, D, $c/\gamma B$, and $D_f/\gamma B$. For the case of surface footings, $D_f/B = 0$, for higher values of $c/\gamma B$, the effect of A_v on the critical acceleration was found to be negligible. For the case of $D_f/B = 1$, the effect of A_v is significant for lower values of D. However, when $D \geq 3.0$, the effect of A_v becomes negligible. Usually a factor of safety of 3.0, $F = 3.0$, is adopted to obtain allowable static bearing capacity.

Driving forces, F_D, which act parallel to the sliding plane ac are,

$$\frac{F_D}{L} = (1 - A_v)(qB + W_A)\sin\alpha_2 + A_h(qB + W_A)\cos\alpha_2 \quad (7.41)$$

where, L is the footing length.

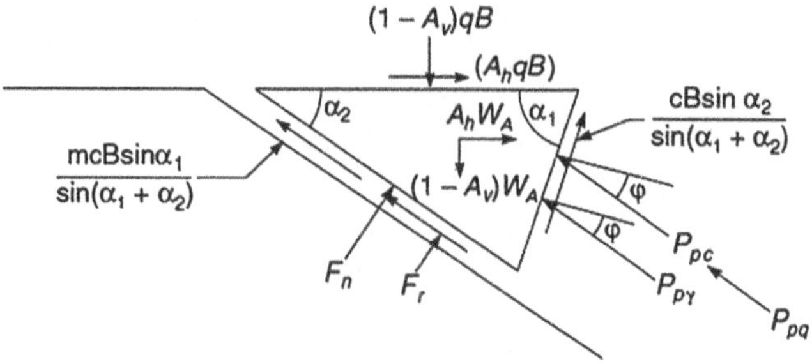

(a)

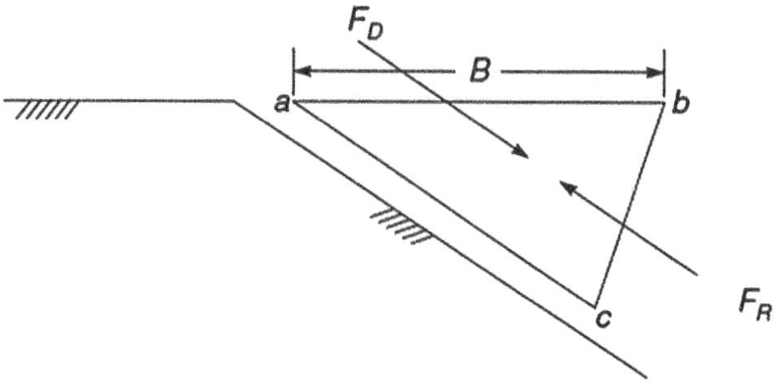

(b)

Fig. 7.27 Sliding-block mechanism (Al-Karni, 1993). (**a**) Forces on the sliding block (**b**) Direction of the driving and resisting forces

The resisting force F_R, which act in the opposite direction to F_D are,

$$\frac{F_R}{L} = (P_{pc} + P_{pq} + P_{p\gamma}) + \frac{mcB\sin\alpha_1}{\sin(\alpha_1+\alpha_2)} + \left[(1-A_v)(qB+W_A)\left(\cos\alpha_2 - \frac{A_h}{1-A_v}\sin\alpha_2\right)\right]\tan\phi \tag{7.42}$$

$P_{p\gamma}$, P_{pq} and P_{pc} are passive pressures and can be calculated using Eqs. (7.43a), (7.43b) and (7.43c).

$$P_{p\gamma} = \frac{\left[\left(N_{\gamma E} + \frac{(1-A_v)H_o}{B}\right)\left(\frac{\gamma B^2}{2}\right) + \frac{H_o\gamma BA_hA_v}{2[\tan(\alpha_2-\phi_m)+A_h]}\right]}{\cos(\alpha_1-\phi) + \frac{\sin(\alpha_1-\phi)-A_h\cos(\alpha_1-\phi)}{\sin(\alpha_2-\phi_m)+A_h\cos(\alpha_2-\phi_m)}\cos(\alpha_2-\phi_m)} \tag{7.43a}$$

$$P_{pq} = \frac{(1-A_v)q_{qE}r_o}{\cos\phi} \times \exp[2(\pi-\alpha_1-\alpha_3)\tan\phi] \times \left(\cos^2\alpha_3 + K_{PE}\sin^2\alpha_3\right) \tag{7.43b}$$

$$P_{\mathrm{pc}} = \frac{2c\sqrt{K_{\mathrm{pE}}}\sin^2\alpha_3}{\cos\phi} \times r_{\mathrm{o}}[\exp 2(\pi - \alpha_1 - \alpha_3)\tan\phi] + \frac{cr_{\mathrm{o}}}{\sin\phi}\{\exp[2(\pi - \alpha_1 - \alpha_3)\tan\phi] - 1\} \tag{7.43c}$$

Additional resisting forces that develop on the sides of the sliding block are,

$$F_{\mathrm{s}} = (2\sigma_{\mathrm{y}}\tan\phi)\frac{1}{2}B^2\frac{\sin\alpha_1\sin\alpha_2}{\sin(\alpha_1+\alpha_2)} \tag{7.44}$$

where, σ_{y}, is the normal stress acting on the sides of the sliding block. Since the problem is considered as a plane strain problem, σ_{y}, may be computed as,

$$\sigma_{\mathrm{y}} = 0.37(1 - A_{\mathrm{v}})\frac{(qB + W_{\mathrm{A}})L}{B \times L}(1 + K_{\mathrm{AE}}) \tag{7.45}$$

The side friction forces are then,

$$\frac{F_{\mathrm{S}}}{L} = 0.74(1 - A_{\mathrm{v}})(qB + W_{\mathrm{A}})(1 + K_{\mathrm{AE}})\tan\phi\frac{B\sin\alpha_1\sin\alpha_2}{2L\sin(\alpha_1+\alpha_2)} \tag{7.46}$$

An addition of Eqs. (7.42) and (7.46) gives,

$$\frac{F_{\mathrm{R}}}{L} = (P_{\mathrm{pc}} + P_{\mathrm{pq}} + P_{\mathrm{p}\gamma}) + \left[(1 - A_{\mathrm{v}})(qB + E_{\mathrm{A}})\left(\cos\alpha_2 - \frac{A_{\mathrm{h}}}{1 - A_{\mathrm{v}}}\sin\alpha_2\right)\right]\tan\phi + \frac{mcB\sin\alpha_1}{\sin(\alpha_1+\alpha_2)} + 0.74\frac{B\sin\alpha_1\sin\alpha_2}{2L\sin(\alpha_1+\alpha_2)} \times (1 - A_{\mathrm{v}})(qB + W_{\mathrm{A}})(1 + K_{\mathrm{AE}})\tan\phi \tag{7.47}$$

From Newton's Law of Motion, the acceleration of the active wedge, abc, relative to plane surfaces, ac, is

$$\frac{(1 - A_{\mathrm{v}})(qB + W_{\mathrm{A}})L}{g}\ddot{x} = F_{\mathrm{D}} - F_{\mathrm{R}} \tag{7.48}$$

where, x is relative displacement measured downward along the slope, ac. Substituting F_{D} and F_{R} in Eq. (7.48), we get,

$$\frac{\ddot{x}}{g} = \left(1 + \frac{A_{\mathrm{h}}}{1 - A_{\mathrm{v}}}\right)\sin\alpha_2 + \left(\frac{A_{\mathrm{h}}}{1 - A_{\mathrm{v}}} - \tan\phi\right)\cos\alpha_2 - \frac{(P_{\mathrm{pc}} + P_{\mathrm{pq}} + P_{p\gamma})}{(1 - A_{\mathrm{v}})(qB + W_{\mathrm{A}})} - \frac{mcB\sin\alpha_1}{\sin(\alpha_1+\alpha_2)(1 - A_{\mathrm{v}})(qB + W_{\mathrm{A}})} - 0.74\chi(1 + K_{\mathrm{AE}})\tan\phi \tag{7.49}$$

The block will start moving when the applied acceleration exceeds the critical acceleration, A_{c} (Fig. 7.28). The solution of Eq. (7.49) will depend on the variation of A_{h} with time above the critical acceleration, A_{c}. Sarma (1975) investigated the effect

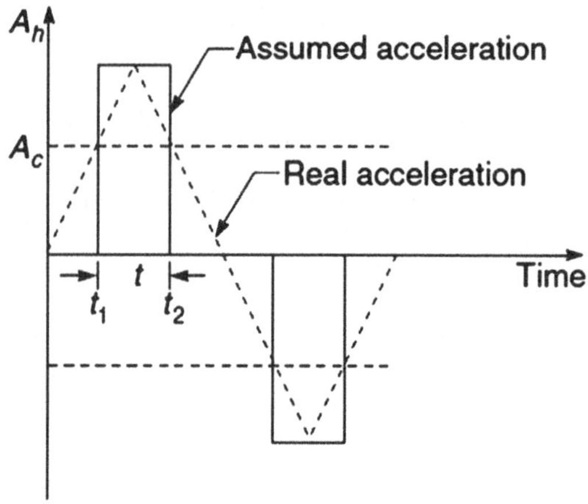

Fig. 7.28 Variation of acceleration with time (Al-Karni, 1993)

of different types of acceleration functions on the calculate displacement by using the sliding block model. Acceleration functions are triangular, rectangular and sinusoidal. On comparison with field data, Sarma (1975) found that the rectangular acceleration function gives a good approximation of the sliding block displacement, when the ratio A_c/A_h is smaller than 0.5. For simplicity, the variation of A_h with time may be assumed as a rectangular function as shown in Fig. 7.28.

By integrating Eq. (7.49) with time and applying the two boundary conditions, namely (i) $t = 0, \dot{x} = 0$ and (ii) $t = 0, x = 0$, the displacement of the block along the side ac for time t is,

$$\frac{2x}{g(\kappa t^2)} = \left(1 + \frac{A_h}{1 - A_v} \tan\phi\right) \sin\alpha_2 + \left(\frac{A_h}{1 - A_v} - \tan\phi\right) \cos\alpha_2 - A_1 - \frac{mcB \sin\alpha_1}{\sin(\alpha_1 + \alpha_2)(1 - A_v)(qB + W_A)} - 0.74\gamma(1 + K_{AE}) \tan\phi \quad (7.50)$$

For evaluating seismic settlement, calculate the allowable static bearing capacity for a desired factor of safety and the ultimate seismic bearing capacity. The seismic settlement will be zero, if the calculated allowable static bearing capacity is less than the seismic bearing capacity. The following procedure should be adopted if the seismic bearing capacity is lesser than the allowable static bearing capacity:

1. Find the critical acceleration, A_c, as explained earlier.
2. Calculate passive pressures P_{pc}, P_{pq} and $P_{p\gamma}$ from Eqs. (7.42a), (7.42b) and (7.42c)respectively, and coefficient of seismic active earth pressure K_{AE} from Eq. (7.31).
3. Value of κ may be assumed to be 0.143; however, its exact value shall be obtained experimentally.
4. Compute time t for half cycle of loading by,

$$t = \frac{1}{2f}\left[1 - \frac{k_c}{k_h}\right] \quad (7.51)$$

where, f is frequency of loading.

5. Substitute the obtained value of t in Eq. (7.50) to get x, the relative displacement of the sliding block in the direction of the slip plane.
6. The seismic vertical displacement for half cycle is computed as,

$$\delta_{sv} = x \sin \alpha_2 \quad (7.52)$$

7. The seismic vertical displacement for half cycle is computed as,

$$\delta_{sh} = x \cos \alpha_2 \quad (7.53)$$

8. Add the calculated vertical settlement to D_f and return to step 1, to obtain the settlement for second half of the cycle.
9. Repeat steps 1–7 for the number of cycles desired.

4 Design Recommendations

For satisfactory performance of retaining walls and shallow foundations, firstly these may be designed using a pseudo-static approach considering adequate factor safety. A displacement analysis should be carried out to estimate seismic settlement.

4.1 Retaining Wall

1. Stability analysis of a wall retaining a dry/moist backfill may be performed using the method proposed by Saran and Gupta (2003).
2. Stability analysis of a wall retaining having a partly submerged backfill may be performed using the method proposed by Saran and Rangwala (2012).
3. The procedure suggested by Richards and Elms (1979) may be used based on permissible wall displacement may be adopted to satisfy the displacement criteria.

4.2 Shallow Foundation

1. Seismic bearing capacity may be computed using the method proposed by Richards et al. (1993)
2. Seismic settlement may be determined using the method suggested by Al-Karni (1993).

References

Al-Atik, L., & Sitar, N. (2010). Seismic earth pressures on cantilever retaining structures. *Journal Geotechnical and Geoenviromental Engineering, 136*(10), 1324–1333.

Al-Karni, A. A. (1993). *Seismic settlement and bearing capacity of shallow footings on cohesionless soil*. Ph.D. thesis, University of Arizona, USA.

Budhu, M., & Al-Karni, A. A. (1993). Seismic bearing capacity of soils. *Géotechnique, 43*(1), 181–187.

Choudhury, D., & Rao, K. S. S. (2005). Seismic bearing capacity of shallow strip footings. *Geotechnical and Geological Engineering, 23*(4), 403–418.

Clough, G. W. & Fragaszy, R. F. (1977). A study of earth loadings on floodway retaining structures in the 1971 San Fernando Valley earthquake. *Proceedings of the 6th WCEE, Sarita Prakashan, Meerut, India* (Vol. 3).

Coulomb, C. A. (1776). *Essai sur une Application des Regles des maximis et minimis a quelques Problemes de statique Relatifs a l'Architecture, Académie Royale des Sciences par divers Savans* (Vol. 7). Paris: De l'Imprimerie Royale.

Culmann, K. (1866). *Die graphische statik*. Zurich: Mayr and Zeller.

Das, B. M., & Puri, V. K. (1996). Static and dynamic active earth pressure. *Geotechnical and Geological Engineering, 14*(4), 353–366.

Das, B. M., & Ramana, G. V. (2010). *Principles of soil dynamics* (2nd ed.). Stanford: Cengage Learning.

Davies, T. G., Richards, R., & Chen, K. H. (1986). Passive pressure during seismic loading. *Journal of Geotechnical Engineering, ASCE, 112*(GT4), 479–484.

Dormieux, L., & Pecker, A. (1995). Seismic bearing capacity of foundation on cohesionless soil. *Journal of Geotechnical Engineering, 121*(3), 300–303.

Ebelling, R. M. & Morrison E. E. (1992). *The seismic design of waterfront retaining structures*. US Army Technical Report ITL-92-11 and US Navy Technical Report NCEL TR-939.

Franklind, A.G., & Chang, F.K. (1977). Permanent displacements of earth embankments by Newmark sliding block analysis. *US Army Waterways Experiment Station*, Misc. Paper S-71-17.

Gazetas, G., Psarropoulos, P. N., Anastasopoulos, I., & Gerolymos, N. (2004). Seismic behavior of flexible retaining systems subjected to short-duration moderately strong excitation. *Soil Dynamics and Earthquake Engineering, 24*, 537–550.

Ghosh, P. (2008). Upper bound solutions of bearing capacity of strip footing by pseudo-dynamic approach. *Acta Geotechnica, 3*(2), 115–123.

Ghosh, P., & Choudhury, D. (2011). Seismic bearing capacity factors for shallow strip footings by pseudo-dynamic approach. *Disaster Advances, 4*(3), 34–42.

IS 1893-Part-III. (2012). *Criteria for earthquake resistant design of structures – part 3 bridges and retaining walls*. New Delhi: Bureau of Indian Standards.

Kapila, I. P. (1962). Earthquake resisting design of Retaining walls. *Proceedings of the 2nd Symposium in Earthquake Engineering, University of Roorkee, Roorkee*.

Lai, C. S. (1979). *Behaviour of retaining walls under seismic loading*. New Zealand: M.E. Report, University of Canterbury.

Lambe, T. W., & Whitman, R. V. (2008). *Soil mechanics SI version*. New Delhi: Wiley India Pvt. Limited.

Mononobe, N., & Matsuo, H. (1929). On the determination of earth pressure during earthquake. *Proceedings of the World Engineering Congress, Tokyo, Japan*, 177–185.

Mononobe, N., & Matsuo, H. (1932). Experimental investigation of lateral earth pressure during earthquakes. *Earthquake Research Institute and Research Office of Public Works*, 884–902.

Nadim, F., & Whitman, R. V. (1983). Seismically induced movement of retaining walls. *Journal of Geotechnical Engineering, 109*(7), 915–931.

Nandakumaran, P. (1973). *Behaviour of retaining walls under dynamic loads*. Ph.D. thesis, University of Roorkee, Roorkee, India.

Newmark, N. (1965). Effects of earthquakes on dams and embankments. *Geotechnique, 15*(2), 139–160.

Okabe, S. (1924). General theory on earth pressure and seismic stability of retaining wall and dam. *Journal of the Japanese Society of Civil Engineering, 10*(6), 1277–1323.

Prakash, S., Puri, V. K., & Khandoker, J. U. (1981). Rocking displacements of retaining walls during earthquake. *International Conference on Recent Advances in Geotechanical Earthquake Engineering and Soil Dynamics, St. Louis, U.S.A.* (Vol. III, 1021–1025).

Prakash, S., & Saran, S. (1966). Static and dynamic earth pressures behind retaining walls. *Proceedings of the 3rd symposium on earthquake engineering, Roorkee,* I, 277–288.

Prandtl, L. (1921). Über die Eindringung festigkeit (Härte) plastischer Baustoffe und die Festigkeitvon Schneiden (in German). ZAMM – Journal of Applied Mathematics and Mechanics/Zeitschrift für Angewandte Mathematik und Mechanik, 1(1), 15–20.

Reddy, R. K., Saran, S., & Viladkar, M. N. (1985). Prediction of displacements of retaining walls under dynamic conditions. *Bulletin of Indian Society Earthquake Technology.*, Paper No. 239, *22*(3), 101–115.

Richards, R., Jr., & Elms, D. G. (1979). Seismic behavior of gravity retaining walls. *Journal of Geotechnical Engineering, ASCE, 105*(GT4), 449–464.

Richards, R., Elms, D. G., & Budhu, M. (1993). Seismic bearing capacity and settlements of foundations. *Journal of Geotechnical Engineering, 119*(4), 662–674.

Richards, R., & Shi, X. (1994). Seismic lateral pressures in soils with cohesion. *Journal of Geotechnical and Geoenvironmental Engineering, 120*(7), 1230–1251.

Saran, S., & Gupta, R. P. (2003). Seismic earth pressure behind retaining walls. *Indian Geotechnical Journal, 33*(3), 195–213.

Saran, S., & Prakash, A. (1971). Seismic pressure distribution in earth retaining walls. *Proceedings of the Third European Symposium on Earthquake Engineering, Bulgaria Academy of Science, Bulgaria* , 355–362.

Saran, S., & Prakash, S. (1968). Dimensionless parameters for static and dynamic earth pressures behind retaining walls. *Journal, Indian National Society of Soil Mechanics and Foundation Engineering, 7*, 295–310.

Saran, S., & Rangwala, H. M. (2012). Seismic earth pressure in yielding walls gravity retaining partially submerged cohesionless backfill. *International Journal of Geotechnical Engineering, 6* (3), 309–318. Accepted for publication in July Issue.

Saran, S. K., & Rangwala, H. (2011). Seismic bearing capacity of footings. *International Journal of Geotechnical Engineering, 5*(4), 475–483.

Sarma, S. K. (1975). Seismic stability of earth dams and embankments. *Geotechnique, 25*(4), 743–761.

Sarma, S. K., & Iossifelis, I. S. (1990). Seismic bearing capacity factors of shallow strip footings. *Géotechnique, 40*, 265–273.

Seed, H. B., & Whitman, R. V. (1970). Design of earth retaining structures of dynamic loads. *Proceedings of the Speciality Conference on Lateral Stresses in the Ground and Design of Earth Retaining Structures, ASCE,* 103–147.

Shukla, S. K., Gupta, S. K., & Sivakugan, N. (2009). Active earth pressure on retaining wall for c-ϕ soil backfill under seismic loading condition. *Journal of Geotechnical and Geoenvironmental Engineering, 135*(5), 690–696.

Shukla, S. K., & Habibi, D. (2011). Dynamic passive pressure from c-ϕ soil backfills. *Soil Dynamics and Earthquake Engineering, 31*(5–6), 845–848.

Shukla, S. K., Nagaratnam, S., & Das, B. M. (2011). Analytical expression for dynamic passive pressure from c–ϕ soil backfill with surcharge. *International Journal of Geotechnical Engineering, 5*(3), 357–362.

Shukla, S. K., & Zahid, M. (2011). Analytical expression for dynamic active earth pressure from c-ϕ soil backfill with surcharge. *International Journal of Geotechnical Engineering, 5*(2), 143–150.

Soubra, A. H. (1999). Upper-bound solutions for bearing capacity of foundations. *Journal of Geotechnical and Geoenvironmental Engineering, 125*(1), 59.

Steedman, R. S., & Zeng, X. (1990). The influence of phase on the calculation of pseudo-static earth pressure on a retaining wall. *Géotechnique, 40*(1), 103–112.

Zarrabi, K. (1979). *Sliding of gravity retaining wall during earthquakes considering vertical acceleration and changing inclination of failure surface*. M.S. thesis, MIT, USA.

Richards, R. Jr. and Elms, D.G. (1979). Seismic behavior of gravity retaining walls. *Journal of Geotechnical Engineering, ASCE,* 105(GT4), 449–464.

Saran, S. (2012). Analysis and design of foundations and retaining structures subjected to seismic loads. *I K International Publishing House Pvt. Ltd, New Delhi.*

Chapter 8
Engineering of Ground for Liquefaction Mitigation

A. Murali Krishna and M. R. Madhav

Notations

a	Radius of the granular pile
b	Radius of the unit cell
d	Diameter of granular pile
d_c	Dilation coefficient
RGP	Rammed granular pile
k_h	Horizontal permeability of untreated ground
k_s	Permeability of ground
$k_h(r)$	Horizontal permeability of treated ground
k_w	Permeability of drain (well)
L	Length of pile
$L_w = \frac{32}{\pi^2}\frac{k_s}{k_w}\left(\frac{H}{d_w}\right)^2$	drain resistance
m_v	Coefficient of volume change of untreated ground
$m_v(r)$	Coefficient of volume compressibility/volume change of treated ground
N_{eq}	Equivalent number of uniform stress cycles induced by earthquake
N_l	Number of uniform stress cycles required to cause liquefaction
R	Non-dimensionalized radial distance, r/b
R_{ma}	Normalized coefficient of volume change at near end, $m_v(a)/m_v$
R_{mb}	Normalized coefficient of volume change at farthest end, $m_v(b)/m_v$
R_{ka}	Normalized coefficient of permeability at near end, $k_h(a)/k_h$

A. Murali Krishna (✉)
Department of Civil Engineering, Indian Institute of Technology Guwahati, Guwahati, India
e-mail: amurali@iitg.ernet.in

M. R. Madhav
Department of Civil Engineering, I.I.T. Hyderabad & JNTU, Hyderabad, India

M. L. Sharma et al. (eds.), *Advances in Indian Earthquake Engineering and Seismology*, https://doi.org/10.1007/978-3-319-76855-7_8

R_{kb}	Normalized coefficient of permeability at farthest end, $k_h(b)/k_h$
r	Radial distance measured from the center of granular pile
r_N	Cyclic ratio
s	Diameter of unit cell
S	Spacing between the granular piles
SCP	Sand compaction piles
SPT N	Standard penetration test number
T	Normalized time, t/t_d
t	Time
$T_{ad} = \left(\frac{k_h t_d}{\gamma_w m_v}\right) \cdot \frac{1}{a^2}$	dimensionless time factor
$T_{bd} = \left(\frac{k_h t_d}{\gamma_w m_v}\right) \cdot \frac{1}{b^2}$	dimensionless time factor
t_d	Duration of earthquake
u	Excess hydrostatic pressure
u_g	Excess hydrostatic pressure generated by earthquake shaking
W or r_u *or* Ru_z	Pore pressure ratio ($= u/\sigma'_{v0}$)
W_{max}	Maximum pore pressure ratio W throughout the layer at a given T

1 Introduction

Earthquakes are constantly posing risk to life and infrastructure facilities. Among the various seismic hazards, liquefaction is probably the most disastrous one leading to huge damage to structures and human life. Liquefaction, the state under which soil deposit loses its strength and flows as fluid, is a major cause for damage during earthquakes. If the ground surface is inclined (sloping ground surface), liquefaction also leads to lateral spreading. Liquefaction is manifested by the formation of sand boils and mud spouts at the ground surface, by seepage of water through ground cracks or in some cases by the development of quicksand conditions over substantial areas (Seed and Idriss, 1982). The 1819 Rann of Kutch earthquake and the recent 2001 Bhuj/Kutch earthquakes have witnessed large areas that suffered damages due to liquefaction (Shankar, 2001).

Bhuj 2001 earthquake boosted the research activities in India on geotechnical earthquake engineering particularly on liquefaction, its evaluation and mitigation. Various themes of research activities in connection with liquefaction can be grouped into: laboratory evaluation of liquefaction initiation and post-liquefaction strength (Sitharam et al., 2004); effects of liquefaction on foundations (Bhattacharya, 2006; Maheshwari et al., 2008), and liquefaction mitigation (Murali Krishna et al., 2006). Sitharam et al. (2004) conducted experimental study on the undrained shear behavior of silty sands collected from the locations of Bhuj earthquake focusing on the effects of silt content and confining stress on the post-liquefaction residual shear strength.

Basically, three approaches are available to reduce liquefaction hazards when designing and constructing new structures (JGS, 1998). The first one is to avoid construction on liquefaction-susceptible soils; the second one is to build liquefaction-resistant structures. The third possibility is to strengthen the ground, by improving the strength, density, and drainage characteristics of soils, using a variety of soil improvement techniques. The third one is considered as the best strategy and is the most preferred choice. In short, ground engineering techniques are commonly employed to mitigate liquefaction and other seismic hazards. Resistance to liquefaction can be improved by increasing the density, modifying the grain size distribution, stabilizing the soil fabric, reducing the degree of saturation, dissipation of the excess pore pressures generated and intercepting the propagation of excess pore pressures, etc. The most common methods to achieve the above and to improve the engineering properties of the soils can be classified as densification, reinforcement, grouting/mixing and drainage.

This chapter presents a review of the fundamental aspects of the liquefaction and its evaluation followed by discussions on the application of ground engineering methods to mitigate liquefaction hazards. Recent developments on the application of granular inclusions in the form of stone columns/granular piles as general ground improvement method as well as seismic hazard mitigation measure are presented.

2 Liquefaction

One of the major consequences of earthquakes is the phenomenon of "liquefaction." A significant amount of work has been performed in the last few decades on liquefaction and its evaluation and remediation (Martin et al., 1975; Seed and Booker, 1977; Seed, 1979; Ishihara, 1993; Seed et al., 1975, 1976, 2003; Youd et al., 2001; Cetin et al., 2004; Sawicki and Mierczynski, 2006; Madabhushi, 2007; Murali Krishna and Madhav, 2008; Idriss and Boulanger, 2008; Murali Krishna, 2011; etc.). Liquefaction occurs in saturated loose sandy deposits if subjected to seismic forces. Deposits of loose granular soils get affected by ground vibrations induced by an earthquake, resulting in large total and differential settlements of the ground surface. In cases of deposits with high water table, the tendency to get densified may result in the development of excess hydrostatic pore water pressures of such magnitude to cause liquefaction of the soil, resulting in excessively large settlements and tilting of structures.

The pore pressure in the soil increases under repeated earthquake forces. Consequently, the effective stress decreases and if the value of the latter approaches zero, the deposit loses its strength completely. Even if the effective stresses do not reduce to zero, the ground becomes very soft and large strains, deformations, and lateral flows result. Castro (1975) distinguishes two different phenomena that were referred to as "liquefaction" and "cyclic mobility." Liquefaction was referred to as complete loss of shear strength and can occur in loose sands. Cyclic mobility is the gradual increase of cyclic strains without complete loss of shear strength. It could occur in loose to medium dense sands, silty sands, etc. The different types of ground failures caused by liquefaction may be defined or identified and vary in intensity, placement in environment, and consequence. They are sand boils, lateral spreads, flow failures, loss of bearing strength and ground oscillations (Ishihara, 1993). Field observations of various forms of liquefaction under different in situ stress conditions and soil behavior are listed in Table 8.1.

Table 8.1 Different forms of liquefaction field observations

In situ Stress Condition	Soil Behavior	Typical Field Observation
No driving shear stress	• Volume decrease • Pore pressure increase	• Ground settlement • Sand boils and ejection from surface fissures
Driving shear greater than residual strength	• Loss of stability • Liquefaction	• Flow slides, sinking of heavy buildings • Floating of light structures
Driving shear less than residual strength	• Limited shear distortion • Soil mass remains stable	• Slumping of slopes, settlement of buildings, lateral spreading

2.1 Liquefaction-Susceptible Soils

It has been considered that relatively "clean" sandy soils having small percentages of fines are susceptible to seismically induced liquefaction. However, significant liquefaction experiences over the years expanded the limits of the soil types that are susceptible to liquefaction and associated phenomena. Figure 8.1 shows the grain size distribution of soil having the possibility of liquefaction (PIANC, 2001).

It was well documented that coarse, gravelly soils were also subjected to liquefaction (Andrus, 1994; Seed et al., 2003) under certain state and drainage conditions. In addition, cohesive soils were also identified as potentially liquefiable soils under some conditions based on their fines content, plasticity, liquid limit and natural water content (Andrews and Martin, 2000). Seed et al. (2003) recommend the zones of potentially liquefiable soils with significant fines content as shown in Fig. 8.2, and the recommendations are "*(1) Soils within Zone A are considered potentially susceptible to 'classic' cyclically induced liquefaction, (2) Soils within Zone B may be liquefiable, and (3) Soils in Zone C (not within Zones A or B) are not generally susceptible to 'classic' cyclic liquefaction, but should be checked for potential sensitivity (loss of strength with remoulding or monotonic accumulation of shear deformation).*"

2.2 Liquefaction Evaluation

The liquefaction potential of any given soil deposit is determined by a combination of the properties of soils, environmental factors and characteristics of the earthquake to which they may be subjected to. Simplified techniques based on in situ test measurements are commonly used to assess seismic liquefaction potential. Liquefaction potential can be evaluated based on various in situ tests like Standard Penetration Test (SPT), Cone Penetration Test (CPT), shear wave velocity measurement, and Becker Penetration Test (BPT) for gravely soils. Most of the simplified

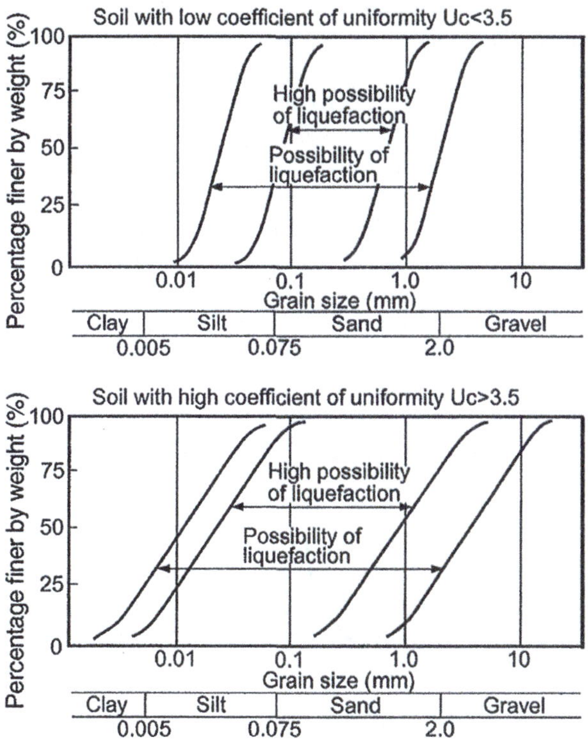

Fig. 8.1 Range of grain size distribution for soils susceptible to liquefaction (PIANC, 2001)

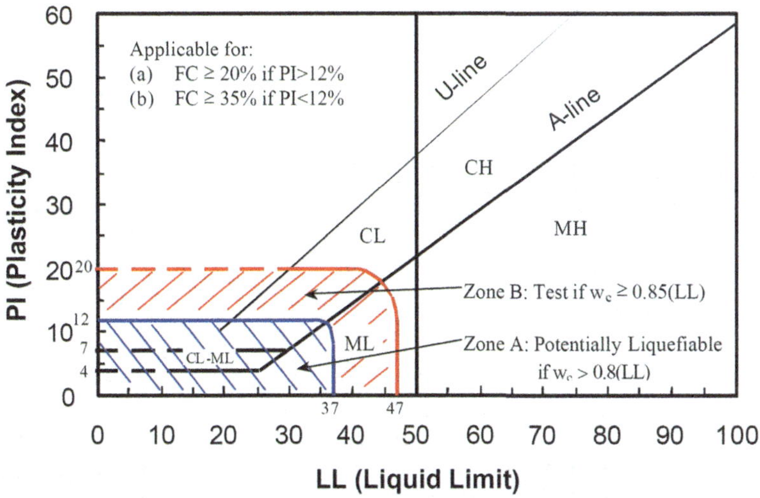

Fig. 8.2 Zones of potentially liquefiable soils with significant fines content (after Seed et al., 2003)

charts and equations rely on the analyses of liquefaction case histories. Using empirical, simple regression, and statistical methods, a boundary (liquefaction curve) or classification technique is obtained to separate the occurrence or nonoccurrence of liquefaction. Figure 8.3 shows such boundary curves between liquefiable and nonliquefiable soils based on different in situ tests.

There are different probabilistic and deterministic approaches also in practice for liquefaction hazard assessment. These methods can be grouped into simplified stress-based or strain-based methods, laboratory tests: stress-controlled or strain-controlled methods, etc. A detailed summary on the different liquefaction evaluation methods and information on magnitude of scaling factors, correction factors for overburden pressure, slope of the ground and input values for earthquake magnitude

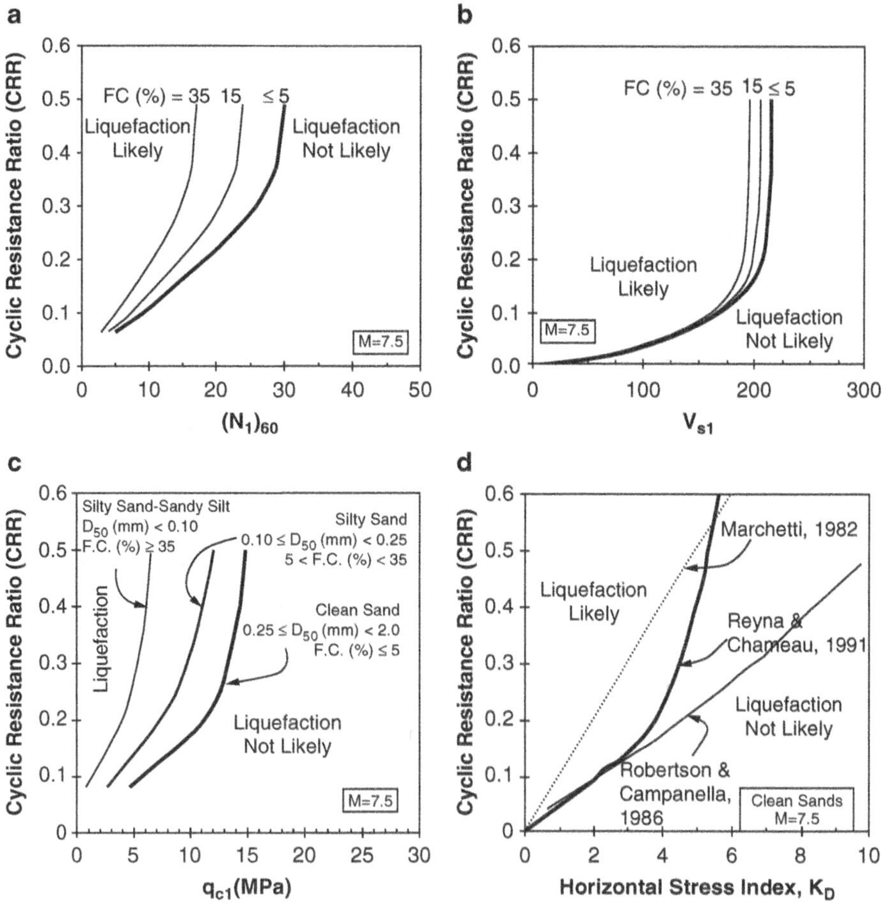

Fig. 8.3 Observed boundaries for "Liquefy–No Liquefy" curves from various in situ tests: (**a**) SPT (NCEER, 1997), (**b**) shear wave velocity (Andrus and Stokoe, 1997), (**c**) CPT (Stark and Olson, 1995), and (**d**) dilatometer test (Reyna and Chameau, 1991)

and peak acceleration, etc., is given by Youd et al. (2001), Seed et al. (2003) and Idriss and Boulanger (2008). Assessment of liquefaction, using probabilistic methods based on SPT (Cetin et al., 2004) and CPT (Moss et al., 2006), is also available.

3 Liquefaction Mitigation Methods

The basic approach for earthquake disaster mitigation can be broadly classified into two major categories, viz., (1) Preventing or minimizing the probability of liquefaction (ground improvement) and (2) Minimization of damages in the event of liquefaction (structural improvement). The basic strategy for the liquefaction mitigation is demonstrated in Fig. 8.4 (PIANC, 2001). The former, soil improvement/ engineering of ground, can further be divided into measures that will improve (a) Ground characteristics and (b) Condition of stress, deformation and pore pressure in the ground. The ground can be engineered to withstand the expected intensity of earthquake by; (i) densification, (ii) solidification, (iii) replacement of part of the soil, or (iv) lowering of the ground water table. The in situ stress and pore water pressure

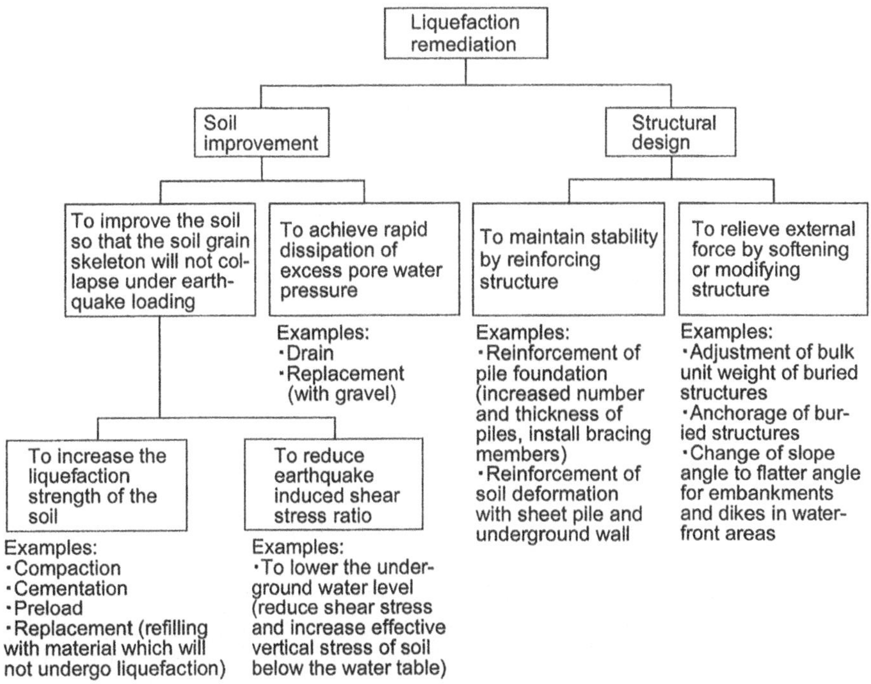

Fig. 8.4 Strategy to be considered for liquefaction remediation (PIANC 2001)

conditions can be controlled by (v) increasing the effective stress, (vi) permitting the dissipation of excess pore pressure, or (vii) restricting the shear deformations.

If liquefaction cannot be prevented, its effect on damages to the structures founded on such ground that possibly can get liquefied can be minimized by (1) Strengthening the ground by pile foundations, (2) Strengthening the foundation itself, (3) Reducing lift-off of underground structures, (4) Providing flexible structure that will absorb ground deformations, and (5) Limiting or controlling post-liquefaction ground displacements.

4 Ground Engineering for Liquefaction Mitigation

Various types of remediation methods against liquefaction have been developed in the last few decades and are being applied to structures. Remediation methods may need to be applied to new structures as well as old existing structures. Various ground engineering methods can be grouped based on the mechanisms to mitigate the liquefaction potential and damage. They are mainly: densification, solidification, replacement, drainage, and others include lowering of ground water table and shear strain restraint method. Figure 8.5 shows the various principles and techniques under each principle for liquefaction mitigation (JGS, 1998). Among the various techniques shown in Fig. 8.5, more widely used methods for liquefaction mitigation are: Vibro methods (Vibro-rod, Vibro-compaction, Vibro-replacement), Deep dynamic compaction, Compaction grouting, Deep soil mixing, Jet grouting, Drainage, Permeation grouting, Explosive compaction and removal and replacement (Idriss and Boulanger, 2008). Figure 8.6 (Mitchell, 2008) shows the applicability of various ground engineering methods for the soils most susceptible to liquefaction.

Among these different ground engineering methods that are applicable to liquefiable soils, the most commonly used methods, like stone columns/granular piles/drains, sand compaction piles, deep soil mixing, and dynamic compaction are discussed in the following sections.

4.1 Stone Columns/Granular Piles

Granular inclusions in the form of granular drains/piles or stone columns are the most widely preferred alternative, among the various ground engineering options, due to several advantages associated with them. Rammed Granular Piles (RGP) function as drains and permit rapid dissipation of earthquake-induced pore pressures by virtue of their high permeability. The generated pore water pressure due to repeated loading may get dissipated almost as fast as they are generated. In addition, they tend to dilate as they get sheared during an earthquake event. Seismic forces which tend to generate positive pore pressures in these deposits cause an opposite effect of dilation in dense granular piles. One of the chief benefits of ground

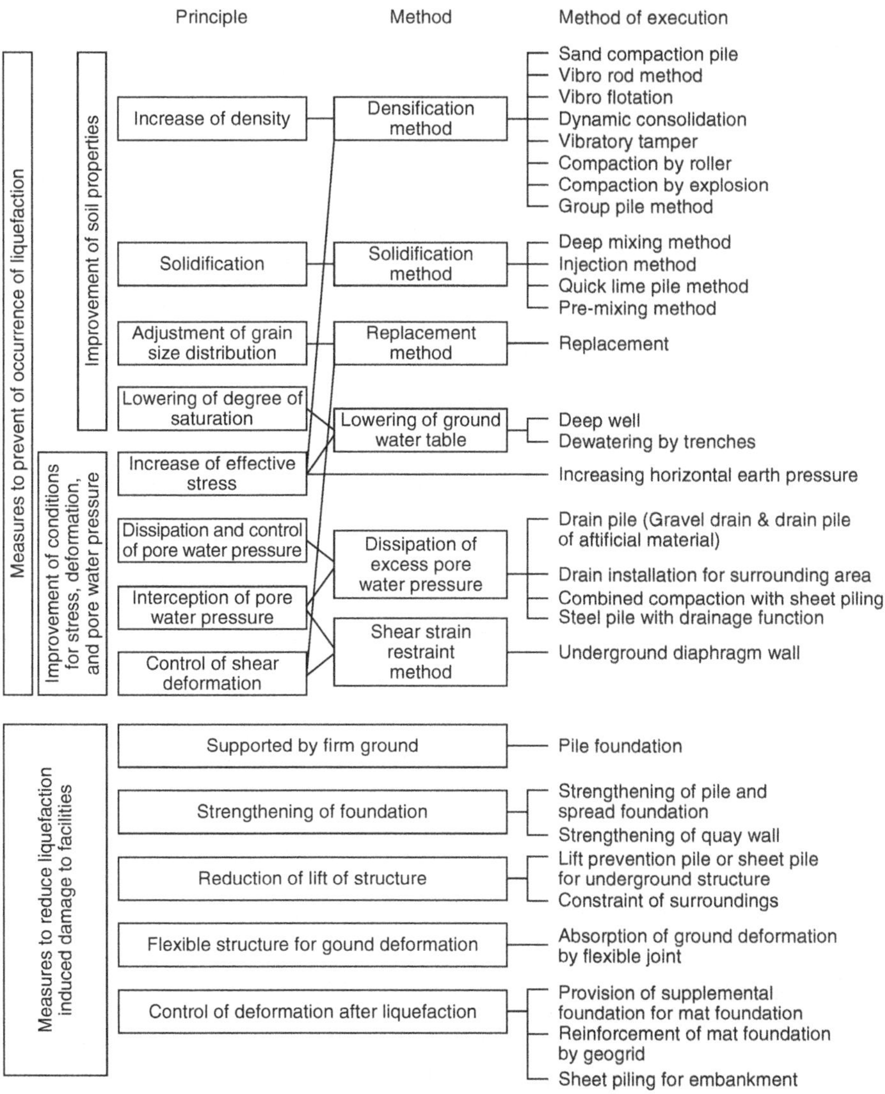

Fig. 8.5 Principles and their techniques of ground engineering for liquefaction remediation (after JGS 1998)

treatment with granular piles is the densification of in situ ground by which the in situ properties of the ground get modified to mitigate the seismic risks, especially, liquefaction potential. Densification by RGP causes increase in deformation moduli and decrease in the coefficients of permeability and volume change. The densification effect decreases with distance from the center of the compaction point and may become negligible at the periphery of unit cell. Further, the very high deformation modulus and stiffness of the granular pile material provide reinforcement for the in

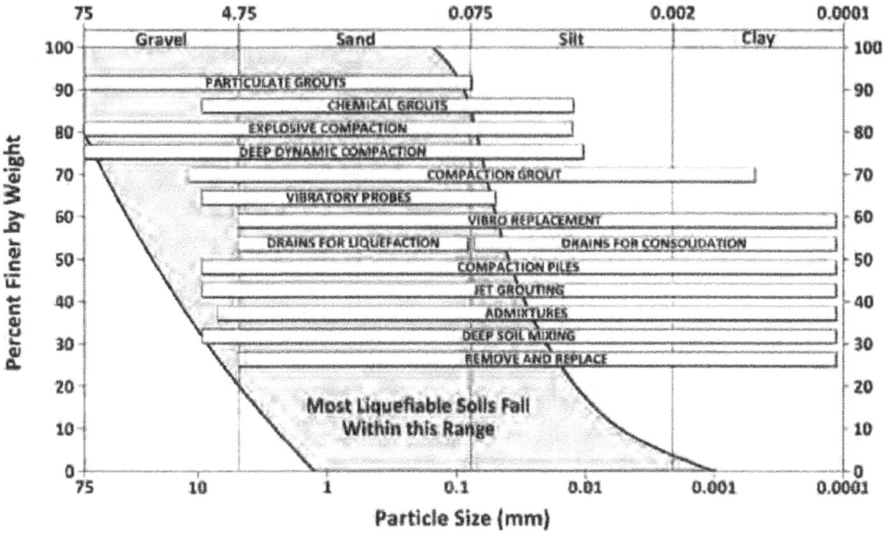

Fig. 8.6 General applicability of ground improvement methods for most liquefiable soils (Mitchell, 2008)

situ soil and offer another mechanism to mitigate liquefaction and participate in resisting lateral spreading. Thus, different mechanisms operate in the functioning of stone columns/granular piles in liquefaction mitigation. These mechanisms can be stated as drainage, storage, dilation, densification and reinforcement. It is observed that granular piles are effective in mitigating liquefaction damage due to the reinforcement effect and drainage facility. Provision of gravel drains/granular piles/stone columns is the most commonly adopted ground treatment methodology for liquefaction mitigation and it has proved its effectiveness in many instances (Mitchell and Wentz, 1991). Though granular piles are efficient in many ways in mitigating the seismic risks, liquefaction mitigation is the main advantage as it is associated with the drainage function which is a very special feature as granular drains. Adalier and Elgamal (2004) summarized some of the field case histories on the use of stone columns as liquefaction countermeasure. An overview of the various mechanisms that function by granular piles in mitigating the liquefaction was presented by Madhav and Murali Krishna (2008) and Murali Krishna (2011).

Various techniques of installation have been conceived for various types of columnar inclusions in a wide variety of soils such as loose sandy to soft compressible soils depending on technical ability, efficiency and local conditions. Granular piles are installed by vibro-compaction, vibro-replacement, cased borehole (rammed stone columns/RGP), or simple auger boring methods (Datye and Nagaraju, 1975, 1981; Balaam and Booker, 1981; Barksdale and Bachus, 1983; JGS, 1998). Figure 8.7 shows the common methods followed for construction of stone columns.

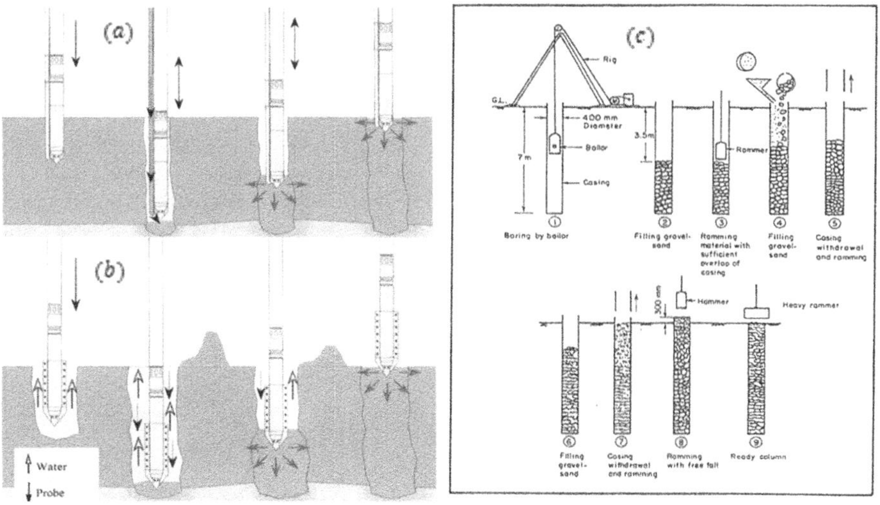

Fig. 8.7 (**a**) Dry bottom-feed; (**b**) Wet top-feed method process (JGS, 1998), (**c**) Cased-borehole method (after Datye and Nagaraju, 1975)

The diameter of the granular pile, spacing between the granular piles, pattern of the pile installation, information about the length of the pile (termination depth) and gradation details of the granular material etc. need to be defined in the design process. Barksdale and Bachus (1983) presented design aspects to be considered while designing the stone columns. Priebe and Keller (1995) demonstrated the design concepts for vibro-replacement. IS 15284 and JGS 1998 are the other two good resources for granular pile design. Design curves developed by Seed and Booker (1977) are used in general for sizing the granular/gravel drains as a liquefaction remediation measure considering the drainage mechanism. To consider the densification effect in connection to liquefaction mitigation, improved SPT N values or cone tip resistance values could be correlated to cyclic resistance ratio and design accordingly (Baez, 1995).

Stone columns function in three fundamental mechanisms as a seismic counter measure. Installation of stone columns densifies the surrounding in situ soil (Murali Krishna and Madhav, 2007, 2009; Murali Krishna et al., 2007), thereby improving the resistance to earthquake-induced shear stresses. Further, stone columns share a significant proportion of the seismic shear stresses due to their higher shear resistance as a reinforcement mechanism. Besides these basic seismic resistance mechanisms, if the in situ soil and existing ground water level conditions lead to liquefaction condition, the stone columns function as drains in dissipating excess pore water pressures developed during seismic loading.

In many instances, it is required to design the stone columns to prevent liquefaction. Design charts developed by Seed and Booker (1977) are being extensively used for this purpose (Fig. 8.8). Figure 8.8 shows the variation of the greatest pore pressure ratio W_{max}, developed as a function of the spacing ratio, *a*/*b*, for values of

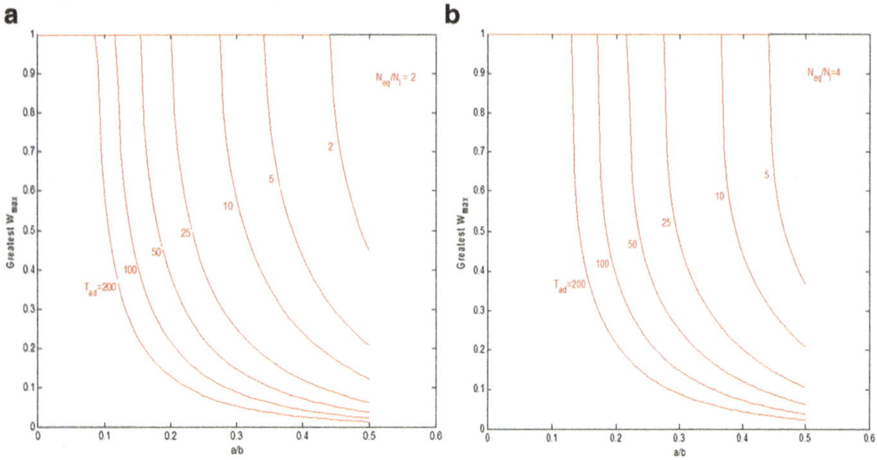

Fig. 8.8 Relationship between greatest pore pressure ratio and drain system parameters for (**a**) $N_{eq}/N_l = 2$ and (**b**) $N_{eq}/N_l = 4$ (redrawn after Seed and Booker, 1977)

N_{eq}/N_l equal to 2 and 4 and for a range of values of the dimensionless time factor $T_{ad} = (k/\gamma_w)[t_d/(m_v a^2)]$. For any particular soil and a selected diameter of stone column, N_{eq}/N_l and T_{ad} will be known, and thus, the value of a/b corresponding to allowable value of W_{max} can be read directly from the curves.

Over the years, the original analytical framework went through various modifications incorporating other complementary effects of granular inclusions in addition to the drainage effect. Stone columns improve the ground by reinforcement, densification of the surrounding soil apart from providing drainage; they dilate during seismic events generating negative pore water pressure for small instances and provide some storage. Thus, the various mechanisms of gravel drains/granular piles in liquefaction mitigation can be listed as drainage, storage, dilation, densification and reinforcement.

Design diagrams by Iai and Koizumi (1986) and Onoue (1988) incorporated the effects of drain resistance in the analyses of Seed and Booker (1977) (Fig. 8.9). Baez and Martin (1992) presented an evaluation of the relative effectiveness of stone columns for the mitigation of liquefaction of soil. Pestana et al. (1998) analyzed the development of excess pore pressure in a layered soil profile, accounting for vertical and horizontal drainage with a nonconstant "equivalent hydraulic conductivity", and head losses due to horizontal flow into the drain and the presence of a reservoir directly connected to the drain were considered. Boulanger et al. (1998) evaluated the drainage capacity of stone columns or gravel drains for mitigating liquefaction hazards. Dilation effect of the stone columns, due to densification around and within the stone columns, on the drainage function of granular piles was studied by Madhav and Arlekar (2000) by extending the Seed and Booker model (1977). It was shown that the dilation effect on pore pressure dissipation by granular piles for the range of parameters considered is marginal. Poorooshasb et al. (2000) demonstrated the

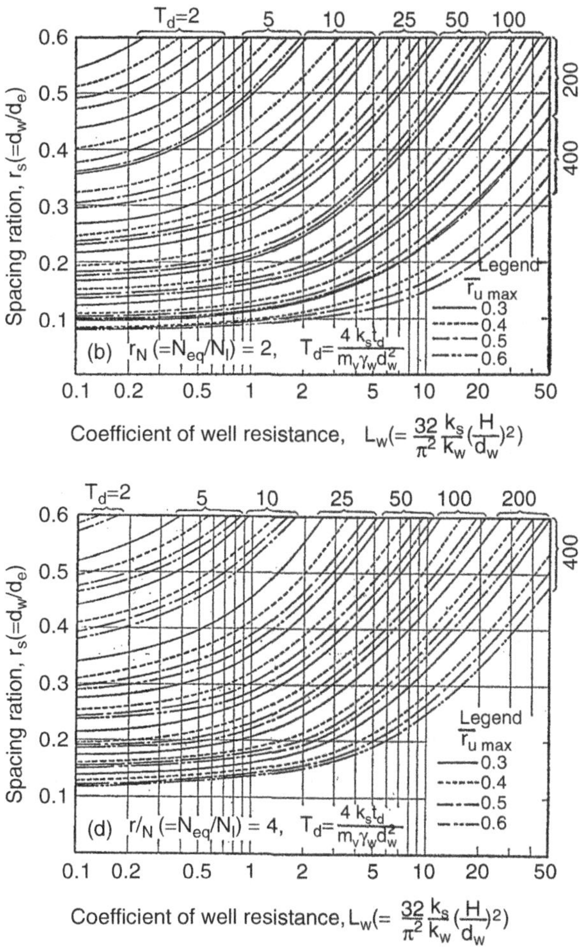

Fig. 8.9 Diagrams for designing the gravel drains considering well resistance (after Onoue, 1988)

effectiveness of inclusion of stone columns in reducing the risk of liquefaction of very loose to loose sandy and silty sand layers using the concept of equivalent permeability.

Poorooshasb et al. (2006) and Noorzad et al. (2007) demonstrated the reinforcement effect of stone columns while analyzing their performance during an earthquake. They proposed that the seismic load imposed on the soil is shared between the stone column and the surrounding ground and stone column carries the major load. Murali Krishna et al. (2006) studied the densification effect with respect to the coefficients of permeability and volume change at the near and at the farthest ends of the granular pile, individually and together on maximum pore pressure variations during an earthquake event (Fig. 8.10).

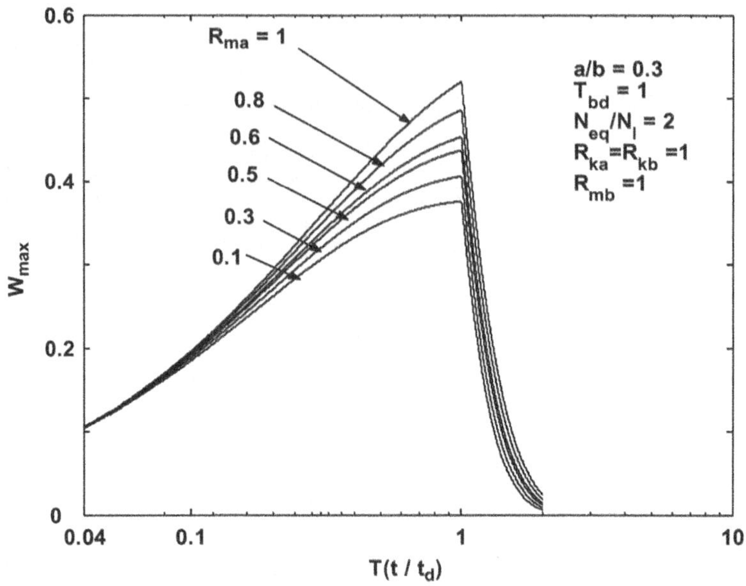

Fig. 8.10 Effect of densification at near end (R_{ma}) on W_{max} (after Murali Krishna et al., 2006)

Murali Krishna and Madhav (2008) combined both the densification and dilation effects and incorporated them in the analysis of pore pressure generation and dissipation (Fig. 8.11). They also quantified the effect of the type of variation (linear or exponential) of the parameters with distance on maximum pore pressure ratios and concluded that the pore pressure ratios are not sensitive to the type of variation of permeability with distance (Fig. 8.12). Recently, Bouckovalas et al. (2009, 2011) considered sand fabric evolution effects on drain design for liquefaction mitigation. They proposed that overlooking the shakedown effects of fabric evolution during cyclic loading underestimates the effectiveness of gravel drains.

Figure 8.13 presents the maximum pore pressure ratio variation with the time for different effects, namely: (1) No densification and no soil fabric effect, similar to the original Seed and Booker (1977) model; (2) Densification effect only (considering the change in flow parameters at the near only with $R_{ka} = R_{ma} = 0.8$); (3) Soil fabric evolution effect only; and (4) densification and soil fabric effects together. $T_{bd} = 1$, $a/b = 0.5$, and $N_{eq}/N_l = 2$ are considered for the purpose. The maximum pore ratios are about 0.64, 0.85, 0.43, and 0.53 for the above effects considered in that order, respectively. This implied that the densification effect alone raised the maximum pore water pressure (W_{max}) by about 30% and soil fabric effect reduced the maximum pore water pressure by about 32% whereas the combined effects effectively reduced the W_{max} by about 17%. It is recommended to use the combination of various effects and mechanisms in the analysis and design of stone columns as seismic risk mitigation elements.

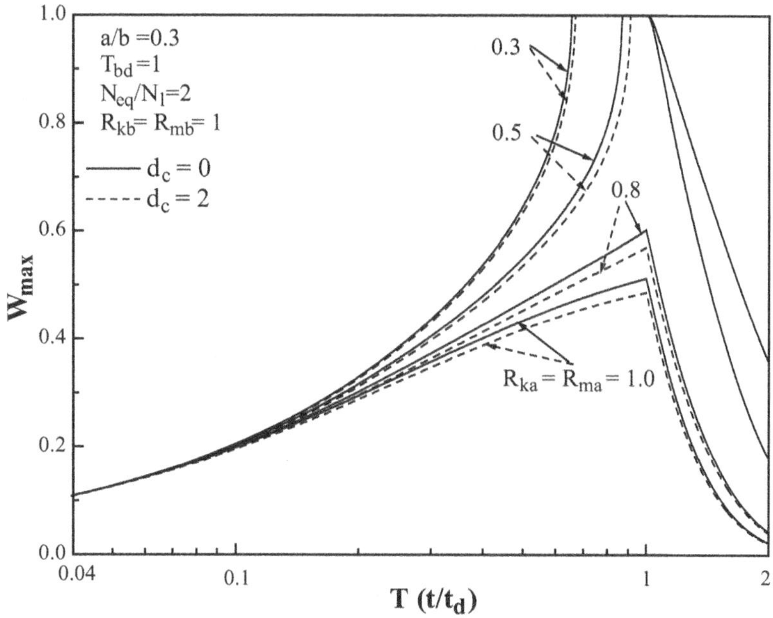

Fig. 8.11 Effect of densification with respect to R_{ka} and R_{ma} and dilation on W_{max} (after Murali Krishna and Madhav, 2008)

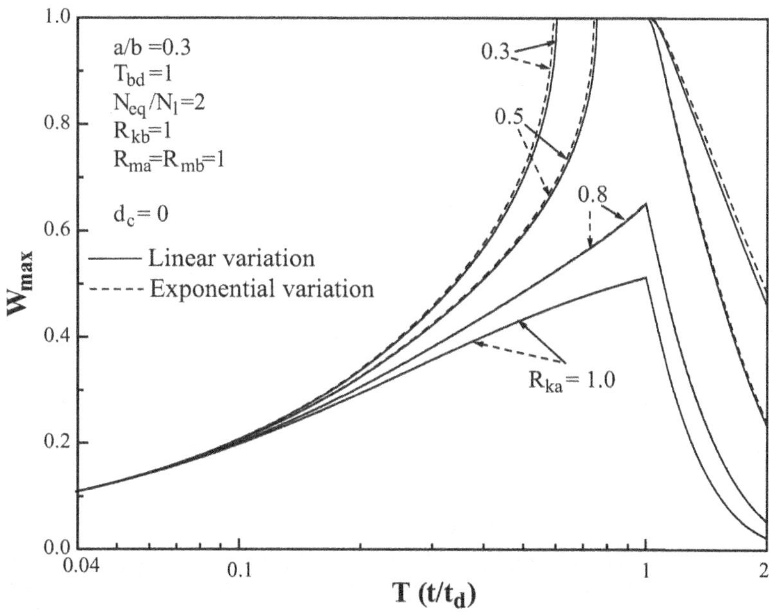

Fig. 8.12 Effect of R_{ka} on W_{max} for linear and exponential variations

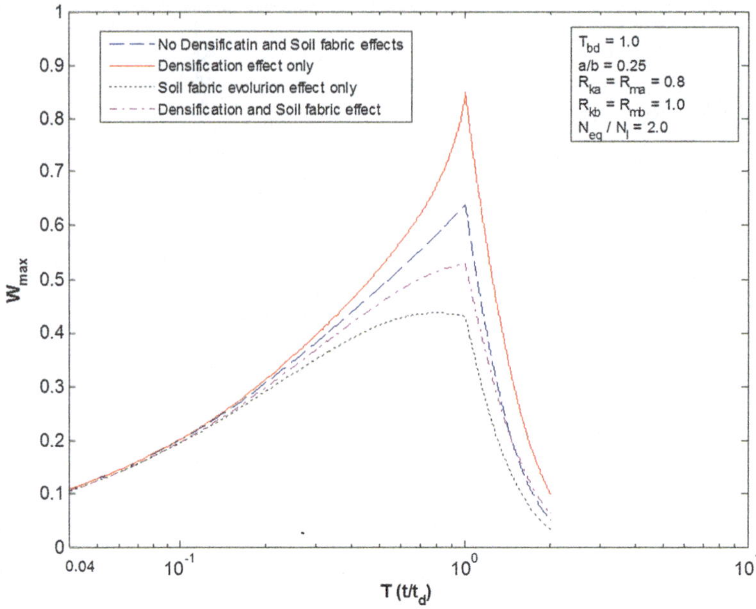

Fig. 8.13 Comparison of the maximum excess pore water pressure ratios for different effects

4.2 Sand Compaction Piles

One of the most commonly used methods under densification principle is sand compaction pile (SCP) method to densify loose sandy deposits and to improve their capacity in terms of liquefaction resistance. Typical SCP construction procedure is demonstrated schematically in the Fig. 8.14 (Okamura et al., 2006). A casing pipe of about 0.4 m dia. is penetrated and sand is placed in the casing pipe from top. After reaching final penetration depth, casing pipe is withdrawn by 0.5 m and sand is released under pressure and then compressed vertically so that the final diameter of the SCP to be about 0.7 m. The procedure continues until completion of SCP up to the full height. They are installed in triangular or square/rectangular pattern. The spacing between the SCPs ranges between 1 and 2.5 m having replacement ratios varying within 10–25%. Design of SCPs involves the diameter of SCP, spacing and pattern of installation. Data obtained from earlier case studies or the trail studies in terms of improved penetration resistance form the basis for the design. JGS (1998) and Kitazume (2005) present the methods and procedures to be followed for designing the SCPs in sandy and clayey soils. Various methods that can be adopted in sizing the SCPs are shown in Fig. 8.15.

Installation of SCP in loose sands has four fundamental advantages: (a) Reducing the void ratio (densification) by compaction, (b) Increasing shear resistance and horizontal resistance, (c) Change in earth pressure at rest, and (d) Formation of uniform ground (JGS, 1998). Further installation of SCP minimizes the degree of

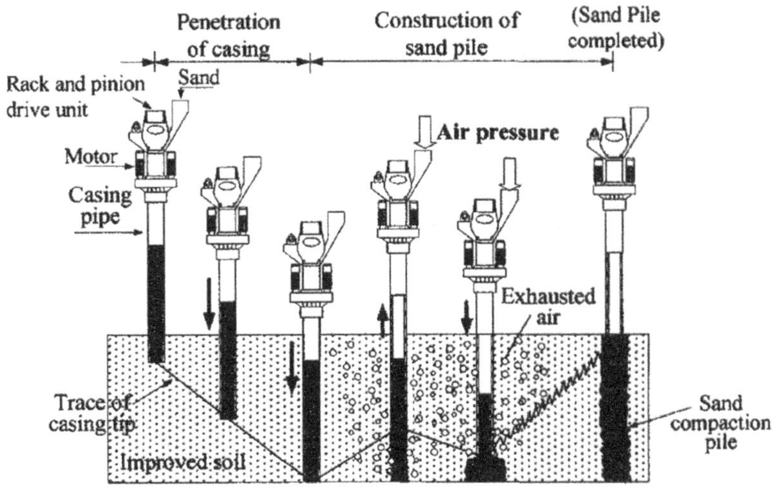

Fig. 8.14 Installation of sand compaction piles (after Okamura et al., 2006)

saturation levels which alter the rate of excess pore water pressure development during seismic loading, thus enhance the liquefaction resistance. The effect of densification is manifested through an increase in the coefficient of earth pressure at rest (Massarsch and Fellenius, 2002) and in the values of modulus of deformation of the soil (Ohbayashi et al., 1999). Ohbayashi et al. (1999) summarized measured values of Swedish Weight Sounding, (N_{sw}), SPT N, and CPT (q_c) at different sites wherein the increase in the measured parameters are presented as a function of the distance from the center of the compaction point.

The densification effect (Tsukamoto et al., 2000) becomes negligible at a distance of about 2.0 m from the center of the SCPs but the increase depends on the fines content (Ohbayashi et al., 1999). Figure 8.16 shows the relation between liquefaction resistance (cyclic stress ratio) and penetration resistance with different saturation levels and elevations (Okamura et al., 2003, 2006).

4.3 Deep Mixing Method (DMM)

Solidification principles are applied to mitigate liquefaction disasters where other more conventional methods are not viable for depth and economical reasons. Among various methods of solidification, like Deep mixing, Compaction grouting (Injection method), Premixing methods, DMM is widely used as a countermeasure against liquefaction for road embankments, river dikes, excavations and open cuts and protection of tank foundations on soft reclaimed ground (Porbaha et al., 1999). The uses of DMM for liquefaction mitigation include liquefaction prevention, reinforcing the liquefied soil, and reduce the excess pore pressure generation.

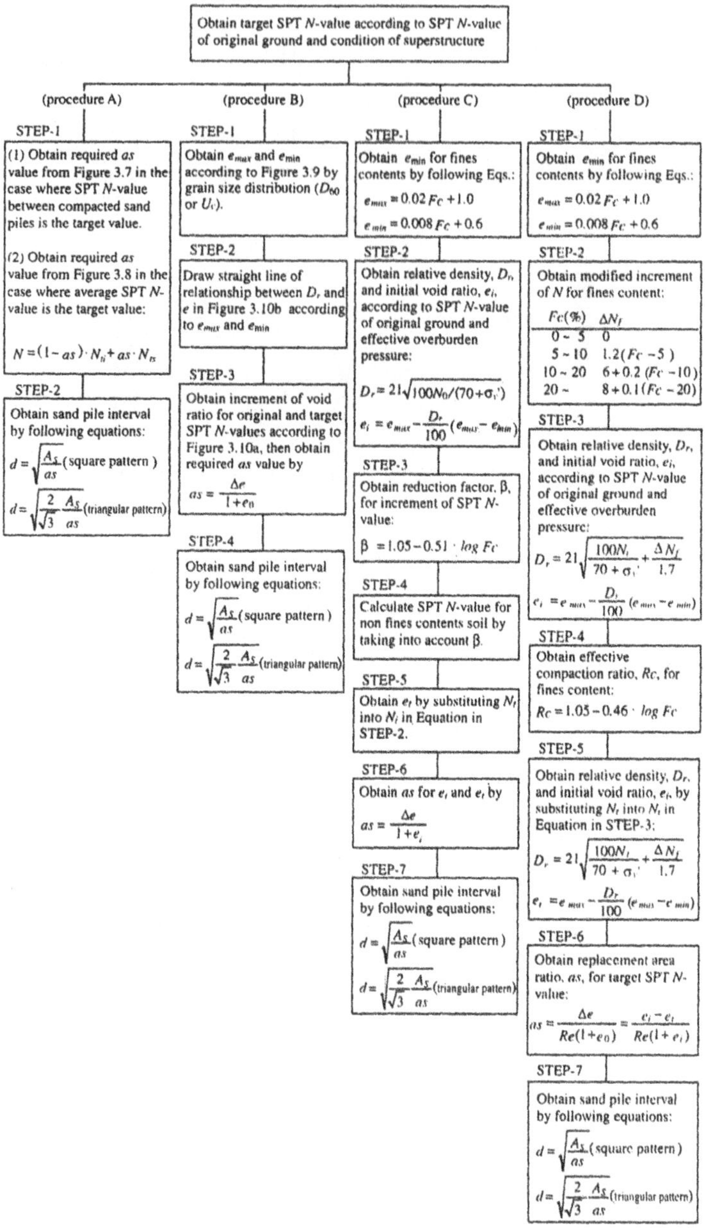

Fig. 8.15 Design procedures for determining sand pile arrangement (after Kitazume, 2005)

In DMM, the in situ soil will be mixed with a cementitious reagent injected as slurry (wet method) or powder (dry method) to improve its engineering behavior. The typical reagent materials are cement, lime, bentonite or other stabilizing agents. DMM is generally implemented in column, block, wall, or lattice/grid patterns as

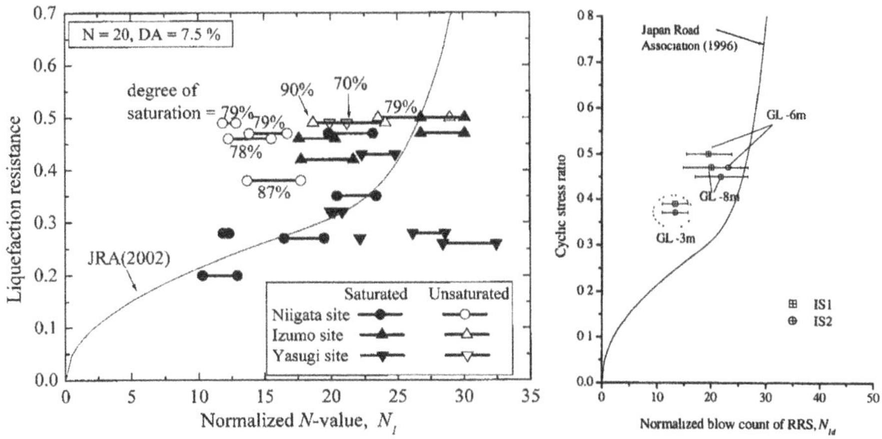

Fig. 8.16 Change in liquefaction resistance (cyclic stress ratio) with *N*-value for (**a**) change in degree of saturation (Okamura et al., 2006) and (**b**) change in elevation (Okamura et al., 2003)

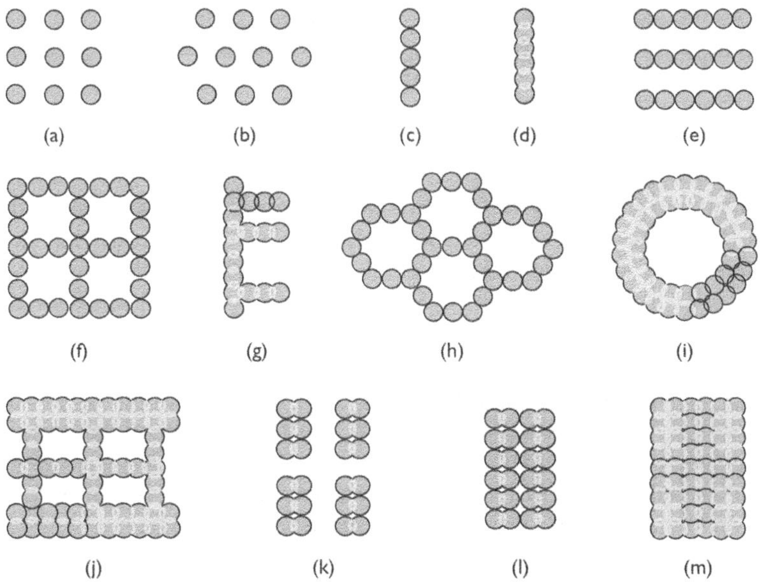

Fig. 8.17 Different patterns of deep soil mixing application: (**a**) column-type (square arrangement), (**b**) column-type (square and triangular arrangement); (**c**) tangent wall; (**d**) overlapped wall; (**e**) tangent walls; (**f**) tangent grid; (**g**) overlapped wall with buttresses; (**h**) tangent cells; (**i**) ring; (**j**) lattice; (**k**) group columns; (**l**) group columns in-contact; (**m**) block (after Topolnicki, 2004)

shown in Fig. 8.17 (Topolnicki, 2004). Typical construction procedure for implementing the DMM is described in Fig. 8.18. Use of a grid or lattice pattern is relatively effective due to its ability to engage the entire treated area as a unit, and fully mobilize the compressive strength of the cemented soil volume.

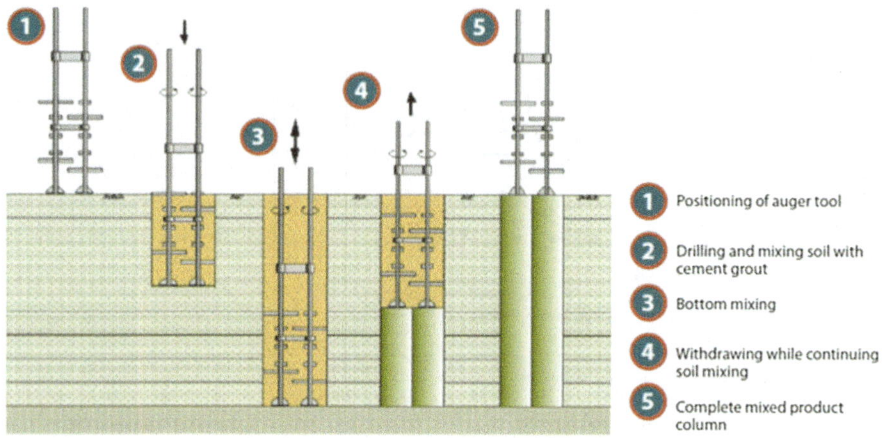

Fig. 8.18 DMM implementation (Jafecusa, 2012)

The mechanism of such lattice pattern was threefold; "(a) The "cells" absorb shear stresses and reduce the amplitude of the lateral granular movement and the development of excess pore pressure, (b) The confinement prevents lateral spreading, and (c) The compressive strength of the columns minimizes settlement" (FHWA, 2000).

Design of DM application involves: selection of suitable DM process (wet or dry) and the construction method; selection of reagent and strength of stabilized soil in specific ground conditions (mix design) and selection of the installation pattern and dimension of the improved ground (Topolnicki 2004). The improved ground is considered to be an underground structure with higher stiffness than the surrounding soil that receives earthquake motion from the underlying ground. Different forces are to be considered on such underground rigid body as: (a) body and inertia forces; (b) earth and water pressured arising from the surrounding soil; and (c) surcharge pressure. Typical steps to be followed in the design are shown in Fig. 8.19 (BS EN 14679 2005). A design procedure to prevent liquefaction-induced damage using the deep mixing was presented by Matsuo et al. (1996). Porbaha et al. (1999) and Siddharthan and Porbaha (2008a, b) discussed the concepts of deep mixing technology in the light of seismic response and liquefaction mitigation.

4.4 Dynamic Compaction

Among various methods available for densification of in situ soils, dynamic compaction is generally used method to densify large area. In this method a heavy tamper, having a mass of 5.4–27.2 Mg, is repeatedly raised and dropped with a single cable from varying heights, in the range 12–30 m, to impact the ground (FHWA, 1995). The tamping is carried out in grid pattern so that the energy

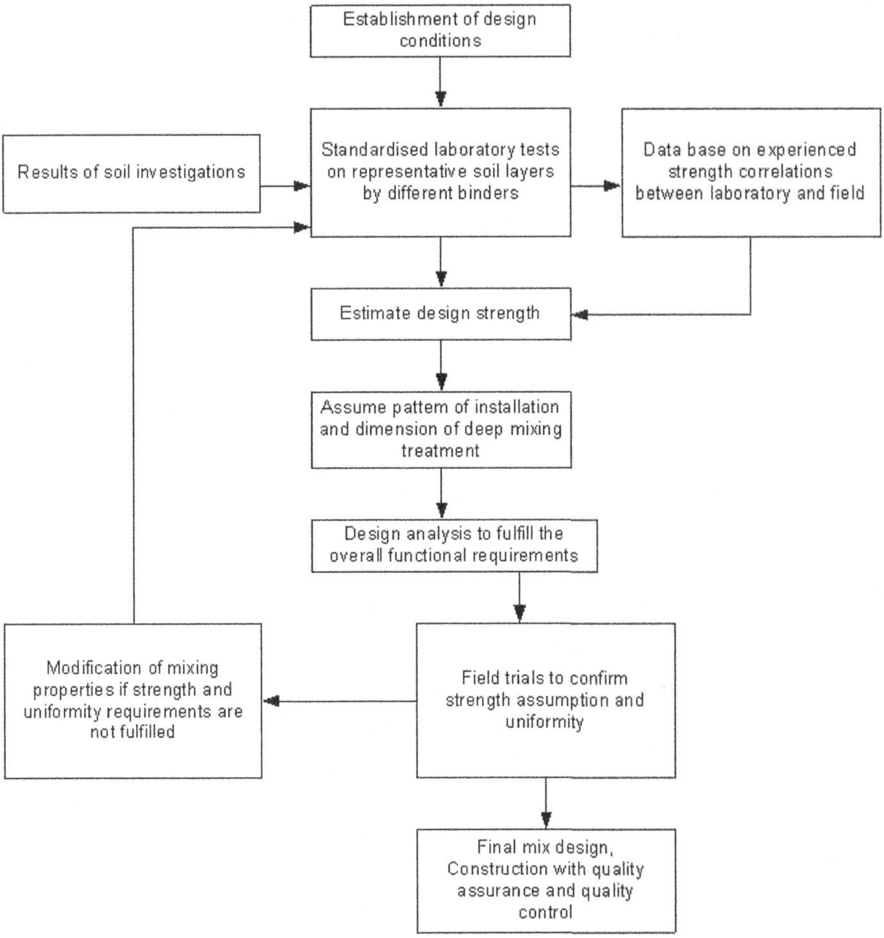

Fig. 8.19 Iterative design process for deep soil mixing (after BS EN 14679 2005)

is applied in phases over the entire area using either single or multiple passes. The craters formed in each pass of tamping are either leveled with a dozer or filled with granular fill material before the next pass of energy is applied. The energy released will densify the in situ soil up to certain depth. The effective depth of treatment (*D*) thus depends on energy released, i.e., mass of the tamped and its dropping height (Fig. 8.20). Dynamic compaction was used for liquefaction mitigation in several occasions (Dise et al., 1994; Shenthan et al., 2004; Sanjay et al., 2010; Kumar, 2001).

Design aspects of the dynamic compaction involves: (a) Selection of the area to densify; (b) Determining the applied energy to be used over the project site to result in the targeted improvement; (c) Selection of the tamper mass and drop height to correspond to the required depth of improvement; (d) Determination of the grid

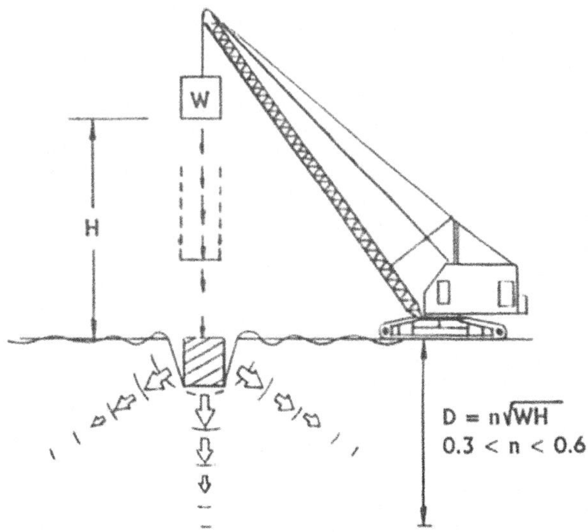

Fig. 8.20 Dynamic compaction method (after FHWA, 1995)

spacing and number of phases; and (e) The need for a surface stabilizing layer (FHWA, 1995).

Improvement in the soil characteristics due to densification can be monitored by any in situ testing method and by comparing such test data before and after dynamic compaction. With the improved soil behavior the liquefaction potential could be minimized to the desired levels. Figure 8.21 shows such data pertaining to liquefaction remediation for a construction site in Greater Noida, Uttar Pradesh, using dynamic compaction method (Sanjay et al., 2010). The figure clearly depicts the levels of improvement in terms of SPT and SCPT test values and the resulted improvement in the liquefaction resistance in terms of stress ratios and factors of safety at different elevations, before and after the dynamic compaction.

5 Comparative Assessment of Engineering of Ground Methods

Yasuda et al. (1996) analyzed the various cases of liquefied and nonliquefied ground in the two reclaimed islands, Rokko and Port in Kobe, following the Great Hanshin-Awaji (1995 Hyogoken-Nambu) earthquake. The reclaimed ground was 15–20 m thick. A remarkable phenomenon was observed and reported. Some zones in the two artificial islands did not liquefy even though large surrounding areas liquefied and caused heavy distress to the buildings and structures built on them. At these islands, the ground was improved by preloading, vertical drains, vibro-compaction and sand compaction piles. Figure 8.22 summarizes the SPT *N*-values before and after ground

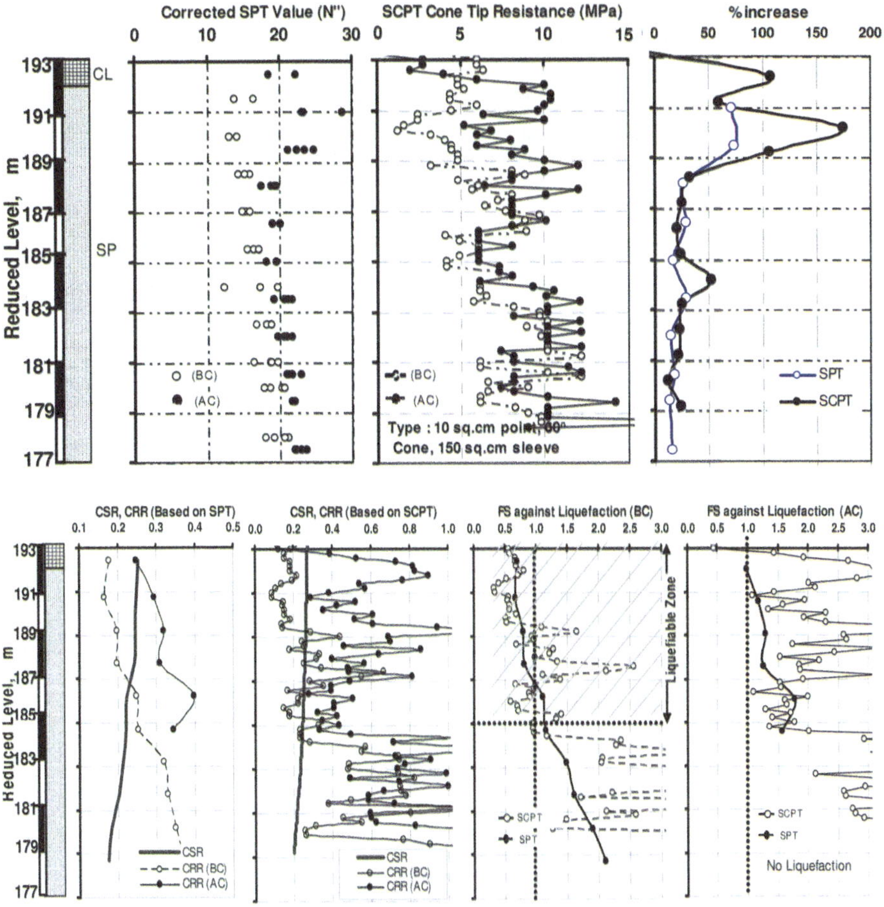

* BC = Before Compaction, AC = After Compaction

Fig. 8.21 In situ test (SPT and SCPT) results and liquefaction analysis results before and after the dynamic compaction (after Sanjay et al., 2010)

improvement. The *N*-values ranged between 8 to 15 in the untreated areas, but increased to 25 or more and to 14–18 in the Port and Rokko island sites, respectively.

The improvement has been nearly 100% and 200%, respectively, in the two areas. The relative ground subsidences of ground for untreated, ground treated by the above listed treatment methods are depicted in Fig. 8.23. The subsidence of the untreated ground ranged within 30–100 cm with an average value of 50 cm. It got reduced to less than 5 cm in case of rod- or sand compaction pile methods. No or very little subsidence, ejection of water or damage to buildings was observed in areas treated by these two methods.

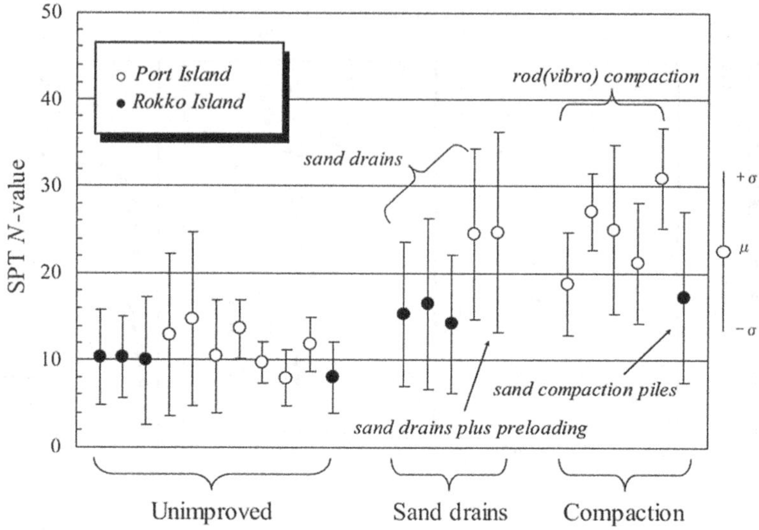

Fig. 8.22 SPT test values before and after improvement by different methods (after Yasuda et al., 1996)

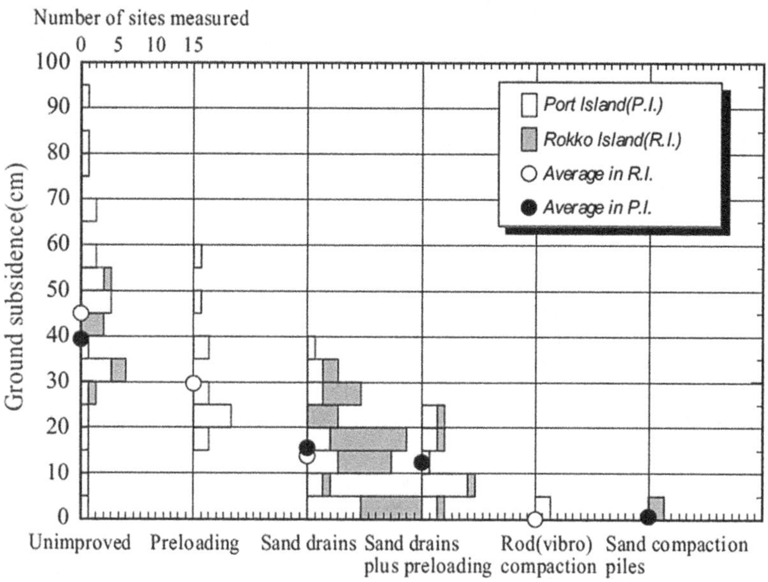

Fig. 8.23 Comparison of ground subsidence in zones treated with different methods (after Yasuda et al., 1996)

6 Conclusions

Liquefaction is the most disastrous feature during an earthquake that causes huge loss and damage to various structures built on or in the ground. Ground improvement methods are extensively used to enhance the in situ ground performance under various loading conditions and also as liquefaction countermeasures. Various types of ground engineering methods employed for liquefaction mitigation and remediation can be grouped into different principles: Densification, Solidification, Reinforcement and Drainage. Among the various options available, the most widely used method for mitigating the liquefaction hazard is the installation of stone columns/granular piles/drains. This method covers the three main principles of ground engineering and employs different mechanism in improving the liquefaction resistance. Mechanisms such as reinforcement, densification, storage, dilation along with the drainage mitigate the damages due to liquefaction. The other methods widely used for the purpose in the order of priority would be sand compaction piles, deep soil mixing and dynamic compaction methods. Selection of a particular method depends on various engineering, economical, and feasibility factors. One should work on different options and evolve the best possible method considering such factors.

References

Adalier, K., & Elgamal, A. (2004). Mitigation of liquefaction and associated ground deformations by stone columns. *Engineering Geology, 72*(3–4), 275–291.

Andrews, D.C.A. and Martin, G.R. (2000). Criteria for liquefaction of silty soils, Proc. 12th World Conference on Earthquake Engineering Auckland, New Zealand, Paper 0312.

Andrus, R. D. (1994). *In situ characterization of gravelly soils that liquefied in the 1983 Borah Peak Earthquake*. Ph.D. thesis, University of Texas at Austin, Austin, TX.

Andrus, R. D., & Stokoe, K. H. (1997). *Liquefaction resistance based on shear wave velocity* (Tech. Rep. NCEER-97-0022). National Center for Earthquake Engineering Research, State University of New York at Buffalo, Buffalo, NY, 89–128.

Baez, J. I. (1995). *A design model for the reduction of soil liquefaction by vibro-stone columns*. Ph. D thesis, University of Southern California, Los Angeles, CA.

Baez, J. I., & Martin, G. R. (1992). *Quantitative evaluation of stone column technique for earthquake liquefaction mitigation* 10th World Conference on Earthquake Engineering, Madrid, Spain, 1477–1483.

Balaam, N. P., & Booker, J. R. (1981). Analysis of rigid raft supported by granular piles. *International Journal for Numerical and Analytical Methods in Geomechanics, 5*(4), 379–403.

Barksdale, R. D., & Bachus, R. C. (1983). *Design and construction of stone columns* (Rep. No. FHWA/RD-83/026), U. S. Department of Transportation, Federal Highway Administration, Washington, DC, 194.

Bhattacharya, S. (2006). Safety assessment of existing piled foundations in liquefiable soils against buckling instability. *ISET Journal of Earthquake Technology, 43*(4), 133–147.

Bouckovalas, G. D, Papadimitriou, A. G, Niarchos, D. (2009). *Gravel drains for the remediation of liquefiable sites: The Seed & Booker (1977) approach revisited.* Proceedings of International

Conference on Performance-Based Design in Earthquake Geotechnical Engineering—From Case History to Practice, IS-Tokyo 2009.

Bouckovalas, G. D., Papadimitriou, A. G., Niarchos, D., & Tsiapas, Y. Z. (2011). Sand fabric evolution effects on drain design for liquefaction mitigation. *Soil Dynamics and Earthquake Engineering, 31*, 1426–1439.

Boulanger, R., Idriss, I., Stewart, D., Hashash, Y., & Schmidt, B. (1998). Drainage capacity of stone columns or gravel drains for mitigating liquefaction, *Proceedings of Geotechnical Earthquake Engineering and Soil Dynamics III*, ASCE Geotechnical Special Publication *75*(1), 678–690.

BS EN 14679 (2005). Execution of special geotechnical works—deep mixing. British Standards (English version).

Castro, G. (1975). Liquefaction and cyclic mobility of saturated sands. *Journal of the Geotechnical Engineering Division, ASCE, 101*(GT6), 551–569.

Cetin, K. O., Seed, R. B., Der Kiureghian, A., Tokimatsu, K., Harder, L. F., Jr., Kayen, R. E., & Moss, R. E. S. (2004). Standard penetration test-based probabilistic and deterministic assessment of seismic soil liquefaction potential. *Journal of Geotechnical and Geoenviromental Engineering, 130*(12), 1314–1340.

Datye, K. R., & Nagaraju, S. S. (1975). *Installation and testing of rammed stone columns*. Proceeding of the IGS Specialty Session, 5th Asian Regional Conference on SMFE, Bangalore, India, 101–104.

Datye, K. R., & Nagaraju, S. S. (1981). *Design approach and field control for stone columns*. Proceedings of 10th International Conference on Soil Mechanics and Foundation Engineering, Stockholm, Sweden. 3, 637–640.

Dise, K., Stevens, M. G., & Von Thun, J. L. (1994). *Dynamic compaction to remediate liquefiable embankment foundation soils*. ASCE Geotechnical Special Publication No. 45, In-Situ Deep Soil Improvement l–25.

FHWA. (1995). *Dynamic compaction* (FHWA Rep. No. FHWA-SA-95-037). U.S. Federal Highway Administration, Washington, DC

FHWA. (2000). *An introduction to the deep soil mixing methods as used in geotechnical applications* (FHWA Rep. No. FHWA-RD-99-138). Prepared by Geosystems (D.A. Bruce) for US Department of Transportation, Federal Highway Administration, 143.

Iai, S., & Koizumi, K. (1986). *Estimation of earthquake induced excess pore water pressure for gravel drains*. Proceedings of 7th Japan Earthquake Engineering Symposium, 679–684.

Idriss, I. M, & Boulanger, R. W. (2008). *Soil liquefaction during earthquake*. Monograph-12. Oakland, CA: Earthquake Engineering Research Institute.

IS 15284 (Part 1). (2003). *Design and construction for ground improvement — Guidelines, Part 1 Stone Columns, analysis*. New Delhi: Bureau of Indian Standards.

Ishihara, K. (1993). Liquefaction and flow failure during earthquakes. *Geotechnique, 42*(3), 351–415.

Jafecusa. (2012). Retrieved April 1, 2012 from http://jafecusa.com/?page_id=2796.

JGS. (1998). Remedial measures against soil liquefaction. In A. A. Balkema (Ed.). Rotterdam: The Japanese Geotechnical Society.

Kitazume, M. (2005). *The sand compaction pile method*. Taylor & Francis Group Publisher, London: Balkeme Publishers.

Kumar, S. (2001). Reducing liquefaction potential using dynamic compaction and construction of stone columns. *Geotechnical and Geological Engineering, 19*, 169–182.

Madabhushi, S. P. G. (2007). Ground improvement methods for liquefaction remediation. *Ground Improvement, 11*(4), 195–206.

Madhav M. R., Arlekar, J. N. (2000). *Dilation of granular piles in mitigating liquefaction of sand deposits* (No: 1035, CD-ROM). 12th World Conference Earthquake Engineering, Auckland.

Madhav, M. R., & Murali Krishna, A. (2008). Liquefaction mitigation of sand deposits by granular piles-an overview. In H. L. Liu, A. Deng, & J. Chu (Eds.), *Geotechnical engineering for disaster mitigation & rehabilitation*, Nanjing, China. 66–79.

Maheshwari, B. K., Nath, U. K., & Ramasamy, G. (2008). Influence of liquefaction on pile-soil interaction in vertical vibration. *ISET Journal of Earthquake Technology, 45*, 1–12.

Martin, G. R., Finn, W. D. L., & Seed, H. B. (1975). Fundamentals of liquefaction under cyclic loading. *Journal of the Geotechnical Engineering Division, ASCE, 101*(GT5), 425–438.

Massarsch, K. R., & Fellenius, B. H. (2002). Vibratory compaction of coarse-grained soils. *Canadian Geotechnical Journal, 39*(3), 695–709.

Matsuo, O., Shimazu, T., Goto, Y., Suzuki, Y., Okumura, R., & Kuwabara, M. (1996). *Deep mixing method as a liquefaction prevention measure. Grouting and deep mixing*. Proceedings of IS-Tokyo '96. The Second International Conference on Ground Improvement Geosystems, Tokyo, May 14–17, 521–526.

Mitchell, J. K. (2008). Mitigation of liquefaction potential of silty sands. In J. E. Laier, D. K. Crapps, M. H. Hussein (Eds.), *From research to practice in geotechnical engineering*. Geotechnical Special Publication 180, ASCE, Reston, VA.

Mitchell, J. K., & Wentz, F. J. (1991). *Performance of improved ground during the Loma Prieta earthquake* (Report. No. UCB/EERC-91/12). Berkeley, CA: Earthquake Engineering Research Center, University of California.

Moss, R. E. S., Seed, R. B., Kayen, R. E., Stewart, J. P., Der Kiureghian, A., & Cetin, K. O. (2006). CPT-based probabilistic and deterministic assessment of in situ seismic soil liquefaction potential. *Journal of Geotechnical and Geoenvironmental Engineering, 132*(8), 1032–1051.

Murali Krishna, A. (2011). Mitigation of liquefaction hazard using granular piles. *International Journal of Geotechnical Earthquake Engineering, 2*(1), 44–66.

Murali Krishna, A., & Madhav, M. R. (2007). Equivalent deformation properties of ground treated with rammed granular piles. *International Journal of Geotechnical Engineering, 1*(1), 31–38.

Murali Krishna, A., & Madhav, M. R. (2008). Densification and dilation effects of granular piles in liquefaction mitigation. *Indian Geotechnical Journal, 38*(3), 295–316.

Murali Krishna, A., & Madhav, M. R. (2009). Treatment of loose to medium dense sands by granular piles: Improved SPT N1 values. *Geotechnical and Geological Engineering, 27*, 455–459.

Murali Krishna, A., Madhav, M. R., & Madhavi Latha, G. (2006). Liquefaction mitigation of ground treated with granular piles: Densification effect. *ISET Journal of Earthquake Technology, 43*(4), 105–120.

Murali Krishna, A., Madhav, M. R., & Madhavi Latha, G. (2007). Densification effect of granular piles on settlement response of treated ground. *Ground Improvement, 11*(3), 127–136.

NCEER. (1997). National Center for Earthquake Engineering Research (NCEER) (1997). In T. L. Youd, & I. M. Idriss (Eds.), *Proceedings of the NCEER Workshop on Evaluation of Liquefaction Resistance of Soils* (Tech. Rep. NCEER-97-022).

Noorzad, A, Poorooshasb, H. B., & Madhav, M. R. (2007). *Performance of partially penetrating stone columns during an earthquake*. Proceedings of the Tenth Symposium on Numerical models in Geomechanics (NUMOG X), Rhodes, Greece, 503–508.

Ohbayashi, J., Harda, K., & Yamamoto, M. (1999). Resistance against liquefaction of ground improved by sand compaction pile method. In Seco a Pinto (Ed.), *Earthquake Geotechnical Engineering, Balkama, Rotterdam* (312p).

Okamura, M., Ishihara, M., & Oshita, T. (2003). Liquefaction resistance of sand improved with sand compaction piles. *Soils Foundations, 43*(5), 175–187.

Okamura, M., Ishihara, M., & Tamura, K. (2006). Degree of saturation and liquefaction resistances of sand improved with sand compaction pile. *Journal of Geotechnical and Geoenvironmental Engineering, 132*(2), 258–264.

Onoue, A. (1988). Diagrams considering well resistance for designing spacing ratio of gravel drains. *Soils and Foundations, 28*(3), 160–168.

Pestana, J. M., Hunt, C. E., Goughnour, R. R., & Kammerer, A. M. (1998). *Effect of storage capacity on vertical drain performance in liquefiable sand deposits*, Proceedings: Second International Conference on Ground Improvement Techniques, Singapore, 373–380.

PIANC. (2001). *Seismic design guidelines for port structures.* Working Group No. 34 of the Maritime Navigation Commission International Navigation Association, Tokyo: A.A. Balkema Publishers.

Poorooshasb, H. B., Noorzad, A., Miura, N., & Madhav, M. R. (2000). *Prevention of earthquake induced liquefaction of sandy deposits using stone columns.* Proceedings of the International Symposium on Lowland Technology 2000, Saga. 213–220.

Poorooshasb H. B., Madhav M. R., & Noorzad, A. (2006). *Performance of stone columns subjected to a seismic base excitation.* Proceedings of the International Symposium on Lowland Technology 2006, Saga, 1189–1194.

Porbaha, A., Zen, K., & Kobayashi, M. (1999). Deep mixing technology for liquefaction mitigation. *Journal of Infrastructure Systems, 5*(1), 21–34.

Priebe, H. J., & Keller, G. (1995). Design of vibro replacement. *Ground Engineering, 28*(10), 31–37.

Reyna, F., and Chameau, J.L. (1991), *Dilatometer based liquefaction potential of sites in the Imperial Valley.* Proceedings: Second International Conference on Recent Advances in Geotechnical Earthquake Engineering and Soil Dynamics, St. Louis.1, 385–392.

Sanjay, G., Shankar, R., & Sorabh, G. (2010). *Dynamic compaction to mitigate liquefaction potential.* Proceedings of the Indian Geotechnical Conference 2010, December 16–18, 2010, 783–786.

Sawicki, A., & Mierczynski, J. (2006). Developments in modeling liquefaction of granular soils, caused by cyclic loads. *Applied Mechanics Reviews, Transactions of the ASME, 59*(3), 91–106.

Seed, H. B. (1979). Soil liquefaction and cyclic mobility evaluation for level ground during earthquakes. *Journal of the Geotechnical Engineering Division, ASCE, 105*(2), 201–255.

Seed, H. B., & Booker, J. R. (1977). Stabilization of potentially liquefiable sand deposits using gravel drains. *Journal of the Geotechnical Engineering Division, 103*(7), 757–768.

Seed, H. B., & Idriss, I. M. (1982). *Ground motions and soil liquefaction during earthquakes.* Oakland, CA: Earthquake Engineering Research Institute.

Seed, H. B., Idriss, I. M., Makdisi, F., & Bannerjee, N. (1975). *Representation of irregular stress time histories by equivalent uniform stress series in liquefaction analyses* (Rep. No. UCB/EERC/75-29). Berkeley, CA: Earthquake Engineering Research Center, University of California.

Seed, H. B., Martin, P. P., & Lysmer, M. J. (1976). Pore-water pressure changes during soil liquefaction. *Journal of Geotechnical Engineering Division, ASCE, 102*(GT4), 323–346.

Seed, R. B., Cetin, K. O., Moss, R. E. S., Kammerer, A. M., Wu, J., Pestana, J. M., Riemer, M. F., Sancio, R. B., Bray, J. D., Kayen, R. E., & Faris, A. (2003). *Recent advances in soil liquefaction engineering: A unified and consistent framework.* 26th Annual ASCE Los Angeles Geotechnical Spring Seminar, Long Beach, CA, April 30, 2003, 1–71.

Shankar, R. (2001). Seismotectonics of Kutch rift basin and its bearing on the Himalayan seismicity. *ISET Journal of Earthquake Technology, 38*(2–4), 59–65.

Shenthan, T., Nashed, R., Thevanayagam, S., & Martin, G. R. (2004). Liquefaction mitigation in silty soils using composite stone columns and dynamic compaction. *Earthquake Engineering and Engineering Vibration, 3*(1), 39–49.

Siddharthan, R. V., & Porbaha, A. (2008a). Seismic response evaluation of deep mixed improved ground part I: Proposed approach. *Ground Improvement, 131*(G13), 153–162.

Siddharthan, R. V., & Porbaha, A. (2008b). Seismic response evaluation of deep mixed improved ground part II: Verification. *Ground Improvement, 131*(G13), 163–169.

Sitharam, T. G., Raju, L. G., & Murthy, B. R. S. (2004). Cyclic and monotonic undrained shear response of silty sand from Bhuj region in India. *ISET Journal of Earthquake Technology, 41* (2–4), 249–260.

Stark, T. D., & Olson, S. M. (1995). Liquefaction resistance using CPT and field case histories. *Journal of Geotechnical Engineering, ASCE, 121*(GT12), 856–869.

Topolnicki, M. (2004). In situ soil mixing. In M. P. Moseley & K. Kirsch (Eds.), *Ground improvement* (2nd ed.). New York: Spon Press, Taylor & Francis Group.

Tsukamoto, Y., Ishihara, K., Yamamoto, M., Harada, K., & Yabe, H. (2000). Soil densification due to static sand pile installation for liquefaction remediation. *Soils and Foundations, 40*(2), 9–20.
Yasuda, S., Ishihara, K., Harada, K., & Shinkawa, N. (1996). Effect of soil improvement on ground subsidence due to liquefaction. Soils and Foundations, Special Issue, 99–107.
Youd, T. L., Idriss, I. M., Ronald, D. A., Arango, I., Castro, G., Christian, J. T., Dobry, R., Finn, W. D. L., Harder, L. F., Jr., Hynes, M. E., Ishihara, K., Koester, J. P., Liao, S. S. C., Marcuson, W. F., III, Martin, G. R., Mitchell, J. K., Moriwaki, Y., Power, M. S., Robertson, P. K., Seed, R. B., Stokoe, I. I., & H, K. (2001). Liquefaction resistance of soils: Summary report from the 1996 NCEER and 1998 NCEER/NSF workshops on evaluation of liquefaction resistance of soils. *Journal of Geotechnical and Geoenvironmental Engineering, 124*(10), 817–833.

Chapter 9
Recent Advances in Soil Dynamics Relevant to Geotechnical Earthquake Engineering

T. G. Sitharam, K. S. Vipin, and Naveen James

1 Introduction

Earthquake hazards are one of the worst natural disasters, causing huge loss to human life and destruction of structures. The influence of local soil condition on seismic hazard has been known for many decades. It's a well-known concept that the characteristics of ground level motion like amplitude, frequency and duration are strongly influenced by topography and material properties of soil overlying the bedrock. This change in the ground-motion characteristics when the waves travel from bedrock to ground surface is termed as local site effect. Due to the local site effects, the earthquake motion at the surface will be entirely different from that at the bedrock level. The study of this variation is very important for shallow-founded structures, geotechnical structures like foundations, retaining walls and dams, floating piles and underground structures. Much of the earthquake damages due to shaking, lateral spreading and liquefaction can be attributed to local site effects. For proper assessment of local site effects, the characteristics and thickness of soil conditions are to be identified based on borings and in situ geophysical and geotechnical tests. In addition to this, the geometry of soil profile is also very essential.

Assessment of local site effects can be carried out at three major scales – macro-scale, micro-scale, and site-specific level. Site effect assessment at macro-scale (macrozonation) is carried out at very small scale like 1:10,00,000 and above

T. G. Sitharam (✉)
Department of Civil Engineering, Indian Institute of Science (IISc), Bangalore, India
e-mail: sitharam@iisc.ac.in

K. S. Vipin
Earthquake Specialist, Swiss Reinsurance Company (Swiss Re), Bangalore, India

N. James
Department of Civil Engineering, Indian Institute of Technology Ropar, Rupnagar, Punjab, India

M. L. Sharma et al. (eds.), *Advances in Indian Earthquake Engineering and Seismology*, https://doi.org/10.1007/978-3-319-76855-7_9

(grid size higher than 5 km × 5 km). An example of macroscale work includes zonation of a country or a state. At this scale, it is extremely difficult to obtain the geotechnical data for the vast study area and hence the site characterization is mainly done based on the geology of the region (Borcherdt and Gibbs, 1976; Shima, 1978; Midorikawa, 1987 and Youd and Perkins, 1978).

Micro-level hazard zoning is carried out for cities and towns with scale varying from 1:10,00,000 to 1:5,000 (grid size of 2 km × 2 km). Geotechnical and geophysical tests are done at each grid point for site characterization purpose. Once the spatial distribution of site classes are known, the site amplification and liquefaction potential can be evaluated based on empirical relationships. The assessment of local site effects at micro-scale is an excellent tool for town planning and urban development (Anbazhagan and Sitharam, 2008).

The site-specific assessment of site amplification and liquefaction assessment are carried out at very large scale (less than 1:5,000 or grid size of 1 km × 1 km or less). This type of hazard zoning is carried out for critical installations like nuclear power plants, dams, etc. At this scale, the subsurface profile of site is accurately determined by geotechnical and geophysical tests. Equivalent linear or nonlinear site response analyses are carried out for prediction of earthquake motion at ground surface. Liquefaction potential can be estimated using the deterministic methodology suggested by Seed and Idriss (1971) or based on probabilistic methods.

2 Effect of Local Site Conditions on Ground Motion

The analysis of strong-motion records has found out that the difference of stiffness between the overlying soil and the underlying bedrock will affect the amplitude, frequency and duration of seismic waves. The changes in these parameters will depend on the local site conditions (properties and the geometry of the overlying soil) (Fig. 9.1). Many historical evidences of the influence of local geological

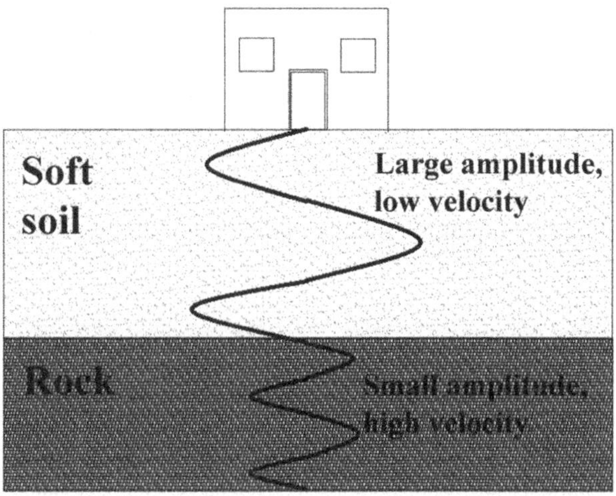

Fig. 9.1 Effect of local site conditions

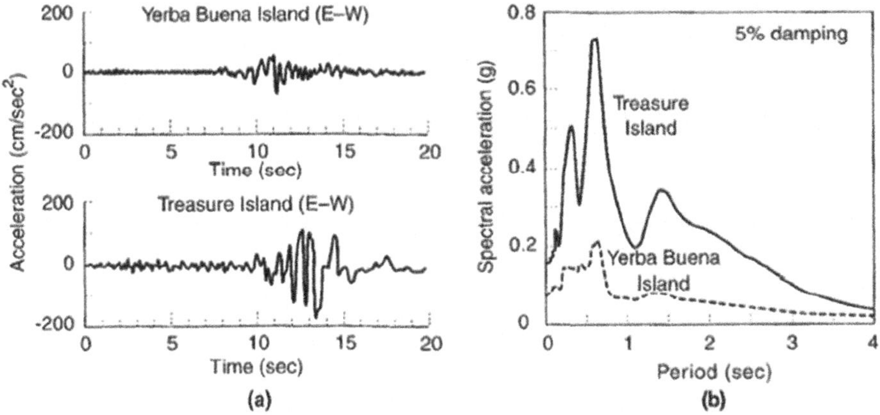

Fig. 9.2 Site amplification during Loma Prieta earthquake (Seed and Harder, 1990)

characteristics on the intensity of ground shaking and damage caused due to an earthquake are found. However, it was not until the early 1970s that the local site conditions were included in building codes and provisions. In addition to theoretical evidence, amplification functions developed from measurements of surface and bedrock motions at the same location and comparisons of surface-motion characteristics from nearby sites with different subsurface conditions, all confirm the effects of local site conditions of earthquake ground motions.

The effects of site amplification were evident at some locations during the Mexico (1985), Loma Prieta (1989), and Bhuj (2001) Earthquakes. The earthquake ground acceleration records for two sites at Loma Prieta (Fig. 9.2)—rock site (Yerba Buena) and soft soil site (Treasure Island).

The major factors which will influence the local site effects are discussed in the following sections.

2.1 *Effect of Topography*

Local site effects produce significant amplifications of the ground motion during an earthquake. One of the important causes of local site effects is topographical effects. Irregular topography can substantially affect the amplitude and frequency characteristics of seismic motion. Macro seismic observations of destructive earthquakes often show higher damage intensity at the top of hills, ridges and canyons than at lower elevations and on flat areas.

Two general types of topography have to be distinguished:

- Surface topography, mainly characterized by mountainous features, such as the presence of rock ridges or steep soil slopes.
- Subsurface (or subsoil) topography, is either caused by lateral heterogeneities of the subsoil layers or by sharp basin geometry (Fäh et al., 1997; Mayer-Rosa and Jimenez, 1999).

2.2 *Effect of Ground Water*

Groundwater can also play a significant role in how earthquakes affect the ground surface when an earthquake occurs. The most well-known effect is liquefaction.

2.3 *Deep Soil Effects*

The local site effects observed at Mexico and Loma Prieta brought the importance of assessment of Deep Soil Effects. Both the cities were underlain by thick soil deposits and based on the records of these two earthquakes, Idriss (1990) suggested methods for evaluating the surface level acceleration values for sites with deep soil deposits.

2.4 *Estimation of Local Site Effects*

There are different techniques available for the estimation of local site effects. Broadly they can be categorized as empirical, experimental and numerical methods.

2.4.1 Empirical Methods

Several researchers have developed empirical relations between surface geology and various ground-motion parameters. These relationships are developed from a particular set of data where both earthquake observations and information on surface geology are available, which can be applied at other sites where only geological information is known. The site amplification will depend on various other factors also and hence these methods should be used when there is not enough geotechnical data to evaluate the site amplification.

1. *Earthquake Intensity Increment Based on Geology*: The relationship between surface geology and seismic intensity increments has been developed by various researchers. The correlations given by Medvedev (1962), Evernden and Thomson (1985), Kagami et al. (1988) and Astroza and Monge (1991) are based on the seismic observations in Asia, California, Japan and Chile respectively.
2. *Relative Amplification Based on Geology*: Borcherdt and Gibbs (1976) proposed the concept of relative amplification which can be used to evaluate the effect of local geology quantitatively. Relative amplification is defined as the amplification with respect to reference site. They measured ground motions generated by nuclear explosions at sites with various geological conditions to obtain the spectral amplifications of the motions with respect to granitic rock sites.
3. *Relative Amplification Based on Geotechnical Parameters*: The most important geotechnical parameters, which can be used to estimate the amplification factors,

are average shear wave velocity and SPT "N" value. Shima (1978) found that the analytically calculated amplification factor is linearly related with the ratio of shear wave velocity of the surface layer to that of bedrock. When the bedrock shear wave velocity is found to be relatively constant over a wide area, the relative amplification in each locality can be obtained from the shear wave velocity of the surface layer.

4. *Amplification Based on Surface Topography*: Topographical effects can play an important role on ground motion characteristics and this has been highlighted by many researchers (Geli et al., 1988; Faccioli, 1991; Chávez-García et al., 1996; Reinoso et al., 1997; Athanasopoulus et al., 1999). Theoretical and experimental studies on topographical effects on ground motion have been made by Aki (1988) who showed the effects of topography using a simple structure of a triangle-edge model. Faccioli (1991) used Aki's model as ridge–valley topography and addressed the relative amplification at the crest of the ridge compared to the base and also the deamplification in the valley.

2.4.2 Experimental Methods

These methods are developed using different kinds of data based on microtremor measurements, weak seismicity survey, and strong-motion data. Detailed description for estimating site effects using the above methods is given below:

1. *Microtremor Data*: The site effects are often expressed by the amplification factor and resonance/fundamental frequency. Usually there are various vibrations in the ground which are caused by natural or ambient noise like wind, sea waves, traffic, industrial machinery, etc. The vibrations that have comparatively small periods of less than 1 sec are called microtremors and those that have a larger period range is called microseisms. Applications of this method include site response analysis, natural frequency of structures etc. In India, Mukhopadhyay et al. (2002) conducted microtremor studies in Delhi and compared the resonance frequency obtained by microtremor with that estimated from strong-motion records.
2. *Weak-Motion Data*: Weak-motion data are the records from small to moderate, natural or artificial seismic events (small-magnitude earthquakes, aftershocks of big events, mine or quarry blasts, nuclear tests). Such data can be recorded by digital, high-sensitive instruments identical to those used by seismologists for microseismicity and seismotectonic studies. Field and Jacob (1993) quoted that the greatest challenge in the estimation of site response from such instrumental recordings is removing the source and path effects. Two techniques are developed depending on whether or not they need a reference site with respect to which the particular effects at other sites are estimated.

    (a) *Reference Site Technique*: The identification and removal of the effects of source, path and the site characteristics is the greatest challenge in evaluating the site response. The simplest method to evaluate the site response is to divide the response spectrum obtained at the site with that of the bedrock (reference site). If the recording in the rock is at a close distance to the soil site, then the three governing factors, which will affect the ground motion, will be the same for both the soil site and the rock.

(b) *Nonreference Site Technique*: In practice, adequate reference sites are not always available. For this reason, different methods without reference sites have been developed. It consists of taking the spectral ratio between the horizontal and the vertical components of the shear wave part. This technique is a combination of Langston (1979) receiver function method for determining the velocity structure of the crust from the horizontal-to-vertical spectral ratio (HVSR) of teleseismic P waves and the Nakamura (1989) method.

3. *Strong-Motion Data*: The development of strong-motion arrays makes it possible to evaluate site effects in mega cities like Los Angeles, Tokyo, Taipei, and Mexico City using the strong-motion data. While using this method, even the nonlinear site effects are included in the recordings. Recent studies show that there is a fairly good agreement between the old and new techniques.

2.4.3 Numerical Methods

The most commonly used methods for one, two, and three dimensional analysis of site response are discussed below:

1. *One-dimensional site response analysis*: The basic assumption of one-dimensional analysis is that all the boundaries (between different soil layers) are horizontal and the response is caused by the wave propagating in a vertical direction.

   The linear approach has been implemented in the following procedures, which are commonly used for ground response analysis (Kramer, 1996):

   - Transfer functions: It is a frequency-dependent function, which describes the ratio of the displacement amplitudes (for particular frequency) at any two points in a soil layer. This will provide the amplification or deamplification factor to arrive at the surface-level motion.
   - Equivalent linear approximation of nonlinear response: Since soil is a nonlinear material, the linear approximation will not give accurate results for the problems involving ground response analysis. Hence, in these cases, equivalent linear analysis is preferred to give more accurate results. The shear modulus and damping ratio of soil depend upon the level of strain to which soil is subjected. In equivalent linear analysis, this variation in soil parameters is taken into account by an iterative procedure.
   - Deconvolution: It is the technique used to evaluate the earthquake motion at the bedrock level from the measured/recorded surface-level ground motions. It uses transfer function technique to back calculate the bedrock motions. This technique can include strain-dependent modulus and damping properties similar to equivalent linear approach.

Nonlinear Approach: The linear approach is very simple, and it is easy to compute, but it cannot evaluate the nonlinear response of the soil precisely. This issue can be overcome by using the nonlinear response of soil using direct numerical integration (in small time intervals) in time domain. The integration of motion in small time intervals will permit the use of any linear or nonlinear stress-strain models. There are many types of software, which can incorporate the nonlinear response of soils such as PLAXIS, SASSI2000, FLAC, QUAKE/W, etc.

2. *Two-dimensional site response analysis*: The one-dimensional analysis may not give very accurate results in cases where the soil profile is level or gently sloping. In the case of sites where embedded structures like pipelines or tunnels are there, one-dimensional analysis will not yield the desired results. The two-dimensional analysis can be done either based on frequency-domain or time-domain methods. This analysis can be done using dynamic finite element methods adopting either equivalent linear approach or nonlinear approach (Kramer, 1996). Numerical modeling software like PLAXIS, FLAC, QUAKE/W, etc. can be used for modeling two-dimensional cases.
3. *Three-dimensional site response approach*: There may be cases in which there is variation in soil profile in three dimensions and the two-dimensional approach may not be adequate. This is ideal for studying the response of three-dimensional structures. The methods and the approaches adopted are similar to the two-dimensional approach. The important approaches adopted are equivalent linear finite element approach, nonlinear finite element approach, etc. Software like ABAQUS, PLAXIS, FLAC, SASSI2000, etc. can be used for the three-dimensional modeling purpose.

3 Evaluation of Dynamic Properties

The response of the soil to cyclic loading strongly influences the earthquake damage. Hence the evaluation of the dynamic properties (shear modulus and damping ratio) of soils is of utmost importance in geotechnical earthquake engineering. The dynamic properties can be evaluated based on field and lab tests, and these tests can be broadly divided into two – High-strain and Low-strain tests. Since soil properties are highly nonlinear, the properties which influence wave propagation, stiffness, damping, Poisson's ratio etc. need to be evaluated at low strain. The high-strain tests are most commonly used to measure the soil strength, which need to be evaluated at higher-strain levels.

3.1 Field Tests

The field tests or the in situ tests measure the dynamic soil properties without altering the chemical, thermal or structural condition of the soil. The field test can be broadly divided into two—low-strain and large-strain tests.

3.1.1 Low-Strain Tests

The strain levels in these types of tests will be around 0.0001%. Some of the important low-strain tests are discussed below:

1. *Seismic Reflection Test:* This test is used to evaluate the wave propagation velocity and the thickness of soil layers. The test setup will consist of a source producing a seismic impulse and a receiver to identify the arrival of seismic waves and the travel time from source to receiver is measured. Based on these measurements, the thickness of soil layer can be evaluated. Even though it is more commonly used than seismic reflection test, its major application is for delineation of major stratigraphic units.
2. *Seismic Refraction Test*: The successful application of seismic refraction and reflection for profiling depends upon how well we modeled the wave propagation in the surface layers. The propagation of seismic waves through near surface deposits is very complex. The particulate, layered and fractured nature of the ground means that waves undergo not only reflection and refraction but also diffraction, making modeling of seismic energy transmission impractical. This test will use the arrival time of the first seismic wave at the receiver. Using the results obtained from this test, the delineation of major stratigraphic units is possible.
3. *Seismic Cross-Hole Test*: Cross-hole testing provides useful information on dynamic soil properties required for site-specific ground response analyses, liquefaction potential studies, and dynamic machine foundation design. Perhaps it is the best in situ method used for obtaining the variation of low-strain shear wave velocity with depth. Unlike MASW/SASW testing, it does not rely on any indirect methods to determine the wave velocities. Dynamic soil parameters, such as moduli and Poisson's ratios, can be easily determined from the measured shear and compressional wave velocities. The only disadvantage with this method is that it requires drilling of boreholes for its testing.
4. *Down-Hole and Up-Hole Tests*: Procedure for testing of down-hole seismic method is similar to the seismic cross test discussed in the previous section. The bore-hole drilled for cross-hole testing can be used for this test. Three orthogonally oriented geophones are lowered in to the borehole to different depths. At each depth, geophones are clammed to the borehole and are used in recording arrival times of seismic waves. Source is placed at the ground surface. Test setup for uphole is just opposite to downhole. In uphole test, source is lowered, and receiver is placed at the surface. Rest of the procedure is similar.
5. *Steady-State Vibration Test*: In this test the wave propagation velocities are measured from steady-state vibration characteristics. However these tests can be useful for determining the near surface shear wave velocity and they fail to provide the details of highly variable soil profiles.
6. *Multichannel Analysis of Surface Wave (MASW)*: Shear wave velocity (V_s) is an essential parameter for evaluating the dynamic properties (Shear Modulus and Modulus Reduction Curve) of soil and also for site classification. A number of

geophysical methods have been proposed for near-surface characterization and measurement of shear wave velocity by using a great variety of testing configurations, processing techniques, and inversion algorithms. The most widely used technique is MASW (Multichannel Analysis of Surface Waves). MASW is a geophysical method, which generates a shear wave velocity (V_s) profile (i.e., V_s vs. depth) by analyzing Raleigh surface waves on a multichannel record. MASW identifies each type of seismic waves on a multichannel record based on the normal pattern recognition technique that has been used in oil exploration for several decades. The MASW method was first introduced in Geophysics by Park et al. (1999). It is a very conventional mode of survey using an active seismic source (e.g., a sledge hammer) and a linear receiver array, collecting data in a roll-along mode. The MASW is used in geotechnical engineering for the measurement of shear wave velocity and dynamic properties, identification of subsurface material boundaries and spatial variations of shear wave velocity.

The captured Rayleigh wave is further analyzed using suitable software to generate shear wave velocity (V_s) data. This is being done in three steps: (a) preparation of a multichannel record (sometimes called a shot gather or a field file), (b) dispersion-curve analysis, and (c) inversion. MASW has been effectively used with highest signal-to-noise ratio (S/N) of surface waves.

The generation of a dispersion curve is a critical step in MASW method. A dispersion curve is generally displayed as a function of phase velocity versus frequency. Phase velocity can be calculated from the linear slope of each component on the swept-frequency record. A dispersion curve generated for a particular location in Karnataka (India) is shown in (Fig. 9.3). The shear wave velocity profile obtained by inversion of this dispersion curve is shown in Fig. (9.4).

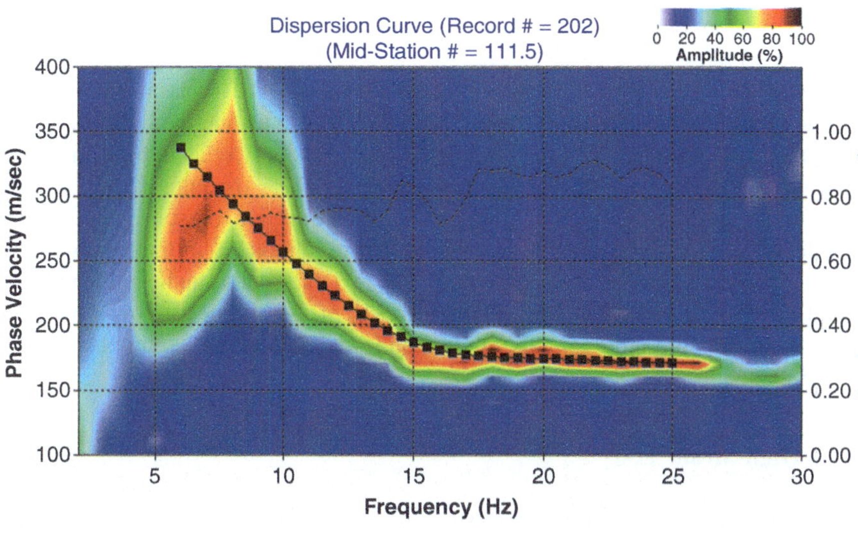

Fig. 9.3 Dispersion curve for a particular location in Karnataka

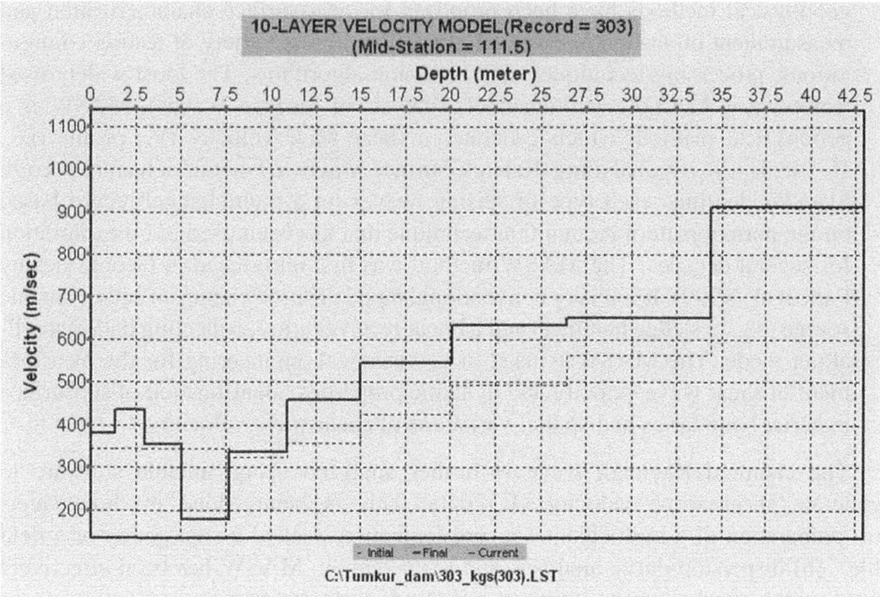

Fig. 9.4 Shear wave velocity profile obtained from the above dispersion curve

Hunter et al. (2002) compared the shear wave velocity measured through borehole method and MASW technique for several sites in the Fraser River Delta and found that the difference between the results was only 9%. Under normal conditions, MASW can provide the profile up to 30–80 m. However, with special equipment, it can go up to 250–300 m, as depth of profile depth mainly depends upon the source energy and geophone spacing. These tests are very useful when the rock depth is more than 30 m.

Advantages and Disadvantages of V_s Measurements: The V_s measurements are possible in very stiff and gravelly soils. The V_s values can be obtained in sites where borings are not permitted. The shear wave velocity is directly related to the mechanical property of soil and it is related to small-strain shear modulus G_{max}. No soil sample is obtained in the test and hence visual inspection of soils may not be possible. Very thin layer of loose soil strata may go unnoticed in the test. Various models are assumed for interpretation of results, and the results can be influenced by ground water conditions, presence of clay. The V_s measurement is a low-strain test and the pore pressure buildup and liquefaction initiation are high-strain behaviors.

3.1.2 High-Strain Tests

The most commonly used in situ soil tests are high-strain tests. There are correlations available between the results of high-strain tests and low-strain tests, and based on these correlations, one can easily calculate in situ shear modulus. Some of the commonly used high-strain tests are given below:

1. *Standard Penetration Test* (*SPT*): The standard penetration test is done using a split spoon sampler in a borehole/auger hole. This sampler consists of a driving shoe, a split barrel of circular cross section (longitudinally split into two parts) and a coupling. The standard penetration test is performed at every 0.75 m interval in a borehole. Even though it is one of the most common tests, the SPT N value obtained has to be corrected before it can be used for any of the empirical relations. Energy correction becomes very important considering the use of various types of equipments/hammers being used in India. Generally the 60% energy correction is recommended worldwide.

Use of SPT values in evaluating the site effects: The major use of SPT data is in evaluating various soil parameters, soil profiling, and liquefaction potential evaluation. There are lots of correlations available for obtaining other soil properties from SPT values. These include correlations to evaluate the relative density, angle of internal friction, specific weight, unconfined compressive strength, etc. Readers can refer to Bowles (1997) for more details.

2. *Cone Penetration Test* (*CPT*): Cone penetration test (CPT) is an in situ test done to determine the soil properties and to get the soil stratigraphy. This test was initially developed by the Dutch Laboratory for Soil Mechanics (in 1955), and hence it is sometimes known as the Dutch cone test. There are different correlations available in the literature relating CPT values with other soil properties. Using the CPT results, the soil type can be identified from the soil behavior type index (I_c).
3. *Seismic Cone Penetration Test* (*SCPT*): The seismic cone penetration test uses a standard cone penetrometer with two geophones. One set of geophones is located behind the friction sleeve, and the other set is located 1 meter above the first set. It can also be fitted with a geophone or an accelerometer fitted above the friction sleeve. The test method consists of measuring the travel time of seismic waves propagating between a wave source and ground surface.
4. *Dilatometer Test*: Dilatometer consists of a stainless steel blade with a thin flat circular expandable steel membrane on one side. Dilatometer is advanced into a bore-hole from the ground surface and tests are conducted at an interval of 10–20 cm. At each interval, the dilatometer is stopped and membrane is inflated under gas pressure. The readings of inflation of the membrane and corresponding gas pressure are recorded. There are correlations available to relate the test results with low-strain soil stiffness and liquefaction resistance of soil.
5. *Pressuremeter Test*: A pressuremeter is a cylindrical device that uses flexible membrane for application of uniform pressure to the walls of a borehole. It measures the stress-deformation behavior of the soil by measuring volume of fluid injected into the flexible membrane and the corresponding pressure applied. This is the only in situ test that can measure stress-strain as well as strength behavior of in situ soil.

Table 9.1 Comparison of three major in situ tests (Youd et al., 2001)

Features	SPT	CPT	V_s
Past measurements at liquefaction sites	Abundant	Abundant	Limited
Type of stress-strain behavior influencing test	Partially drained, large strain	Drained, large strain	Small strain
Quality control and repeatability	Poor to good	Very good	Good
Detection of variability of soil deposits	Good for closely spaced tests	Very good	Fair
Soil types in which test is recommended	Non-gravel	Non-gravel	All
Soil sample retrieved	Yes	No	No
Test measures index or engineering property	Index	Index	Engineering

6. *Field Vane Shear Test*: Vane shear test is done on fully saturated clays for evaluation of undrained shear strength. This is very suitable for soft clays whose shear strength (less than 100 KPa) will be changed considerably by sampling. The equipment consists of stainless steel vane (with four blades) connected to the end of a high tensile rod. The rod is enclosed by a sleeve packed by grease. The typical dimension of the vanes is usually 50 mm by 100 mm or 75 mm by 150 mm and the diameter of the rod should not exceed 12.5 mm (BIS-4434, 1978). After pushing the vane and the rod into the clay, torque is gradually applied to the top end of the rod till the clay fails in shear. The shear strength can be calculated based on the torque applied and the test is repeated at the desired depths.

The advantages and disadvantages of three commonly used in situ tests are listed in Table 9.1 (Youd et al., 2001).

3.1.3 Laboratory Tests

These tests are usually performed on very small representative samples. The success of the laboratory testing depends heavily on simulating the actual field conditions (initial state of the sample and the loading conditions). Since the dynamic soil properties depend on lots of factors, the preparation of soil sample for testing needs to be done with utmost care. Some of the laboratory tests used for evaluating dynamic properties of soil are given below:

1. *Resonant Column Test* (*low-strain test*): This is the most common low-strain test which is being used to evaluate the dynamic properties of soil. In this test the soil sample, either solid or hollow cylindrical samples are subjected to torsional or axial loading using an electromagnetic loading system. The fundamental frequency of the sample can be evaluated and this in turn will give the value of the

shear modulus of the soil specimen. Even though the resonant column test is very good in evaluating the damping and strain-dependent properties of soils, the response will depend on the response of the apparatus also.

2. *Bender Element Test* (*low-strain test*): In this test the shear wave velocity of laboratory specimen can be measured using a piezoelectric bender element. A transmitter and receiver elements (piezoelectric) are placed at each end of the sample. There will be a change in dimension of these piezoelectric elements when subjected to change in voltage. An electric pulse applied to the transmitter causes it to deform rapidly and produce a stress wave that will travel through the specimen toward the receiver. When the stress wave reaches the receiver, it generates a voltage pulse and this is measured. The wave speed is calculated from the arrival time and the known distance between transmitter and receiver.
3. *Cyclic Triaxial Tests* (*high-strain test*): The test device consists of the standard triaxial testing equipment with a cyclic axial loading unit. In some cases, the cell pressure is also applied cyclically and it is possible to simulate isotropic or anisotropic initial stress conditions. The values of the shear modulus and damping ratio can be obtained from the stress-strain response of the samples. Cyclic triaxial test is very useful in determining the liquefaction potential of the soil.
4. *Cyclic Direct Simple Shear Test* (*high-strain test*): The cyclic direct simple shear test can simulate the earthquake loading more precisely than the cyclic triaxial test and hence this is one of the tests which is commonly used for liquefaction testing. When a cyclic shear stress is applied to the top or bottom of the specimen, the deformation is similar to that of a soil element in which there is a vertical propagation of S wave.
5. *Cyclic Torsional Test:* In cyclic torsional test, cylindrical soil specimen is subjected to torsion and hence the reversibility in shear stress can be achieved when compared to the cyclic triaxial and cyclic simple shear tests. This test is the most suitable and the most commonly used one to measure stiffness and damping characteristic for wide range of strain level. While testing on solid specimen, it was noted that the shear strain is maximum at the outer edge and close to zero along the axis of the specimen. In order to achieve the radial uniformity of shear strains, hollow cylindrical specimens are preferred over solid cylindrical specimens.

The above tests have been used for evaluating the modulus reduction curves – G/G_{max} with shear strain and the damping ratios with shear strains.

4 Seismic Site Characterization Methods

One of the most important steps in earthquake hazard evaluation is site characterization. This involves acquisition, synthesis and interpretation of qualitative and quantitative information about the site of interest. The scale (number of tests per grid points) at which the insitu field testing can be carried out for site response studies is also an important parameter. The number of field tests will vary with the scale of study and heterogeneity of the soil. When the study areas are very large, guidelines of geotechnical testing for railway lines, tunnels, and highways becomes the yard stick. Seismic site characterization can be carried out based on the following methods.

4.1 Based on Geology

The first step in a site characterization is the geological investigation of the region to determine the geological settings. Geological maps, aerial photographs and field reconnaissance performed by geologists can be used for this purpose.

4.2 Based on Geotechnical Data

Site response and ground failure are strongly influenced by the properties of soil. Geotechnical reports for the sites may be available from various governmental or nongovernmental agencies. The traditional methods like drilling and undisturbed sampling can provide adequate stratigraphic details and estimates of geotechnical parameters. But they cannot provide useful estimates of hydrogeological conditions which are very important in estimating the seismic hazards like liquefaction, landslides, etc.

Dynamic response of soils under dynamic loads depends on the cyclic stress-strain characteristics of the soil in shear. The small-strain shear modulus (G_{max}), the shear modulus ratio (G/G_{max}) and the cyclic shear strain amplitude are the basic characteristics of soil deformation that play an important role in dynamic response analysis (Sun et al., 1988; Seed et al., 1986). During earthquakes, soils are subjected to irregular dynamic loads that cause degradation of stiffness and shear strength with respect to number of cycles.

The other important parameter controlling the cyclic stress–strain characteristics of soils is the number of cycles. This parameter plays a crucial role especially in analyzing the behavior of soil layers under earthquake loads.

The insitu tests generally conducted to identify the soil stratification and engineering properties of the soil layers are penetration tests. Two methods that have been widely used are the Standard Penetration Test (SPT) and Cone Penetration Test (CPT). SPT is generally used to investigate cohesionless or relatively stiff soil deposits, whereas CPT is used to identify soil properties in soft soil deposits (Lunne et al., 1997).

4.3 Based on Geophysical Investigations

Seismic tests are classified into borehole (invasive) and surface (non invasive) methods. They are based on the propagation of body waves and surface waves, which are associated to very small strain (<0.001%). The shear wave velocity (V_s) is a soil property used to determine the shear modulus (G) of the soil as below:

$$G = \rho V_s^{\,2} \tag{9.1}$$

Seismic tests are also used to determine the material damping ratio by measuring the spatial attenuation of body or surface waves:

$$D_o = \alpha V / 2\pi f \ (D_o < 10\%) \tag{9.2}$$

where D_o is the small strain damping ratio, α- attenuation coefficient, V- velocity respectively, of P, S, or R waves and f- frequency.

Recently most of codes like Eurocode-8 (2003), NEHRP (BSSC, 2003), International Building Code (IBC, 2009), etc. specify the site classification based on the average shear wave velocity values in the top 30 m (V_s^{30}). The amplification of shear waves mainly depends on the density and the shear wave velocity of the overlying soil layer. Since the variation in density of soil is comparatively less, the amplification depends heavily on the shear wave velocity near the earth surface. There are two methods to denote the near-surface shear wave velocity (V_s)—depth corresponding to one quarter wave length of the period of interest and the average shear wave velocity in the top 30 m. The main disadvantage with quarter wavelength V_s is that the depths associated with this will be very deep. Hence the classification based on V_s^{30} is being used more commonly now days. It is calculated using the equation

$$V_s^{30} = \frac{30}{\sum_{i=1}^{N}\left(\frac{d_i}{v_i}\right)} \tag{9.3}$$

where d_i is the thickness of the ith soil layer in meters, v_i is the shear wave velocity for the ith layer in m/s, and N is the no. of layers in the top 30-m soil strata which will be considered in evaluating V_s^{30} values.

Eurocode-8 and NEHRP: A site classification scheme based on V_s^{30} values was proposed by Borcherdt (1994) and a similar scheme was adopted by the National Earthquake Hazard Reduction Program (NEHRP) also. The NEHRP (BSSC, 2003) site classification scheme is given in Table 9.2. Eurocode-8 (2003) has also classified the site based on V_s^{30}, standard penetration test (SPT) and cone penetration test (CPT) values. The classification given by Eurocode-8 is given in Table 9.3. Even though both the schemes use similar methods to identify the site classes, the range of V_s^{30} values specified for each site class is different in both the methods. However IS code BIS-1893 (2002) addresses only three site classes (hard rock, medium soil and soft soil) based on SPT "N" values.

Table 9.2 Site classification as per NEHRP scheme: (BSSC, 2003)

NEHRP Site Class	Description	V_s^{30}
A	Hard rock	>1500 m/s
B	Firm and hard rock	760–1500 m/s
C	Dense soil, soft rock	360–760 m/s
D	Stiff soil	180–360 m/s
E	Soft clays	<180 m/s
F	Special sandy soils, e.g., liquefiable soils, sensitive clays, organic soils, soft clays >36 m thick	

Table 9.3 Site classification adopted by Eurocode–8 (2003)

Ground type	Description of stratigraphic profile	Parameters		
		V_s^{30} (m/s)	SPT	C_U (KPa)
A	Rock or other rock-like geological formation, including utmost 5 m of weaker material at the surface	>800		
B	Deposits of very dense sand, gravel, or very stiff clay, at least several tens of meters in thickness, characterized by a gradual increase of mechanical properties with depth	360–800	>50	>250
C	Deep deposits of dense or medium-dense sand, gravel, or stiff clay with thickness from several tens to many hundreds of meters	180–360	15–50	70–250
D	Deposits of loose-to-medium cohesionless soil (with or without some soft cohesive layers), or of predominantly soft-to-firm cohesive soil	<180	<15	<70
E	A soil profile consisting of a surface alluvium layer with V_s^{30} values of type C or D and thickness varying between about 5m and 20 m, underlain by stiffer material with V_s^{30} > 800 m/s			
S1	Deposits consisting, or containing a layer at least 10 m thick, of soft clays/silts with a high plasticity index (PI > 40) and high water content	<100 (indicative)		10–20
S2	Deposits of liquefiable soils, sensitive clays, or any other soil profile not included in types A–E or S1			

A modified site classification system based on geotechnical data was proposed by Rodriguez-Marek et al. (2001). In this system the stiffness of soil was also taken into account for the site classification. This system is presented in Table 9.4. The main advantage of this system is that it correlates the V_s^{30} values with the geotechnical and surface geological features.

Table 9.4 Classification based on geotechnical features (Rodriguez-Marek et al. 2001)

Site	Description	Comments
A	Hard rock	Crystalline bedrock; $V_s^{30} \geq 1500$ m/s
B	Competent bed rock	$V_s^{30} > 600$ m/s or <6 m of soil. Most unweathered California rock cases
C1	Weathered rock	$V_s^{30} \sim 300$ m/s increasing to >600 m/s, weathering zone >6 m and <30 m
C2	Shallow stiff soil	Soil depth > 6 m and <30 m
C3	Intermediate depth stiff soil	Soil depth > 30 m and <60 m
D1	Deep stiff Holocene soil	Soil depth > 60 m and <200 m
D2	Deep stiff Pleistocene soil	Soil depth > 60 m and <200 m
D3	Very deep stiff soil	Soil depth > 200 m
E1	Medium thickness soft clay	Thickness of soft clay layer 3–12 m
E2	Deep soft clay	Thickness of soft clay layer >12 m
F	Potentially liquefiable sand	Holocene loose sand with high water table, $Z_w \leq 6$ m

Table 9.5 Slope ranges for various NEHRP site classes (after Wald and Allen, 2007)

		Slope Range (m/m)	
Class	V_s^{30} Range (m/s)	Active Tectonic	Stable Continent
E	<180	$<1.0 \times 10^{-4}$	$<2.0 \times 10^{-5}$
D	180–240	1.0×10^{-4} to 2.2×10^{-3}	2.0×10^{-5} to 2.0×10^{-3}
	240–300	2.2×10^{-3} to 6.3×10^{-3}	2.0×10^{-3} to 4.0×10^{-3}
	300–360	6.3×10^{-3} to 0.018	4.0×10^{-3} to 7.2×10^{-3}
C	360–490	0.018 to 0.050	7.2×10^{-3} to 0.013
	490–620	0.050 to 0.10	0.013 to 0.018
	620–760	0.1 to 0.138	0.018 to 0.025
B	>760	>0.138	>0.025

4.4 Based on Topographic Slope

Recent studies by Wald and Allen (2007) have shown that topographic variations can be an indicator to near-surface geomorphology and lithology to the first order. The slope of surface is having good correlation with the site conditions – with steep mountains indicates rocky terrain, nearly flat basins indicates soil and intermediate slopes indicates a transition from rock to soil. Matsuoka et al. (2005) have found a good correlation between V_s^{30} and the slope of the topography for Japan. Chiou and Youngs (2006) have correlated V_s^{30} with the terrain elevation for Taiwan. Wald and Allen (2007) computed V_s^{30} for the entire California region based on slope map. Table 9.5 presents a correlation of V_s^{30} values with the terrain slope for the California region as proposed by Wald and Allen (2007). Slope map at a macroscale can be obtained from a digital elevation model for that region; hence site characterization based on topographic slopes is very useful for local site estimation at macro-scale.

5 Evaluation of Ground Response Spectra and PGA Values for Different Site Classes

When the study area is very vast, it is extremely difficult to classify the region into different site classes based on the geotechnical or geophysical data. Hence in those cases the PGA values can be evaluated for different NEHRP or Eurocode site classes. Based on site investigation, we can identify the site class at any location can be determined and for the appropriate site class, the PGA or S_a values can be obtained from the respective maps.

The response of a structure to the earthquake ground motion is given by response spectra, and hence it has evolved as the backbone of earthquake structural-resistant design. Despite some of the shortcomings of this method, it remains popular among the practicing engineers. There are various provisions adopted by codes for smoothing the response spectra. A comparison of smoothed response spectra obtained using various codal provisions for Mumbai is shown in Fig. 9.5. The details of the calculation are available at Vipin and Sitharam (2010).

6 Liquefaction Potential Evaluation

During an earthquake, soil can fail due to liquefaction with devastating effect such as land sliding, lateral spreading, or large ground settlement. The phenomenon of seismic soil liquefaction had been observed for many years, but was brought to the

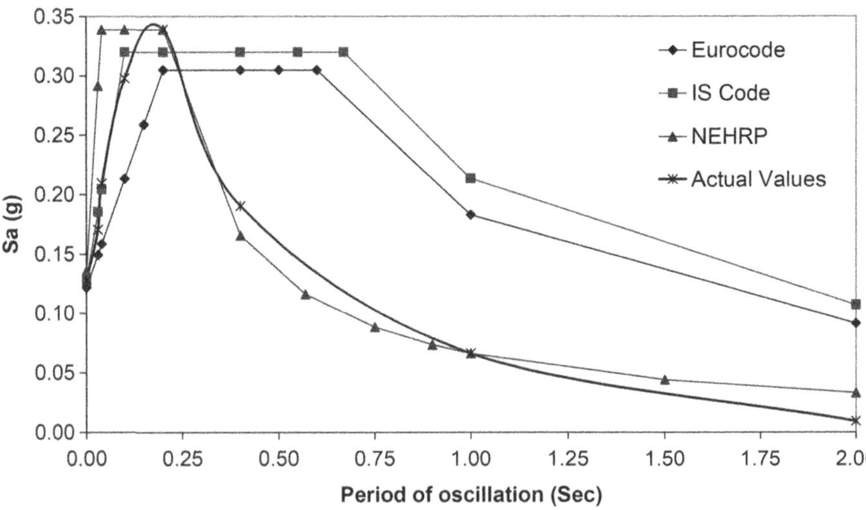

Fig. 9.5 Comparison of actual (simulated) and design response spectra for site class D (Mumbai) based on different codal provisions (return period, 475 years)

attention of researchers after Niigata (1964) and Alaska (1964) earthquakes. Liquefaction is a phenomenon in which the strength and stiffness of a soil is reduced by earthquake shaking or by other sudden dynamic loading. This is due to the reduction in effective stress of the soil due to the sudden earthquake loading. In the case of saturated soils, the sudden loading will cause an increase in pore fluid pressure, and this will reduce the effective stress. In the case of dry soils, liquefaction can occur with increase in pore air pressure (Lohse et al., 2004).

6.1 Evaluation of Liquefaction Susceptibility

The liquefaction susceptibility is evaluated by considering the soil properties alone, without considering the earthquake loading. If a soil at a particular site is susceptible to liquefaction then only is it prone to the liquefaction hazards. The important factors that will decide the susceptibility of soil liquefaction at the site are:

- Type of soil (index properties of soil)
- Shape of soil particles
- Permeability of soil
- Presence of seismic waves
- Depth of groundwater table
- Historical environment
- Age of soil
- Confining pressure
- Relative density of soil
- Natural soil deposits in water bodies

6.2 Evaluation of Liquefaction Potential

There are a number of approaches available to evaluate the liquefaction potential of the soils such as cyclic strain approach, cyclic stress approach, energy dissipation approach, effective stress-based response analysis approach and probabilistic approach. Out of these approaches, cyclic stress approach is most commonly used due to its simplicity and robustness to accurately model earthquake-induced stresses within the ground. Over the years many design charts and correlations were developed based on cyclic stress approach for the estimation of liquefaction resistance of soils through laboratory as well as in situ tests. However the cyclic strain approach is also becoming popular.

The evaluation of liquefaction potential involves two stages – (1) evaluation of earthquake loading and (2) evaluation of soil strength against earthquake loading. The earthquake loading on soil is expressed using the term cyclic stress ratio (CSR), and the soil strength (capacity of soil) to resist liquefaction is expressed using cyclic resistance ratio (CRR).

One of the first methods to quantify the liquefaction potential of soils is the simplified procedure developed by Seed and Idriss. The "simplified method" suggested by Seed and Idriss to evaluate the cyclic stress ratio (CSR) values is

$$\mathrm{CSR} = 0.65 \frac{a_{\max}}{g} \frac{\sigma_{vo}}{\sigma'_{vo}} \frac{r_d}{\mathrm{MSF}} \tag{9.4}$$

where $a_{\max}$ is the peak ground acceleration (at surface level), σ_{vo} and σ'_{vo} are the total and effective over burden pressure, r_d is the depth reduction factor, and MSF is the magnitude scaling factor. The above relation was developed for an earthquake of magnitude M_w -7.5, and if the magnitude of an earthquake is different from this, it is being taken care by the MSF.

Evaluation of the liquefaction resistance can be obtained from both laboratory and field tests. The former method is seldom used because of the cost and the difficulties involved in getting undisturbed samples.

6.2.1 Liquefaction Resistance of Soils from Field Tests

The difficulties and the high cost involved in the laboratory test make the insitu test a convenient method for evaluating liquefaction potential. The four major in situ test methods which are considered for the liquefaction potential evaluation are given below:

1. The standard penetration test (SPT)
2. The cone penetration test (CPT)
3. Measurement of in situ shear wave velocity (*Vs*)
4. The Becker penetration test (BPT)

The curves between cyclic resistance ratio (CRR) and the corrected SPT values suggested by Youd et al. (2001) for liquefaction potential evaluation are given in (Fig. 9.6). Another commonly adopted method to assess the liquefaction potential for clean sand from CPT data was suggested by Robertson and Wride (1998) and is shown in (Fig. 9.7). The liquefaction potential can also be assessed based on shear wave velocity values. The correlation between CRR and the corrected shear wave velocity was proposed by Andrus and Stokoe (2000), and shown in (Fig. 9.8).

Of the above mentioned tests, the most commonly used tests are based on SPT. A brief description of some of the important methods adopted for liquefaction potential evaluation based on SPT values is discussed in this chapter.

The most widely used methods to evaluate the liquefaction potential based on SPT values were proposed by Seed and Idriss, Seed and Peacock (1971) and Seed et al. (1984). The NCEER workshops in 1996 and 1998 (Youd et al., 2001) resulted in a number of suggested revisions to the SPT-based procedure. Idriss and Boulanger (2004) presented a revised curve between CSR and modified SPT value based on the reexamination of the available field data.

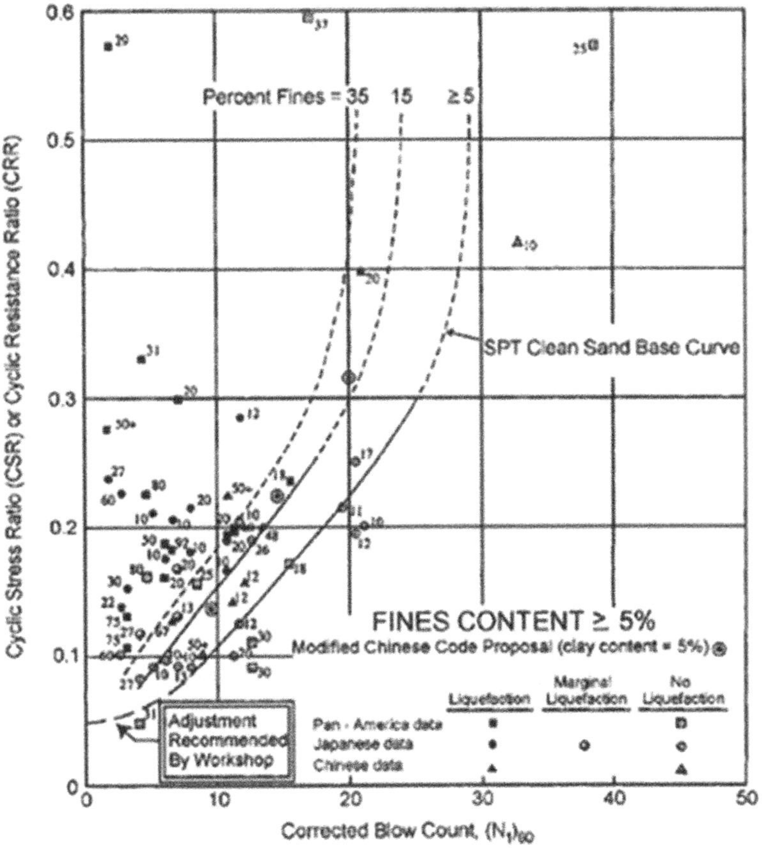

Fig. 9.6 SPT-based deterministic cyclic resistance curves proposed by Youd et al. (2001)

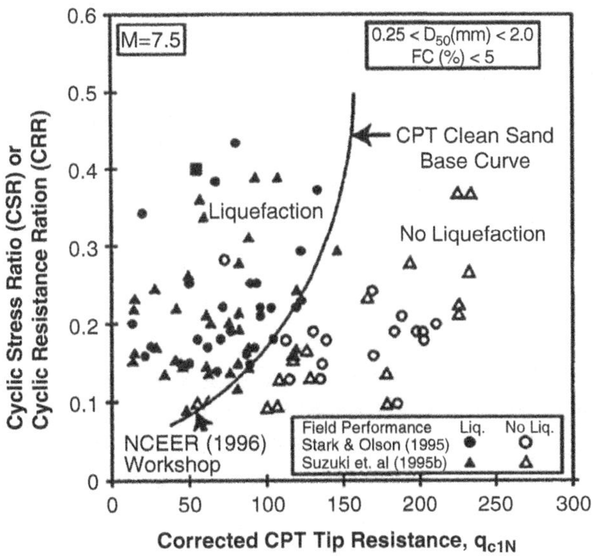

Fig. 9.7 CPT-based liquefaction triggering correlation proposed by Robertson and Wride (1998) for clean, dry sands

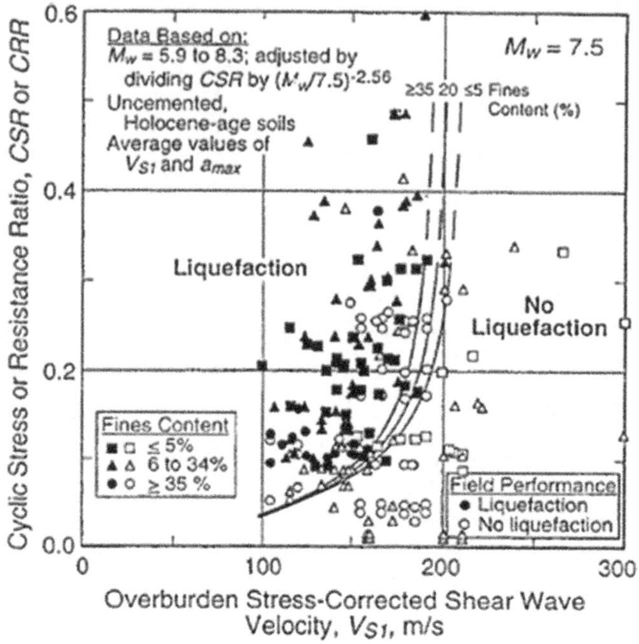

Fig. 9.8 V_S-based liquefaction triggering correlation as proposed by Andrus and Stokoe (2000)

6.2.2 Probabilistic Methods

A number of researchers have developed correlations based on probabilistic methods to evaluate the liquefaction potential. One of the first to develop a relation was Liao et al. (1988) and subsequently other relations were suggested by Youd and Noble (1997), Toprak et al. (1999), etc. The relationships provided by these researchers are in the form of probability contours (probability of triggering soil liquefaction). A recent and comprehensive work in the area of probabilistic liquefaction potential evaluation was done by Cetin et al. (2004). The probability of liquefaction can be evaluated using the procedure suggested by Cetin et al. (2004). A probabilistic performance-based approach for liquefaction potential evaluation was suggested by Kramer and Mayfield (2007). In this approach, the uncertainty in the earthquake loading for the initiation of liquefaction is explicitly included. Readers can refer Vipin et al. (2010) and Vipin and Sitharam (2011) for more details of this analysis.

The liquefaction potential of Gujarat was evaluated using the performance-based approach, in terms of SPT values required to prevent liquefaction. The liquefaction hazard curve for selected cities in Gujarat is shown in (Fig. 9.9). These curves give the variation of corrected SPT values ($N_{1,60,cs}$) required to prevent liquefaction against the mean annual rate of exceedance. Based on a geotechnical investigation

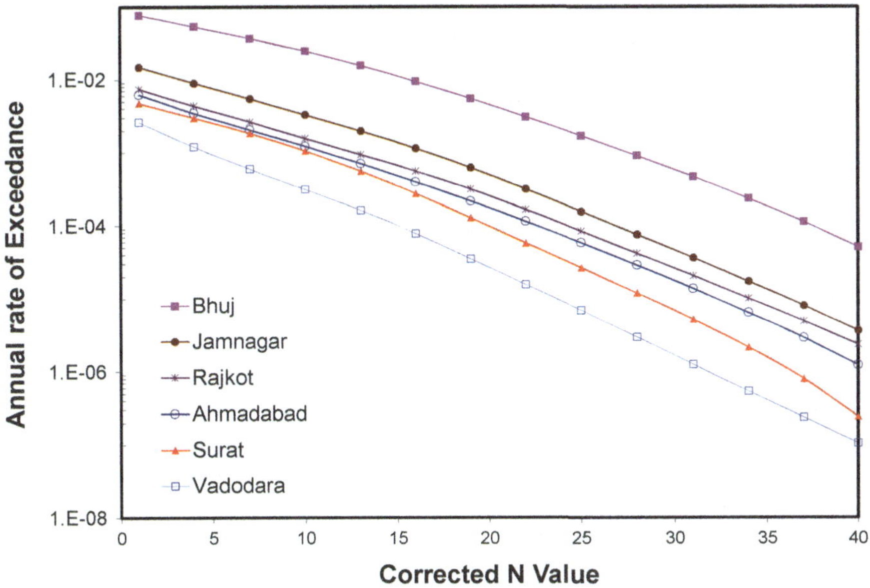

Fig. 9.9 Liquefaction hazard curve for some of the selected cities in Gujarat

the corrected SPT values need to be evaluated for any of these locations. If the corrected SPT value is higher than the SPT value obtained from the curve, for a given return period, then the site is safe against liquefaction for that given return period. For example, if the corrected SPT value for a site at Bhuj is 20, then it can be seen that the location is safe against liquefaction for a return period of 100 years (approximately), whereas for the same SPT value at Vadodara, the liquefaction return period is of the order of $5x10^4$ years. This clearly indicates that the Bhuj region is more vulnerable to liquefaction than Vadodara.

7 Conclusions

An overview of various methods adopted for site characterization studies and evaluation of dynamic soil properties are discussed in this chapter with an emphasis on the recent advances in these techniques. Selection of any particular method also depends upon the scale at which seismic hazards need to be assessed. Even though the accuracy of site characterization and local site effect assessment at macroscale is low, still it is acceptable for preliminary hazard estimation and demarcating areas of high hazard. Site-specific analysis is a rigorous procedure where complete categorization of soil profile needs to be carried out based on lager number of geotechnical and geophysical test data. Hence it is economically nonviable for small scale level of infrastructure developments and normally recommended for sites having critical installations like nuclear power plant or dam. For all the critical sites, geotechnical and geophysical tests are to be carried out at very close interval. Based on the soil profile data (obtained from bore logs) and shear wave velocity profile (obtained from geophysical tests), the site amplification factor can be obtained using equivalent/non linear ground response analysis. For vast area, the initial assessment of liquefaction potential needs to be done based on the SPT or CPT data. For much more in-depth analysis in critical cases, laboratory testing of the soil samples need to be done.

References

Aki, K. (1988). Local site effects on strong ground motion. *Proceedings of the Earthquake Engineering and Soil Dynamics II, Park City, Utah, June 27–30,* 103–155.

Anbazhagan, P., & Sitharam, T. G. (2008). Seismic microzonation of Bangalore, India. *Journal of Earth System Science, 117,* 833–852.

Andrus, R. D., & Stokoe, K. H., II. (2000). Liquefaction resistance of soils from shear wave velocity. *Journal of Geotechnical and Geoenvironmental Engineering, 126*(11), 1015–1025.

Astroza, M., & Monge, J. (1991). Seismic microzones in the city of Santiago. Relation damage-geological unit. *Proceedings of the Fourth International Conference on Seismic Zonation, Earthquake Engineering Research Institute, Stanford, CA, USA, August 25–29,* 3, 595–601.

Athanasopoulus, G. A., Pelekis, P. C., & Leonidou, E. A. (1999). Effects of surface topography on seismic ground response in the Egion (Greece) 15 June 1995 earthquake. *Soil Dynamics and Earthquake Engineering, 18*(2), 135–149.

BIS-1893. (2002). *Criteria for earthquake resistant design of structures, part 1 – general provisions and buildings.* New Delhi: Bureau of Indian Standards.

BIS-4434. (1978). *Code of practice for in-situ vane shear test for soils (first revision).* New Delhi: Bureau of Indian Standards.

Borcherdt, R. D. (1994). Estimates of site-dependent response spectra for design (methodology and justification). *Earthquake Spectra, 10*(4), 617–653.

Borcherdt, R. D., & Gibbs, J. F. (1976). Effects of local geological conditions in the San Francisco Bay region on ground motions and the intensities of the 1906 earthquake. *Bulletin of the Seismological Society of America, 66,* 467–500.

Bowles, J. E. (1997). *Foundation analysis and design.* Singapore: McGraw-Hill.

BSSC. (2003). *NEHRP recommended provisions for seismic regulations for new buildings and other structures (FEMA 450), part 1: provisions.* Washington, DC: Building Seismic Safety Council for the Federal Emergency Management Agency.

Cetin, K. O., Seed, R. B., Kiureghian, D. A., Tokimastu, K., Harder, L. F., Kayen, R. E., et al. (2004). Standard penetration test-based probabilistic and deterministic assessment of seismic soil liquefaction potential. *Journal of Geotechnical and Geoenvironmental Engineering, 130* (12), 1314–1340.

Chávez-García, F. J., Sanchez, L. R., & Hatzfeld, D. (1996). Topographic site effects and HVSR: A comparison between observations and theory. *Bulletin of the Seismological Society of America, 86,* 1559–1573.

Chiou, B. S.-J., & Youngs, R. R. (2006). PEER-NGA empirical ground motion model for the average horizontal component of peak acceleration and pseudo-spectral acceleration for spectral periods of 0.01 to 10 seconds. *Interim Report for USGS Review, 23,* 219 pp. http://peer.berkeley.edu/lifelines/repngamodels.html.

Eurocode-8. (2003). *BS-EN 1998-1, "Design of structures for earthquake resistance—Part 1: General rules, seismic actions and rules for buildings"*. Brussels: European Committee for Standardization.

Evernden, J. F., & Thomson, J. M. (1985). Predicting seismic intensities. In J. I. Ziony (Ed.), *Evaluating earthquake hazards in the Los Angeles region —An earth-science perspective,* U.S. Geological Survey Professional Paper, 1360, 151–202.

Faccioli, E. (1991). Seismic amplification in the presence of geological and topographic irregularities. In S. Prakash (Ed.), *Proceedings of the second international conference on recent advances in geotechnical earthquake engineering and soil dynamics, St. Louis, Missouri,* Rolla, MO: University of Missouri-Rolla, 2, 1779–1797.

Fäh, D., Rüttener, E., Noack, T., & Kruspan, P. (1997). Microzonation of the City of Basel. *Journal of Seismology, 1*, 87–102.

Field, E. H., & Jacob, K. H. (1993). The theoretical response of sedimentary layers to ambient seismic noise. *Geophysics Research Letters, 20*, 2925–2928.

Geli, L., Bard, P. Y., & Jullen, B. (1988). The effect of topography on earthquake ground motion: A review and new results. *Bulletin of the Seismological Society of America, 78*, 42–63.

Hunter, J. A., Benjumea, B., Harris, J. B., Miller, R. D., Pullan, S. E., & Burns, R. A. (2002). Surface and downhole shear wave seismic methods for thick soil site investigations. *Soil Dynamics and Earthquake Engineering, 22*, 931–941.

IBC. (2009). *International building code*. Washington: International Code Council.

Idriss, I. M. (1990). Response of soft soil sites during earthquakes. *Proceedings of Memorial Symposium to Honor Professor H. B. Seed. Berkeley, CA.*

Idriss, I. M., & Boulanger, R. W. (2004). Semi-empirical procedures for evaluating liquefaction potential during earth-quakes. *Proceedings of 11th Int. Conf. on Soil Dynamics & Earth-quake Engineering & 33d Int. Conf. on Earthquake Geotechnical Engineering, Berkeley,* 32–56.

Kagami, H., Okada, S., & Ohta, G. (1988). Versatile application of dense and precision seismic intensity data by an advanced questionnaire survey. *Proceedings of 9th World Conference on Earthquake Engineering,* 8, 937–942.

Kramer, S. L. (1996). *Geotechnical earthquake engineering*. Englewood Cliffs, NJ: Prentice Hall Publishers.

Kramer, S. L., & Mayfield, R. T. (2007). Return period of soil liquefaction. *Journal of Geotechnical and Geoenvironmental Engineering, 133*(7), 802–813.

Langston, C. A. (1979). Structure under mount rainier, Washington, inferred from teleseismic body waves. *Bulletin of the Seismological Society of America, 84*(B9), 4749–4762.

Liao, S. S. C., Veneziano, D., & Whitman, R. V. (1988). Regression models for evaluating liquefaction probability. *Journal of Geotechnical Engineering, 14*(4), 389–411.

Lohse, D., Rauhé, R., Bergmann, R., & van der Meer, D. (2004). Granular physics: Creating a dry variety of quicksand. *Nature, 432*, 689–690.

Lunne, T., Robertson, P. K., & Powell, J. J. M. (1997). *Cone penetration testing in geotechnical practice*. London: Blackie Academic and Professional.

Matsuoka, M., Wakamatsu, K., Fujimoto, K., & Midorikawa, S. (2005). Nationwide site amplification zoning using GIS-based Japan Engineering Geomorphologic Classification Map. *Proceedings of 9th International Conference on Structural Safety and Reliability,* 239–246.

Mayer-Rosa, D., & Jimenez, M.-J. (1999). *Seismic zoning, recommendations for Switzerland.* Vorbereitung: Landeshydrologie und -Geologie, Geologischer Bericht.

Medvedev, J. (1962). *Engineering seismology* (Vol. 260). Moscow: Academia Nauk Press.

Midorikawa, S. (1987). Prediction of isoseismal map in the Kanto plain due to hypothetical earthquake. *Journal of Structural Engineering, 33B*, 43–48.

Mukhopadhyay, S., Pandey, Y., Dharmaraju, R., Chauhan, P. K. S., Singh, P., & Dev, A. (2002). Seismic microzonation of Delhi for ground shaking site effects. *Current Science, 82*, 877–881.

Nakamura, Y. (1989). A method for dynamic characteristics estimation of subsurface using microtremor on the ground surface. *Quarterly Report of Railway Technical Research Institute, 30*(1), 25–33.

Park, C. B., Miller, R. D., & Xia, J. (1999). Multi-channel analysis of surface waves. *Geophysics, 64*(3), 800–808.

Reinoso, E., Wrobel, L. C., & Power, H. (1997). Three-dimensional scattering of seismic waves from topographical structures. *Soil Dynamics and Earthquake Engineering, 16*, 41–61.

Robertson, P. K., & Wride, C. E. (1998). Evaluating cyclic liquefaction potential using the cone penetration test. *Canadian Geotechnical Journal, 35*(3), 442–459.

Rodriguez-Marek, A., Bray, J. D., & Abrahamson, N. A. (2001). An empirical geotechnical seismic site response procedure. *Earthquake Spectra, 17*(1), 65–87.

Seed, H. B., & Idriss, I. M. (1971). Simplified procedure for evaluating soil liquefaction potential. *Journal of the Soil Mechanics and Foundations Division, 97*, 1249–1273.

Seed, H. B., & Peacock, W. H. (1971). Test procedures for measuring soil liquefaction characteristics. *Journal of the Geotechnical Engineering Division ASCE, 97*(8), 1099–1119.

Seed, H. B., Tokimatsu, K., Harder, L. F., & Chung, R. M. (1984). The influence of SPT procedures in soil liquefaction resistance evaluations. In *Report no. UCB/EERC-84/15*. Berkeley, CA: Earthquake Engineering Research Center.

Seed, H. B., Wong, R. T., Idriss, I. M., & Tokimastu, K. (1986). Moduli and damping factors for dynamic analysis of cohesionless soils. *Journal of Geotechnical Engineering, ASCE, 112*(11), 1016–1032.

Seed, R. B., & Harder, L. F. (1990). SPT-based analysis of cyclic pore pressure generation and undrained residual strength. In *H.Bolton seed memorial symposium proceedings* (Vol. 2). Vancouver, BC: BiTech Publishers Ltd.

Shima, E. (1978). Seismic microzoning map of Tokyo. *Proceedings of 2nd International Conference on Microzonation*, 1, 433–443.

Sun, J. I., Golesorkhi, R., & Seed, H. B.. (1988). Dynamic moduli and damping ratios for cohesive soils. *EERC Report No.UCB/EERC-88/15.*

Toprak, S., Holzer, T. L., Bennett, M. J., & Tinsley, J. C. III. (1999). CPT and SPT based probabilistic assessment of liquefaction potential. *Proceedings of 7th U.S.-Japan Workshop on Earthquake Resistant Des. of Lifeline Facilities and Countermeasures Against Liquefaction, Technical Report MCEER-99-0019, Multidisciplinary Center for Earthquake Engineering Research, Buffalo, NY,* 69–86.

Vipin, K. S., Anbazhagan, P., & Sitharam, T. G. (2010). Probabilistic evaluation of seismic soil liquefaction potential based on SPT data. *Natural Hazards, 53*, 547–560.

Vipin, K. S., & Sitharam T. G. (2010). Development of site specific design response spectrum based on different codal provisions. *14th Symposium on Earthquake Engineering, IIT Roorkee*, 212–223.

Vipin, K. S., & Sitharam, T. G. (2011). Evaluation of liquefaction return period based on local site classes: Probabilistic performance based logic tree approach. *International Journal of Geotechnical Engineering, 5*, 245–254.

Wald, J. D., & Allen, I. T. (2007). Topographic slope as a proxy for seismic site conditions and amplification. *Bulletin of the Seismological Society of America, 97*(5), 1379–1395.

Youd, T. L., Idriss, I. M., Andrus, R. D., Arango, I., Castro, G., Christian, J. T., et al. (2001). Liquefaction resistance of soils: Summary report from the 1996 NCEER and 1998 NCEER/NSF workshops on evaluation of liquefaction resistance of soils. *Journal of Geotechnical and Geoenvironmental Engineering, 127*(10), 817–833.

Youd, T. L., & Noble, S. K. (1997). Liquefaction criteria based on statistical and probabilistic analyses. *Proceedings of the NCEER Workshop on Evaluation of Liquefaction Resistance of Soils, Technical Report NCEER-97-0022, Multidisciplinary Center for Earthquake Engineering Research, Buffalo, New York.*

Youd, T. L., & Perkins, D. M. (1978). Mapping liquefaction-induced ground failure potential. *Journal of the Geotechnical Engineering Division, ASCE, 104*, 443–446.

Part III
Structural Dynamics

Chapter 10
Historical Development and Present Status of Earthquake Resistant Design of Bridges

Mahesh Tandon

1 Introduction

A bridge collapse during earthquakes may not result in many lives lost as compared to that in a building collapse, but the economic loss can be truly substantial. Bridges often provide a vital link to earthquake-ravaged areas and hence have a vital post-disaster purpose. Critical bridges must remain functional after the event at all costs to provide relief and also for security and defense purposes. This concept is the basis of the so-called performance-based approach towards which all present-day international codes are veering towards.

For far too long there has been a misleading impression that earthquake forces are a low-level static force. In fact they are a high magnitude dynamic force induced by mass inertia due to severe ground shaking. Thankfully, the gap between the impressions of the past and the reality which has always been present seems now to be narrowing.

When there was not enough knowledge about into the subject of earthquake engineering the discipline was merely thought of as an extension of civil or structural engineering. It was the geologists, seismologists and others who provided the insight into the generation of earthquakes and the probability of their occurrence. Seismic zonations and micro-zonation can be said to be almost entirely the contribution of their expertise.

M. Tandon (✉)
Tandon Consultants Pvt. Ltd., New Delhi, India
e-mail: mahesh.tandon@tcpl.com

M. L. Sharma et al. (eds.), *Advances in Indian Earthquake Engineering and Seismology*, https://doi.org/10.1007/978-3-319-76855-7_10

2 Early Developments in India

Pioneering work in the field of earthquake engineering was done by the Geological Survey of India after the 1897 Assam Earthquake (Jain and Nigam, 2000; Tandon and Srivastava, 1974). Apart from making an attempt to put the knowledge about earthquakes on a scientific basis and making a historical catalogue of earthquakes in various parts of India, an earthquake-resistant type of housing was developed which is still in evident in Northeast India.

The pioneering work was revived at the institution rechristened as IIT Roorkee with the establishment of the School of Research and Training in Earthquake Engineering (SRTEE) in 1961. This department was amongst the few of its kind in the world which were devoted exclusively to the teaching, research, training and consultancy in the field of earthquake engineering. Apart from creating a country wide awareness of the discipline, it must also be credited with formulating the first edition of IS:1893 in 1962, a code which is currently in its fifth revision (Tandon and Srivastava, 2002).

3 Early Days of the International Scenario

Research and codification for earthquake-resistant design of buildings has always preceded that for bridges.

The Italians were the pioneers for developing a criteria for horizontal forces to be applied to structures to resist the onslaught of earthquakes. This was in the wake of the 1908 Messina earthquake ($M = 7.5$). The criteria were developed at the instance of the Government by a special panel appointed by it that consisted of practitioners and academicians. They suggested a static horizontal force of $0.125 \times W$ to be applied to the structure, where $W =$ total vertical load of structure.

The Japanese, after the 1923 Kanto earthquake ($M = 7.9$), when many bridges and buildings were destroyed, followed suit with a similar static horizontal force criteria of $0.1 \times W$.

The Los Angeles City in the United States woke up only after the devastation caused by Santa Barbara (1925), Long Beach (1933) and other earthquakes. They also came up with a similar horizontal force criteria, i.e., $0.08 \times W$, for all buildings except schools, which were to be designed for $0.10 \times W$. To their credit it must be said that the Los Angeles City code was the first to introduce changes in the code in 1943 which indirectly took into account the dynamic properties of the structure by relating the forces to the height of the building. Then in 1957, it incorporated directly the flexibility of the structure in the code.

On the bridges front, in Japan, the draft Specifications for Highway Bridge, 1939 specified a design seismic coefficient of 0.20 g. The 1956 specifications revised it to a range of 0.1–0.35 g. Despite this large increase in the static horizontal force, failures were observed during the 1964 Niigata earthquake ($M = 7.5$) and the 1968 Tokachioki earthquake ($M = 7.9$).

California State in the western part of the United States of America is amongst the most seismically active areas in the world. Caltrans (previously called the California State Highways Department), from 1943 when the first criteria was published upto 1965, the static horizontal seismic force was still considered at about 0.06 g level. In 1971, during the San Fernando earthquake ($M = 6.6$) several freeway structures, apart from buildings collapsed. This event resulted in a sharp increase of 2–2.5 times in the design forces in the Caltrans jurisdiction.

The devastation caused by the San Fernando earthquake led to considerable introspection amongst the engineering community in California and could be said to have triggered finally a changeover to an inelastic design criteria for bridges. It was realized that merely upgrading the capacity demand every time there were damages due to earthquake was not the answer. The AASHTO Interim Specifications (1975) indicated a ground acceleration up to 0.5 g with response reduction factors and ductility requirements. The Japanese followed soon thereafter and in 1981 published the criteria of inelastic design and ductility requirements with ground accelerations reaching 1.0 g.

As can be concluded, since the last four or five decades a dynamics approach taking account of the flexibility of the structure has become universally accepted as the appropriate method of analysis for earthquakes. And three decades ago, it was realized that structural response went well into the inelastic range and that it was necessary to introduce ductility into the structural system.

Historically speaking (Housner, 1997; Beckett and Alexandrou, 1997) we can identify three major milestones in the discipline of earthquake engineering as far as structural analysis is concerned:

1. Development of criteria of horizontal static forces to simulate the effect of earthquakes (1908).
2. Accounting for the flexibility of the structure while evaluating the strength demand placed by earthquakes (1943).
3. Recognizing that structures enter the inelastic (or post-elastic) stage during severe ground shaking and that their safety depended on ductility (1975).

4 Differences with Respect to Buildings

While considerable research and codification of seismic design has been done in relation to buildings worldwide, the same quantum of effort and interest is not apparent for bridges. The structural form of bridges is different from those of buildings, including the nature of loading, degree of indeterminacy and the potential of dissipating energy. Whereas the most vulnerable portion of a building is its superstructure, it is the bearings, substructure, and foundation that are most susceptible to damage in a bridge during an earthquake. The design life of a building is almost universally accepted as 50 years, while that for bridges it is 100 years. These factors must be taken into account while formulating the codes of practice for the two types of structures.

5 Present Status

5.1 Major Shift in Approach

With the occurrence of every major earthquake, there has been in the past, almost a worldwide tendency to increase the capacity demand of the structure to counteract such events. It is only in the last few years that new strategies have been successfully developed to handle this problem economically.

The current international practice has shifted toward a performance-based engineering design, wherein the accent is on serviceability and safety under different levels of magnitude of earthquakes for structures of varying importance. Also there is an increasing realization that apart from techniques for improving ductility, the structural engineer's tool box should include energy-dissipating and energy-sharing devices and those that can control the response of the system.

There have also been further advances on appropriate methods and devices of preventing "dislodgement" or "unseating" of the superstructure in the event of severe ground shaking.

5.2 Plastic Hinging and Ductility

As mentioned earlier, there is a marked difference in seismic design aspects of bridges and buildings. The reduced degree of indeterminacy of bridge structures leads to reduced potential of dissipating energy and load redistribution. In bridges, the substructures (piers and abutments) are the main structural elements which provide resistance to seismic action. For energy dissipation, ductile behavior is necessary during flexure of these structural elements under the lateral seismic loads.

This essentially means that the formation of plastic hinges or "flexural yielding" is allowed to occur in these elements during severe shaking to bring down the lateral design forces to acceptable levels. Since yielding would lead to damage, plastic hinging is localized by design at points accessible for inspection and repair, i.e., parts of the substructure that lie upwards of the foundation, (Fig. 10.1). No plastic hinges are of course allowed to occur in the foundations and nor in the bridge deck.

5.3 Superstructure Dislodgement Prevention and Integral Bridges

Bearings, in general, are comparatively fragile and brittle elements (Fig. 10.2). Usual bearings of various types (metallic, elastomeric, pot, etc.) can be designed to have the capacity of sustaining lateral forces of about 25% of their vertical load carrying capacity (Indian Roads Congress 2014). For larger lateral forces, it is more suitable

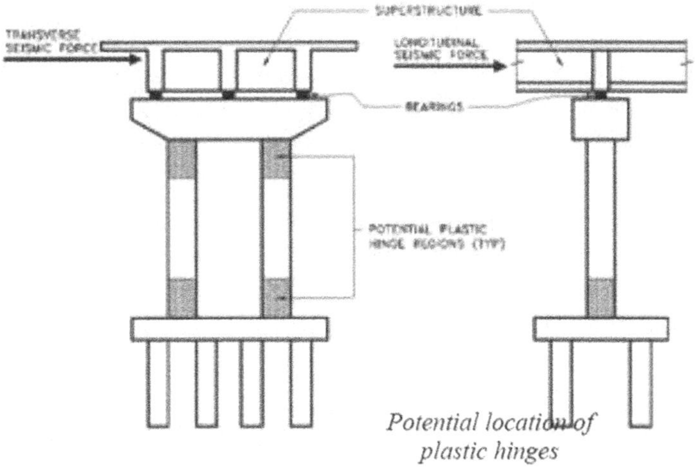

Fig. 10.1 Well-designed structures dissipate seismic energy by inelastic deformations in localized zones of selected members

Fig. 10.2 Surajbari old bridge: Metallic bearings destroyed during earthquake

and economical to provide resistance to these forces separately by some other structural element, as discussed later in this paper.

Superstructures, by themselves, usually have adequate strength to resist seismic forces. In many earthquakes, it was noticed that the superstructure was dislodged and had fallen onto the ground or was damaged due to loss of support caused by large

Fig. 10.3 Girder shifted in the longitudinal direction with loss of seating during shaking

displacements of elastomeric bearings, (Fig. 10.3), or due to out of phase displacement of adjacent piers of large height difference.

Use of elastomeric bearings as a means for transferring loads other than vertical loads must be examined carefully. The so called "response reduction factors" applicable for ductile behavior of reinforced concrete substructures are not to be used for arriving at displacements, for which structures on such bearings are specially vulnerable.

To counteract un-seating failures the following countermeasures are suggested:

1. Provide "reaction blocks" or other types of seismic restrainers for preventing dislodgement of superstructure at pier/abutment cap level.
2. Provide adequate support lengths for superstructure on pier/abutment cap.
3. Design and construct "Integral" bridges whereby the substructure and superstructure can be made monolithic.

Countermeasures for (1) and (2) have been discussed in detail in a previous paper (Tandon, 2001), and examples indicative of these features have also been included in Tandon (2005a). In some recent projects with continuous concrete bridges, "reaction blocks" were provided for longitudinal seismic forces as depicted in (Figs. 10.4 and 10.5). The provision of elastomeric pads on a vertical face, not only ensures even distribution of the applied lateral force to the reaction block but also acts as a "shock absorber," thereby introducing more damping into the system.

For more details of actual projects the reader's attention is drawn to Ref. (Indian Roads Congress 2000).

Coming to the proposition (3), integral bridges can be made earthquake-resistant more conveniently than bridges with bearings. Apart from obviating the necessity of

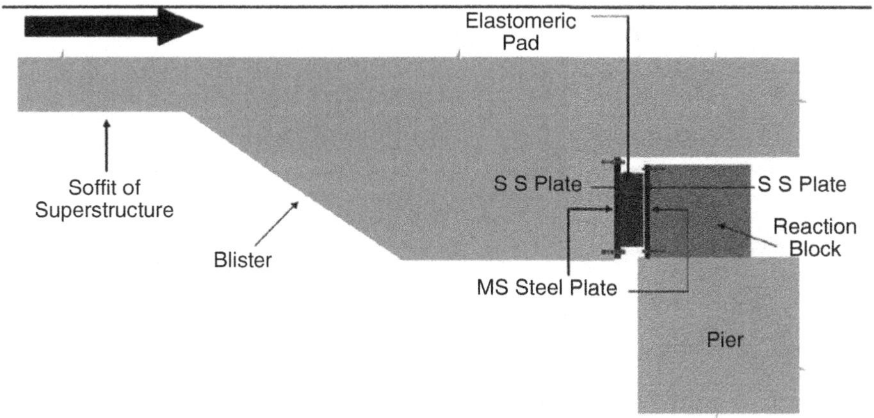

Fig. 10.4 Longitudinal seismic restrainer (vertical elastomer pad introduces damping to longitudinal forces)

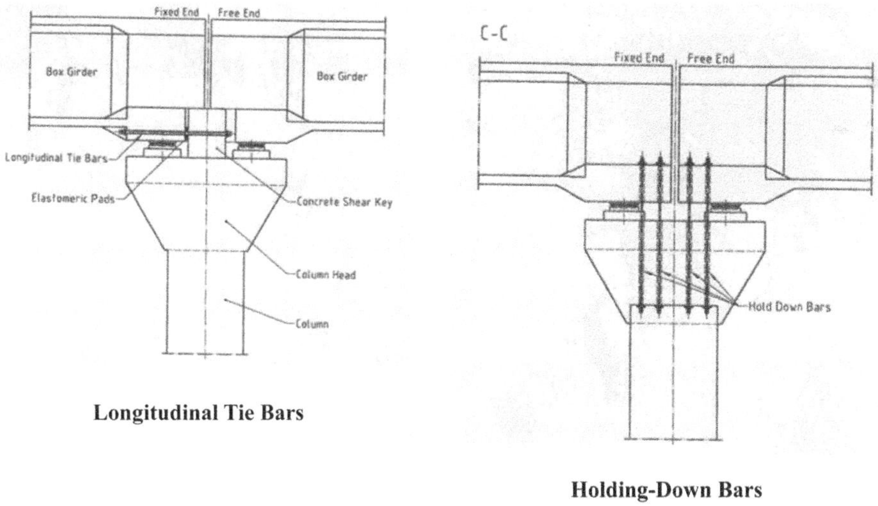

Fig. 10.5 Prevention of dislodgement by steel bars. (**a**) Longitudinal Tie Bars, (**b**) Holding-down bars

providing seismic restrainers and/or wide support lengths for the superstructure, the number of potential locations of plastic hinges can be increased, and ductility of a high order introduced into the system. A good example of an integral bridge is shown in (Figs. 10.6 and 10.7), wherein the potential seismic performance was greatly enhanced using the concept.

Fig. 10.6 Panchsheel Club Flyover, New Delhi, side view

Fig. 10.7 Panchcheel Club Flyover, New Delhi piers and deck are monolithic

5.4 Concepts of "Energy Dissipation" and "Energy Sharing"

The concepts of "Energy Dissipation" and "Energy Sharing" by using devices external to the structure have been increasingly adopted in recent years. Let us see what these concepts mean and how they have been utilized in actual projects.

5.4.1 Energy Dissipation

The use of special devices that reduce the seismic forces can be effectively utilized in the structure. By de-coupling the structure from seismic ground motions, it is possible to reduce the earthquake-induced forces in it. This can be done in two ways:

1. Increase natural period of the structure by base isolation.
2. Increase damping of the system by energy-dissipating devices.

The central issues are to limit the seismic energy entering into the structure from the ground in the first place and then to dissipate as much of it as possible by damping devices.

As mentioned earlier, the usual elastomeric bearings can be rarely used for moderate to high seismic zones except for transmitting vertical loads. Also, abutments and piers supporting bridge decks on elastomeric bearings should in general be designed to remain within the elastic range (Tandon, 2005b). Some special devices incorporating rubber (Fig. 10.8), can be used advantageously for energy dissipation as they limit displacement by high damping. Even though they have a good potential, these special devices have so far been used only in buildings in India.

Large viscous fluid dampers, (Fig. 10.9), are also being used for major bridge construction, though their efficacy in real earthquakes still requires to be validated.

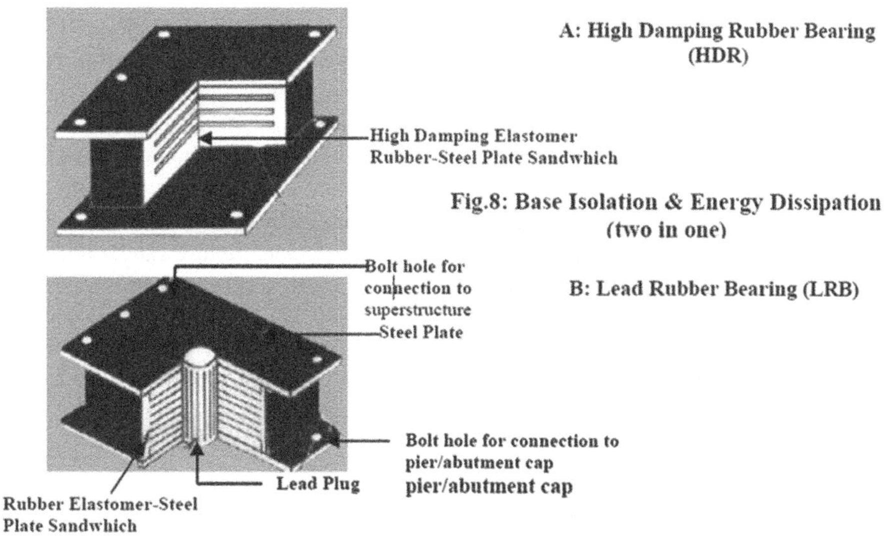

Fig. 10.8 Base isolation and energy dissipation (two in one): (**a**) high damping rubber bearing (HDR) and (**b**) Lead Rubber Bearing (LRB)

Fig. 10.9 Example of seismic viscous fluid damper for major bridge

5.4.2 Energy Sharing

Sometimes it is advantageous that the seismic energy entering from the ground into the structure does not get localized. Special devices exist which can avoid significant energy accumulation and ensure its distribution to various structural elements. Here, the idea is not to reduce the total seismic energy entering into the structure but to judiciously distribute it amongst the designated resisting elements. Such devices go by the name of shock transmission units (STUs). Their action is shown in Fig. 10.10, the behavior being similar to a car seatbelt. As structure A and structure B move slowly relative to each other, the fluid is able to migrate through narrow orifices from one side of the piston to the other. For rapid movements (e.g., earthquakes), the transfer of fluid is not possible, thereby locking the piston to its cylinder. In such circumstances the device acts as a rigid link between Structure A and Structure B. In bridge structures the inertial force from the complete superstructure can be transmitted to designated sub-structures. Application of STUs to a 1.0 km long bridge over Ganga near Allahabad is shown in (Figs. 10.11 and 10.12). The expansion joints are located only at the abutments and at the central pier as shown in (Fig. 10.12), wherein the seismic forces are transmitted to three piers in each of the two halves of the structure.

5.4.3 Passive and Active Measures

There is thus scope for both "passive" control by prescribed ductile detailing procedures and "active" control by specific devices for earthquake-resistant bridges. The judicious use of these ideas can lead to economical and safe bridge structures.

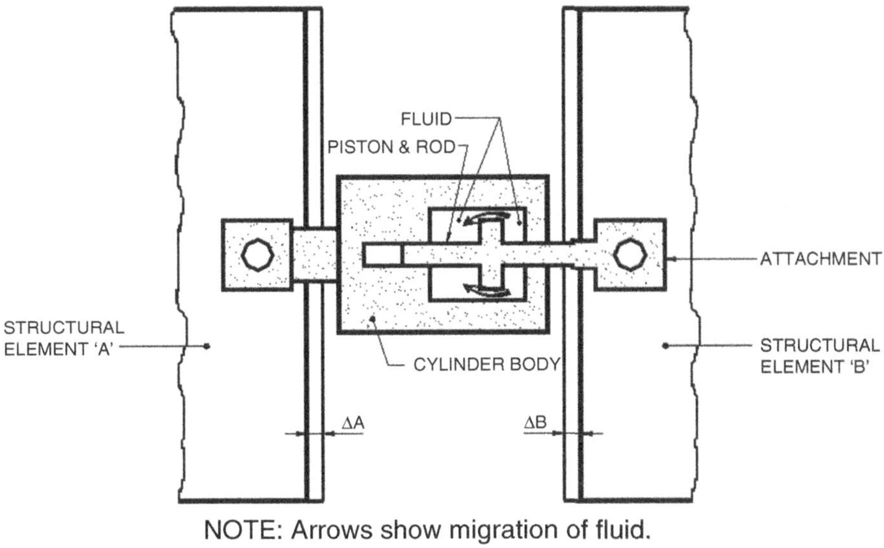

Fig. 10.10 Shock transmission unit—the principle

Fig. 10.11 Allahabad bypass bridge over river Ganga, Allahabad

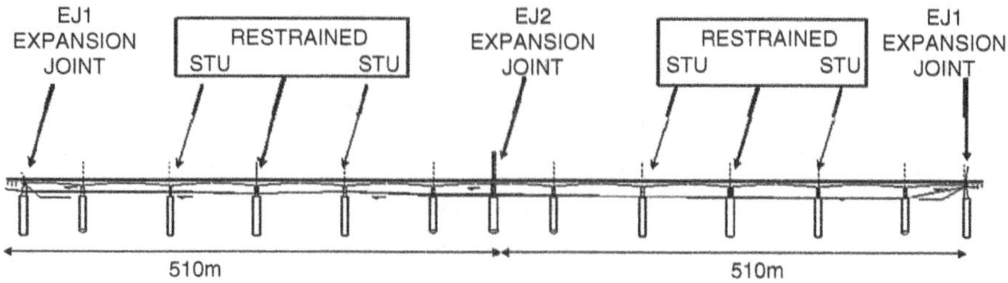

Fig. 10.12 NHAI's Ganga bridge at Allahabad showing application of STUs

References

Beckett, D., & Alexandrou, A. (1997). *Introduction to Eurocode 2: Design of concrete structures (including seismic actions)*. E& FNSPON.

Beckett, D., & Alexandrou, A. (2002). Standard specifications and code of practice for road bridges, Section IX. In *Bearings, Parts I, II and III. IRC: 83*. New Delhi: Indian Roads Congress.

Housner, G. W. (1997). *Connections – The EERI oral history series*. Oakland, CA: Earthquake Engineering Research Institute.

Jain, S. K., & Nigam, N. C. (2000). Historical developments and current status of earthquake engineering in India. *12th World Conference in Earthquake Engineering*.

Tandon, M. (2000). Standard specifications and code of practice for road bridges, Section II. Loads and stresses. In *IRC:6*. New Delhi: Indian Roads Congress.

Tandon, M. (2001). Earthquake resistant design of bridges. *Journal of Indian Roads Congress*, New Delhi, 479.

Tandon, A. N., & Srivastava, H. N. (1974). *Earthquake occurrence in India*. Earthquake Engineering: Jai Krishna Sixtieth Birth Anniversary Commemoration Volume. Savita Prakashan.

Tandon, A. N., & Srivastava, H. N. (2002). *Criteria for earthquake resistant design of structures, part 1 – general provisions and buildings*. In *IS:1893 (Part 1):2002*. New Delhi: Bureau of Indian Standards.

Tandon, M. (2005a). *Economical design of earthquake resistant bridges*. Annual Lecture of Indian Society for Earthquake Technology, Roorkee.

Tandon, M. (2005b). *prEN 1998-2 Eurocode 8, Part 2. 'Design of structure of earthquake resistance: Bridges' final draft*.

Chapter 11
Developments in Earthquake Resistant Design of Reinforced Concrete Buildings

S. K. Thakkar

1 Introduction

The discipline of earthquake engineering has developed in the last over 150 years. The loss of lives, damage of buildings and economic loss has been the biggest motivating factors in this development. The primary goal of the development has been to reduce earthquake risk on people and built environment. The occurrences of earthquakes have been reasonably well understood with the explanations provided by the theory of plate tectonics. Though the regions where earthquakes are likely to occur are known but their size, location and time of occurrence, that is earthquake prediction cannot be made with any degree of certainty. The inability to predict earthquakes has not discouraged the research and development of techniques that can protect structures from damage. Though it is not possible to predict earthquakes and prevent earthquake occurrence but it is possible to prevent loss of life, injuries to people and structural damage caused by strong ground shaking by following well developed principles and procedures of earthquake resistant design of structures. The protection of buildings and civil infrastructures from earthquake damage is an engineering problem; that is earthquake effect on foundations and structures needs to be scientifically analyzed and best estimates of induced forces and deformations in the structure and foundation is made, then designed to withstand these effects. The subject of earthquake protection and safety falls in the domain of earthquake engineering which has developed as a discipline of civil engineering. The earthquake motion induces inertia forces and deformations in the structure. The magnitude of earthquake force that acts on a structure depends on period of a structure, stiffness, energy-dissipating characteristics of a structure, amplitude and frequency characteristics of a ground motion. The developments of basic concepts in

S. K. Thakkar (✉)
Department of Earthquake Engineering, Indian Institute of Technology, Roorkee, Uttarakhand, India

M. L. Sharma et al. (eds.), *Advances in Indian Earthquake Engineering and Seismology*, https://doi.org/10.1007/978-3-319-76855-7_11

earthquake engineering have largely occurred on the basis of observed behavior of structures in past earthquakes. The testing of structural models on shaking tables under simulated ground motions, the testing of models under quasi-static loading and pseudo-dynamic tests have led to verification of analytical methods and created deeper understanding of earthquake behavior. The safety of structures in earthquakes has always posed challenges to researchers and designers. There are issues which continue to be researched in order to develop more rational methods of seismic analysis, safe and economical methods of design. In particular estimation of earthquake hazard, earthquake effects and soil effects on structures has been most challenging. There has been significant research in the methods of seismic analysis, earthquake resistant design and earthquake protection technology in the last five decades.

Researchers (Otani, 2004; Villavarde, 2009) have traced the history of earthquake engineering and earthquake resistant design and defined the scope of earthquake engineering and its developments. Bertero (Bertero and Bozorgnia, 2004) has highlighted the growth of earthquake engineering, its definition and its goal to the present state. Some researchers have highlighted the importance of design for controlled damage of structures under earthquakes so as to reduce cost of post earthquake repairs (Priestley, 2000; Otani, 1997). Otani (2004) has highlighted that controlling inter storey drift through the use of structural walls or structural control devices and improving method to fasten non-structural elements to the structure may reduce damage to partition walls. Kawashima (Kawashima, 1997) has drawn attention to the need to consider residual displacement in defining damage control performance level. It has been highlighted that damage control and maintenance of building function will become a major issue in the future (Otani, 2004).

The earthquake resistant design is different than the designs for gravity and other forces as it includes the aspect of stiffness, strength and ductility in the design. For economic reasons structures are not designed elastically but are allowed to be stressed in inelastic range for design earthquake that is certain amount of damage is acceptable in the seismic design. With the developments in performance based seismic design, it has become possible to design the structures for controlled damage under design earthquakes.

The state of the art in earthquake engineering has come of an age now where it is possible to design structures to withstand earthquake effects without getting them significantly damaged while existing seismically deficient structures can be retrofitted to withstand safely the effect of future earthquakes. However there are issues of concern like significant damage is observed in structural and non structural elements of engineered buildings in moderate earthquakes requiring enormous post earthquake repairs and retrofitting. There are undesirable structural responses and collapse observed in some structures. There are large numbers of existing structures that were designed prior to the existence of modern codes which are found seismically deficient according to current standards; it may require retrofitting for their safety in future earthquakes. The above concerns clearly demonstrate that the current seismic design methods have not adequately addressed the issue of damage control in the event of moderate earthquakes. The purpose of this chapter is to focus on the basic concepts developed in earthquake engineering over a period of time, lessons learned from past earthquakes and project challenges for future research and development in order to mitigate effects of earthquakes on structures with the emphasis on reinforced concrete buildings.

2 Earthquake Engineering

The earthquake engineering has developed as a multidisciplinary subject largely centered on civil engineering which has drawn inputs from seismology, geology and various engineering disciplines. The knowledge of structural engineering, geotechnical engineering, instrumentation, computer modeling, soil dynamics and structural dynamics has largely contributed in developments of methods of designing structures and foundations in with standing effects of earthquakes. The seismology, geology and probability theory has enabled understanding of occurrence of earthquakes and in assessment of earthquake hazard while structural dynamics in particular have enabled rational evaluation of earthquake forces.

The earthquake engineering deals with many other studies such as:

1. Post-earthquake damage investigations of structures, foundations and soil.
2. Deployment of strong-motion instrument network in seismic zones to record "free-field" ground motions and response of structures in earthquakes.
3. Micro-earthquake recording at project sites.
4. Microzonation studies.
5. Determination of earthquake hazard and seismic risk.
6. Shake table testing of scaled models/structures under simulated earthquake motion; quasi-static and pseudo-dynamic testing of models; seismic qualification studies; field testing of full scale structures and ambient vibration testing.
7. Structural response control techniques.
8. Structural health monitoring and damage detection.
9. System identification studies.
10. Vulnerability assessment, repair and retrofitting of existing structures.
11. Formulation of engineering decision models.
12. Development of seismic guidelines and codes for earthquake-resistant design, earthquake protection systems and retrofitting of structures.

3 Fundamental Concepts

The fundamental concepts in earthquake engineering are the foundation to the development of any earthquake-resisting system and design method which can lead to safe performance in earthquakes. The following concepts have emerged as a result of theoretical background, laboratory testing under simulated conditions, analytical and experimental studies of seismic response of structures and their observed behavior in earthquakes:

3.1 Earthquake Forces

- The earthquake forces are the result of inertia effect caused by vibration of structure due to ground motion. The structural damage is mainly caused by horizontal inertia forces associated with lateral vibration of structure. The vertically induced forces are generally small. The inertia forces are resisted by elastic/inelastic resistance of the structure. As inertia forces are directly proportional to mass, possibility to reduce mass of the structure should be explored in the design process.
- The flexibility in a structure causes dynamic amplification of ground acceleration to occur along the height as a result the acceleration at higher floors in a multistory building become larger than the base acceleration.
- As the earthquake ground motion largely consists of frequencies lying in the range of 0–33 Hz, there cannot be resonance at a single frequency of structure like in a sinusoidal motion; however, there can be a quasi-resonance at dominant frequencies. Majority of structures possess frequencies in the same range as that of an earthquake and are thus affected by earthquake.
- The damping inherent in the structure has a significant contribution in reducing induced forces and displacements in the structure due to earthquakes.
- The earthquake forces that develop in a structure are dependent on its period of vibration, damping and characteristics of ground motion; the forces can be rationally computed by dynamic analysis.
- One of the most important concepts developed in earthquake engineering is that of "Response Spectrum" which is a relationship between time period of a structure with certain damping and spectral values of acceleration, velocity and displacement for a given earthquake motion. This relationship has enabled rational determination of seismic force in a structure.
- The structure with higher fundamental period of vibration develops smaller seismic forces compared to structures of shorter periods, this follows from the shape of response spectrum. Possibility to elongate fundamental time period of a structure should be explored while making choice of the structural system in order to reduce the seismic forces.

3.2 Structural System

- The simple and symmetrical plans develop smaller forces. Symmetry reduces development of torsion forces while unsymmetry increases torsion effects. The symmetry in both plan and elevation are preferred design features.

- The other forms of irregularities such as raising building on floating columns/walls, sudden change of stiffness in adjacent storeys, missing columns and beams in structural system induce discontinuity in load path are undesirable features.
- The nonstructural elements such as infill partition walls and stair cases have significant contribution in resisting earthquake forces, their adequate consideration in analysis, design and detailing is necessary.
- A properly selected structural configuration and system in the conceptual design can minimize undesirable response and best resist the lateral forces generated by earthquake.
- The concentration of a mass at one particular level is not preferred from seismic view point.
- The uniform distribution of mass and stiffness along the height is preferred.
- The distribution of lateral earthquake resisting elements in plan in two principal directions is a preferred feature of seismic design.
- The buildings constructed too close to each other are susceptible to damage by "pounding".

3.3 Design Concepts

- The earthquake resistant design is achieved by design of structure both for strength and ductility.
- The design forces are much smaller than those that would be required if the structure were to remain elastic. The ductile behavior of the structure enables lowering of design seismic forces.
- The traditional design is based on, "Force based design" in which the design forces for a structural component are obtained by dividing forces computed using elastic dynamic analysis by "Response reduction factor" to account for inelastic behavior.
- The strong column weak beam design leads to ductile behavior and enables sound earthquake resistant design.
- As per the observed behavior in earthquakes, the traditional soft storey design is vulnerable to earthquake damage. These design needs provision of adequate stiffness, strength and ductility.

3.4 Energy Dissipation

- The ductile detailing is recognized as an important concept as it leads to energy dissipation in inelastic behavior.
- The ductile detailing in columns, beams and beam-column joints are preferred features.

- The ductility demand distribution along the height should be as uniform as possible.
- The reinforced concrete structures and steel structures are designed to demonstrate ductile behavior following principles of ductile detailing.
- The ductility demand is less than the ductility capacity. This can be verified in design by non-linear methods of analysis.

3.5 Soil and Foundation

- The structures are normally considered fixed at the level of foundation. There is no effect of soil structure interaction when structure rests on rocky or firm foundation. The soil structure interaction effect may be important in certain situations particularly when stiff structures are founded on soft soils. Soft soil deposits are also known to amplify ground motions where certain frequencies are filtered out while others are amplified as a result the base motion is dominated for a range of frequencies.
- The liquefaction of soil is one of the major concerns for earthquake safety. The sites with potential of liquefaction, instability of slopes and landslides should be avoided as far as possible. The engineering solutions to deal with these problems are required where these problems are encountered.
- The foundation should not be normally allowed to yield or lift under seismic conditions.

The desired features mentioned above are required for a good engineering practice. Where the desirable features are not met due to functional constraints, elaborate dynamic analyses are required in order to incorporate their effects in design. It is necessary to appreciate the importance of fundamental concepts in developing conceptual design for reducing earthquake risk.

4 Development of New Design Concepts

4.1 Capacity Design

Capacity design is another design concept developed to achieve ductile failure and avoid brittle failure mode in structures. This is an integrated design procedure developed for reinforced buildings by Paulay T (Paulay, 1970) which was aimed to realize formation of intended plastic mechanism. The capacity design is used to ensure that an appropriate hierarchy of resistance exists within the various structural components. This objective is achieved by designing all the members intended to remain elastic against all brittle modes of failure using "Capacity design effects". Such effects are computed from equilibrium conditions of intended plastic mechanism, when all flexural hinges have developed plastic moment capacity including their over strength.

4.2 Performance Based Design

Performance-based design (Priestley, 2000) is a design concept developed to obtain acceptable level of performance for expected levels of seismic ground motions. Both performance levels and corresponding levels of ground motions are specified. The background of this development lies in the, "Vision 2000 document" (OES, 1995) prepared by the Structural Engineers Association of California. The objective of the design approach is to ensure that the structural system and components have enough capacity to withstand the deformation imposed by design motion. It was realized that displacements rather than the forces as measurement of damage allow a structure to fulfill this condition. A target displacement could be specified, the analysis performed and then design strength and stiffness determined as the end products of design for a structure. This method of design has been known as displacement-based design (Priestley et al., 2007). In performance-based design/displacement-based design, the designer needs to define acceptable deformation/displacement based on post-performance requirement and available deformation capacity. In the performance-based design, the emphasis is on performance. This method of design has certain merits over much used "Force-based design" which has been comprehensively discussed elsewhere (Priestley, 2000). The performance-based design method has a greater potential of meeting limit states of design and limiting the damage in structures as compared to force-based design. The main issues remain to develop a design method readily adaptable by code, a define design limit states such as serviceability, ultimate and damage control and corresponding level of earthquake motions, their return period and probability of exceedance. The updating of design methods is a continuing process which is based on the experience gained from performance of structures, deficiencies uncovered in design methods from the observed behavior in earthquakes and research and development.

In India, seismic design is based on force-based design using response reduction factors. There is a need to bring emphasis on "performance" in Indian seismic codes in current seismic design method; alternately performance-based design may be adopted. Further, there is a need to adopt non-linear time history methods of dynamic analysis in the design process in codes as these are considered to be most rational as of date compared to other available methods.

5 Response Reduction Factor

The response reduction factor is an important concept developed in earthquake engineering that enables explicit reduction of elastic seismic forces to acceptable design-level forces. It is based on the research work of Veletsos and Newmark (1960). This factor is also termed as response modification factor "R" and behavior factors (q-factor) and force reduction factor in some codes. Depending on period range, three concepts have emerged.

5.1 Equal Displacement

The maximum displacements of elastic and elastic-plastic, single-degree-of-freedom system under earthquake motion is seen to be equal in the long-period range, consequently the response reduction factor turns out to be:

$$R = 1/\mu$$

where, μ = ductility factor of the system, the ratio of ultimate deformation to yield deformation.

5.2 Equal Energy

The strain energy of elastic and elastic-plastic, single-degree-of-freedom-system under earthquake motion is seen to be conserved in short period range; consequently the response reduction factor turns out to be:

$$R = 1/\surd(2\mu - 1)$$

5.3 Equal Acceleration

For zero period, that is, for a theoretically rigid structure, the maximum acceleration for elastic and elastic-plastic system is equal; consequently the response reduction factor turns out to be:

$$R = 1$$

The response reduction factors primarily reflect ductility capacity of a structure, but when these factors are employed in codes, these also included other effects that may effectively result in response reduction such as structural over strength, redundancy and energy dissipation capacity.

The response reduction factors are also differentiated among buildings with different materials and structure types. They are semi-empirical in nature and are also based on performance of buildings in past earthquakes and analytical and experimental studies in laboratory. The reduction factors/ductility factors used in traditional design methods are not normally verified at the end of the design in seismic codes (IS: 1893 2002); these are deemed to be available in the structure with the provisions of ductile detailing. It is also difficult to rationally evaluate these factors for a structure as these depend on several other parameters besides period, such as hysteretic curve, structural geometry and foundation compliance. The designer should ensure by his experience and judgment that not too big or too small values are adopted as response reduction factors so as to render structure unsafe or may lead to over conservative design. Though the concept of reduction in response is well appreciated from the building performance in earthquakes but the magnitude of the response reduction factors is found to be varying; it is often challenged because of the above reasons. Conceptually the response reduction factors are applied globally to a structure and are primarily based on displacement ductility capacity. In certain structures such as "Bridges" different reduction factors are recommended for different components such as superstructure, substructure, foundations, bearings etc. It should be appreciated that the use of response reduction factor in design is somewhat overstretched.

6 Lessons from Earthquakes

The performance of structures in the past earthquakes have taught many useful lessons with regard to deficiencies in current design methods, materials of construction, type of structural system, ductile detailing, role of nonstructural elements, pounding, foundation and soil effects etc. These lessons have enabled development of concepts that are useful in preliminary design and planning of structures and in removing the shortcomings of design methods recommended in seismic codes. The other important aspect highlighted is that the shortcomings of earlier designs should not be repeated in the future design and construction. Some typical lessons are highlighted here:

6.1 Soil Amplification

Acceleration is an important parameter in earthquake engineering. It is acceleration versus time that is recorded for three components of ground motion by strong motion instruments for collection of ground motion data. Acceleration responses at certain locations due to local site characteristics are amplified and known to possess dominant periods in a certain range. The dominant periods are reflected by "peaks" in Fourier spectrum of ground motion. The buildings possessing periods in this range are subjected to large inertia forces due to quasi-resonance and are susceptible to damage. Such a behavior is observed even at distances of 300–400 km from the epicenter where local soil has caused amplification of motion and induced damage to tall buildings. Local soil amplification effects where likely to occur should be kept in view in seismic design.

6.2 Ductile Behavior

Ductility is a key parameter in seismic design. It has a dual role; on one hand it reduces design forces to an acceptable level while on other hand it provides energy dissipation in inelastic range; this effectively increases damping and further decreases the response. The buildings with ductile detailing in accordance with provisions of seismic codes have normally performed well without collapse but with some structural damage. The detailing rules in respect to transverse reinforcement in columns in plastic hinge regions, detailing of beam-column joints, splice detailing are found to be important in prevention of collapse and also in reducing seismic damage. Many of the failures in columns have occurred because of larger spacing of transverse reinforcement. The ductile behavior is proven to be an effective safety measure particularly in collapse prevention in earthquakes.

6.3 Soft Storey Failure

The first soft storey is common in multistory buildings where the open space is required for commercial facility or garages. The building with first soft storey is observed to be vulnerable to damage and collapse due to inadequate strength and ductility. There are innumerable examples where collapse or severe damage occurred in the first storey due to lack of strength and ductility. The columns in first storey are required to resist large base shears and should be designed for ductile behavior. The avoidance of open first storey is often not possible. The shear failure of soft storey can be protected by following capacity design principles.

6.4 Torsional Behavior

In buildings unsymmetrical in plan, for example, where walls are provided on one side while on other side there are open frames, the eccentricity between centre of mass and rigidity causes torsional vibration during earthquake. Large damage occurs in members away from the centre of rigidity. It is often possible in the initial stages of planning to reduce eccentricity by adjusting locations of walls.

6.5 Mode of Failure

It is important to understand mode of failure of building components in earthquake and the consequences of member failure in structural performance. The failure of vertical members often leads to collapse of building. The failure types of beams, columns, walls, and beam-column joints may be different. Some of the most common modes of failure are flexure compression failure of columns, shear failure of columns, failure of beam-column joint, bond-splitting failure, splice failure of longitudinal reinforcement and anchorage failures. The deficiency in lateral load-resisting system has led to total collapse of structures. The structural failures can be prevented by proper choice of structural system, adequate design for strength and ductility and following ductile detailing provisions.

6.6 Non-structural Elements

The damage to nonstructural elements such as partition walls used as infill in frames, windows, doors, mechanical facilities and stair cases is observed to be significant in earthquakes. There are numerous cases of damage to infill walls and staircases in earthquakes. The cost of repair work is heavy on account of damage to such components and also building use can be greatly disrupted because of such damages. The prevention of damage of such elements is considered to be very important. The nonstructural elements are often disregarded in structural analysis although they contribute significantly to the stiffness of the building as a whole. The perfect solutions for their safety are not yet found, but various options are available for reduction of such damages: (i) structural separation of non-structural elements from structural components with suitable detailing provisions, (ii) adequate consideration of stiffness of non-structural elements in structural analysis and their design to resist induced seismic forces, and (c) controlling of inter-storey drift in design process.

6.7 Pounding of Buildings

Pounding of two adjacent buildings causes structural damage. Adequate structural separation between two buildings is the solution to this problem. The separation distance required between two adjacent buildings can be estimated on the basis of displacements calculated from dynamic analysis of buildings for design earthquake motion.

6.8 Seismic Design Method

The force-based method of design currently adopted by majority of seismic codes of designing structures for small design forces is well proven on the basis of lessons from earthquakes. It is also brought out from the observed behavior that the structures designed using this method may be subjected to significant damage in certain situations indicating shortcoming of this method. However further update in this design method of controlling damage by introducing displacement check have been introduced by researchers followed with subsequent developments in a range of design methods falling in the category of displacement-based design/performance-based design.

6.9 Age Effect

Deterioration of material strength with age and aggressive environment is a well known feature. This decreases the seismic resistance of the building. Earthquake damages also produce the similar effect. There are no rational methods developed to determine loss of stiffness and strength due to age. It is important to maintain the structure at regular intervals by repair and strengthening so as to keep control of durability and strength.

6.10 Foundation Issues

The failure of foundation is caused by: (i) liquefaction of soil, (ii) landslides, (iii) compaction of soil, (iv) differential settlement and (v) fault rupture. It is difficult to design safe foundation for fault rupture. The repair and retrofit measures in foundation is expensive and extremely difficult. It is advisable to reduce the possibility of foundation failure by proper choice of location and conservative design.

6.11 Quality Issue

The quality of construction has a significant impact on the performance of buildings in earthquakes. Reinforced concrete buildings with poor quality of construction were subjected to greater damage while buildings with good quality of construction have performed well. Monitoring of quality of construction work and material strength during construction by experienced people is necessary to maintain the high standards of workmanship and construction material.

7 Structural Response Control

Structural response control devices consist in adding certain features to a building which alters its dynamic characteristics so as to reduce its response under earthquake motions. This is termed as "earthquake protection" in contrast to the term "earthquake resistance" used in traditional design for strength and ductility. In the traditional method of design, the seismic demand computed from dynamic analysis is met in the structure by introducing necessary capacity into it by design; there is no control of seismic demand in this approach. While in the structural control method the seismic demand in the structure is substantially reduced by modifying dynamic characteristics of the structure, consequently reduced capacity is to be built into the structure to meet the demand enabling better control of seismic behavior and damage. There are two types of structural control, passive and active. In the passive control such as base isolation technique the structure is placed on bearing pads such as elastomeric bearings which has low horizontal stiffness but high vertical stiffness to transfer vertical loads. The horizontal flexibility in bearings causes elongation of period of structure which is far away from dominant period of earthquake motion. This is termed as detuning the frequency of the structure with respect to frequency of excitation. Typical period of isolated building may be kept as 2.0 s while dominant period of excitation may range between 0.05s–0.1 s. This difference in periods of structure and base motion does not allow dominant energy of ground motion to be transmitted to the structure. As a result deformation is confined to the bearing level and structure oscillates as a rigid body leading to earthquake protection. Certain amount of damping is desired in the system to control displacements at the bearing level. This may be available inherently in the bearing; alternatively it may be provided by an external damping device placed in parallel with the isolation bearing. Another concept that is often employed in response control of tall buildings is by Passive Energy-Dissipating (PED) devices. This may consist in adding damping device in the building, usually in the locations of diagonals of bay such as, friction damper, viscous damper, viscoelastic damper, hysteretic damper, etc. Tuned mass dampers are yet another set of devices consisting of a mass, flexible element and a damper. The natural period of damper is tuned with the fundamental period of the building as a result it counters the motion of the building and reduces amplitude of vibration of building caused by wind or earthquake. Tuned mass dampers are placed on top of the building to control the vibrations of the building. In active control systems, actuators are used to apply forces that counter the earthquake forces in the structure. As earthquake motion is random and changes rapidly, actuator control systems are controlled by the computer to determine the magnitude and direction of counteracting forces. A variety of mass dampers and passive damping devices are now available for structural control of buildings in earthquakes. The structural response control techniques are applicable for design of new structures as well as for retrofitting.

Though wide range of structural control devices have been developed and incorporated in selected buildings and structures. Many buildings with base isolation systems and other passive and active devices have been subjected to earthquakes and these structures have performed very well demonstrating their effectiveness, yet their use in civil structures has been limited. One of the reasons of its less use could be its greater cost relative to traditional designs based on strength and ductility. There is a tremendous scope of reducing damages in structures in moderate earthquakes with this alternate method of earthquake protection. Further it could be the most suitable method to retrofit structures where modification in building superstructure and addition of structural walls for strengthening may not be possible due to aesthetic constraints and limited capacity of existing foundations, respectively. More comprehensive research to cut down on the cost of structural control technology will be required.

8 Repair and Retrofitting of Structures

Despite advances in earthquake engineering, a large number of existing buildings and other structures were constructed using techniques and seismic codes that be may be far below the current standards. Majority of such structures are vulnerable in future earthquakes and may be subjected to damage. Such structures may need seismic assessment according to current codes, and these may need retrofitting in order to survive the future earthquakes. The important issues in repair and retrofitting of structures are: (i) Assessment of seismic capacity of existing structures, (ii) Development of economical methods of retrofitting, and (iii) Testing the effectiveness of retrofitting. Seismic repair and retrofitting have become an important issue and a challenge for disaster prevention, considering the vulnerability of large number of existing structures in seismic zones.

9 Challenges in Earthquake Engineering

The following is the summary of advances in earthquake engineering and seismic design methods that are currently employed in practice:

- Force-based design with response reduction factors.
- Performance-based design for planned performance.
- Seismic base isolation with primary aim to elongate time period of structure.
- Passive energy dissipation devices in the form of dampers with primary aim to add damping to the system.
- Active control devices with primary aim to apply forces to counter earthquake effect.
- Tuned mass dampers on top of the structure with the aim to control vibrations caused by wind and earthquake.
- Repair and retrofit techniques for existing structures.

With advances in earthquake engineering as indicated above it is now possible to tackle the problem of earthquake-resistant design of new structures as well as retrofit of existing structures more rationally than in the past. However, there are issues in earthquake engineering which need further research and development to meet the challenges of the future:

- Structures are still subjected to considerable damage in structural and nonstructural elements even in moderate earthquakes; this implies that design methods should be upgraded to control the damage.
- The development of appropriate structural systems that minimizes earthquake effects and are capable of resisting vertical and lateral loads effectively.
- Development of performance-based design methods and specifying design limit states which can limit the damage to an acceptable level.
- Development of cost-effective structural control technology to reduce post-earthquake damages in structures.
- Techniques for seismic assessment of existing structures considering loss of stiffness and strength.
- Techniques for economical retrofitting of structures without overstressing of foundation.
- Verification methods of effectiveness of retrofit techniques.
- Techniques of damage detection and determining seismic capacity of earthquake-damaged structures.

10 Conclusions

This chapter briefly reviews fundamental concepts, lessons learned from earthquakes, advances in earthquake engineering, design methods and issues that need further research and development. There should be emphasis on conceptual design at the planning stage of the project to minimize earthquake effects. The state of the art in earthquake engineering has reached a stage where protection of human lives is possible in well designed engineered buildings. The existing buildings can be retrofitted to withstand future earthquakes. Performance-based engineering is one of the major developments which has made possible the design of structures for a given performance under design earthquake(s). The further challenges are in the development of appropriate structural systems that can best provide the earthquake resistance upgrading of design methods to control the level of damage in earthquakes and to design economical retrofit measures for existing structures. Structural control technology has a tremendous scope to control earthquake damages in structures which needs to be explored by further research. Indian seismic design codes have yet to adopt performance-based design methodology; also there exists a need to adopt non-linear dynamic methods of time history analysis in the design process in the codes.

References

Bertero, V. V., & Bozorgnia, Y. (Eds.). (2004). *The early years of earthquake engineering and its modern goal: Earthquake engineering, from engineering seismology to performance based engineering*. Boca Raton, FL: CRC Press.

IS: 1893(Part 1). (2002). *Indian standard, criteria for earthquake resistant design of structures, part 1, general provisions and buildings, fifth revision*. New Delhi: Bureau of Indian Standards.

Kawashima, K. (1997). The 1996 Japanese seismic design specifications of highway bridges and the performance based design. *Proceedings of international conference at Bled, Slovenia. A.A. Balkema, Rotterdam/Brookfield* (371–382).

OES. (1995). *California office of emergency services, vision 2000: Performance based seismic engineering of buildings*. Sacramento, CA: Structural Engineers Association of California.

Otani, S. (1997). Developments of performance based design methodology in Japan. *Proceedings of International Conference, Bled, Slovenia. A.A. Balkema, Rotterdam/Brookfield* (59–68).

Otani, S. (2004). Earthquake resistant design, of reinforced concrete buildings, past and future. *Journal of Advanced Concrete Technology, Japan Concrete Institute, 2*(1), 3–24.

Paulay, T. (1970). Capacity design of reinforced concrete ductile frames. *Proceedings, Workshop on Earthquake-Resistant Reinforced Concrete Building Constructions, University of California at Berkeley,* III, 1043–1075.

Priestley, M. J. N. (2000). Performance based design. *Proceedings of Twelfth World Conference on Earthquake Engineering, Paper no. 2831, New Zealand.*

Priestley, M. J. N., Calvi, G. M., & Kowalsky, M. J. (2007). Displacement based seismic design of concrete structures. *Proceedings of 6th National Conference on Earthquake Engineering, Turkey* (113–137).

Veletsos, A. S., & Newmark, N. M. (1960). Effects of inelastic behavior on the response of simple systems to earthquake motions. *Proceedings, Second World Conference on Earthquake Engineering, Tokyo-Kyoto,* II, 895–912.

Villavarde, R. (2009). *Fundamental concepts of earthquake engineering*. Boca Raton: CRC Press.

Chapter 12
Earthquake Resistant Design of Masonry Buildings

Anand Swarup Arya

1 Introduction

In the past severe earthquake all over the world, the masonry buildings have generally been damaged the most because of their heavy weight, small or no tensile strength, small shearing resistance, lack of proper bonding between longitudinal and cross walls and poor workmanship. Yet such buildings are being put up in the conventional way for reasons of climatic suitability, low-cost, local availability of materials, widespread knowledge of methods of construction etc. It is quite apparent that it will not be possible to do away with this kind of construction in the seismic areas particularly in India and other developing countries. Census of Housing in India has shown that the number of housing units using burnt brick as the wall material was 111.9 million in 2001 which has increased to 146.5 million in 2011. Census showing an increase of 31% in 10 years. Therefore, finding effective methods of strengthening such buildings is of paramount importance so that the danger to life and property during future earthquake is minimized. An important requirement of such methods is that they should be cheap and easily applicable.

We at the Department of Earthquake Engineering, IIT Rooree (then, SRTEE) had considered the importance of the earthquake safety of masonry building as early as 1960, and the first ME thesis on the subject was submitted in 1962. The collaborative work at SRTEE and Central Building Research Institute (CBRI), Roorkee between the author and Dr. S. M.K. Chetty during the period from 1962 to 1966 resulted in the formulation of the first standard IS:4326-1967 on the subject. The main provisions in regard to the design of earthquake resistant masonry buildings since then have remained basically the same except some additions requiring marginal

A. S. Arya (✉)
Earthquake Engineering, IIT, Roorkee, Uttarakhand, India

State Disaster Management Authority, Bihar, India
e-mail: anandsarya@gmail.com

M. L. Sharma et al. (eds.), *Advances in Indian Earthquake Engineering and Seismology*, https://doi.org/10.1007/978-3-319-76855-7_12

improvements in the safety requirements. The main aim of this chapter is to explain the basis of the design of Seismic Bands at various levels and the vertical bars at critical points of the masonry buildings so that the so called method of calculation of the reinforcement is clearly understood by the profession and underlying assumptions are fully appreciated. It will also clarify that the "*prescriptive*" provisions were indeed based on detailed "pre-engineering" calculations which were fully supported by the tests carried out at the Department of Earthquake Engineering, IIT Roorkee from time to time.

2 Structural Action

During ground motion, an inertia force acts on the mass of the building due to which the building tends to remain stationary while the ground moves. If the building is rigidly fixed into ground, the inertia forces will cause horizontal shears in the building, the magnitude of which will be a function of the ground motion and stiffness and damping characteristics of the building.

As a result of the shaking, the roof tends to separate from the supports, the roof covering tends to be dislodged, walls tend to tear apart, and if unable to do so the walls tend to shear off diagonally in the direction of motion. If filler walls are used within steel, concrete, or timber framing, they may fall out of the frame bodily unless properly tied to the framing members. Thus on the whole it is indicated that *a necessary condition for earthquake resistance is that the various parts of the building should be adequately tied together* (Arya, 1967) and Ch. 10 in (Arya, 2007).

Consider the structural elements and structures shown in Fig. 12.1. In (a), wall A is free standing and the ground motion is acting transverse to it. The force acting on

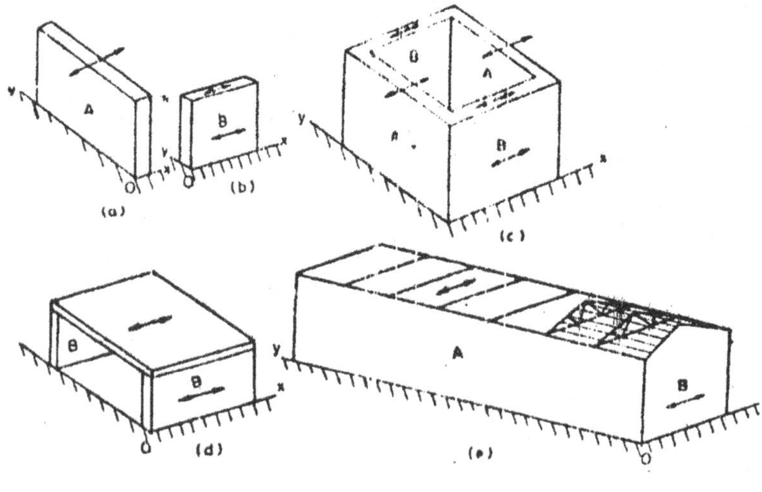

Fig. 12.1 Various wall positions under lateral earthquake force

the mass of the wall tends to overturn it. The resistance of the wall is obviously very small. In (b), the free standing wall B is subjected to ground motion in its own plane. It is clear that in this case, because of its large depth in the plane of bending, the wall will offer great resistance. Such a wall is termed as *shear wall.* It may be noted that for ground motion parallel to y-axis, wall A will act as shear wall and wall B will topple over. Now consider the combination of walls A and B as shown in (c). For the x-direction of motion as shown, walls B act as shear walls and besides themselves they offer resistance against the collapse of walls A as well. Walls A now act as vertical plate supported on two vertical sides and bottom and subjected to the inertia force on their own mass. Near the vertical edges, the wall will carry bending moments in the horizontal plane for which the masonry has little strength. Consequently cracking and separation of the walls may occur. If however, a horizontal *bending member* (like a timber beam or reinforced concrete runner stiffer than the slab) was inserted at a suitable level in wall A and continued in wall B, this will take care of the bending tensions in the horizontal plane. So far as vertical bending is concerned, the wall gets *pre-compression* due to self weight along with that or floor or roof supported by the wall and can be made to take care of vertical bending tensions. The situation will be the same for walls B for ground motion along y-axis. Thus the *bending member* will be required in walls B also. Therefore, a horizontal runner, strong in bending, is required around the enclosure. Such a runner is termed as seismic *band* and depending upon at which level it is used, it is called *roof band, lintel band* or *gable band.*

In Fig. 12.1(d), a roof slab is resting on walls B and earthquake is along the x-axis. Assuming that there is enough adhesion between the slab and the walls, the slab transfers its inertia force at the top of walls B, causing shearing and overturning actions in them. To be able to transfer its force on to the walls, the slab must have enough strength in bending in the horizontal plane. Whereas Reinforce Concrete (RC) slabs shall possess such strength inherently, other types of roofs or floors such as timber joists with timber plank or brick tile covering will have to be connected together and fixed to the walls suitably so that they are able to transfer their inertia force to the walls. After this load transfer, the walls B must have enough strength as shear walls to withstand the applied inertia force and its own inertia force. It is quite clear that the structure shown at (d), when subjected to ground motion along y-axis, will collapse very easily because walls B have very little bending resistance in the y-direction. Hence bracing arrangement, either by shear walls or by diagonal braces, must be provided in the y-direction.

Lastly, consider the barrack type structure carrying roof trusses as shown at (e). The trusses rest on walls A and walls B are gabled to receive the purlins of the end bays. The ground motion is along x-axis. The inertia forces will be transmitted from sheeting to purlins to trusses and from trusses to walls A. The end purlins will transmit some force directly to gable ends. In order that the structure does not collapse, the following arrangements must be made:

1. The trusses must be anchored into the walls by holding down bolts,
2. Walls A, which do not get much support from the walls B in this case, must be made strong enough in vertical bending as cantilever, or

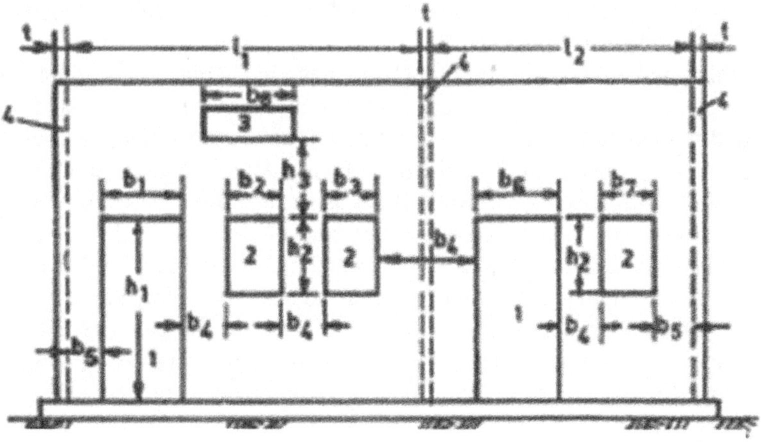

Fig. 12.2 Dimensions of openings and piers for recommendations in Table 12.3

3. Some suitable arrangement must be made to transmit the force horizontally to end walls B which are well suited to resist the force by shear wall action.

Strengthening of walls A in vertical direction may be done by providing adequate pilasters under the roof trusses. For horizontal transfer of forces, a *band* may be provided at top of wall to which the roof trusses are connected, or *diagonal bracing* may be provided in plan at the level of main ties of the trusses which should extend from one gable end to the other end and be suitably connected to the end walls B (see Fig. 12.2). The inertia force of top half height of walls A may be assumed to be resisted by such a *band or bracing system.*

Now if ground motion is along y-direction, walls A will be in a position to act as shear walls, and all forces may be transmitted to them. In this case, the purlins act as ties and struts and transfer the inertia force of roof to the gable ends. A sloping band is required at the top of gable ends to transfer this inertia force from gable ends to the walls A and where this band joins roof band at the top of walls A.

The above discussion leads to the following requirements for structural safely of masonry buildings during ground motions:

1. *A free standing wall must be designed as vertical cantilever.*
2. *A shear wall must be capable of resisting all horizontal forces due to its own mass and those transmitted to it.*
3. *The roof or floor elements must be tied together and be capable of transferring their inertia force to shear walls by bending in horizontal plane as diaphragms.*
4. *The trusses must be anchored to supporting walls and have an arrangement of transferring their inertia force to end walls, say by cross bracing.*

5. *The shear walls must be present along both the axes of the building and reinforced vertically near there ends.*
6. *The walls must be effectively tied together to avoid separation at vertical joints due to shaking.*
7. *Horizontal bands must be provided for tying the walls together and carrying horizontal bending tension and transferring the inertia load horizontally to the shear walls.*

3 Main Design Provisions in IS:1326-1993

3.1 Categorization of Buildings (IS:4326-1967)

From the earthquake safety point of view, all design issues are based on probable *seismic intensity* at the site, the stiffness of the *site soil* on which the building will stand and the *importance* of the building. The seismic intensity is based on the *seismic zones* and the *importance of the building* according to IS: 1893 (Part- 1) 2002. The soil is classified as soft, medium and stiff. In the earlier version of IS:4326, categorization of the buildings for design purposes was based on the three factors: intensity as per the zone, importance factor as per the use of building and the softness of the soil defined in IS:1893. However, the earthquake design spectra in IS:1893-2002 now are flat at one value for the three types of soils for time periods of less than 0.4 s. in which the masonry buildings of one to four stories lie. Therefore, the building categories are now defined only based on seismic zones and the importance of the buildings as shown in Table 12.1 for Zones II to V (Zone I was merged into Zone II in IS:1893-2002, hence category A also does not appear in the table).

3.2 Masonry Units

Masonry units are of burnt bricks, solid concrete blocks, hollow concrete blocks or squared stone blocks. These have to be corresponding to various Indian Standard Codes and should have minimum crushing strength not less than 3.5 MPa.

Table 12.1 Building categories for earthquake-resistant design

Importance factor	Seismic zone II	III	IV	V
1.0	B	C	D	E
1.5	C	D	E	E

3.3 Mortar (IS:4326-1967)

The mortar used in masonry wall construction is specified on the basis of *building category* in Table 12.2. These are as per the mortar specified in IS:1905. Mortar of higher quality and strength is used so as to achieve high compressive and tensile strength of the wall and designed adequate performance of even important buildings in various seismic zones.

3.4 Openings in theWalls (IS:4326-1967 and Ch. 16 in (Arya, 2007)

Door and window openings in walls reduce their lateral load resistance both in-plane as well as out-of-plane earthquake forces. Therefore, they should preferably be small and more centrally located. This subject formed an issue of research at SRTEE, starting with the size and position of a single opening in the wall and then through detailed lateral load analysis of three and four storeyed buildings using shear wall analysis. As a result of such researches, the constraints on the size of the openings and the piers between the openings were specified as shown in Fig. 12.2 and Table 12.3. The other main provisions in the standard are presented in paras 4 and 5 below:

4 Provisions of Seismic Bands (IS:4326-1967)

IS:4326 requires that all masonry buildings should be strengthened by using seismic bands at various level as required depending on the building type and the building category. For example, for a building with flat roof resting on soft soil, the seismic bands are required at the *plinth level*, the *lintel level* of the door and the window openings in each storey (see Fig. 12.3).

Also, if the floor or roof consists of pre-cast or pre-fabricated units such as wooden beams, RC or steel joists placed in parallel and carrying pre-fabricated

Table 12.2 Recommended mortar mixes

Category of building from Table 12.1	Proportion of Cement–Lime–Sand
Any ordinary	M_2 (Cement–sand 1:6) or M_3 (Lime–cinder 1:3) or richer
B, C	M_2 (Cement–lime–sand 1:2:9 or Cement–sand 1:6) or richer
D, E	H_2 (Cement–sand 1:4) or M_1 (Cement–lime–sand 1:1:6) or richer

Note: Though the equivalent mortar with lime will have less strength at 28 days, their strength after 1 year will be comparable to that of cement mortar

Table 12.3 Size and position of openings in bearing walls

S. no.	Position of opening	Details of Opening for Building Category		
		A and B	C	D and E
1.	Distance b_5 from the inside corner of outside wall, min	Zero mm	230 mm	450 mm
2.	For total length of openings, the ratio $(b_1 + b_2 + b_3)/l_1$ or $(b_6 + b_7)/l_2$ shall not exceed:			
	(a) One-storeyed building	0.60	0.55	0.50
	(b) Two-storeyed building	0.50	4.46	0.42
	(c) 3 or 4 storeyed building	0.42	0.37	0.33 for D only
3.	Pier width between consecutive openings b_4. *min*	340 mm	450 mm	560 mm
4.	Vertical distance between two openings one above the other h_3, *min*	600 mm	600 mm	600 mm
5.	Width of opening of ventilator b_8, *max*	900 mm	900 mm	900 mm

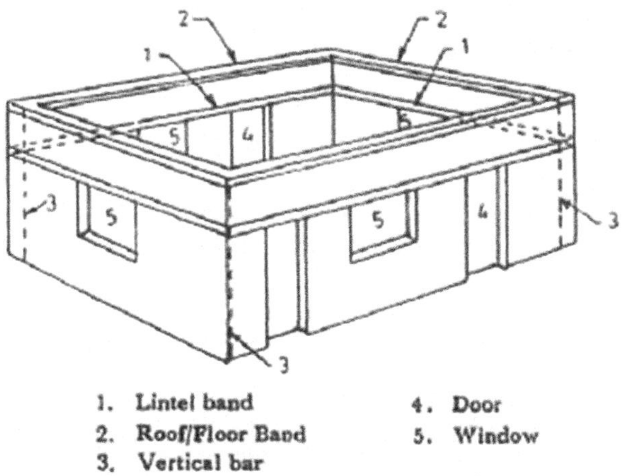

Fig. 12.3 Overall arrangement of reinforcing masonry buildings

elements like brick tiles, concrete, panel or stone slabs, etc., spanning across them and covering material like earth-fill or compacted lime concrete flooring and roofing, a band will also be needed at the ceiling level of such roofs or floors or at the same level of the roof or floor on all the walls.

In the case of pitched roof, bands will also be needed at the *eave level*, for enclosing the *gable end walls* and on top of ridge walls where such wall is provided. All such *bands* have to be constructed on all external and internal walls in continuous manner providing full continuity at the corners and T-junctions of the walls (see Fig. 12.4).

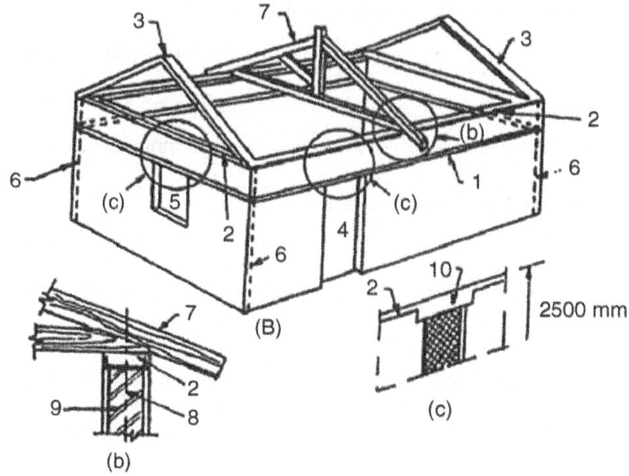

1. Lintel band
2. Eave level (Roof) band
3. Gable band
4. Door
5. Window
6. Vertical steel bar
7. Rafter
8. Holding down bolt
9. Brick/Stone wall
10. Door lintel integrated with roof band

a) Perspective view
b) Details of truss connection with wall
c) Details of integrating door lintel with roof band

Fig. 12.4 Overall arrangement of reinforcing masonry building having pitched roof

4.1 Method of Design of Seismic Bands (Arya, 1967 and Ch. 10 in Arya, 2007)

The following assumptions were made in developing the designs as shown in Table 12.4.

- Reinforced concrete *bands* were assumed to provide support to the plate action of the wall while bending out of plane in view of the reinforced concrete being stiffer than brickwork in the ratio of 8 to 10, (ratio of their modulus of elasticity).
- A *seismic band* at door and window common level was considered most appropriate to divide the wall height into two parts. The seismic inertia of the wall on the middle height between the *plinth band* and the *rood band* was assumed to be supported by the *lintel band* going over all the walls with full bending continuity at the corners and T-junctions.
- A seismic coefficient *four* times of the seismic coefficient prescribed for the seismic zone in IS:1893 was taken for the design of the RC section of the band, in which the reinforcing steel stress was taken as its *yield stress* or *0.2% proof stress*. It was assumed that if the effective seismic force exceeded this design value, the steel will go into the ductile deformation providing plastic energy absorption.

Table 12.4 Recommended longitudinal steel in reinforced concrete bands

	Building Category B		Building Category C		Building Category D		Building Category E	
Span m	No. of Bars	Dia (mm)	No. of Bars	Dia (mm)	No. of Bars	Dia (mm)	No. of Bars	Dia (mm)
5 or less	2	8	2	8	2	8	2	10
6	2	8	2	8	2	10	2	12
7	2	8	2	10	2	12	4	10
8	2	10	2	12	4	10	4	12

1. Span of wall will be the distance between center lines of its cross walls or buttresses. For spans greater than 8 m, it will be desirable to insert pilasters or buttresses to reduce the span or special calculations shall be made to determine the strength of wall and section of band.
2. The number and diameter of bars given above pertain to high strength deformed bars.
3. The width of RC, band is assumed to have the same thickness as the wall. Wall thickness shall be 200 mm minimum. A clear cover of 20 mm from the face of the wall will be maintained.
4. The vertical thickness of RC band shall be kept at 75 mm minimum, where two longitudinal bars are specified, one on each face and 150 mm, where four bars are specified.
5. Concrete mix shall be of grade M20 of IS 456 or 1:1.5:3 by volume.
6. The longitudinal steel bars shall be held in position by steel links or stirrups 6 mm diameters spaced at 150 mm apart.

- In view of the continuity planned at the corners and T-junctions of walls, the bending moment in the band was taken as WL/10 where W was the total seismic load and L the span of the band between the corner and T-junctions. The required area of steel reinforcement was defined by steel beam theory. In view of the fact that the longitudinal steel in the band is to be specified to be the same on both faces to take case of the reversible direction of the seismic force.
- The storey height for the *band* calculations was taken as 3.5 m. as usually adopted in school buildings and usually a less height is adopted in residential buildings.

A large number of calculations were carried out jointly by the author and CBRI scientist before adopting the final design value in the standard IS:4326-1967.

5 Provision of Vertical Steel Reinforcing in Masonry Walls (Arya, 1967 and Ch. 10 in Arya, 2007)

For working out the vertical steel bar requirements in the masonry walls, the method of calculation as suggested in PCA 1955 for analyzing masonry shear wall was adopted. Accordingly, the reinforcement as given in Table 12.5 to be provided at the corners and T-junctions of the walls was adopted in IS:4326 for which a large number of hand calculations were carried out by the author and the CBRI scientist working jointly. The following assumptions were made:

Table 12.5 Vertical steel reinforcement in masonry walls

No. of Storeys	Storey	Diameter of HSD Single Bar in MM at Each Critical Section			
		Category B	Category C	Category D	Category E
One	–	Nil	Nil	10	12
Two	Top	Nil	Nil	10	12
	Bottom	Nil	Nil	12	16
Three	Top	Nil	10	10	12
	Middle	Nil	10	12	16
	Bottom	Nil	12	12	16
Four	Top	10	10	10	Four-storeyed building not permitted
	Third	10	10	12	
	Second	10	12	16	
	Bottom	12	12	20	

1. The diameters given above are for H.S.D. bars. For mild steel plain bars, use equivalent diameters as given under Table 12.5.
2. The vertical bars will be covered with concrete M20 or mortar 1:3 grade in suitably created pockets around the bars. This will ensure their safety from corrosion and good bond with masonry.

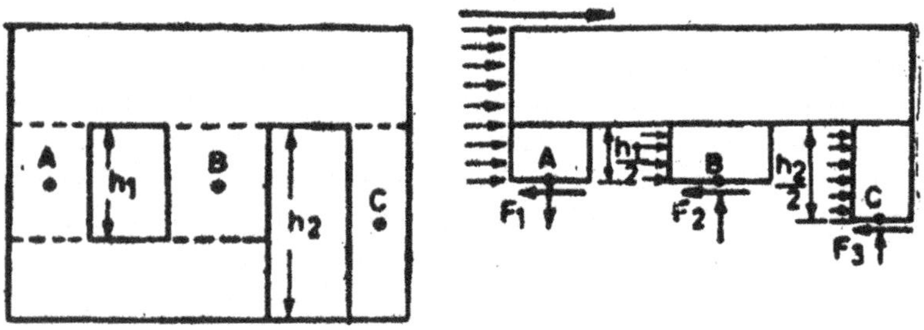

Fig. 12.5 Vertical steel reinforcement in mansory walls

- The masonry above the lintel band has very little deformation as compared with the in-plane lateral deformations in the masonry piers existing between the openings, hence that could be neglected. So also, the masonry below the window opening could be taken as rigid as compared with the piers.
- When the shear wall deflects laterally, the piers will behave like flat ended at the top and bottom with a point of contra flexure at the mid height of the pier (see Fig. 12.5).
- The piers will be subjected to lateral shear force in the ratio of their *shear stiffnesses*. The shear acting at the point of contra flexure will create bending moments of opposite nature at the two ends of the pier.
- For calculating the possible tensile stresses in the piers, a seismic coefficient *four* times the specified coefficient in the code for the zone was adopted namely, "0.16 for zone III, 0.20 for zone IV and 0.32 for zone V" as per IS:1893-1962.

Each pier will, therefore, be subjected to a shear force, V, bending moments in the plane of the shear walls at both ends, top and bottom, causing bending stresses at the ends, $\pm p_b$. It will also carry a vertical compression due to the weights supported by the pier coming from the floors and roof above, p_o. In addition, the whole building will be subjected to overturning effect due to the applied lateral seismic loads by which the "leeward" piers will carry additional compression p_c and the "windward" piers will be subjected to additional tension, $-p_t$. Thus, every pier will be subjected to stresses as follows:

Shear stress V/bd, where 'b' represents the thickness of wall and 'd' the width of the pier.

Bending stress $\pm p_b$

Direct stress p_o

Overturning stress $+p_c$ or $-p_t$

Combining these stresses, one edge of the pier may come in tension requiring tensile reinforcement. Large number of calculations were made and the total tension as worked out at the edge of the end piers as well as at the inner edge of the door or window openings, was used to determine the area of the steel bar required using *yield stress* or *0.2% proof stress*. It was assumed that where a larger tension may occur, the steel bar may go into ductile state and provide plastic energy absorption.

6 Experimental Verification (Arya, 1976; Qamaruddin et al., 1978; Arya et al., 1980; Arya, 1981; Arya, 1984a; Arya, 1984b)

A number of experimental test have been carried out on wagon-shock table at Department of Earthquake Engineering, IIT Roorkee, using *half size* single room specimens subjected to moderate to large shocks and measuring the accelerations at the base as well as the top of the specimen. All the tests have shown that those specimen which were reinforced according to the provisions in IS:4326-1967, although cracked at higher level of acceleration, did not collapse in any of the tests. At high levels of acceleration imparted to the specimens diagonal cracking did occur in piers showing the deficiency under shear strength for which providing bar reinforcement was not practical.

7 Theoretical Checks on NBO Supported Experimental Low-Cost Buildings (Arya et al., 1982 and Ch. 16 in Arya, 2007)

The National Buildings Organisation had put up four and five storeyed buildings with single brick load bearing constructions and using different experimental components in seismic zones III and IV of India during 1974 to 1982. The main objectives of the new techniques employed in the buildings were economy and speedy construction. The buildings included: (a) at Roorkee, Zone IV, four-storeyed, (b) at Lucknow, Zone III, four-storeyed, (c) at Parwanoo, Zone IV, four-storeyed, (d) at Delhi, Zone IV, four-storeyed, and (e) at Kolkata, Zone III, five-storeyed. A study was undertaken to examine the suitability of such buildings in the seismic environment. The results of the study are summarized below:

No adverse behavior of one brick thick load bearing walls observed in the 3 and 4 storeyed buildings in zones III and IV. For the normal frequent earthquake condition the compressive stresses in all walls and piers remain within the enhanced permissible stresses for this condition and the reinforcement provided in these buildings is found to be adequate to take care of the tensile stresses generated in the large earthquake condition. Provision of vertical steel, though in a somewhat smaller cross-sectional area, at corners and junctions of walls in *four*-storeyed buildings even in Zone III would be desirable.

The five-storeyed building at Calcutta showed weakness in its longitudinal direction for the codal value as well as 4 times this value as probable maximum seismic co-efficient used in the analysis, mainly because the provision of a series of openings consisted about 2/3 of the wall length instead of less value as specified in IS:4326. The reinforcement, which should have been provided in the piers around openings was also not provided. It was recommended that all specifications of IS:4326 must be adhered to for seismic safety including the number of storeys.

8 Observations in Real Earthquake

A real life proof of the adequacy of the seismic resistant measures specified in IS:4326 came out during the Uttarkashi Earthquake in 1991 from the performance observed of ITBP colony buildings constructed at Mahitanda village very close to Uttarkashi. The buildings were one and two storeyed residential buildings constructed using solid concrete blocks with RC slab roofing in which all provisions of IS:4326, namely, seismic bands as well as vertical bars were used by Central Public Works Department (CPWD), Government of India. During the earthquake none of the buildings was even moderately damaged.

9 Conclusion

From the above presentation, it will be clearly observed that the earthquake resistant provisions in IS:4326 are based on a large number of engineering calculations based on realistic assumptions and given in very simple tabulated form for making it convenient for the professionals as well as other users for constructing earthquake-resistant masonry building. For very important or historical buildings, calculations can be carried out by qualified engineers in the similar manner so that dimensions of the seismic bands and the reinforcement could be designed as found appropriate. The design data provided in IS:4326 may not be taken as "imaginary thumb rules or prescription" but used with confidence since those are based on preengineering calculations and verified by tests and real life earthquake performance.

References

Arya, A. S. (1967). Design and construction of masonry building in seismic area. *Bulletin of the Indian Society of Earthquake Technology (ISET), Roorkee, 4*(24), 25–37.

Arya, A. S. (1976). *Experimental evaluation of a seismic strengthening method of brick buildings, VI Symposium on earthquake engineering, Roorkee: University of Roorkee,* I, 353–359,

Arya, A. S. (1981). A new concept for resistance of masonry buildings in severe earthquake shocks. *Journal of Institution of Engineers (India), 61*(Cl.6), 302–308.

Arya, A. S. (1984a). Dynamic testing of brick building models. *Proceedings of Institution of Civil Engineers, U.K. Part-2, Paper 8788, 77,* 253–365.

Arya, A. S. (1984b). Seismic analysis of experimental buildings. In *International conference on low cost housing for developing countries, Roorkee: Central Building Research Institute.* 623–636.

Arya, A. S., Chandra, B., Gupta, S. P., & Pankaj, B. (1982). Evaluation of seismic performance of multi-storeyed brick buildings constructed under experimental housing scheme of national buildings organization. In *Earthquake engineering studies, EQ 82-14.* Roorkee: Department of Earthquake Engineering, University of Roorkee.

Arya, A. S., Chandra, B., & Qamaruddin, M. (1980). *A new system of brick buildings for improved behaviour during earthquake.* Seventh World Conference on Earthquake Engineering, Istanbul, Turkey (p. 4/2250).

Arya Anand, S. (2007). *Earthquake disaster reduction: Masonry buildings, design and construction.* New Delhi: KW Publishers Pvt. Ltd.

IS:4326-1967 & 1993. (1967). *Code of practice for earthquake resistant construction of buildings.* New Delhi: ISI.

Portland Cement Association. (1955). *Analysis of small reinforced concrete buildings for earthquake forces.* Portland Cement Association, Chicago.

Qamaruddin, M., Arya, A. S., & Chandra, B. (1978). Experimental evaluation of aseismic strengthening method of brick building. In *Sixth symposium on earthquake engineering,* Roorkee, 353–360.

Chapter 13
Seismic Design Philosophy: From Force-Based to Displacement-Based Design

Devdas Menon, A. Meher Prasad, and Jiji Anna Varughese

1 Introduction

The frequent occurrence of major and minor earthquakes and the associated damages have led the structural engineering community to think about the need for a rational design method which can control the possible structural and nonstructural damages and can give an economical yet safe design. In recent years, this has resulted in the development of several displacement-based design methods, which are philosophically different from the traditional force-based design methods.

The early SEAOC Blue Book (SEAOC, 1959) has expressed the basic seismic design philosophy as follows:

Structures should be designed to resist:

- A *minor* earthquake without damage
- A *moderate* earthquake without structural damage, but with some nonstructural damage
- A *major* earthquake without collapse, with some structural and nonstructural damages

Seismic design guidelines, such as FEMA 356, FEMA P695, ATC 40, etc., provide useful information for designing the structures for such multiple performance objectives. IS 1893:2002 also aims at the above philosophy of attaining multilevel performance objective, as stated in cl.6.1.3. However, the design provisions in the code are limited in scope to analysis under a single *design basis earthquake*.

D. Menon (✉) · A. Meher Prasad
Department of Civil Engineering, Indian Institute of Technology Madras, Chennai, TN, India
e-mail: dmenon@iitm.ac.in

J. A. Varughese
Department of Civil Engineering, GEC Barton Hill, Thiruvananthapuram, Kerala, India

M. L. Sharma et al. (eds.), *Advances in Indian Earthquake Engineering and Seismology*, https://doi.org/10.1007/978-3-319-76855-7_13

Different seismic analysis methods are currently in use, which vary in complexity, designer input, sophistication in modelling and field of applicability. The more important among them are briefly discussed here, with special focus on displacement-based design methods.

2 Types of Seismic Analyses

In general, for simple and regular structures, a static linear analysis is sufficient whereas for more complex and irregular structures, dynamic analysis, either linear or nonlinear, is required based on the degree of irregularity. In seismic analysis, a certain level of inelasticity is allowed, to make the design more economical, by taking advantage of the ductility capacity of the structure. In linear methods, inelasticity is accounted for indirectly, by adopting response reduction factor (R) for calculating the design base shear (V_B). But, in nonlinear analyses, both material and geometric nonlinearities are considered explicitly. This calls for proper knowledge of nonlinear material modelling and seismic demand modelling.

2.1 Equivalent Static Analysis (Seismic Coefficient Method)

Designers are familiar with the conventional force-based design in which the lateral load-resisting system is designed for an equivalent static load, which is calculated based on the design peak ground acceleration (PGA), permissible ductility (μ), type of soil and importance of the structure (I). Usually, the equivalent static approach is adopted when the building configuration is regular and there is no significant mass or stiffness irregularity.

2.2 Response Spectrum Method

Linear dynamic analysis, also known as response spectrum analysis (RSA), considers the effect of several modes, and hence it is appropriate for all buildings and irregular buildings. Undamped free vibration analysis of the entire building is performed using appropriate masses and elastic stiffness of the structural system, to obtain the natural periods (T) and mode shapes (φ). Responses are calculated for each mode separately and are combined together to get the overall response. The modal masses of the considered modes should be at least 90% of the total seismic mass. Different modal combination rules are given in IS 1893:2002 to get the peak response quantities like member forces, displacements, storey forces, storey shears, base reactions, etc. Figure 13.1 shows the first three mode shapes and the associated spectral acceleration coefficients.

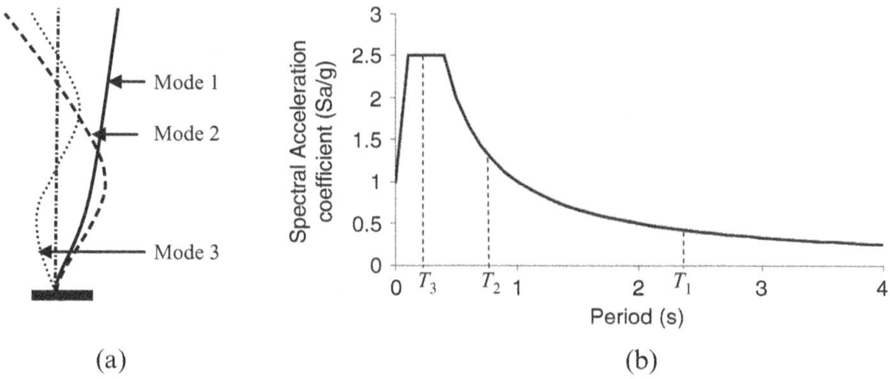

Fig. 13.1 Spectral acceleration for response spectrum analysis

2.3 Performance Assessment Methods

Performance assessment of existing structures is required to be done in the context of the frequent occurrence of seismic hazards, so that proper strengthening measures can be adopted. The most popular among the assessment techniques is the *pushover analysis*, which gives useful information regarding the capacity of the structure. When the capacity curve is plotted along with the demand curve, a *performance point* can be located. If there is no performance point, the capacity of the structure or the damping has to be increased so as to get a performance point. Another performance assessment tool is the *nonlinear time history analysis*, which is a dynamic analysis performed for a suite of ground motion time histories.

2.3.1 Pushover Analysis

Nonlinear analyses are more advanced methods which give useful information on the behavior of structures in the nonlinear range. Nonlinear static analysis, commonly known as pushover analysis is performed to evaluate the capacity of an existing structure. ATC 40 gives useful information on evaluating the performance of RC structures.

In pushover analysis, the magnitude of lateral load is increased monotonically maintaining a predefined distribution pattern along the height of the building (Fig. 13.2). The building is displaced till the *control node* reaches target displacement or till the building collapses. Usually, the center of gravity of roof is taken as the control node. Pushover analysis may be carried out twice: first, till the collapse of the building to estimate target displacement, and second, till the target displacement, to estimate the seismic demand.

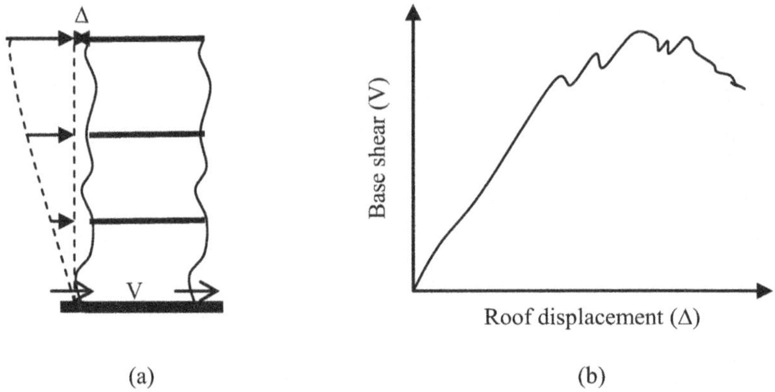

Fig. 13.2 Pushover analysis. (**a**) Building model, (**b**) Pushover curve

Target displacement is the displacement demand (roof level) for the building at the control node subjected to the ground motion under consideration. There are two approaches to calculate target displacement:

1. FEMA 356 Coefficient Method
2. Capacity Spectrum Method

As per FEMA 356, target displacement (δ_t) can be estimated as

$$\delta_t = C_0 C_1 C_2 C_3 S_a \left(\frac{T_e}{2\pi}\right)^2 \tag{13.1}$$

where T_e is the effective fundamental period of the building in seconds in the direction under consideration, S_a is the spectral acceleration at the effective fundamental period, C_0 is a shape factor to convert the spectral displacement of equivalent single-degree-of-freedom (SDOF) system to the displacement at roof, C_1 is the ratio of expected displacement for an inelastic system to the displacement of a linear system, C_2 is a factor that accounts for pinching in the load deformation relation due to strength and stiffness degradation and C_3 is a factor to adjust geometric nonlinearity (P–Δ) effects.

In the capacity spectrum method (Freeman, 1978), the pushover curve is plotted in acceleration–displacement response spectrum (ADRS) format, which is called the *capacity spectrum* (Fig. 13.3). Simultaneously, the acceleration and displacement spectral values as calculated from the corresponding response spectrum for a certain damping (say 5% initially) and are plotted. The equivalent damping and natural period increase with the increase in nonlinear deformation of the components. The spectral values of the acceleration and displacement for various damping levels can be plotted using the multiplication factors for the 5% damping curve, as given in IS 1893:2002. Thus the instantaneous spectral acceleration and displacement point

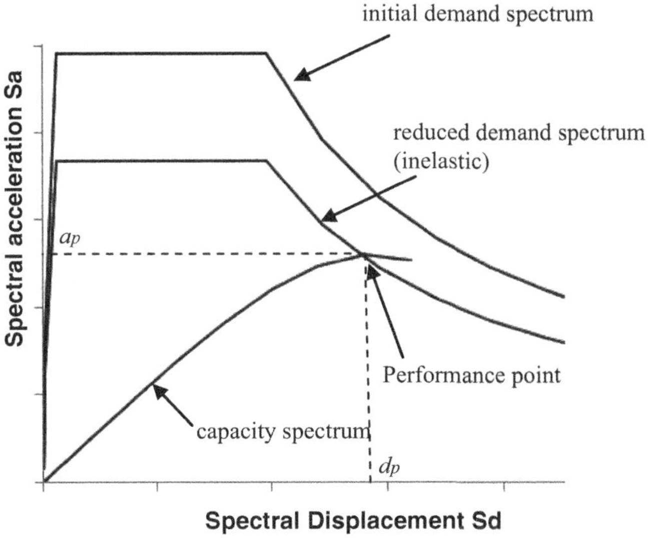

Fig. 13.3 Typical demand and capacity spectra in ADRS format

(demand point) shifts to a different response spectrum for higher damping. The locus of the demand points in the ADRS plot is referred to as the *demand spectrum.*

The *performance point* is the point where the capacity spectrum crosses the demand spectrum. If the performance point exists and the damage state at this point is acceptable, then the building is considered to be adequate for the design earthquake. But, the analysis cannot predict accurately the higher mode responses of a flexible building.

Capacity spectrum method (CSM) and displacement coefficient method (DCM) have been updated by the project ATC 55, which is principally aimed to improve the accuracy of the peak SDOF displacement estimations. This is documented in the FEMA 440:2005 report. The improved DCM modifies C_1 and C_2 coefficients, to improve the expected elastoplastic oscillator deformation estimations from their elastic counterparts (C_1) and modify these estimations for cyclic degradation (C_2). This procedure suggests elimination of the coefficient (C_3), which accounts for P-delta effects. Instead, it establishes a limit on the lateral strength to avoid dynamic instability.

The improved CSM proposes new effective damping and period relationships for a wide variety of cyclic behavior and post-yield stiffness to predict the nonlinear SDOF deformation demands through an equivalent linear system. The improved CSM determines the equivalent linear parameters (equivalent period, T_{eq}, and equivalent damping, β_{eq}) through a statistical analysis that minimizes the extreme differences between the maximum response of an actual inelastic SDOF system and its equivalent linear counterpart.

Modal Pushover Analysis developed by Chopra and Goel (2002) is an improved procedure to calculate target displacement. The force distribution is expressed as a combination of modal contributions. The peak value of roof displacement due to n^{th} mode is expressed as

$$u_{\text{no, roof}} = \Gamma_n \phi_{n,\text{roof}} D_n \tag{13.2}$$

where $u_{\text{no,roof}}$ is the target displacement of the building at roof due to n^{th} mode, Γ_n is the nth mode participation factor, $\varphi_{n,\text{roof}}$ is the nth mode shape coefficient at roof, and D_n is governed by the equation of motion. The peak modal responses from all the modes considered are combined according to appropriate modal combination rule, such as square root of sum of squares (SRSS), complete quadratic combination (CQC), etc. This application of modal combination rules to inelastic systems lacks a theoretical basis. However, it seems to be reasonable as it resembles the RSA analysis procedure.

Adaptive Pushover Analysis takes into account the current level of local resistance and updates the forcing function accordingly. The lateral load pattern is not kept constant, but it is continuously updated, based on a combination of the instantaneous mode shapes and spectral amplifications corresponding to the inelastic period of the structure. Bracci et al. (1997) first developed adaptive pushover analysis procedure using the inelastic forces of the previous equilibrated load step to update the lateral load pattern. A displacement-based adaptive pushover analysis was developed by Papanikolaou et al. (2005) where displacement was used as the forcing function. The applied displacement at each step is determined by modal analysis or any other method that explicitly accounts for the structural characteristics at the current level of inelasticity, in a way that approximates the expected dynamic deformations. The procedure is not popular in practice because of its computational difficulties.

2.3.2 Nonlinear Time History Analysis

This is generally the most accurate method of analysis and is adopted by researchers for evaluating the performance of a structure designed using any of the standard design methods. A large number of subjective modelling decisions have to be taken for performance assessment (Priestley et al., 2007). The structural engineer should have appropriate experience in nonlinear time history analysis, a good knowledge of nonlinear material modelling, and suitable selection and scaling of ground motions.

Most of the early work in the inelastic analysis of concrete structures was based on bilinear systems. However, it was soon realized that RC elements do not offer the large energy dissipation capacity which is inherent in a bilinear system. A more general stiffness-degrading model for reinforced concrete was first introduced by Clough and Johnson (1966). This model has the advantage over the bilinear model that the loading stiffness is modified as peak rotation increases. Takeda et al. (1970) developed a nonlinear model which can closely reproduce the behavior of RC

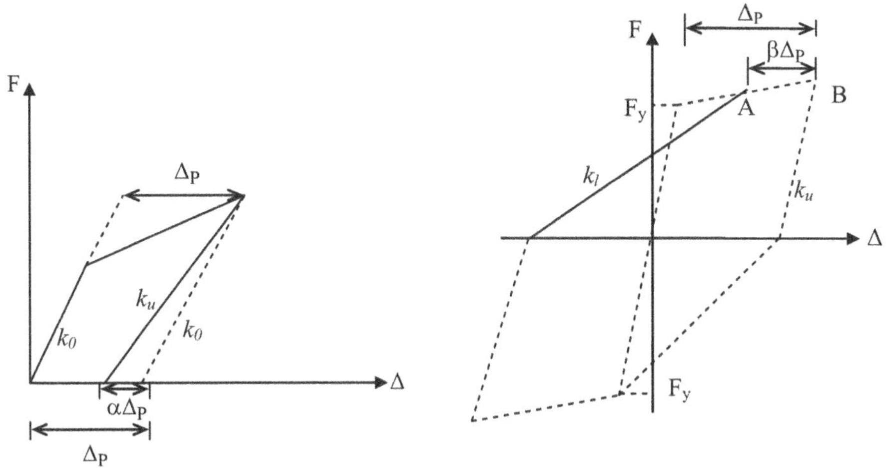

Fig. 13.4 Modified Takeda model

elements in flexure. The model has a bilinear envelope curve, and it is designated to dissipate energy at low cycles once the cracking point is exceeded. Emori and Schnobrich (1978) later introduced modifications into the Takeda model to take into account the slippage and shear pinching effects.

The moment–rotation relationship for the modified Takeda model is shown in Fig. 13.4. The primary curve for the model is a bilinear curve which changes slope at the point of yielding. Two other modifications were introduced by Litton (1975). The first is for stiffness degradation in the unloading part. Instead of unloading with initial slope (k_0), the parameter α is used to modify the unloading stiffness k_u. The second modification is for reloading stiffness k_l. Instead of loading toward the point of maximum (B), another point (A), which is set by parameter β, is aimed at. Parameter α decreases the unloading stiffness and parameter β increases the reloading stiffness.

Most seismic design codes prescribe general guidelines for selecting the type of earthquake records required for nonlinear dynamic analysis purposes. The period range for spectral matching varies among code provisions. Moreover, the minimum number of records required for structural analysis is three in all cases and when a set of at least seven ground motions is used, the structural engineer is allowed to compute the mean structural response. Otherwise, only a maximum response value is computed if three to six recordings are used. As per ASCE7-05 (2005), for 2-D analysis, the ground motions shall be scaled such that the average value of the 5% damped response spectra for the suite of motions is not less than the design response spectrum for the site for periods ranging from 0.2T to 1.5 T where T is the natural period of the structure in the fundamental mode for the direction of response being analyzed.

2.4 Incremental Dynamic Analysis

Usually the performance assessment is done deterministically, based on discrete hazard and performance levels. The Pacific Earthquake Engineering Research (PEER) center has focused for several years on the development of procedures for a comprehensive probabilistic seismic performance assessment of buildings and bridges (Krawinkler & Miranda, 2004). An important issue in performance-based earthquake engineering is the estimation of structural performance under seismic loads, in particular, the estimation of the mean annual frequency (MAF) of exceeding a specified level of structural demand (e.g., the maximum of peak inter-story drift ratio θ_{max}) or a certain limit state capacity (e.g., global dynamic instability).

A promising method that has recently risen to meet these needs is the *Incremental Dynamic Analysis* (IDA), which involves performing nonlinear dynamic analyses of the structural model under a suite of ground motion records, each scaled to several intensity levels designed to force the structure all the way from elasticity to final global dynamic instability (Vamvatsikos & Cornell, 2004). The important steps of IDA are as follows:

1. Choosing suitable ground motion intensity measures (e.g., Peak Ground Acceleration, PGA) and representative Damage Measures, DM (e.g., storey drift index)
2. Suitable scaling of ground motions
3. Employing proper interpolation
4. Summarizing techniques for multiple records to estimate the probability distribution of the structural demand given the seismic intensity
5. Defining limit-states, such as immediate occupancy (IO), collapse prevention (CP), and global dynamic instability (GI)

The results can be integrated with conventional probabilistic seismic hazard analysis (PSHA) to estimate mean annual frequencies of limit-state exceedance.

A probabilistic design/assessment approach to performance-based seismic engineering has been proposed by Cornell and Krawinkler (2000). The basis for assessing the adequacy of the structure will be a vector of certain key decision variables (DV) such as the exceedance of one limit state (e.g., collapse). These can only be predicted probabilistically. Therefore, the specific objectives of engineering assessment analysis are in effect quantities such as the mean annual frequency of collapse λ. The proposed strategy involves the expansion of the mean annual frequency in terms of the structural Damage Measures, DM, and ground motion Intensity Measures, IM, which can be written symbolically as

$$\lambda(\mathrm{DV}) = \int\int G(\mathrm{DV}|\mathrm{DM})dG(\mathrm{DM}|\mathrm{IM})d\lambda(\mathrm{IM}) \tag{13.3}$$

Here, G(DV|DM) is the probability that the decision variable exceeds specified values given that the engineering damage measures (e.g., the maximum storey drift index) are equal to particular values (fragility curves). G(DM|IM) is the probability that the damage measure exceeds these values given that the intensity measure

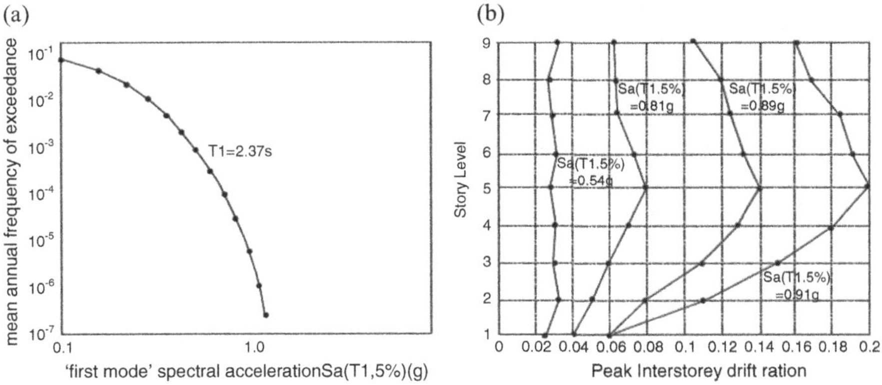

Fig. 13.5 (**a**) Hazard curve (**b**) the median peak inter-storey drift ratio (Ref: Vamvatsikos & Cornell, 2004)

(PGA) equals particular values. λ(IM) is the mean annual frequency of the intensity measure.

The results of IDA in terms of MAF and storey drift at various ground motion IM are shown in Fig. 13.5.

3 Displacement-Based Design

Structural design is different from performance assessment. It can be viewed as an iterative assessment process that starts with a judgmental conceptual design for which performance assessment is carried out and the design is improved in successive iterations until the performance targets are met (Krawinkler & Miranda, 2004).

Displacements are rather a direct measure of damage than forces and hence it will be more reasonable if displacements are considered at the very start of the design process and the limit states are defined accordingly. Displacement-based design methods aim for multiple performance levels. For ordinary buildings, it can be immediate occupancy (IO) for serviceability earthquakes (SE) and life safety (LS) under design-basis earthquake (DBE). For important buildings like school or hospital buildings, one can go for higher levels of performance objective, i.e., IO under DBE and LS under maximum considered earthquake (MCE). Analysis should be done for each performance objective separately, and structural design should be done for the worst condition.

Any method which starts with an objective of designing the structure to achieve a predefined displacement or drift can be generally considered as a Displacement-Based Design (DBD) method. In DBD, a target displacement is assumed corresponding to the required performance level. The structural design is aimed to achieve this displacement while ensuring that the displacement is not exceeded. The permissible values of inter-storey drift for different performance levels are given in ATC 40 and FEMA 356 which help in estimating the target displacement.

3.1 Direct Displacement-Based Design

Direct Displacement-Based Design (DDBD) can be performed in three major steps (Priestley et al., 2007): (1) Estimation of design base shear and its distribution, (2) Design of yielding members, and (3) Design of non-yielding members *using* capacity design. The design should be followed by proper performance verification using well-established nonlinear analysis techniques. Thus, unlike the conventional force-based design, DBD calls for a detailed look on the local as well as the global behavior of structure thus helping in reducing damage during an earthquake.

The given structure is converted into an equivalent SDOF system having effective mass (m_{eff}), effective stiffness (k_{eff}), and effective damping (ξ_{eff}) based on a target displacement. The design base shear (V_B) is calculated using the secant period (T_{eff}) (calculated from the displacement spectrum scaled based on the design ductility) corresponding to the maximum displacement. The lateral loads acting at various floor levels are obtained based on a distribution which depends on the seismic mass and the inelastic displacement.

The structure is analysed for the lateral loads, and the designated yielding members (beams) are designed accordingly. The designated non-yielding members (columns) are designed based on capacity-based design considering the overstrength of beams and dynamic amplification due to higher mode effects.

A comparison of direct displacement-based methods proposed by Priestley and Kowalsky (2000) and Chopra and Goel (2001) is given in Table 13.1.

In the direct displacement-based design (Priestley & Kowalsky, 2000), the secant stiffness of the equivalent SDOF system corresponding to target displacement (Δ_T) is calculated from the elastic displacement spectrum suitably scaled for equivalent

Table 13.1 Comparison of DDBD by Priestley and Kowalsky (2000) and Chopra and Goel (2001)

DDBD (Priestley & Kowalsky, 2000)	DDBD (Chopra & Goel, 2001)
• Assume critical drift (θ_c) • Calculate inelastic displacement (Δ_i) • Obtain equivalent SDOF system properties and the design displacement (Δ_d) • From the displacement spectrum for equivalent damping, find out effective secant period (T_e) and secant stiffness (K_e) • Design base shear $V_B = K_e\Delta_d$ • Distribute the base shear to various floor levels using $F_i = F_t + 0.9V_B \frac{m_i\Delta_i}{\sum_{i=1}^{n} m_i\Delta_i}$ where m_i is the seismic mass at i^{th} floor level and $F_t = 0.1\ V_B$ at roof and $F_t = 0$ at all other floors • Design the structure for the lateral load	• Assume allowable plastic rotation (θ_p) • Design displacement, • $\Delta_d = \Delta_y + H_e\theta_p$ • Δ_y, Yield displacement; H_e, effective height • Use inelastic displacement spectrum to estimate the initial elastic fundamental period, T_i, and the stiffness, K_i • Design base shear $V_B = K_i\Delta_y$ • Distribute as per Priestley's method and design the frame • Calculate yield displacement and yield force (V_y) of the frame • If $V_B = V_y$, and Δ_y = assumed Δ_y, adopt the design • Otherwise, change Δ_d as per the new Δ_y and redesign

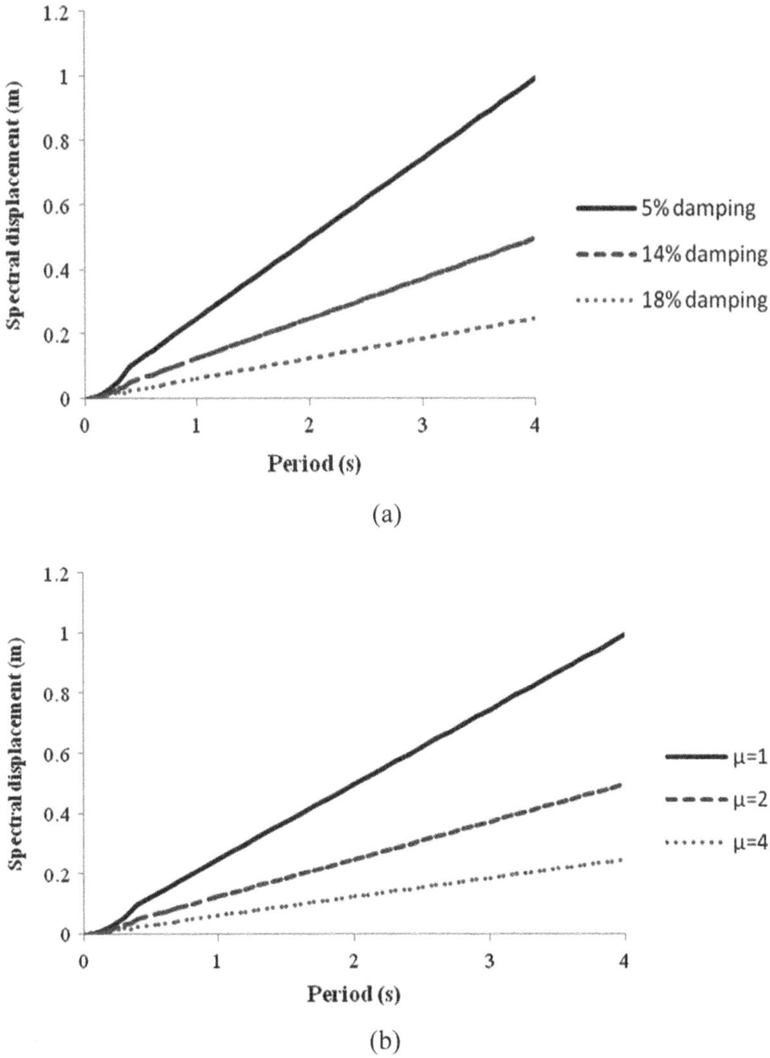

Fig. 13.6 Typical displacement spectra. (**a**) Elastic spectra for equivalent damping, (**b**) inelastic spectra

damping (Fig. 13.6). But, in DDBD method by Chopra and Goel (2001), the inelastic displacement spectrum for various ductility levels is adopted. The design base shear for the two methods is obtained as

$$V_{\mathrm{B}} = \begin{cases} K_{\mathrm{secant}} \times \Delta_{\mathrm{T}} \ (\text{as per Priestley's method}) \\ K_{\mathrm{initial}} \times \Delta_{\mathrm{y}} \ (\text{as per Chopra's method}) \end{cases} \tag{13.4}$$

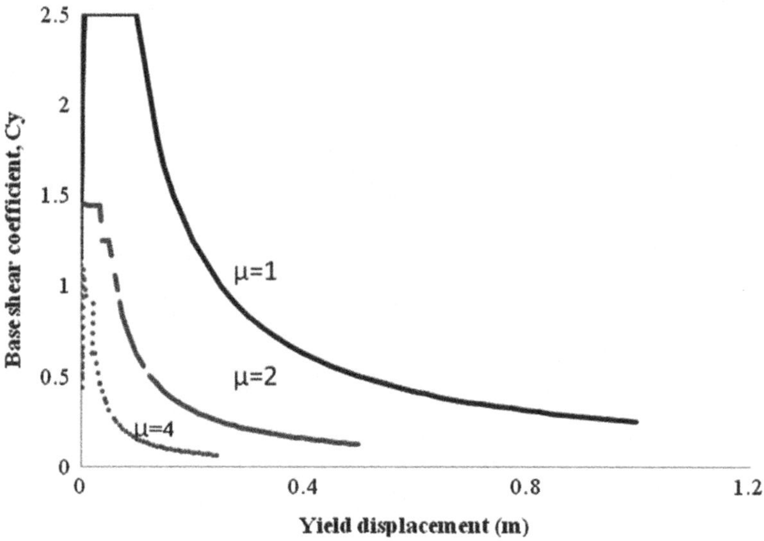

Fig. 13.7 Typical yield point spectra

3.2 Yield Point Spectrum Method

In the yield point spectrum method (Aschheim & Black, 2000), yield point spectra (Fig. 13.7) for various ductility levels are used to estimate the design base shear. Yield Point Spectrum (YPS) plots spectral response data for oscillators having constant displacement ductility on the axes of base shear coefficient and yield displacement. The curve for elastic response ($\mu = 1$) is identical to the spectral demand curve of capacity spectrum method (CSM). For higher values of ductility, the curves diverge, since the CSM explicitly plots Δ_u while YPS explicitly plots Δ_y. For both the methods, periods are constant along radial lines extending from the origin. Yield point spectrum can be plotted for an elastic response spectrum or for a given ground motion.

Target displacement (Δ_T) is determined to satisfy the drift limit for desired risk event. For this target displacement, the drift control branch is drawn satisfying $\mu\Delta_y = \Delta_T$, and the permissible design region is identified corresponding to allowable ductility. A suitable Δ_y for the structure is selected and used to determine required yield strength coefficient (C_y). Base shear can be determined as $V_b = m_{eff}C_y g$.

3.3 Performance-Based Plastic Design Method

One of the recently developed energy methods is the Performance-Based Plastic Design (PBPD) proposed by Goel et al. (2010). Even though it uses work-energy

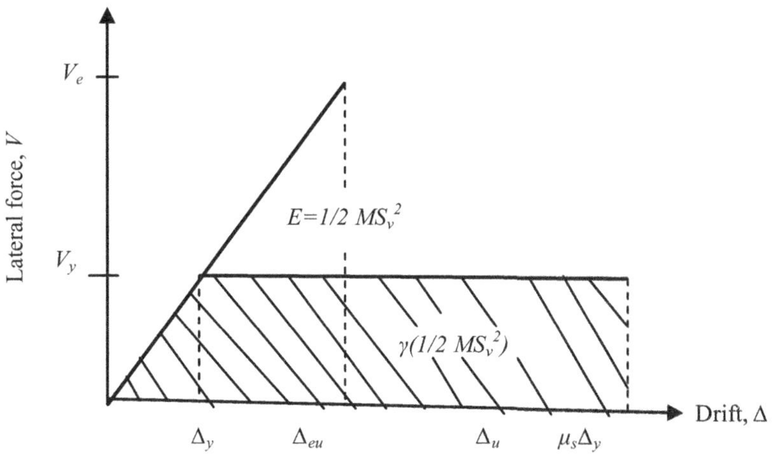

Fig. 13.8 PBPD concept

principle for calculating the design base shear, it considers target drift as the design constraint and hence can be grouped under DBD methods. The method has been successfully applied to a variety of common steel framing systems and more recently to RC frames. PBPD method uses pre-selected target drift and yield mechanisms as key performance objectives. The design base shear for a specified hazard, which is generally given as design spectrum in codes, is calculated by equating the work needed to push the structure monotonically up to the target drift to the energy required by an equivalent elasto-plastic (EP) SDOF system to achieve the same state (Fig. 13.8).

A new distribution of lateral design forces (Chao et al., 2007) is used which is based on relative distribution of maximum storey shears consistent with inelastic dynamic response results. Plastic design is then performed to detail the frame members and connections in order to achieve the intended yield mechanism and behavior. Thus, determination of design base shear, lateral force distribution and plastic design are the three main components of PBPD method.

Assuming an idealized EP force-deformation behavior of the system, the work-energy equation can be written as

$$(E_e + E_p) = \gamma\left(\frac{1}{2}MS_v^2\right) = \frac{1}{2}\gamma M\left(\frac{T}{2\pi}S_a g\right)^2 \qquad (13.5)$$

where (E_e + E_p) is the elastic and plastic components of energy needed to push the structure up to the target drift, S_v is the design pseudo-spectral velocity, S_a is the design pseudo-spectral acceleration coefficient, T is the natural period of the structure, M is the total mass of the system, and γ is the energy modification factor (Eq. 13.6) which depends on structural ductility factor (μ_s) and ductility reduction factor (R_μ).

$$\gamma = \frac{2\mu_s - 1}{R_\mu^2} \tag{13.6}$$

The work-energy equation can be rewritten in the following form:

$$\frac{1}{2}\frac{W}{g}\left(\frac{T}{2\pi}\frac{V_y}{W}g\right)^2 + V_y\left[\sum_{i=1}^{N}\lambda_i h_i\right]\theta_p = \frac{1}{2}\gamma\frac{W}{g}\left[\frac{T}{2\pi}S_a g\right]^2 \tag{13.7}$$

The required design base shear coefficient is

$$\frac{V_y}{W} = \frac{-\alpha + \sqrt{\alpha^2 + 4\gamma \mathrm{Sa}^2}}{2} \tag{13.8}$$

where α is a dimensionless parameter given by

$$\alpha = \left[h^* \frac{\theta p 8\pi^2}{T^2 g}\right] \tag{13.9}$$

The term θ_p is the plastic component of target drift ratio and $h^* = \sum_{i=1}^{N} \lambda i h i$

where λ_i is the base shear distribution factor and h_i is the height of ith floor above the base.

In PBPD method, energy calculations are done using monotonic force-displacement relations; advanced methods considering the energy dissipated under cyclic loading are yet to be developed (Bertero & Bertero, 2002).

3.4 *Deformation-Controlled Design Method*

The deformation-controlled design (Panagiotakos & Fardis, 1999) aims to integrate a displacement-based design approach within the entire structural design process, including the effects of gravity load. In this method, the longitudinal reinforcement in beams and at base sections of columns is proportioned using force-based design for the ultimate limit state in bending, due to the combination of the serviceability earthquake (SE) with the quasi-permanent gravity loads. This is a force-based design with a response reduction factor ranging from 2.5 to 7.5. Capacity-based design is done for proportioning reinforcement of columns at all locations except at base and for shear forces in all the members. Peak inelastic member deformation demands are estimated for life safety earthquake demand. The reinforcement in critical end regions of beams and columns is proportioned and detailed to meet these inelastic chord rotation demands.

3.5 *Deformation-Based Design Method*

A deformation-based design method concerning analysis of 3-D RC buildings is proposed by Kappos and Stefanidou (2010) which involves the use of advanced analysis tools, i.e., response history analysis for appropriately scaled input motion, for multiple levels of earthquake action. It is expected that the design will develop inelastic behavior in the structure for the serviceability earthquake also. The method can be applied to regular as well as irregular RC 3-D frame buildings. Comparison of the design with that of the same building designed as per EC8 provisions shows more economical detailing of transverse reinforcement in the members which develop very little inelastic behavior under even very strong earthquakes.

4 Comparison of Force-Based and Displacement-Based Design

In force-based design, force-based acceptance criteria are adopted. For instance, if the capacity-demand ratio is acceptable, the design is considered as acceptable. It is based on the practice of designing for other loadings like gravity and wind loads. But, the displacement-based design approach is different. It specifies a drift level and designs the structure to have the specified displacement capacity. It also ensures that the displacements are within allowable limits. Hence, it can provide improved protection for nonstructural components.

In force-based design, the design base shear is calculated based on a factor called response reduction factor (R), which depends on the design ductility, overstrength, and redundancy of the structure. Since, the values of the above parameters are not known at the start of the design process, the value of R is not realistic, and hence, the structure is designed for an unknown ductility level. But, in displacement-based design, since the ultimate displacement is predefined and the yield displacement can be estimated based on the size of the critical yielding elements, the structure is designed for a known ductility level.

In force-based design, the structure is designed for a single seismic hazard level, that is *design basis earthquake* (DBE) which can be calculated from the *maximum considered earthquake* (MCE) and is assumed that the structure will satisfy the *life safety* performance objective. But the satisfaction of one performance objective, say, "*life safety* under DBE" does not guarantee that the structure will satisfy other performance objectives (*operational* under *serviceability earthquake*). But, the displacement-based design methodology offers the advantage of designing the structures for multiple performance levels, and hence, it is very much useful for the design of important and lifeline structures.

5 Conclusions

The seismic design philosophy including that of the traditional force-based design and the recently developed displacement-based seismic design is discussed in the present chapter. Static methods are applicable for regular structures, whereas for highly irregular structures, a detailed nonlinear analysis is required to be performed. Some of the analysis techniques can be used for performance verification of existing buildings and the assessment of new design methods. The reliability of the assessment methods depends on the proper representation of seismic hazard, accurate modelling of material and geometric nonlinearity, and the appropriate selection of element models.

References

ASCE7-05. (2005). *Minimum design loads for buildings and other structures*. Reston, Virginia: American Society of Civil Engineers.

Aschheim, M. A., & Black, E. F. (2000). Yield point spectra for seismic design and rehabilitation. *Earthquake Spectra, 16*(2), 317–336.

ATC 40. (1996). *Seismic evaluation and retrofit of existing concrete buildings*. Redwood City, CA: Applied Technology Council.

Bertero, R. D., & Bertero, V. V. (2002). Performance-based seismic engineering: The need for a reliable conceptual comprehensive approach. *Earthquake Engineering and Structural Dynamics, 31*, 627–652.

Bracci, J. M., Kunnath, S. K., & Reinhorn, A. M. (1997). Seismic performance and retrofit evaluation of reinforced concrete structures. *Journal of Structural Engineering, 123*(1), 3–10.

Bureau of Indian Standards IS 1893 (Part 1). 2002. *Indian Standard Criteria for earthquake resistant design of structures- General provisions and buildings*.

Chao, S. H., Goel, S. C., & Lee, S. S. (2007). A seismic design lateral force distribution based on inelastic state of structures. *Earthquake Spectra, Earthquake Engineering Research Institute, 23* (3), 547–569.

Chopra, A. K., & Goel, R. K. (2001). Direct displacement-based design: Use of inelastic vs. elastic design spectra. *Earthquake Spectra, 17*(1), 47–65.

Chopra, A. K., & Goel, R. K. (2002). A modal pushover analysis procedure for estimating seismic demands for buildings. *Earthquake Engineering and Structural Dynamics, 31*, 561–582.

Clough, R.W. and S.B. Johnston (1966) Effect of stiffness degradation on earthquake ductility requirements. Proceedings of second Japan National Conference on Earthquake Engineering. 227–232.

Cornell, A., & Krawinkler, H. (2000). Progress and challenges in seismic performance and assessment. *PEER Center News, 3*(2), 1–3.

Emori, K. and Schnobrich, W.C. (1978) Analysis of Reinforced Concrete Frame Wall Structures for Strong Motion Earthquakes, Civil Engineering Studies, Structural Research Series No. 457, University of Illinois, Urbana.

FEMA 356. (2000). *Pre standard and commentary for the seismic rehabilitation of buildings*. Washington, DC: Federal Emergency Management Agency.

FEMA 440. (2005). *Improvement of non-linear static procedures*. Washington, DC: Applied Technology Council (ATC 55).

FEMA P695 (2009). Quantification of building seismic performance factors, Applied Technology Council (ATC-63), Washington D.C.

Freeman, S.A., (1978). Prediction of response of concrete buildings to severe earthquake motion, ACI Special Publication SP-55, 589–605.
Goel, S. C., Liao, W. C., Mohammed, R. B., & Chao, S. H. (2010). Performance-based plastic design method for earthquake-resistant structures: An overview. *The Structural Design of Tall and Special Buildings, 19*(1–2), 115–137.
Kappos, A. J., & Stefanidou, S. (2010). A deformation-based seismic design method for 3D R/C irregular buildings using inelastic dynamic analysis. *Bulletin of Earthquake Engineering, 8*, 875–895.
Krawinkler, H., & Miranda, E. (2004). Performance-based earthquake engineering. In *Earthquake engineering from engineering seismology to performance-based engineering*. USA: CRC Press.
Litton, R.W. (1975). A contribution to the analysis of concrete structures under a cyclic loading. Ph. D Thesis, Department of Civil Engineering, University of California, Berkeley.
Panagiotakos, T. B., & Fardis, M. N. (1999). Deformation-controlled earthquake resistant design of R.C. buildings. *Journal of Earthquake Engineering, 3*(4), 498–518.
Papanikolaou, V. K., Elnashai, A. S., & Pareja, J. F. (2005). *Limits of applicability of conventional and adaptive pushover analysis for seismic response assessment report.* Urbana-Champaign: Mid-America Earthquake Centre, University of Illinois.
Priestley, M. J. N., Calvi, G. M., & Kowalsky, M. J. (2007). *Displacement-based seismic design of structures*. Pavia: IUSS press.
Priestley, M. J. N., & Kowalsky, M. J. (2000). Direct displacement-based design of concrete buildings. *Bulletin of the New Zealand National Society for Earthquake Engineering, 33*(4), 421–444.
SEAOC Blue Book. (1959). *Recommended lateral force requirements*. Seismology Committee, Structural Engineers Association of California.
Takeda, T., M. A. Sozen and N. N. Nielsen (1970). Reinfoced concrete response to simulated earthquakes. ASCE Journal, 96 (ST12), 2557–2573.
Takayanagi, T., Derecho, A. T. and Corley W. G. (1979). Analysis of Inelastic Shear Deformation Effects in Reinforced Concrete Structural Wall Systems, Nonlinear Design of Concrete Structures, Proceedings of the CSCE-ASCE-ACI-CEB International Symposium, University of Waterloo, Ontario, 545–579.
Vamvatsikos, D., & Cornell, C. A. (2004). Applied incremental dynamic analysis. *Earthquake Spectra, 20*(2), 523–553.

Chapter 14
Evolution of Earthquake Resistant Design Code - A Template for Future

Ashok K. Jain

1 Introduction

Man has been constructing earthquake-resistant buildings from time immemorial. He developed his own set of guidelines based on his experience and judgment from the past earthquakes. This premise is continuing till date and will continue for a couple of more centuries till the seismology of the given place of interest and behavior of materials and structures are fully understood. In the hill states of North India and northeastern states of India, the construction of light weight wooden houses was very much prevalent. In the Northern hills, they used a combination of wooden planks with diagonal braces in the outer walls, whereas, in the northeastern states, they used bamboo construction. In the old Kathmandu Palace, they used wide top and bottom wooden strips over door and window openings that acted as wooden bands at the sill and lintel levels as shown in Fig. 14.1.

Similarly, the concept of using timber bands was prevalent in Kashmir Valley for over 100 years as shown in Fig. 14.2. The lower storey is in stone masonry, whereas the upper storey is in brick masonry. A typical bamboo hut in the Northeast India and a timber construction in North India are shown in Figs. 14.3 and 14.4.

These materials and construction techniques resulted in a very low rate of casualty in the event of a strong earthquake. Naturally, these techniques were based on experience of our ancestors developed over a long period of time. Later, as the men became more developed and gained knowledge about his surroundings, commerce and trade, science and technology, etc., he started exploring stronger building materials, such as, stone masonry and brick masonry. For the past several

A. K. Jain (✉)
Department of Civil Engineering, Indian Institute of Technology Roorkee, Roorkee, Uttarakhand, India

M. L. Sharma et al. (eds.), *Advances in Indian Earthquake Engineering and Seismology*, https://doi.org/10.1007/978-3-319-76855-7_14

Fig. 14.1 Timber bands in Kathmandu Palace

Fig. 14.2 Timber bands in a masonry building, Baramulla (Courtesy Mrs. Amita Sinvhal, Dept. of Earthquake Eng., IIT Roorkee)

decades, he started preferring concrete and steel buildings. These stronger building materials created several problems, owing to lack of understanding about their behaviour, leading to severe loss of life and property during earthquakes. A need was felt to regulate the building construction industry and a set of specifications referred to as the *Code of Practice* were developed and issued by the respective authority in each country for use of all the stakeholders. As the men became more imaginative, so were his desires and requirements. The oldest known code that includes some reference to buildings also is by *Hammurabi of Babylon* origin dating back to 1770 BC.

Fig. 14.3 A typical timber hut with walls made of bamboo mats in North East India

Fig. 14.4 A typical timber construction in the hills of North India- Ground storey for cattle and upper storey for self

Presently, there is a big gap between the research and practice in many countries. In India, this gap seems to be even wider because of a very lax accountability and judicial system. The limit state design code for Reinforced concrete structures was introduced in 1978. The corresponding code for steel structures was introduced only in 2007 after a gap of nearly 30 years. Logically, the Steel Code should have been introduced much earlier than the Reinforced Concrete Code. The Indian Roads Congress has introduced limit state concept in bridge design through IRC 6-2014 and IRC 112-2011. IS 1893 was first published in 1962, and revised in 1966, 1970, 1975 and 1984 (Jain 1980). The IS:1893-1984 code was revised in 2002 after a gap of nearly 18 years. It has now been revised in 2016, after a gap of 14 years. IS:13920 code was published in 1993 and it has been revised in 2016, after a gap of 23 years (Jain, 2017). Still, microzonation does not appear to be even on the radar of codal committee. New Zealand introduced the concept of design for different levels of ductility back in 1991. This is the only code known to have kept pace with the research. In such a state of affairs, the latest innovative solutions cannot be expected to be part of the code. It is therefore but natural that the society is deprived of the fruits of technological developments and innovations. It is expected that during the next few decades, this gap will reduce and there will be more scope to incorporate innovative solutions at a much faster pace. It is expected that the current *prescriptive criteria* in the codes will be replaced by *performance-based criteria* while putting more onus on the engineers.

1.1 Prescriptive Design Versus Performance Design

In the prescriptive design, the code tells the designer what to do at each stage. The final product is expected to respond as the code desired. The designer is not bothered about the final performance of its product. It is presumed to behave as the codes have envisaged without knowing what exactly the codes have desired or intended. whereas in the performance-based design, the code simply tells what performance is expected out of the given component of the structure as well as the structural system. A performance-based seismic code can be expected to be a probabilistic code. The designer has to choose an appropriate option on how to design it so as to achieve the specified performance level and probability. Thus, the designer has to have a very deep understanding of the behavior of various structural materials, components, and structural systems under different loading conditions and their implications and consequences.

Theodoropoulos (FEMA, 454) has provided an excellent summary of the difficulty in introducing a performance-based code in practice at present: *Enforcing a code that specifies multiple levels of design and performance would require a more complex procedure than currently used. Its success will depend upon increased public understanding of the behavior of structures in earthquakes, the limitations of current codes, and the rationale behind a performance-based approach. A performance-based seismic design code format could provide a unified basis for comparison of design alternatives that give decision-makers a consistent means of quantifying risk. That basis will enable design professionals to respond to the needs of all decision-makers and stakeholders concerned with seismic design, including owners, lenders, insurers, tenants, and communities at large.*

This chapter highlights the development of understanding and knowledge during the past 100 years and shape of a futuristic code that would give more freedom and flexibility to the owners, designers, and builders and improve the performance of a structure under a severe seismic loading.

2 Development of Understanding and Knowledge During the Past 100 Years

The initial work on the development of earthquake-resistant design specifications was carried out in America and Japan during the past 100 years and can be summarized as follows (Elsesser, FEMA 454, 2006, Chap. 7; McClure, 1968):

2.1 Seismic Codal Specifications

- Initial seismic designs for buildings were based on wind loads, using static force concepts. This approach started in the late 1800s and lasted to the mid-1900s.
- After the 1906 San Francisco earthquake, concepts of building dynamic response gained interest, and in the early 1930s, initial studies of structural dynamics with analysis and models were initiated at Stanford University. This approach ultimately led to a design approach that acknowledged the importance of dynamic characteristics (period of vibration, mode shape) rather than static design concepts.
- Several engineers from San Francisco went to Japan after the 1923 Japan earthquake, where three major buildings, statically designed for lateral forces of 10% of gravity, showed marked resistant behavior. The Board of Fire Underwriters of the Pacific, State of California, when their earthquake department was established in 1926, was greatly influenced by the observations in Japan 1923 and Santa Barbara 1925 earthquakes and advocated a static design, using a 10% lateral force factor.
- The "Palo Alto" Code and "Uniform Building Code" 1927 were two of the first codes in the United States to incorporate lateral force requirements based on seismic force as a product of the mass times the acceleration. The coefficient of acceleration varies depending on the bearing capacity of the soil (that is, type of soil) but was approximately 10% of the dead load.
- Dynamic design concepts were enhanced by the acceleration spectra method used for design as developed by Professor Housner at the California Institute of Technology in the 1950s. This concept realized that dynamic characteristics of a building play a very significant role in resisting seismic forces. The terms *period of vibration* that depends on mass and stiffness of the structure and *damping* became the most critical parameters.
- While analytical methods were being developed, engineers needed additional knowledge about nonlinear behavior of structural components. Substantial testing of materials and connection assemblies to justify actual behavior was undertaken from 1950 to 1990 at numerous universities in the United States, Japan, New Zealand, and other countries (notably Lehigh University, Bethlehem, on steel structures; University of California, Berkeley, on steel and RC members; University of Illinois, Urbana Champaign, on RC members; University of

Michigan, Ann Arbor, on steel and RC members; University of Texas on RC members; etc.).

- In 1948, a Joint Committee on Lateral Forces of the San Francisco Section of ASCE and the Structural Engineers Association of Northern California (SEAONC) was formed. This committee, after several years of study, issued a report wherein it was recommended that lateral force coefficients be used which were related to the estimated or calculated fundamental period of vibration of the structure. This report, published in 1951, received worldwide acclaim and is the basis for many earthquake codes including the first Earthquake-Resistant Code of India in 1962.
- In 1960, Section (j) was introduced in the SEAOC recommendations which required that "buildings more than 13 stories or one hundred and sixty feet (160′) in height shall have a complete moment resisting space frame capable of resisting not less than 25 percent of the required seismic load for the structure as a whole. The frame shall be made of ductile material or a ductile combination of materials. The necessary ductility shall be considered to be provided by a steel frame with moment resistant connections or by other systems proven by tests and studies to provide equivalent energy absorption." Thus the concepts of "second line of defense" and "ductility" were introduced.
- A commentary was prepared to complement the recommendations in the SEAOC code and explain its broad considerations with reasoning and background information to the designers. It was decided that it would be unreasonable and economically unjustified to attempt to impose a design which would *survive even the strongest earthquake*, without any damage. One of the important statements in the 1960 Commentary is "The Recommended Lateral Force Requirements are intended to provide this protection in the event of an earthquake of intensity or severity of the strongest of those which California has recorded."

  This level of protection was defined as the primary function of a building code and to provide "*minimum standards*" to assure public safety. Requirements contained in such codes are intended to safeguard against *major structural failures* and to provide protection against loss of life and personal injury. The code does not assure protection against nonstructural damage, such as cracked plaster, broken glass and light fixtures, cracked filler walls, overturned equipment, etc.
- This led to the now famous three-paragraph statement that is contained in nearly each and every earthquake-resistant code in the world:

  Structures designed in conformance with the provisions and principles set forth therein should be able to:

  - Resist minor earthquake *without any structural damage but with some nonstructural damage*
  - Resist moderate earthquakes *with very minor structural damage (very fine cracks in beams and columns that may not be visible with naked eyes), but with extensive nonstructural damage*

- Resist major earthquakes *without collapse of the structural system but with extensive structural as well as nonstructural damage*

- The 5th revised edition of IS:1893-2002 code states that "It is not intended in this standard to lay down regulation so that no structure shall suffer any damage during earthquake of all magnitudes. It has been endeavored to ensure that, as far as possible, structures are able to respond, without structural damage to shocks of moderate intensities and without total collapse to shocks of heavy intensities."
- Further, on various seismic zones in the IS:1893-2002 code, it states that "The object of this map is to classify the area of the country into a number of zones in which one may reasonably expect earthquake shaking of more or less same maximum intensity in future. The Intensity as per Comprehensive Intensity Scale (MSK64) broadly associated with the various zones is VI (or less), VII, VIII and IX (and above) for Zones II, III, IV and V respectively. The maximum seismic ground acceleration in each zone cannot be presently predicted with accuracy either on a deterministic or on a probabilistic basis. The basic zone factors included herein are reasonable estimates of effective peak ground accelerations for the design of various structures covered in this standard."

  Prakash et. al. (2006a, 2006b) compared the base shear given by different revisions of IS:1893, and the physical significance of response reduction factor R.

2.2 Analytical Tools

- Since 1965 to the present, sophisticated computer analysis programs have been and continue to be developed to facilitate design of complex structural systems and the study of nonlinear behavior. Solid SAP and later SAP4/SAP90 series of free software developed at the University of California, Berkeley, were the watershed in creating awareness among the structural engineers throughout the world about how to carry out static and dynamic analysis of building frames. DRAIN-2D developed by Kanaan and Powel in 1973 was used extensively to understand the nonlinear response of RC and steel-framed structures. Its simplicity and flexibility to add more modules were simply remarkable. SAP2000, STAAD Pro, ABACUS, ANSYS, OPENSEES plus many others are the current avatars. Pal et.al. (1987) proposed *inelastic response spectra* based on extensive nonlinear dynamic analysis of SDOF systems.

2.3 Full-Scale Testing

- In the late 1970s, the US–Japan collaboration on large-scale/full-scale testing of six-storey RC and steel buildings gave a serious impetus to the development of a wider and deeper practical knowledge base for the development of aseismic codal practices. The concept of "active and passive energy dissipation devices," "post-yield behavior," and "base isolation" gained wide acceptance because of very comprehensive experimental programs (*including very sophisticated instrumentation*).

- The breath taking full-scale testing facilities developed at the Tokyo Institute of Technology (Tokyo Tech) are a landmark facility in the whole world. It helps develop a clear understanding about the behavior of various structural elements and structural systems, detailing of reinforcement, and causes of failure in buildings and bridges during earthquakes.

2.4 Lessons from Failure of Structures During Earthquakes

- The above technological breakthrough was further supported by a detailed survey of wide spread damage to buildings and life line facilities all over the world by various teams of experts drawn from different countries who provided microscopic pictures of the structural damage along with their "own interpretations." Naturally, all these reports were made available to each and anyone interested. This helped in dissemination of the knowledge, understanding, and refinement of the design and construction practices specific to the respective countries. These concepts and structural systems so developed can now be effectively utilized that previously could only be imagined.

3 Acceptable Risk in an Earthquake (FEMA 749, 2010)

The elements contributing to risk (chance of loss) are shown in Fig. 14.5. In simple terms they are understood as follows:

1. *Hazards*—physical effects generated in the naturally occurring event
2. *Location* of the hazards relative to the community at risk
3. *Exposure*—the value and importance of the various types of structures and lifeline systems in the community serving the populace

Fig. 14.5 Elements of risk

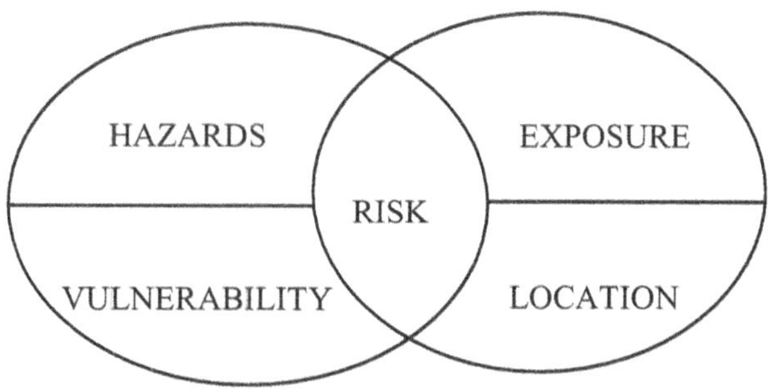

4. *Vulnerability* of the exposed structures and systems to the hazards expected to affect them during their useful life

What is an acceptable risk? It is a million dollar question. Defining acceptable risk is not easy because the risk that is acceptable to one person may be unacceptable to the other. Often a person's perception of an acceptable level of risk depends on whether or not the person believes he or she will be personally affected and how much loss the person is willing to bear, as well as how much the person is willing to spend to avoid the risk. The *NEHRP Recommended Seismic Provisions* has adopted the following target risks as the minimum acceptable for buildings and structures in the United States:

- A small chance (10%) that any structure will experience partial or total collapse as a result of the most intense earthquake ground motion considered by the building codes. These very rare and intense earthquake effects are called risk-targeted *maximum considered earthquake* (MCE) ground motions, and the probability of their occurrence varies across the nation. This collapse-prevention goal is intended as the primary means of ensuring life safety in that most casualties in past earthquakes occurred as a result of structural collapse.
- Limit the chance of collapse (to perhaps 6%) as a result of MCE ground shaking for structures intended primarily for public assembly in a single room or area (e.g., theaters or convention centers), for structures with a very large number of occupants (e.g., high-rise office buildings and sports arenas), and for structures housing a moderately large number of people with limited mobility (e.g., prisons) or who society generally regards as particularly vulnerable and important to protect (e.g., school children).
- For structures that contain a large quantity of toxic materials that could pose a substantial risk to the public (e.g., some chemical plants), provide a small probability that structural damage will result in release of those materials.
- Limit the chance of total or partial collapse as a result of MCE ground motions (3%) for structures deemed essential to emergency response following a natural disaster (e.g., police and fire stations and hospitals), and further limit the chance that earthquake shaking will cause damage to these structures or to their architectural, mechanical, electrical, and plumbing systems sufficient to prevent their post-earthquake use.
- For ordinary structures, the *NEHRP Recommended Seismic Provisions* seeks to provide a probability of 1% or less in 50 years that a structure will experience earthquake-induced collapse.
- For all structures, minimize the risk, in the event of likely earthquake debris generated by damage to cladding and ceilings or mechanical or electrical systems will fall on building occupants or pedestrians.
- To the extent practicable, avoid economic losses associated with damage to structural and nonstructural systems as a result of relatively frequent moderate earthquake events.

4 Structural Non-linear Models (Deierlein et al., NEHRP Brief 4, 2010)

Inelastic structural component models can be differentiated in the manner plasticity is distributed through the member cross sections and along its length. A set of five idealized model types for simulating the inelastic response of beam-columns are shown in Fig. 14.6. Several types of structural members (e.g., beams, columns, braces, and some flexural walls) can be modeled using the concepts illustrated in this figure. Each of these beam-column models may represent very different hysteresis behavior depending upon their stiffness, strength, and detailing of reinforcement (*in RC members*) or member sections and connection details (*in steel members*). Typical hysteresis models are shown in Fig. 14.7.

In modeling the hysteretic properties of actual elements for analysis, the initial stiffness, strength, and post-yield force-displacement response of cross sections should be determined based on principles of mechanics and/or experimental data, considering influences of cyclic loading and interaction of axial, shear, and flexural effects. Both unconfined stress-strain and confined stress-strain curves of concrete, and stress-strain curves of steels of different grades are required to generate moment-curvature relations and plastic hinge properties of various elements (Jain 2016). ASCE 41 provides guidelines for estimation of stiffness, strength, and deformation limits in steel, reinforced concrete, masonry and wood members, base isolators, and energy dissipation components of moment frames, braced frames, shear walls, diaphragms, infills, and foundations.

An idealized force versus deformation relationship for specifying the force and deformation parameters of nonlinear component models is shown in Fig. 14.8 (ASCE 41). These points include effective yield (point B), peak strength (point C), residual strength (point D), and ultimate deformation (point E). Since in most cases the descending slope is more gradual than implied in Fig. 14.8, it may be more

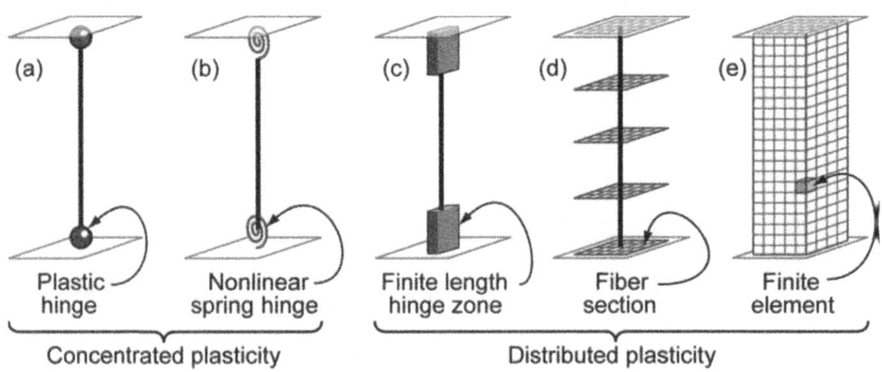

Fig. 14.6 Typical idealization for distribution of plasticity in beam-columns

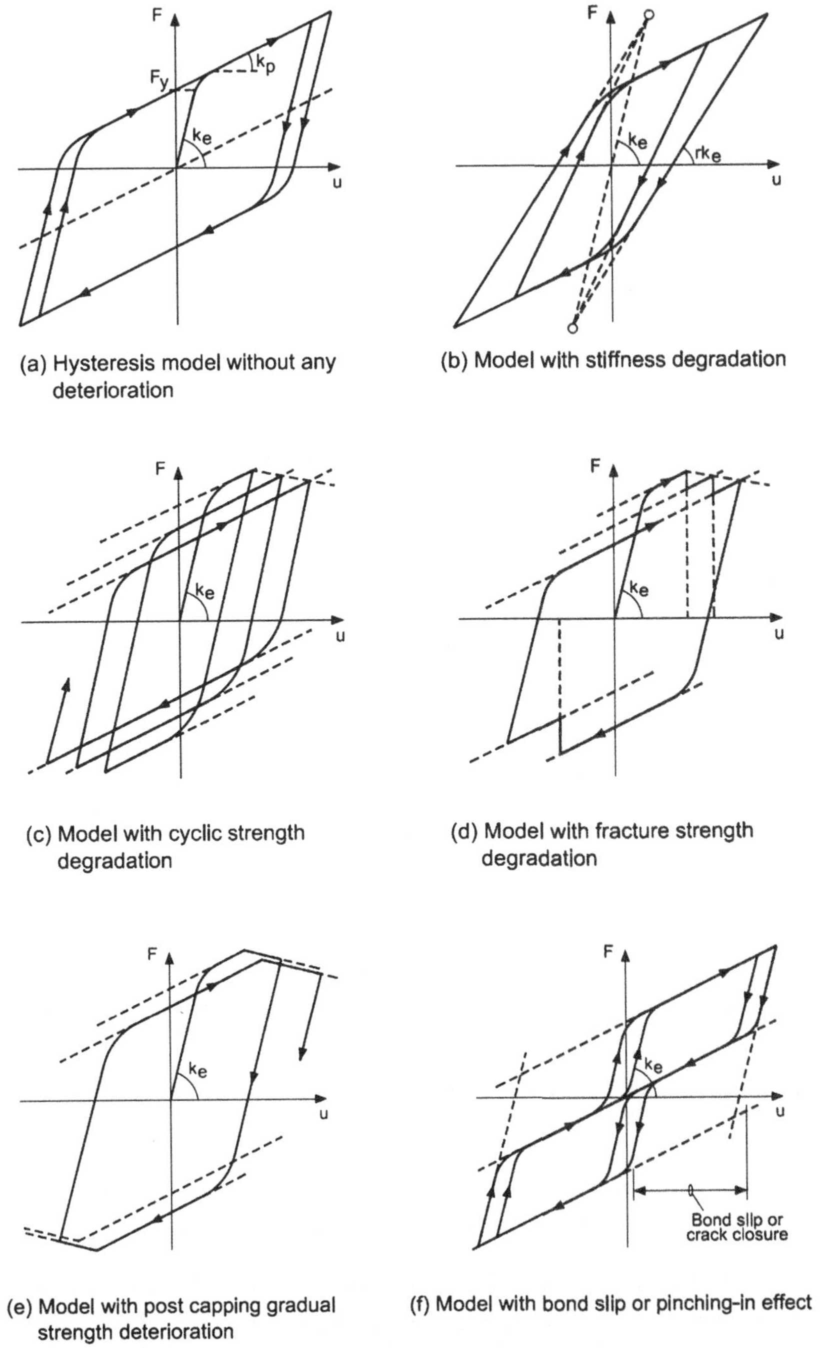

Fig. 14.7 Typical hysteresis models (Deierlein et al., NEHRP Brief 4, 2010)

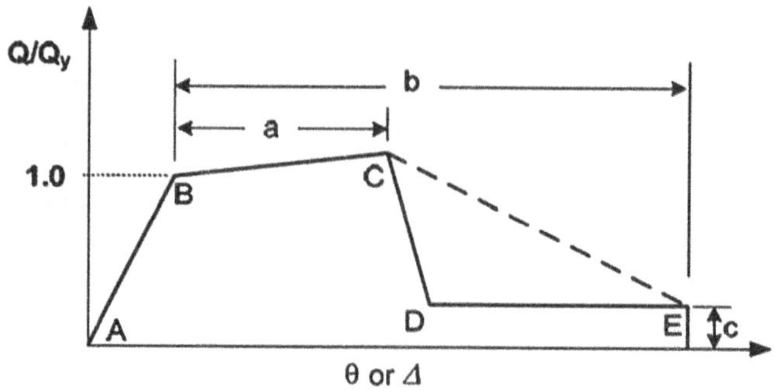

Fig. 14.8 Generalized force-deformation curve (ASCE 41)

reasonable to define the descending branch between point C and E (or to a point between D and E), as illustrated by the dashed line in the figure.

Thus it is now possible to start thinking how to incorporate nonlinear models for different types of structures in the code.

5 Energy-Dissipating Devices

Use of active and passive energy dissipation is an emerging technology that enhances building performance by reducing the demands through addition of damping devices and stiffening elements. There are quite a large number of buildings that have been built in Japan and California using this concept. However, most codes are silent about their design perhaps because it involves use of nonlinear analysis which is fairly complicated and is difficult to codify at present. Therefore, the design of such buildings is left to the experts who can demonstrate the safety of such devices either experimentally or theoretically or both.

Between the two techniques, the passive energy dissipation technique is preferred since it does not require any external source of energy. Moreover, it is too much to rely on computers and power supply in the event of a serious seismic event. In determining the seismic demand parameters, a nonlinear analysis model is required that can capture the forces, deformations, or other demand parameters for the devices. The analysis model of the device should account for its dependence on loading rates, temperature, sustained (gravity and other) loads, and other interactions. Energy dissipation devices are usually modeled as combinations of dashpots and equivalent elastic or hysteretic springs. The resulting models incorporate one or more of the four main categories shown in Fig. 14.9.

In conventional construction, energy dissipation is provided by hysteresis energy dissipated in the members and connections. Hysteretic energy is developed by inelastic action and causes damage. After the event, all such members and connections have to be replaced which is not an easy task. This can be avoided by providing non-destructive means of energy dissipation such as frictional energy dissipaters.

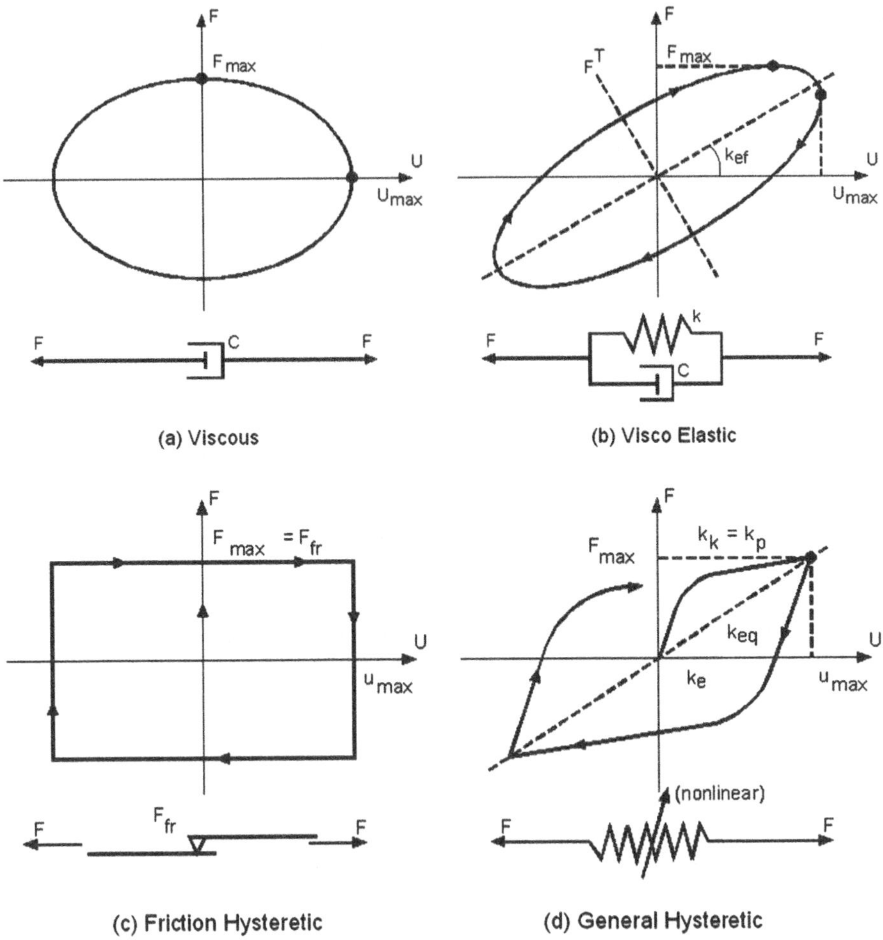

Fig. 14.9 Hysteretic models for Energy-Dissipating Devices (Deierlein et al., NEHRP Brief 4, 2010)

Damage to a structure can be greatly reduced by using such dissipaters. In case of any damage to such devices, they are easy to replace. The trend to using such devices is increasing.

There are two types of passive energy dissipation methods for building systems: viscoelastic and mechanical/hysteretic systems. Fluid dampers are a viscoelastic energy dissipation method, while friction dampers such as slotted bolted connections are mechanical/hysteretic energy dissipation systems. Friction dampers work by dissipating energy by friction between two known surfaces. The resistance to motion is provided by simple friction or Coulomb friction. A slotted bolted connection (SBC) is now extensively used in steel or RC buildings. A typical SBC is shown in Fig. 14.10. In the latest design, the friction occurs between steel and a brass slider plate (Popov et al. 1995).

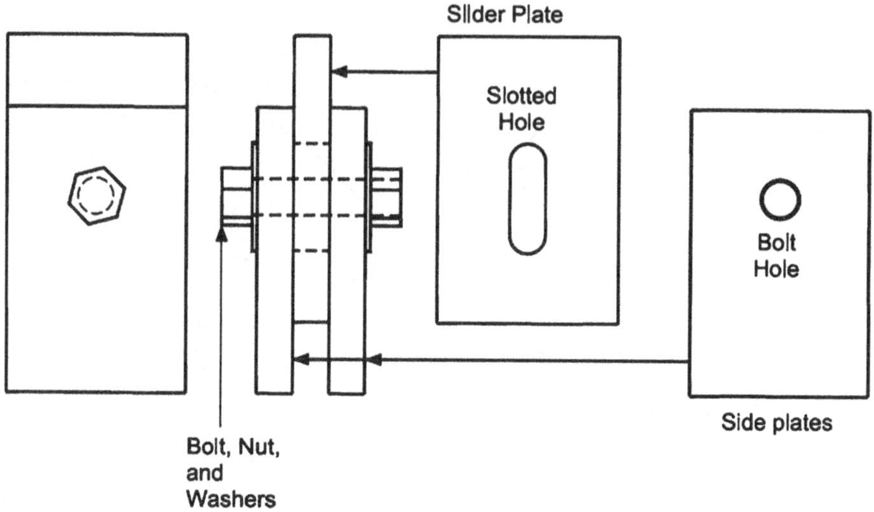

Fig. 14.10 Generic friction damper

6 Seismic Isolators or Base Isolation Devices

Seismic isolators can be generally classified as either elastomeric or sliding, and isolation systems may be composed of one or multiple types of isolators. Elastomeric isolators may be either bearings made of high-damping rubber or bearings with low-damping rubber that have a lead core. Sliding isolators may be flat assemblies or have curved surfaces, such as the friction pendulum system. Nonlinear analysis procedures should explicitly model the force-deflection properties of isolators. The elastomeric bearings may be modeled similarly to the hysteretic model shown in Fig. 14.9d, with the initial stiffness determined by the shear stiffness of the rubber layers and the post-yielding stiffness defined by the characteristics of the rubber and/or lead plug. The sliding isolators are modeled using the friction model shown in Fig. 14.9c, with the friction level defined by the friction coefficient of the sliding surfaces and normal forces due to gravity and other loads acting on the isolator. The variability of model parameters, related to such factors as the rate of loading, temperature dependence, and longevity of device life cycle, is very important and should be accounted in the design (Naeim and Kelly, 1999; ASCE 7, 2016; ASCE 41, 2007).

IS:1893-2002 Code permits the use of base isolation and energy-absorbing devices for earthquake-resistant design. It says:

> Only standard devices having detailed experimental data on the performance should be used. The designer must demonstrate by detailed analyses that these devices provide sufficient protection to the buildings and equipment as envisaged in this standard. Performance of locally assembled isolation and energy absorbing devices should be evaluated experimentally before they are used in practice. Design of buildings and equipment using such device should be reviewed by the competent authority. Base isolation systems are found useful for short period structures, say less than 0.7s including soil-structure interaction.

In many recent buildings, base isolation technique has been successfully applied in buildings having much greater period of vibration.

7 Seismic Design Category (SDC) Concept

The Indian code IS:1893 treats all structures under a single category. It means within the same seismic zone or among the various seismic zones, each structure has to be detailed for the same ductility level irrespective of its risk level. Apparently, this is not rational. It appears to be one of the major reasons for noncompliance of earthquake provisions by the private owners and consultants throughout the country.

The NEHRP *Provisions has introduced* the concept of Seismic Design Category (SDC) to categorize structures according to the seismic risk they could pose. There are six SDCs ranging from A to F with structures posing minimal seismic risk assigned to SDC A and structures posing the highest seismic risk assigned to SDC F. Thus, as the SDC for a structure increases, so do the strength and detailing requirements and the cost of providing seismic resistance. A similar approach is used in Eurocode 1998. Table 14.1 summarizes the potential seismic risk associated with buildings in the various Seismic Design Categories and the primary protective measures required for structures in each of the categories. There is a need to adopt a similar criterion in the Indian code so that analysis and detailing requirements can be made more rational and acceptable to the designers.

8 Non-linear Static (Pushover) Analysis

Pushover analysis is a nonlinear static analysis carried out under conditions of constant gravity loads and monotonically increasing horizontal loads to represent an earthquake loading (Fig. 14.11). It may be applied to verify the structural performance of newly designed and of existing buildings for the following purposes:

(a) To verify or revise the overstrength ratio values.
(b) To estimate the expected plastic mechanisms and the distribution of damage.
(c) To assess the structural performance of existing or retrofitted buildings for the purposes of ECurocode 1998-3, FEMA 440, or ASCE 41.
(d) As an alternative to the design based on Eurocode 1998-linear elastic analysis which uses the behavior factor q. In that case, the target displacement should be used as the basis of the design.

A detailed procedure for carrying out static nonlinear analysis or pushover analysis is available in Applied Technology Council document 40 (ATC 40, 1996), FEMA 440 and Eurocode 1998.

Table 14.1 Seismic design categories, risk, and seismic design criteria (NEHRP)

SDC	Building type and expected MMI	Seismic criteria
A	Buildings located in regions having a very small probability of experiencing damaging earthquake effects	No specific seismic design requirements, but structures are required to have complete lateral force-resisting systems and to meet basic structural integrity criteria
B	Structures of ordinary occupancy that could experience moderate (MMI VI) intensity shaking	Structures must be designed to resist seismic forces
C	Structures of ordinary occupancy that could experience strong (MMI VII) and important structures that could experience moderate (MMI VI) shaking	Structures must be designed to resist seismic forces Critical nonstructural components must be provided with seismic restraint
D	Structures of ordinary occupancy that could experience very strong shaking (MMI VIII) and important structures that could experience MMI VII shaking	Structures must be designed to resist seismic forces Only structural systems capable of providing good performance are permitted Special construction quality assurance measures are required. Nonstructural systems required for *life safety* protection must be demonstrated to be capable of post earthquake functionality.
E	Structures of ordinary occupancy located within a few kilometers of major active faults capable of producing MMI IX or more intense shaking	Structures must be designed to resist seismic forces Only structural systems that are capable of providing superior performance are permitted Many types of irregularities are prohibited. Nonstructural systems required for life safety protection must be demonstrated to be capable of post earthquake functionality. Special construction quality assurance measures are required.
F	Critically important structures located within a few kilometers of major active faults capable of producing MMI IX or more intense shaking	Structures must be designed to resist seismic forces Only structural systems capable of providing superior performance are permitted.Many types of irregularities are prohibited. Nonstructural systems required for *facility function* must be demonstrated to be capable of post earthquake functionality. Special construction quality assurance measures are required.

8.1 ADRS Spectra

Application of the capacity spectrum technique requires that both the demand response spectra and structural capacity (or pushover) curves should be plotted in the spectral acceleration versus spectral displacement domain. Spectra plotted in this format are known as *Acceleration–Displacement Response Spectra* (ADRS) after

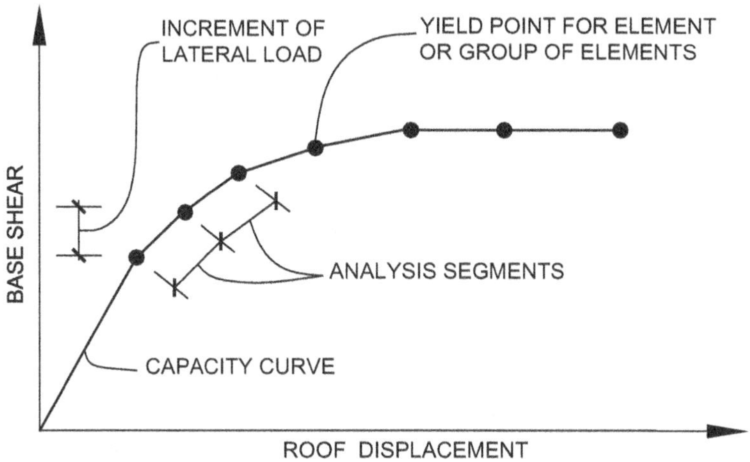

Fig. 14.11 Base Versus Roof Displacement Curve

Mahaney et al. (1993). Every point on a response spectrum curve has associated with it a unique spectral acceleration S_a, spectral velocity S_v, spectral displacement S_d, and period, T. To convert a spectrum from the standard S_a versus T format found in the building code to ADRS format, it is necessary to determine the value of S_{di} for each point S_{ai}, T_i on the curve. This can be done using Eq. 14.1 (Jain, 2012 2016):

$$S_d = \frac{1}{4\pi^2} S_a T^2 \tag{14.1}$$

Standard demand response spectra contain a range of constant spectral acceleration and a second range of constant spectral velocity. Relations between spectral acceleration and displacement at period T are well known. In order to develop the capacity spectrum from the capacity (or pushover) curve, it is necessary to do a point-by-point conversion to first mode spectral coordinates. Any point V_i, Δ_i on the capacity curve is converted to the corresponding point S_{ai}, S_{di} on the capacity spectrum using the equations

$$\begin{aligned} V_b &= \alpha_h W \\ &= (\alpha_i S_{ai}) W \end{aligned} \tag{14.2}$$

$$\text{or} \quad S_{ai} = \frac{V_b/W}{\alpha_i} \tag{14.3}$$

$$\Delta_{roof} = C_{roof}^{(r)} \phi_{roof,1}^{(r)} S_{di}$$

$$\text{or}\quad S_{\rm di} = \frac{\Delta_{\rm roof}}{C^{(r)}_{\rm roof}\phi^{(r)}_{\rm roof.1}} \tag{14.4}$$

$$T = 2\pi\sqrt{\frac{S_{\rm d}}{S_{\rm a}}} \tag{14.5}$$

The response spectra in traditional and ADRS format are given in Fig. 14.12. In the ADRS format, lines radiating from the origin have constant periods.

The capacity spectrum conversion analogy is shown in Fig. 14.13. The capacity spectrum is superimposed over response spectra in traditional and ADRS format in Figs. 14.13 and 14.14. The equal displacement approximation estimates that the inelastic spectral displacement is the same which would occur if the structure remained fully elastic. In the ADRS, the point of intersection of demand spectrum and capacity spectrum is located.

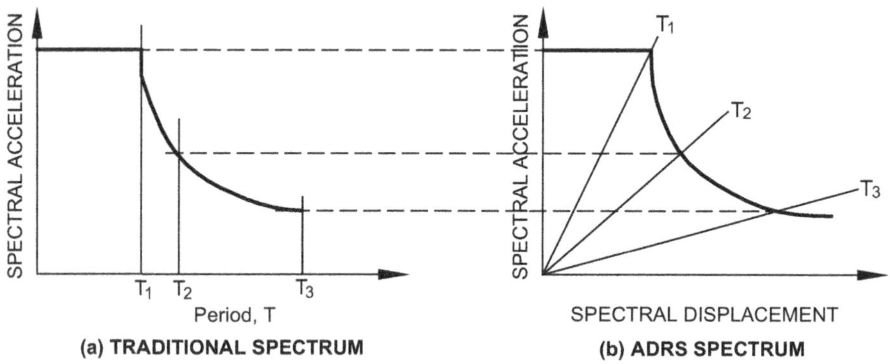

Fig. 14.12 Traditional spectrum versus ADRS Spectrum

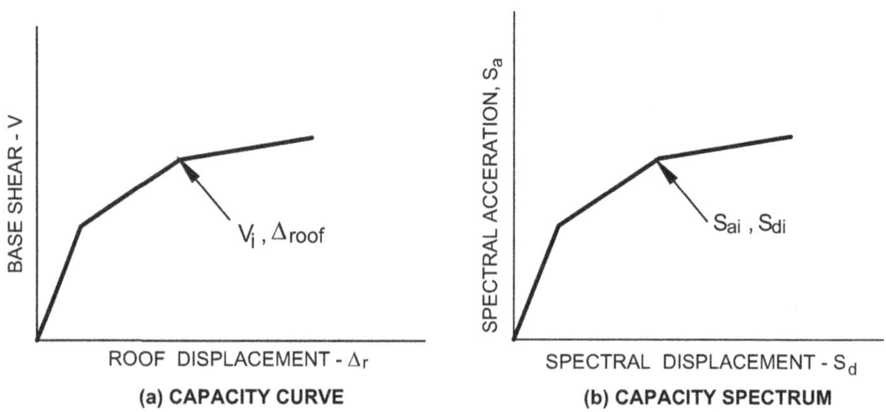

Fig. 14.13 Capacity Curve Versus Capacity Spectrum

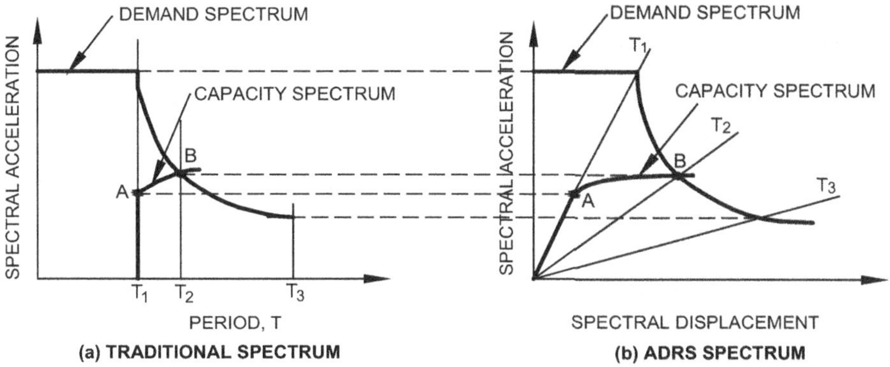

Fig. 14.14 Demand Versus Capacity Spectrum

9 Displacement-Based Design

First let us look at the force-based design and its shortcomings. For force-based design, the steps are essentially as follows (Priestley et al., 2007):

1. Estimate member sizes and stiffness and the structure mass.
2. Calculate the vibration periods.
3. Using an acceleration response spectrum and the vibration periods, calculate spectral accelerations and hence lateral loads.
4. For combined gravity and lateral loads, calculate strength demands on the structural members.
5. Design the members to satisfy these demands.
6. If the member sizes or stiffness are significantly different from the estimates, iterate from step (1).

In a steel frame, member stiffness depends on member size, so if the sizes chosen in step (5) are different from those estimated in step (1), iteration is needed. Hence, this is not strictly a "direct" design. For a reinforced concrete structure, the member stiffness is usually based on the cracked concrete section, and in step (5) only the reinforcement may be designed, with no changes in the estimated member sizes. In this case iteration is not needed, and this is a "direct" design.

9.1 Problems with Force-Based Design

(a) Interdependency of strength and stiffness
(b) Period calculation
(c) Ductility concept and reduction factors

(d) Ductility of structural system
(e) Relationship between strength and ductility demand
(f) Relationship between elastic and inelastic displacement demand
(g) Structures with dual structural systems and the resulting interaction

Priestley et al. (2007) have shown that there are serious problems with each of the above factors that are widely employed in force-based design; hence the force-based design is not very reliable. That is why they strongly recommend the use of the displacement-based design (DBD). This approach is very different from conventional force-based design, and the main steps are as follows (Priestley et al., 2007):

1. Using capacity design principles, choose the components in the structure that are allowed to yield (essentially, choose a plastic mechanism for the structure, e.g., a strong column-weak beam mechanism for a frame).
2. Choose the relative strengths of the yielding components (e.g., the relative strengths of the beams over the frame height).
3. Choose a design displacement (lateral drift) for the structure.
4. Estimate the member sizes and calculate the structure mass.
5. Estimate the yield displacement.
6. Given the design displacement and the yield displacement, get the displacement ductility.
7. Given the displacement ductility, and knowing something about the nonlinear behavior of the structure, estimate the effective damping ratio for the structure (as the amount of yield increases, the amount of dissipated energy increases and hence also the effective damping ratio).
8. Using a displacement response spectrum, the damping ratio, and the design displacement, get an effective period of vibration (note that this is the reverse of the way that an acceleration spectrum is used in force-based design).
9. Given this period and the structure mass, calculate the required effective stiffness for the structure.
10. Multiply the effective stiffness by the design displacement to get the base shear.
11. Given the base shear, use the plastic mechanism, the relative strengths of the yielding components, and equilibrium to calculate the actual strengths of the yielding components.
12. Using capacity design principles, calculate the strengths of the non-yielding components. It may be necessary to iterate from step (4), or to change the choices in steps (1) through (3), but usually the design will be close to a final design. Step (5), which requires estimating the yield displacement, is a key step.

An important assumption in the DBD is that member stiffness is proportional to member strength. This is essentially true for steel beams and columns and also true for reinforced concrete frames and walls. It follows that the yield displacement can be estimated quite accurately knowing the yield strain for steel and the member dimensions. As a consequence, a displacement-based design can be obtained with little or no iteration, and DBD works for both steel and concrete structures.

The DBD process is very different from force-based design, in concept and in detail. The end result of the design process can be a final design. For a large structure, however, the designer will often use this as an initial design to be refined using nonlinear dynamic analysis. An initial design can, of course, be obtained by other methods, including force-based design (as is often done in current practice). The advantage of using DBD is that the initial design is likely to be very close to a final design, and any refinements are likely to be minor. A force-based initial design may be very different from a final design and is likely to require much more effort to refine. The displacement based design process is in its infancy. It needs further clarity and simplification for it to become acceptable to designers.

10 Performance-Based Seismic Design

The objective of the performance-based seismic design (Eurocode, 1998, FEMA 445, Jain 2012, 2016) is to verify the seismic performance of the structure and to evaluate its seismic performance level for the various design limit states. For each limit state, the level of earthquake considered in the design may be determined depending upon the purpose and importance of the structure.

10.1 Performance Index

The performance index should be selected as appropriate means to quantitatively express the performance of a structure subjected to earthquakes. It may include the lateral displacement of the structure, deformation, damage including crack width and force of the members, or the strain and stress of the materials. The lateral displacement of the structure and member force may be used as the performance index.

10.2 Required Performance Index

The required performance index, PI_R, denoting the seismic demand of a structure is defined as the maximum displacement of the structure and the maximum member force during the earthquake for each limit state. The PI_R should be computed using the appropriate analytical methods.

The performance objectives (*Required Performance Indices PI_R*) are specifically stated in terms of the acceptable risk of casualties, direct economic losses, and downtime resulting from earthquake-induced damage. These acceptable risks can be expressed for specific levels of earthquake ground motion intensity, for different earthquake scenarios, or for a specific period of time considering all earthquakes that can occur during the lifetime of the building and the probability of their occurrence. The performance objectives can be split in two parts – *a damage state and a level of seismic hazard*. Seismic performance is described by designating the maximum

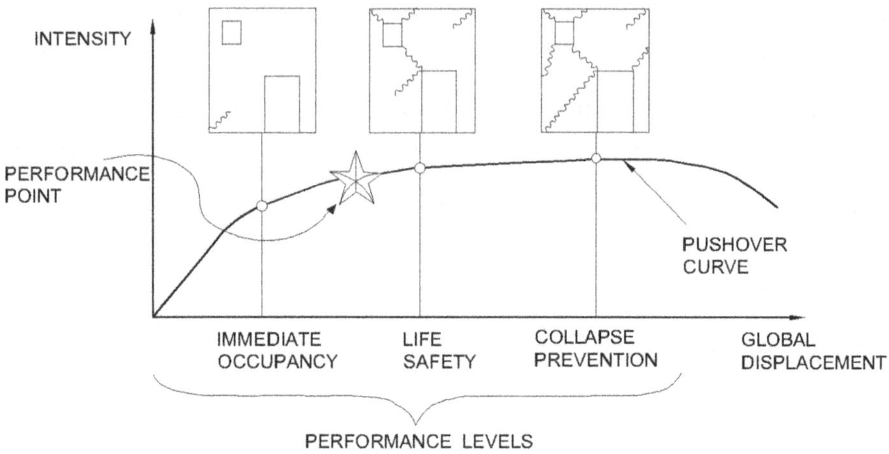

Fig. 14.15 Building damage states

allowable damage state (performance level) for an identified seismic hazard (earthquake ground motion). A performance objective may include consideration of damage states for several levels of ground motion and would then be termed a dual- or multiple-level performance objective. It may be further split into *Structural Performance Level* and *Non-Structural Performance Level*. These may be specified independently; however, the combination of the two determines the overall *Building Performance Level* (Fig. 14.15).

10.3 Structural Performance Levels (ATC 58)

- Immediate Occupancy (SP–1): Continuous service; limited structural damage with the basic vertical and lateral force-resisting system retaining most of their pre-earthquake characteristics and capacities.
- Damage Control (SP–2): Most operations and functions can resume immediately; structure safe for occupancy. Essential operations protected, while nonessential operations are disrupted. This performance level is somewhere between immediate occupancy and life safety.
- Life Safety (SP–3): Significant damage with some margin against total or partial collapse. Injuries may occur with the risk of life-threatening injury being low. Repair may not be economically feasible.
- Structural Stability or Collapse Prevention (SP–4): Substantial structural damage in which the structural system is on the verge of experiencing partial or total collapse. Significant risk of injury exists. Repair may not be technically or economically feasible.

10.4 Possessed Performance Index

The possessed performance index, PI_P, denoting the seismic capacity of a structure is defined as the limit of displacement of the structure specified for each limit state and for each member strength. The member strength should be computed using the material models.

Once performance objectives for a building have been determined, the next step is to design a building so as to allow the building's performance characteristics (*Possessed performance indices*) to be determined. This will include identification of the building's site, size and configuration, occupancy, quality and type of nonstructural systems, the structural system, and estimates of its strength, stiffness, durability, and ductility. A series of simulations (analyses of building response to ground shaking) are performed to estimate the possessed performance indices of the building under various design scenario events.

The process of performance assessment (PI_P) is inherently uncertain and complex. It requires a number of assumptions to be made as to the severity and characteristics of earthquake shaking the building, the condition and occupancy of the building at the time an earthquake occurs, and response of the structural system in the nonlinear range (Aparna and Jain, 2016). It also requires information on the efficiency with which the stakeholders are able to respond and repair the building and restore it to service once damage occurs. The performance assessment process of a building for seismic loads is shown in Fig. 14.16.

These simulation studies provide statistical data on building drifts, floor accelerations, member forces, and deformations, termed demand parameters at different levels of ground motions. This demand data from the structural analyses and data on the building configuration is used to calculate the possible distribution of damage to structural and nonstructural building components and also the potential distribution in casualty and economic and occupancy losses. The performance assessment produces probability distribution functions for casualties, repair costs, and occupancy interruption time. From these distributions it is possible to extract the expected losses at various confidence levels. Further, it is possible to identify the most significant contributors to these losses, to guide design decisions intended to reduce the severity of assessed losses. The seismic performance of the structure should be verified for each limit state by ensuring that:

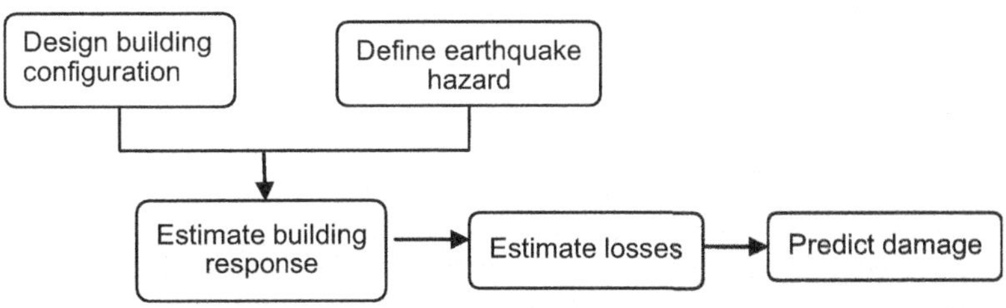

Fig. 14.16 Performance assessment process

$$PI_P > PI_R \tag{14.6}$$

The seismic performance level of the structure should be evaluated as the ratio of PI_P to PI_R. The higher value of this ratio denotes the higher performance level. Following a performance assessment, the building's possessed performance is compared with the required performance indices. In some cases it may not be possible to meet the stated PI_R at reasonable cost, in which case, some relaxation of the PI_R may be appropriate. In other cases, it may be found that the building is inherently capable of performance superior to that required by the performance objectives and that because of other design considerations, it cannot practically be designed to reduce its PI_P. In such cases, the superior performance capability should be documented and accepted.

11 Format of the Futuristic Code

There are two possible options for the format of a futuristic code: (a) a very detailed step-by-step performance-based but prescriptive format and (b) a concise performance-based but philosophical format. In the prescriptive format, it will be a Herculean task to incorporate all the salient features of various new materials, stress-strain curves, nonlinear static analysis, dynamic nonlinear analysis, pushover analysis, base isolation, energy-dissipating devices, etc. besides performance criteria. It will consist of a series of dos and don'ts. No doubt the designers in every corner of the world are used to a prescriptive format as somebody has already made a series of decisions. If, however, the level of understanding and training of an average designer and technician (engaged in the construction industry) is enhanced drastically from the current level, which is no doubt a very difficult but not an impossible task, then it will suffice to simply innumerate the intention and expectation of the code, owner, and the society. In advanced countries, the level of an average designer as well as a technician is already much higher than those of their Indian counterparts. This later format will certainly be manageable and easy to draft as well as understand and implement. It will also give sufficient flexibility to introduce new ideas and innovations. Artficial intelligence can play a significant role in the various stages of planning, design, detailing, construction and later in maintenance of the structure. Nevertheless, the bottom line remains that it will be based on good faith and is liable to be misused and misinterpreted.

11.1 Conflicts Between Intent, Expectations, and Performance

The codes do not and cannot explicitly state their intention about the performance of a building under earthquake force because of the presence and interaction between so many factors. The general public is not expected to appreciate the intent of the code. They assume that once an expert has designed a building in accordance with the latest code, then the building is "absolutely safe." But now in view of failure of so many buildings designed in accordance with the latest specifications in different parts of the world, wider media coverage, and TV discussions, the general public (i.e., owners) is also becoming aware of the ground realities. The structural engineers must explicitly inform the owners of the intention, expectation, and actual performance of the structure during an earthquake. The limitations of the materials, estimation of the seismic force, design philosophy, detailing as well as workmanship, site supervision, etc. must be explained to the owner, architects, and the site engineers so that they may take appropriate steps to face the situation eventually. The Indian insurance companies are also becoming more demanding while taking out the insurance as well as at the time of settlement of the claims.

12 Concluding Remarks

This paper summarizes the state of art of the knowledge in the area of seismic-resistant design of structures developed during the past 100 years and the future direction that the codes are likely to adopt. It is clear that with the explosion of information and understanding of microzonation, nonlinear modeling and behavior of structures, base isolation, energy-dissipating devices, as well as performance-based design concept and artificial intelligence, the future codes are going to be much more demanding and specific in their intentions. The designers should be ready to accept more challenges, responsibilities and risks.

Acknowledgments In the preparation of this chapter, a lot of assistance has been derived from the FEMA and NEHRP publications summarizing the latest enhancements in the knowledge base in the aseismic design. This help is gratefully acknowledged.

References

Aparna, K.P. and Jain, A.K. (2016). Performance evaluation of mutistoreyed RC frames. *Indian Concrete Journal. 90*(6). 38-45.

ASCE 41. (2007). *Seismic rehabilitation of existing buildings, ASCE/SEI Standard 41-06 with supplement 1*. Reston, VA: American Society of Civil Engineers.

ASCE 7. (2016). *Minimum design loads for buildings and other structures, ASCE/SEI Standard ASCE 7-16*. Reston, VA: American Society of Civil Engineers.

ATC 40. (1996). *Seismic evaluation and retrofit of concrete buildings* (Vol. 1). Redwood City, CA: Applied Technology Council.

Deierlein, G. G., Reinhorn, A. M., & Willford, M. R. (2010). Nonlinear structural analysis for seismic design—A guide for practicing engineers, NEHRP Seismic Design Technical Brief No. 4, Oct., NIST, Gaithersburg, Maryland.

Eurocode 1998. (2003). *Design of structures for earthquake resistance*. Brussels: European Committee For Standardization.

FEMA 440 (2005). Improvement of nonlinear static seismic analysis procedures, Washington, D.C.

FEMA 445 (2006). Next generation performance based seismic design guidelines, Washington, D.C.
FEMA 454 (2006). Designing for earthquakes: A manual for architects, Washington, D.C.
FEMA 749 (2010). Earthquake-resistant design concepts, Washington, D.C.
IRC 112. (2011). *Reinforced concrete bridges*. New Delhi: Indian Roads Congress.
IRC 6. (2014). *Loads and stresses*. New Delhi: Indian Roads Congress.
IS 13920. (1993). *Ductile detailing of reinforced concrete structures subjected to seismic forces*. New Delhi: Bureau of Indian Standards.
IS 13920. (2016). *Ductile design and detailing of reinforced concrete structures subjected to seismic forces*. First revision. New Delhi: Bureau of Indian Standards.
IS 1893. (1962). *Recommendations for earthquake resistant design of structures*. New Delhi: Bureau of Indian Standards.
IS 1893. (1966). *Criteria for earthquake resistant design of structures*. First revision, New Delhi: Bureau of Indian Standards.
IS 1893. (1970). *Criteria for earthquake resistant design of structures*. Second revision, New Delhi: Bureau of Indian Standards.
IS 1893. (1975). *Criteria for earthquake resistant design of structures*. Third revision, New Delhi: Bureau of Indian Standards.
IS 1893. (1984). *Criteria for earthquake resistant design of structures*. Fourth revision, New Delhi: Bureau of Indian Standards.
IS 1893. (2002). *Criteria for earthquake resistant design of structures, part 1 General provisions and buildings*. Fifth revision, New Delhi: Bureau of Indian Standards.
IS 1893. (2016). *Criteria for earthquake resistant design of structures, part 1 General provisions and buildings*. Sixth revision, New Delhi: Bureau of Indian Standards.
Jain, A. K. (1980). Review of seismic provisions for concrete buildings. *Indian Concrete Journal, 54*(11), 294–300.
Jain, A. K. (2012). *Reinforced concrete: Limit state design* (7th ed.). Roorkee: Nem Chand & Bros.
Jain, A. K. (2016). *Dynamics of structures with MATLAB applications*, Chennai: Pearson education.
Jain, A. K. (2017). A critical review of IS 13920-2016, *Indian Concrete Journal*, *91*(9), 18-24.
Mahaney, J. A., Paret T. F., Kehoe B. E., & Freeman S. A. (1993). The capacity spectrum method for evaluating structural response during the Loma Prieta earthquake, *National Earthquake Conference*, Memphis.
McClure, F. E. (1968). *Modern earthquake codes: History & development*. Berkeley: Computers and Structures Inc.. reprinted in 2006.
Naeim, F., & Kelly, J. M. (1999). *Design of seismic isolated structures: From theory to practice*. New York, NY: John Wiley and Sons, Inc..
Opensees (2014). Open system for earthquake engineering simulation, PEER, Berkeley: University of California.
Pal, S., Dasaka, S. S., & Jain, A. K. (1987). Inelastic response spectra. Computers and Structures, 25, 335–344.
Popov, E. P., Grigorian, C. E., & Yang, T. S. (1995). Developments in seismic structural analysis and design. *Engineering Structures, 17*(3), 187–197.
Prakash, V., Pore, S. M., & Jain, A. K. (2006a). The role of reduction factor and importance factor in fifth revision of IS:1893, Proceedings of 13th Symposium on Earthquake Engineering, IIT Roorkee, Dec 18–20, 2006, 964–977.
Prakash, V., Pore, S. M., & Jain, A. K. (2006b). Response reduction factors for earthquake resistant design of liquid retaining tanks, Proceedings of 13th Symposium on Earthquake Engineering, IIT Roorkee, Dec 18–20, 2006, 985–995.
Priestley, M. J. N., Calvi, G. M., & Kowalsky, M. J. (2007). *Displacement based seismic design of structures*. Pavia, Italy: IUSS Press.

Chapter 15
Dynamic Parameter Characterization for Railway Bridges Using System Identification

Pradipta Banerji and Sanjay Chikermane

1 Introduction

The problem of assessment of railway bridges with a view to ascertain the remaining service life under the effects of the plying traffic loads is a current research area. For an effective estimate of the service life, the current condition of the bridge, its boundary conditions, and its realistic behavior need to be assessed. With the question of aging infrastructure, another area comes up for question—i.e., the usage of a bridge beyond its design life. For most of these issues, structural health monitoring specifically used for estimation of current condition and remaining life assessment is an imperative bridge management tool (Yanev, 1998).

The Indian railway network is extremely vast comprising more than 125,000 major bridges in addition to numerous culverts and minor bridges. A fair proportion of these have been in use for several decades—most for over 50 years and some approaching 100 years. These old bridges had been designed for axle loads which are significantly lower than those plying the traffic routes today. Also the frequency of trains has increased steadily over the past few decades culminating in an almost exponential rise over the last decade or so (Banerji and Chikermane, 2011). Over the last decade also, the allowable axle loads permitted to run on the bridge has been increased by almost 25% from civil infrastructure; and there have been several papers that have comprehensively reviewed these (Doebling et al., 1996; Ewins, 1984). Of the techniques developed here, depending upon the suitability for the instrumented structure, some of the Indian railway authorities, in view of this increased loadings on the structures initiated a comprehensive structural monitoring program. The present chapter arises out of the

P. Banerji (✉)
Department of Civil Engineering, IIT Bombay, Powai, Mumbai 400076, Maharashtra, India

S. Chikermane
Department of Civil Engineering, Indian Institute of Technology Roorkee, Roorkee, Uttarakhand, India

M. L. Sharma et al. (eds.), *Advances in Indian Earthquake Engineering and Seismology*, https://doi.org/10.1007/978-3-319-76855-7_15

work done by the authors as a response to the railway initiative. There have been several methods used for identifying the dynamic parameters of them have been used in this work.

2 Overview of Bridge

The bridge considered for analysis is a 22 × 30.5 m under-slung steel girder built in 1965. The photograph of the bridge is given in Fig. 15.1. The bridge has a 60-kg rail running on steel channel sleepers which are directly connected with the top chord. Post construction and commissioning, 16-mm-thick additional flange plates were provided in all the girders at the end top chords (U0-U2 and U8-U10), over the existing 8-mm-thick flange plates. There is a rocker bearing on the pier and a roller on the abutment. Pier and Abutment are of mass cement concrete of grade 1:3:6 with open foundation. The instrumentation scheme for this bridge consisted of measuring accelerations, displacements, and strains at several locations as indicated in Fig. 15.2.

Accelerometers are placed at both mid-span and quarter-span to measure the frequencies and global mode shapes of the structure. Displacement transducers are placed at mid-span to measure the vertical deflection of the girder and at the roller end to identify the bearing movements and hence the boundary conditions at the supports. Strain gauges are placed at various locations as shown in Fig. 15.2, to assess the strains at critical locations for various loading conditions.

Fig. 15.1 Photograph of bridge

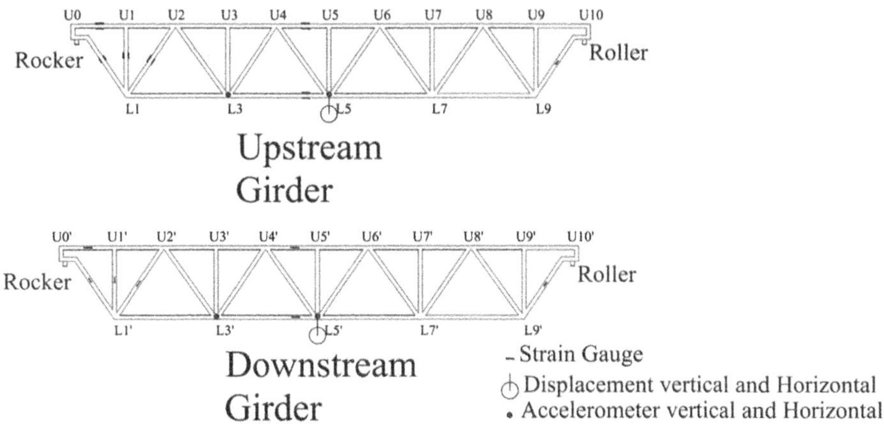

Fig. 15.2 Instrumentation scheme of superstructure

3 System Identification Background

The estimation of parameters from the response data is an important aspect of the entire structural health monitoring paradigm. In this work, we consider five basic methods for estimation of the vibration characteristics from the response data. Three of these are frequency domain methods, and two are time domain algorithms. The methods considered are:

1. Frequency peak picking
2. Frequency domain decomposition
3. Parameter estimation using eigenvalue realization algorithm (ERA)
4. Stochastic subspace identification
5. Parameter estimation using strain data with a statistical formulation

The usage of the ERA and the stochastic subspace identification has been done on the data which is recorded by the accelerometers placed upon the structure. Since the excitation of these responses correspond to the entire structure being excited, these are termed as "global responses". The frequency peak picking has been done with both the accelerometer data and with the frequency transform of the strain data.

In this paper, we introduce the terminology of "global modes" and "local modes." By "global modes" we mean the mode shapes where the entire structure is excited and it has a significant amount of mass participation. By virtue of these characteristics, these are the modes picked up by the accelerometer responses. Against this concept of "global modes", the "local modes" are the ones in an open web girder, where only one or more of the elements behaves like a substructure within the framework of the entire structure. Since these are locally excited, the mass participation of these is low, and also they do not manifest in all the elements. The strain gauges which have been applied on the elements pick up these modes when the element they are applied on corresponds to the excited substructure.

From the accelerometer data, the global modes are picked up. The strain gauges pick up both the local and the global modes. Removing the modes which have been identified by the accelerometer data gives the local modes. It is seen that the frequency of these local modes are sensitive damage features and using these, damage in the open web girder may be identified. The analysis of the local modes as being damage features is not presented here as it has been submitted for publication elsewhere. For brevity, an outline of the formulation is presented here with a more detailed exposition elsewhere (Banerji and Chikermane, 2009a, b).

3.1 Peak Picking

Classical frequency domain approaches are based upon the relationships between the input and the output PSD (Power Spectral Density). For a structure having an input x and an output y, the PSD of the output $G_{yy}(j\omega)$ and the input PSD $G_{xx}(j\omega)$ are related as (Bendat and Piersol, 1980) by:

$$G_{yy}(j\omega) = H(j\omega)*G_{xx}(j\omega)H(j\omega)^T \tag{15.1}$$

where $H(j\omega)$ is the FRF matrix. This can be written in the pole-residue form (Ewins, 1984) as:

$$H(j\omega) = \sum_{i=1}^{N} \left(\frac{R_i}{j\omega - \lambda_i} + \frac{R_i^*}{j\omega - \lambda_i^*} \right); R_i = \phi_i \gamma_i^T \tag{15.2}$$

where ϕ_i, γ_i are the mode shape and the modal participation vector, respectively. When all output measurements are taken as reference, the FRF matrix is square and $\phi_i = \gamma_i$. If the input is considered as white noise, then the input PSD $G_{xx}(j\omega)$ is a constant and the modal decomposition of the output PSD gives the following pole-residue form:

$$G_{yy}(j\omega) = \sum_{i=1}^{N} \left(\frac{A_i}{j\omega - \lambda_i} + \frac{A_i^H}{-j\omega - \lambda_i^*} + \frac{A_i^*}{j\omega - \lambda_i^*} + \frac{A_i^T}{-j\omega - \lambda_i} \right) \tag{15.3}$$

where

$$\begin{aligned} \lambda_i &= -\sigma_i + j\omega_{di} \\ A_i &\approx d_i \phi_i^* \phi_i^T \\ d_r &= \gamma_i^H G_{xx} \gamma_i \end{aligned} \tag{15.4}$$

In the vicinity of a natural modal frequency, Eq. (15.3) can be approximated as only the contribution from the mode under consideration is significant. The approximation is as:

$$\underset{\omega \to \omega_r}{G_{yy}^T(j\omega)} \approx \phi_r \frac{2d_r}{j\omega - \lambda_r} \phi_r^H \tag{15.5}$$

The classical peak picking algorithm is based upon this approximation that in the vicinity of a mode, the power spectral density of the output shows a peak and the column of the PSD matrix at this frequency corresponds to the mode shape.

The peak picking method is a simple technique for estimating a structure's modal frequencies and mode shapes using only system output data. This method stems from the fact that the frequency response function (FRF) of a given system will peak at and around that system's modal frequencies. Assuming that a structure is excited with a white noise, broadband input, the Fourier spectrum of the response data collected at any sensor location can be considered equal to the FRF of the structure at that sensor location. If a structure is lightly damped and has well separated modes, then the imaginary component of an FRF at any modal frequency, for all sensor locations, can be assembled to yield the corresponding mode shape. Since the structures have been sparsely sensed, the mode shape identification has not been attempted.

Spurious noise or measurement errors can also throw up peaks. If, however, the basis of estimation is not a single data set, but multiple data sets over several different train events, the relative significance of this error can be reduced. For the peak picking exercise, only the data from the trailing end of the event has been considered. This is because when a train moves on a bridge, the mass of the train and the excitation may result in blurring the frequency spectrum. The decaying signal after the train has left the span has been considered as being close enough to a free vibration estimate for evaluating the frequency peaks.

3.2 *Frequency Domain Decomposition*

If the modes are not well separated, classical peak picking often fails as it cannot identify closely spaced or repeated modes. Also the mode shapes picked up are the operational deflected shapes. The frequency domain decomposition (FDD) class of methods was developed (Brincker et al., 2000) to overcome these limitations. In these methods, the PSD matrix of the output is decomposed by using singular value decomposition at discrete frequency points $\omega = \omega_{\mathrm{d}}$ in the form of:

$$G_{yy}(j\omega_{\mathrm{d}}) = U_{\mathrm{d}} S_{\mathrm{d}} U_{\mathrm{d}}^{\mathrm{H}} \tag{15.6}$$

At a point where only the rth mode is dominant for a frequency w_r, the PSD matrix approximates to a rank one matrix of the form:

$$\underset{\omega_{\mathrm{d}} \to \omega_r}{G_{yy}(j\omega_{\mathrm{d}})} \approx s_{\mathrm{d}} u_{\mathrm{d}1} u_{\mathrm{d}1}^H \tag{15.7}$$

The first singular vector of the rth frequency is an approximation of the rth mode shape. In the case of repeated modes, the rank of the PSD matrix keeps changing. Modal frequencies can be located by the peaks of the SV plots. From the corresponding singular vectors, mode shapes can be obtained. Since SVD has the ability of separating signal space from noise space, the modes can be indicated from SV plots, and closely spaced modes or even repeated modes can easily be detected.

3.3 Eigen value Realization Algorithm (ERA)

The second-order differential equation of motion is first cast in a state space form as follows (Juang and Phan, 2001):

$$\begin{aligned} M\ddot{y} + C\dot{y} + Ky &= u \\ \ddot{y} + M^{-1}C\dot{y} + M^{-1}Ky &= M^{-1}u \end{aligned} \tag{15.8}$$

Putting $x = [\dot{y}\ y]^T$ this converts the second-order equation into two first-order equations given as

$$\begin{aligned} \dot{x} &= \begin{bmatrix} \ddot{y} \\ \dot{y} \end{bmatrix} = Ax + Bu \\ y &= Cx + Du \end{aligned} \tag{15.9}$$

where the matrices A, B, C, and D are the system matrices. The system identification paradigm tries to identify these matrices. The matrices are given by

$$\begin{aligned} A &= \begin{pmatrix} -M^{-1}C & -M^{-1}K \\ I & 0 \end{pmatrix}; C = (0\ I) \\ B &= \begin{bmatrix} M^{-1} \\ 0 \end{bmatrix}; \qquad D = (0\quad 0) \end{aligned} \tag{15.10}$$

It can be shown that the eigen-values of the system matrix A are identical to the eigen-values of the representative equations of motion (Ewins, 1984). The standard formulation of the ERA (Juang and Phan, 2001) has been considered in this work. The typical length of the acceleration data is a 10-s data length sampled at 200 Hz. On the basis of these Markov parameters, a block Hankel matrix is built up of size $pm \times$ nr, where p is considered as the system order, m is the number of inputs, and r is the number of outputs. In this formulation, n is worked out by considering the maximum matrix size possible considering all the data recorded.

A singular value decomposition of the Markov parameter matrix gives an optimal estimate of the matrices B and C. The algorithm is then repeated for a shifted formulation where the entire recording data is shifted by one sample. The form of the Markov parameter and the optimal form as defined earlier decomposes into an optimal least square estimation of the A matrix. The eigen-values of this matrix are computed, first in the discrete domain, and then converted into the continuous domain.

With this, all the system matrices can therefore, be computed from the available data in an optimal sense. It can also be shown (James III et al., 1993) that the cross-correlation of a sensor giving ambient data with a reference synchronous sensor is analogous to the free vibration of a system. This enables the application of the algorithm based upon the ambient data only.

3.4 *Stochastic Subspace Identification*

The fundamental basis of these methods (Van Overschee and De Moor, 1996) is an estimation of the state sequences directly from the data recorded—without an a priori estimation of the state matrices—using an explicit method, if the state sequences need to be extracted for analysis, or an implicit one if the state matrices are sufficient. The methods used are linear algebraic methods of using an orthogonal or oblique projection of the rows of one set of block Hankel matrices on another. A singular value decomposition then estimates the observability or state sequences followed by a least square solution to estimate the state matrices. The strength of this method lies in the usage of robust tools like singular value decompositions and QR decompositions, which are known to be robust for even ill-conditioned problems.

The state-space solutions given in Eq. (15.9) are first taken into their discrete forms as follows:

$$\begin{aligned} x_{k+1} &= Ax_k + Bu_k + w_k \\ y_k &= Cx_k + Du_k + v_k \end{aligned} \tag{15.11}$$

with

$$E\left(\begin{bmatrix} w_p \\ v_p \end{bmatrix}\begin{bmatrix} w_q^T & v_q^T \end{bmatrix}\right) = \begin{pmatrix} Q & S \\ S^T & R \end{pmatrix}\delta_{pq} \tag{15.12}$$

where E is the expected value operator and δ_{pq} is the Kronecker delta.The vector x is the state vector of the process, and the vectors w and v correspond to the process noise and the measurement noise respectively. The matrix A is the general dynamical system matrix, B is the input matrix which influences how the inputs affect the next state, C is the output matrix which describes how the internal state is transferred to the outside world, and D is the direct transmission matrix. The matrices Q, S, and R are the noise covariance matrices. The matrix pair $\{A,C\}$ is assumed to be observable, which fundamentally means that all the modes can be observed through the output y and are hence identifiable, and the matrix set $\{A,B,Q^{1/2}\}$ is assumed to be controllable, which means that all the modes can be excited through either the input u or the process noise w.

The subspace algorithm works by first assuming that the state sequences information is encapsulated in the output matrix and can be retrieved by doing a projection (either orthogonal or oblique) of the block Hankel matrix of one set of outputs onto the block Hankel matrix of another set of outputs. Once these are extracted, the problem decomposes into a much simpler one as shown below:

$$\underbrace{\begin{pmatrix} x_{i+1} & x_{i+2} & \dots & x_{i+j} \\ y_i & y_{i+1} & \dots & y_{i+j-1} \end{pmatrix}}_{\text{known}} = \begin{pmatrix} A & B \\ C & D \end{pmatrix} \underbrace{\begin{pmatrix} x_i & x_{i+1} & \dots & x_{i+j-1} \\ u_i & u_{i+1} & \dots & u_{i+j-1} \end{pmatrix}}_{\text{known}} + \begin{pmatrix} w_i & w_{i+1} & \dots & w_{i+j-1} \\ v_i & v_{i+1} & \dots & v_{i+j-1} \end{pmatrix} \tag{15.13}$$

The form of Eq. (15.13) is solved by estimating the values of A, B, C, and D which give a solution of

$$\begin{pmatrix} \widehat{A} & \widehat{B} \\ \widehat{C} & \widehat{D} \end{pmatrix} = \frac{\min}{A,B,C,D} \left\| \begin{pmatrix} \widehat{x}_{i+1} & \widehat{x}_{i+2} & \dots & \widehat{x}_{i+j} \\ \widehat{y}_i & \widehat{y}_{i+1} & \dots & \widehat{y}_{i+j-1} \end{pmatrix} - \begin{pmatrix} A & B \\ C & D \end{pmatrix} \begin{pmatrix} \widehat{x}_i & \widehat{x}_{i+1} & \dots & \widehat{x}_{i+j-1} \\ u_i & u_{i+1} & \dots & u_{i+j-1} \end{pmatrix} \right\|^2 \tag{15.14}$$

3.5 Strain Data With a Statistical Formulation

The partial differential equation of a beam, vibrating naturally with a frequency ω is given by:

$$\frac{\partial^2}{\partial x^2}\left[E(x) * I(x)\frac{\partial^2}{\partial x^2}(w(x,t))\right] + \rho A(x)\frac{\partial^2 w(x,t)}{\partial t^2} = 0 \tag{15.15}$$

where ρ is the density, E the Young's modulus, A the area, and I the moment of inertia of a cross section. This can be written in terms of the modal displacement $W(x)$ as

$$\mathrm{EI}\frac{d^4 W(x)}{dx^4} = \rho A \omega^2 W(x) \tag{15.16}$$

Using a weighted function $\delta V(x)$, signifying a displacement field, it is possible to convert the above member level equation into a structural level equation such as

$$\mathrm{EI}\int_{x_k}^{x_l} \frac{d^2 W^*}{dx^2}\frac{d^2 \delta V}{dx^2} dx = \int_{x_k}^{x_l} \rho A \omega^2 W \delta V dx + M_k \frac{d(\delta V)}{dx}(x_k) - M_l \frac{d(\delta V)}{dx}(x_l) \tag{15.17}$$

The member modal strain measured is proportional to the second derivative of the displacement for an Euler–Bernoulli beam. In addition to this, the essential strains which are induced on the structure due to the boundary conditions of the substructure which shows the structural modal effects would also be present. So, for the structure under consideration, if the structural modal effects could be removed, the resulting effects would be due to the member level vibrations.

From these, the boundary conditions and flexural rigidities could be estimated and an estimate made about the condition of the joints.

A peak picking approach which is statistically done over the different sets of individual events measured is undertaken. This approach reduces the dependence or coloring of the frequencies estimated by spikes or noise in the data. Since a large proportion of the strain data evaluated is static in nature, a high-pass filter is applied on the data to remove this portion. The range is then divided into equal ranges of 5 Hz with buffering as described in the peak picking section. So each segment of the frequency axis (except the first 2.5 Hz and the last 2.5 Hz considered) manifests itself in two ranges. A peak is identified as such if it is estimated in either of these ranges. A similar evaluation criterion to pick up the peaks is identified as a peak corresponds to any frequency outlier which lies more than three standard deviation points away from the mean of the range considered.

A correlation of the peaks estimated is done across sensors and also with the accelerometer peaks identified. Since the accelerometer identifies only global peaks and also a global peak would crop up in all the sensors evaluated, there are two independent yardsticks to divide the data into global peaks. For each sensor, the significant peaks which remain after the global peaks have been identified are taken to correspond to the local frequencies.

4 Parameter Estimation

The data was recorded at 200 cycles per second indicating a Nyquist frequency of 100 Hz. Since most structural responses are expected to be well within this range, this was considered as being sufficient for the analysis purposes. Since there is a significant amount of electromagnetic interference around 50 Hz, which corresponds to the AC cycle frequency in India, a band-stop filter was applied on the data with a band of 45–55 Hz. For the application of peak picking and frequency domain decomposition, the free vibration response corresponding to the data set when the train has just left the bridge has been taken, whereas for the ERA and SSI algorithms, the ambient data has been considered.

The results of the global parameters estimated are shown in Table 15.1. In this table the maximum deviation is the deviation between estimation algorithms and the computed average. Figure 15.3 shows the Fourier spectrum of the acceleration data. The mapping of the singular values from frequency domain decomposition is given in Fig. 15.4. It can be seen here that there are indications of closely spaced modes at certain frequencies where the second singular value also shows a peak indicating a

Table 15.1 Results of estimation using different algorithms—global frequencies

Mode	ERA (Hz)	Peak picking (Hz)	FDD (Hz)	SSI (Hz)	Strain based system (Hz)	Average (Hz)	Max. deviation %	Numerical model (Hz)	% Difference between numerical model and average
Mode 1	3.83	3.91	3.52*	4.58	–	3.96	15.7	4.20	6.1%
			6.64*		6.45	6.55	1.5		
Mode 2	7.11	7.42	7.42*	7.94		7.47	6.3	7.64	2.3%
Mode 3	9.96	9.96	9.77	9.71	9.82	9.84	1.4	10.46	6.3%
Mode 4	13.38	–	14.45	14.55	13.33	13.93	4.5	14.92	7.1%
Mode 5	13.91	–	15.63	15.31		14.95	7.0	15.91	6.4%

[a]Indicates closely spaced or repeated modes at this frequency

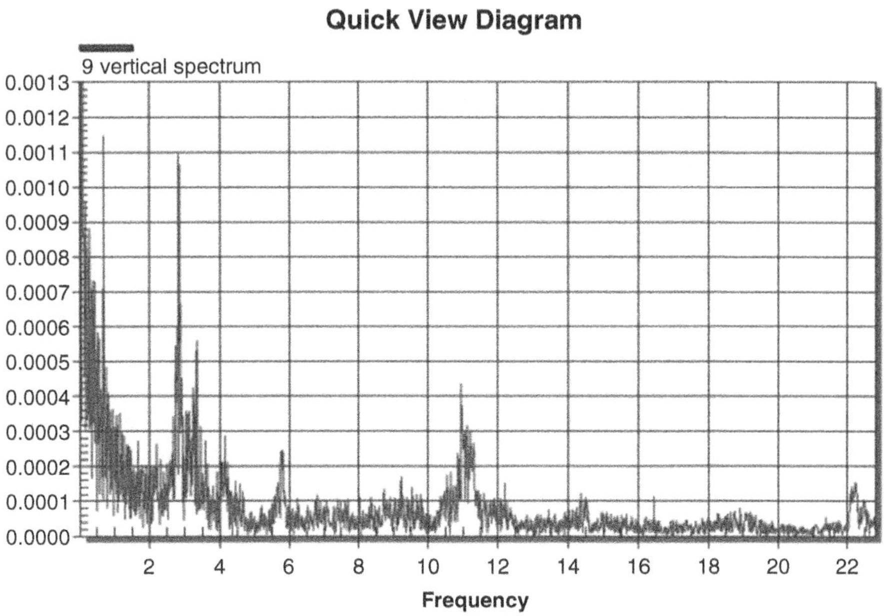

Fig. 15.3 Fourier spectrum for acceleration data—vertical accelerations

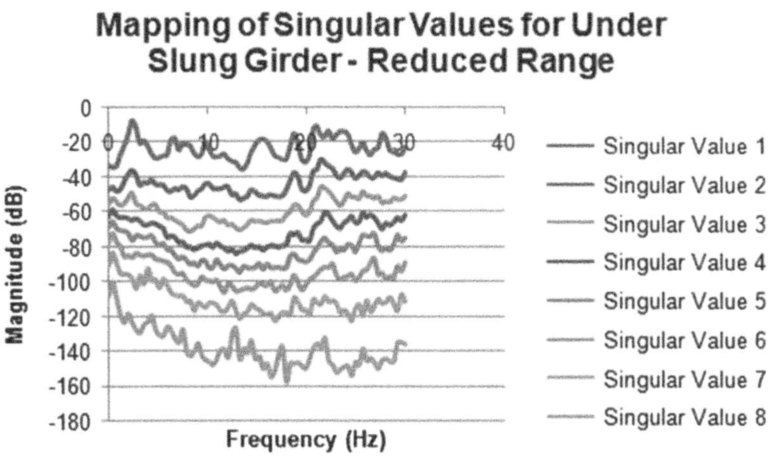

Fig. 15.4 Singular values of girder from frequency domain decomposition

matrix of greater than rank 1 at the frequency under consideration. Figure 15.4 shows a reduced range of frequency. This is done because for most structures, these are the significant frequencies where most of the mass participation occurs and hence are the frequencies which are easily excited.

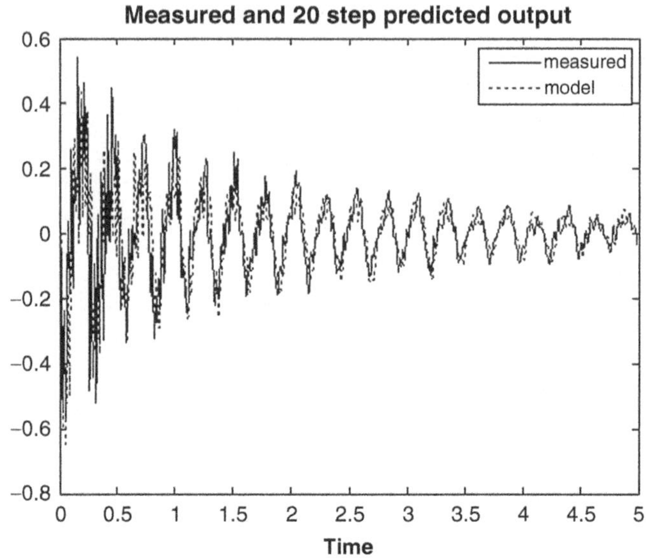

Fig. 15.5 Measured and 20 step predicted response for stochastic subspace iteration

The model order for the bridge is considered as 30 for the ERA implementation. Using this model order, the data sets generated from the results of the eigen-value realization algorithm are compared with classical frequency domain data extraction systems and the numerical model. For the Stochastic Subspace method implementation on the under-slung girder, 26 different data sets were built up from the ambient data measured from site conditions. These data sets were combined together as a single merged data set for running the stochastic subspace algorithm. In addition to the data sets used for building up the model, 18 data sets were used to validate the model. The correlation between the numerical model and the site data for a 20-step predicted output is shown in Fig. 15.5.

Table 15.1 also indicates a comparison with an updated numerical model. The modeling and updating methodology are not indicated here as they have been submitted for publication elsewhere.

The fundamental problem with using strain gauge data is that, unlike vibration data, there is a significant proportion of noise in the strain gauges. Also strain gauges have a large component of static data within them, which interferes in the estimation of dynamic parameters. Since the noise component is extremely high, the ambient strain data cannot be used for realistic parameter estimation as on filtering out noise, very little significant data is left for analysis, so the parameters for strain gauges have been estimated from the forced vibration data.

For this procedure, first a load estimation algorithm was developed, the details of which are not being presented here as they have been submitted for publication elsewhere. Using these load identification procedures the loads applied in the event were first identified. The measured response was then de-convoluted with this load vector to give an estimate of the impulse response. The frequencies were then

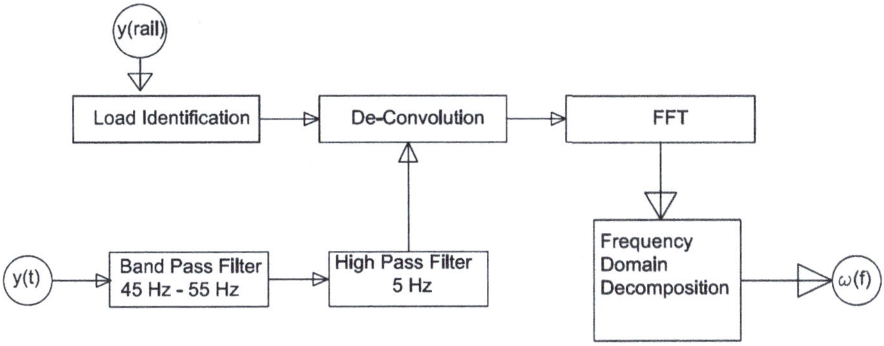

Fig. 15.6 Algorithm for identification of frequencies from strain gauges

Fig. 15.7 Frequency response from pier side diagonal data—20–30 Hz

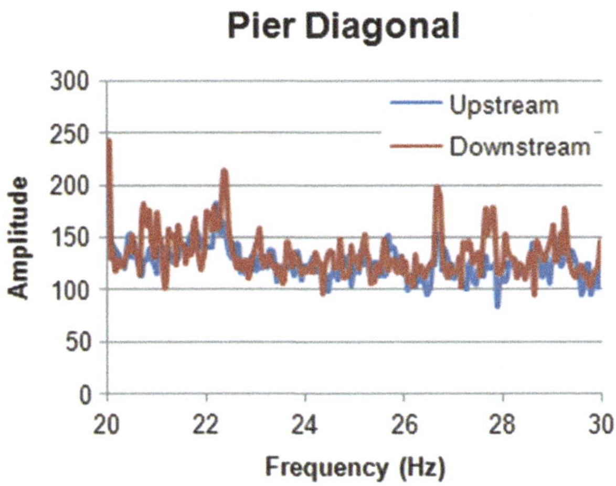

estimated as the Fourier decomposition of this influence line. The entire process is shown below in a block diagram in Fig. 15.6. Sample result sets have been presented for the global results in Figs. 15.7, 15.8, and 15.9 for this bridge. The results from other instrumented locations were seen to be giving similar patterns of results and are presented in Table 15.1. The results for the local or member level frequencies are given in Table 15.2, and the figures for the mode shapes from the updated numerical model are shown in Figs. 15.10 and 15.11.

5 Analysis of Outcomes

As can be seen from Table 15.1, there is a good convergence between the updated numerical model and the globally estimated dynamic parameters. It can also be seen here that although no single algorithm ended up giving a total convergence with the

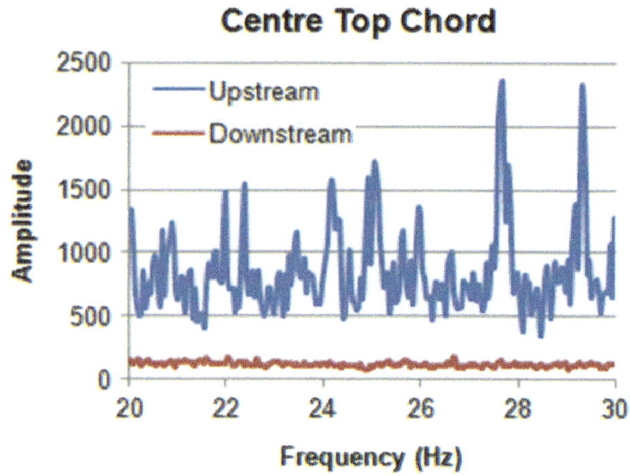

Fig. 15.8 Frequency response from central top chord data—20–30 Hz

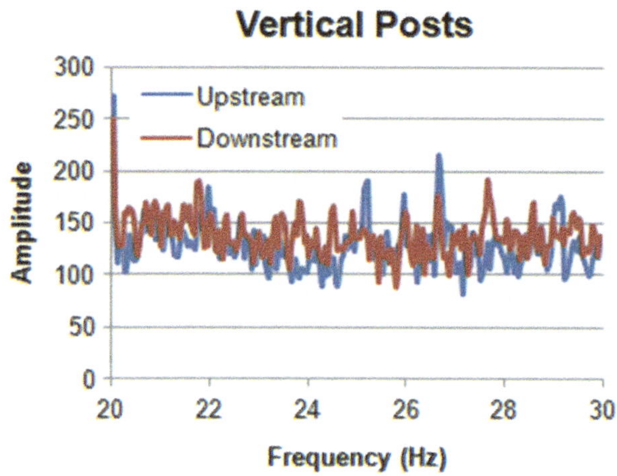

Fig. 15.9 Frequency response from vertical post data—20–30 Hz

model, the averaged values using different algorithms seemed to agree with the numerical model much better. From this table it can also be seen that except for a large deviation in the first mode estimate, most of the other modes were estimated with a reasonable accuracy using all of the methods discussed. The large deviations could be reduced by updating the mass matrix, but as there was no physical validation for that, this activity was not undertaken. For the local modes, it was seen that the mass participation values from the numerical model was very low; hence, since the frequency depended only upon the stiffness and the self-weight of the members, these were estimated with much better accuracy as can be seen from Table 15.2.

The presence and participation of members in the local frequencies was qualitatively estimated. Taking for example, the first local frequency, from Figs. 15.7 and

Table 15.2 Results of estimation using strain data—local frequencies

Number	Numerical model (Hz)	Frequency (Hz)	Remarks
1	23.01	22.22	Occurs in both pier diagonals. Representation of these frequencies is also seen in abutment end rakers and as a bending strain in center top and bottom chords. As it occurs in both the diagonals, it is probably an out-plane bending
2		22.37	
3	24.69	24.52	These occur in vertical posts in upstream and downstream girders. These frequencies are also seen in a minor representation in center top chords. They are possibly in-plane frequencies as the top chord axial members do not show these
4		25.21	
5	25.06	25.647	These are the end raker in-plane strains, as they do not show up in the end raker sensors mapping axial strains. They are also seen in the center top and bottom chord as in-plane bending strains. Some small representation is also seen in the first diagonals. Large representation is also seen in vertical posts
6	39.43	38.3	These are the diagonal in-plane bending strains. They occur in only the upstream diagonal member. They also occur in the vertical posts, bottom chord bending members, and top chord bending members in both the center and rocker top chords

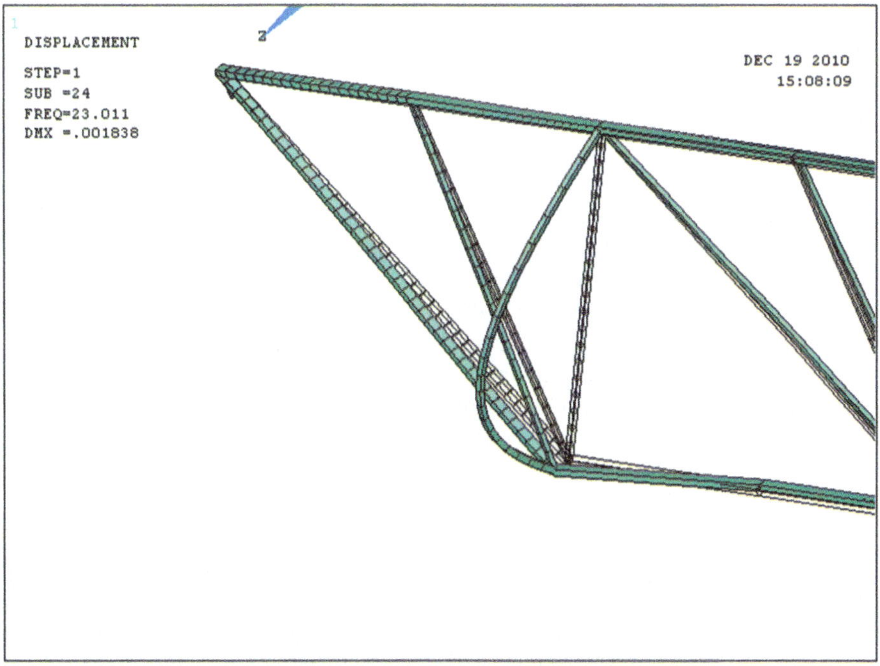

Fig. 15.10 Local mode 1, frequency 23.01 Hz—diagonal dominated

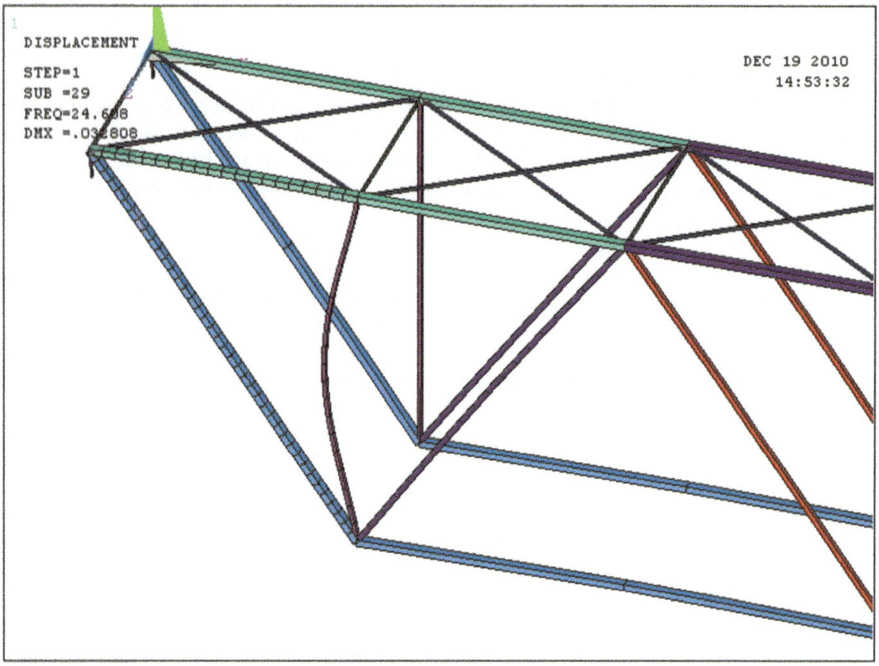

Fig. 15.11 Local mode 2, frequency 24.69 Hz—vertical post with in-plane movement

15.8, it is seen that the diagonal and the central top chord upstream side show this frequency, but it is not seen in the center top chord downstream side or predominantly in the vertical post. Since the instrumentation of the upstream side is on the top and bottom flange, while on the downstream side at the centroidal axis, it can be inferred that the particular frequency is not exhibited as a predominantly axial response for the top chord. This pattern is also exhibited by the numerical model behavior at a similar frequency.

Similarly the next frequency is seen in Figs. 15.8 and 15.9 as being exhibited by both the vertical post and central top chord without any presence in the axial top chord member. The diagonal also does not show any significant sign of this frequency. This result pattern is also similarly exhibited in the numerical model.

It can thus be seen that the frequencies estimated from strain gauges correspond to both types of actions—the structural global behavior and the member local behavior. A synthesis of data can hence be done segregating these.

6 Conclusions

From the analysis of results here, it can be seen that most of the parameter estimation techniques give acceptable results. Time domain techniques tend to give some computational modes which are difficult to identify and remove from the list.

Although most of these correspond to either negative or very large damping values, this need not be always necessary. Computational modes with acceptable damping values could also theoretically exist. Frequency domain techniques, especially the domain decomposition techniques, give very robust results of the estimated parameters with an intuitive accuracy.

It is also seen that the local modes correspond to both in-plane and out-of-plane bending as the main actions. The global modes are used for a validation of the model updating procedure, and it is shown that the updated models give very good estimates of global modes. After a member level updating, the numerical model is seen to agree very well with the local modes as well. Hence it can be seen that if the data is of a good quality and has a sufficient length to ensure frequency resolution, fairly accurate estimates can be made of the system dynamic parameters which can be used to validate the numerical model and seed the model updating process.

References

Banerji, P., & Chikermane, S. (2009a). Structural parameter estimation of two bridges from site data using an Eigen value realization algorithm. 4th International Conference on Structural Health Monitoring on Intelligent Infrastructure (SHMII-4), Zurich.

Banerji, P., & Chikermane, S. (2009b). Structural parameter estimation of two bridges from site data using Kalman filters and stochastic subspace algorithm. IV ECCOMAS Thematic Conference on smart structures and materials (smart'09), Porto.

Banerji, P., & Chikermane, S. (2011). Structural health monitoring of a steel railway bridge for increased axle loads. *Structural Engineering International, 21*(2), 210–216.

Bendat, J. S., & Piersol, A. G. (1980). *Engineering applications of correlation and spectral analysis*. New York, NY: John Wiley & Sons.

Brincker, R., Zhang, L., & Andersen, P. (2000). Modal identification from ambient responses using frequency domain decomposition. Proc. of the 18th Intl. Modal Analysis Conference (IMAC), San Antonio, Texas.

Doebling, S. W., Farrar, C. R., Prime, M. B., & Shevitz, D. W. (1996). Damage identification and health monitoring of structural and mechanical systems from changes in their vibration characteristics: a literature review, Los Alamos National Laboratory Report LA-13070-MS.

Ewins, D. J. (1984). *Modal testing: Theory and practice*. Taunton: Research Studies Press Ltd..

James III, G. H., Carne, T. G. and Lauffer, J. P. (1993). The Natural Excitation Technique (NExT) for modal parameter extraction from operating wind turbines, Sandia Report, SAND92-1666 UC-261.

Juang, J. N., & Phan, M. Q. (2001). *Identification and control of mechanical systems*. Cambridge, UK: Cambridge University Press.

Van Overschee, P., & De Moor, B. (1996). *Subspace identification for linear systems: theory, implementation, applications*. Boston/London/Dordrecht: Kluwer.

Yanev, B. (1998). The management of bridges in New York City. *Engineering Structures, 20*(11), 1020–1026.

Chapter 16
An Overview of Response Spectrum Superposition Methods for MDOF Structures

I. D. Gupta

1 Introduction

The response spectrum superposition methods are used very widely for seismic response analysis of multi-degree-of-freedom (MDOF) structures, due to their ability to predict the maximum structural response in a very simple and efficient way. For a given earthquake acceleration time history as input excitation, a response spectrum represents the maximum response of a single-degree-of-freedom (SDOF) oscillator with a specified viscous damping as a function of its natural period (frequency) (Biot, 1941, 1942; Housner, 1941). Response spectra for different damping values are routinely computed and made available for the strong motion accelerograms obtained during actual earthquakes (Gupta et al., 1993; Chandrasekaran and Das, 1993). The response spectra can be used directly to get an estimate of the maximum response for many structures, which can be idealized as a single-degree-of-freedom (SDOF) system (e.g., elevated water tanks, single-storey buildings, etc.). Under certain approximations, a MDOF structure can also be approximated by an equivalent SDOF system (Clough and Penzien, 1975).

However, the exact time histories of the various response quantities of a MDOF structure to a given ground acceleration time history can be obtained in a simple way using normal mode theory (Biot, 1933). In this approach, the total response of a MDOF structure in a specified degree-of-freedom is expressed by superposition of the responses in its various modes of vibration, where each mode behaves like a SDOF oscillator with a specific natural period and damping ratio. It is much more convenient to compute the response time histories of various modes than directly for the MDOF structure. Also, the response spectrum of input accelerogram can be used to define the absolute maximum response amplitude in each normal mode. But these cannot simply be added to get the total maximum response of a MDOF structure,

I. D. Gupta (✉)
Central Water and Power Research Station (Formerly), Pune, Maharashtra, India

M. L. Sharma et al. (eds.), *Advances in Indian Earthquake Engineering and Seismology*, https://doi.org/10.1007/978-3-319-76855-7_16

because all the modal maxima do not occur at the same time and they are not in the same phase. Also, under certain conditions, the modal responses are characterized by strong correlations. Therefore, several different methods have been proposed over the years to superpose the modal maxima to get directly the absolute maximum response of MDOF structures. A comprehensive review on these methods is provided in Joshi and Gupta (1998) and Gupta and Joshi (1998).

This chapter first describes the normal mode theory for computing the exact time-history response of MDOF structures. The early response spectrum superposition methods (Joshi, 1997), which have also been adopted in some of the design standards, are described the next. Some of these methods neglected the modal correlation (interaction) effect, and the others accounted it in an ad hoc manner. This is followed by later developments based on more rigorous theoretical basis to account for the modal interaction effect. Finally, the mode superposition methods considering explicitly only a few lowest-frequency modes are described. It is impractical to consider explicitly the effect of all the modes for real structures with very large number of degrees of freedom. Example numerical results are presented to illustrate the relative performance of the various methods vis-à-vis the exact results. Though, due to their jagged nature, the response spectra of real earthquakes are not used as it is in actual design applications, the spectra of real earthquakes have been used in the present study to facilitate comparison of the results obtained by various spectrum superposition methods with the corresponding exact solutions.

2 Earthquake Response Spectra

As mentioned before, a response spectrum represents the maximum response as a function of natural period or frequency of a single-degree-of-freedom (SDOF) structure with a specified damping value, when excited by a given acceleration time history of earthquake ground motion. A response spectrum is computed by modeling the SDOF structure as a lumped mass, M, supported on massless columns with total spring constant, K, as shown in Fig. 16.1. A viscous damper with damping constant C models the energy loss. When excited by ground acceleration $\ddot{z}(t)$, the mass M will experience a relative displacement $x(t)$, which is described by the following differential equation of motion:

$$\ddot{x}(t) + 2\zeta\omega_0\dot{x}(t) + \omega_0^2 x(t) = -\ddot{z}(t) \tag{16.1}$$

where ω_0 is the natural frequency in rad/s and ζ the damping ratio of the oscillator as a fraction of the critical damping:

$$\omega_0 = \sqrt{\frac{K}{M}} \quad \text{and} \quad \zeta = \frac{C}{2\sqrt{KM}} \tag{16.2}$$

Assuming zero initial conditions, the time-history solution of Eq. (16.1) is given by the Duhamel integral (Trifunac, 1972):

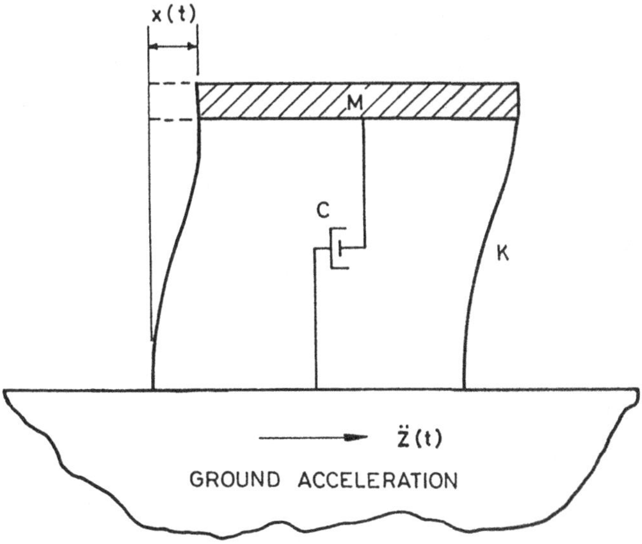

Fig. 16.1 Model of a single-degree-of-freedom (SDOF) system

$$x(t) = \frac{-1}{\omega_0\sqrt{1-\zeta^2}} \int_0^t \ddot{z}(\tau)e^{-\zeta\omega_0(t-\tau)} \sin\omega_0\sqrt{1-\zeta^2}(t-\tau)d\tau \tag{16.3}$$

The expressions for the relative velocity, $\dot{x}(t)$, and the relative acceleration, $\ddot{x}(t)$, responses of the oscillator can simply be obtained by differentiation. However, in practical applications, the absolute acceleration response, $\ddot{x}(t) + \ddot{z}(t)$, is used to get total inertial force acting on mass M of the SDOF oscillator. The maximum absolute values of the relative displacement response $x(t)$, velocity response $\dot{x}(t)$, and the total acceleration response $\ddot{x}(t) + \ddot{z}(t)$ define the exact relative displacement, relative velocity, and absolute acceleration response spectra SD, SV, and SA, respectively. However, the common practice is to compute the relative displacement spectrum SD and approximate the velocity and acceleration spectra by pseudo-spectral velocity (PSV) and pseudo-spectral acceleration (PSA) as follows:

$$\text{SV} \approx \text{PSV} = \frac{2\pi}{T}\text{SD} \quad \text{and} \quad \text{SA} \approx \text{PSA} = \left(\frac{2\pi}{T}\right)^2 \text{SD} \tag{16.4}$$

Two different acceleration time histories are considered to compute the example response spectra and the response of MDOF structures. These are the longitudinal component recorded at 1A foundation gallery of Koyna dam from the main earthquake of December 10, 1967, with M6.5 at closest distance of 17.4 km (termed as Rec # 016) and the S10E component recorded at Nongkhlaw site from the Meghalaya earthquake of September 10, 1986, with M5.2 at closest distance of 63.6 km

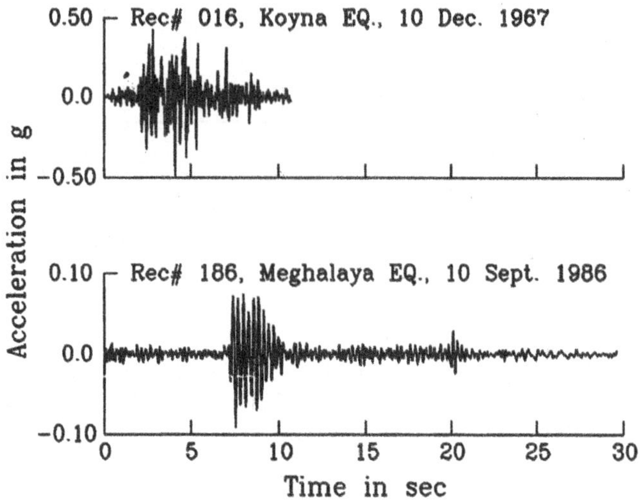

Fig. 16.2 Accelerograms considered to compute the example results

(termed as Rec # 186). These acceleration time histories are plotted in Fig. 16.2, and the various types of response spectra for both the accelerograms are shown in Fig. 16.3. It is seen that PSA approximates SA very well, whereas PSV differs significantly from SV in the longer-period range.

3 Normal Mode Theory

Though some simple types of structures can be assumed to behave like a SDOF system, most real structures of importance like tall buildings, chimneys, towers, etc. need to be described as multi-degree-of-freedom (MDOF) systems. Figure 16.4 shows a simple idealization of multistorey buildings in terms of lumped floor masses m_i connected with massless columns with total storey stiffnesses k_i and damping values c_i.

Assuming the floors to be rigid and the columns to be flexible to lateral deformation, but rigid in the vertical direction, the equations of motion for the whole system under earthquake loading $\ddot{z}(t)$ can be written in the matrix form as:

$$[M]\{\ddot{x}\} + [C]\{\dot{x}\} + [K]\{x\} = -[M]\{I\}\ddot{z}(t) \tag{16.5}$$

In this equation, $\{I\}$ is a column vector with all its elements as unity, and $[M]$, $[K]$, and $[C]$ are the mass, stiffness, and damping matrices, defined in terms of the storey masses, m_i; storey stiffnesses, k_i; and the storey damping values, c_i, as follows:

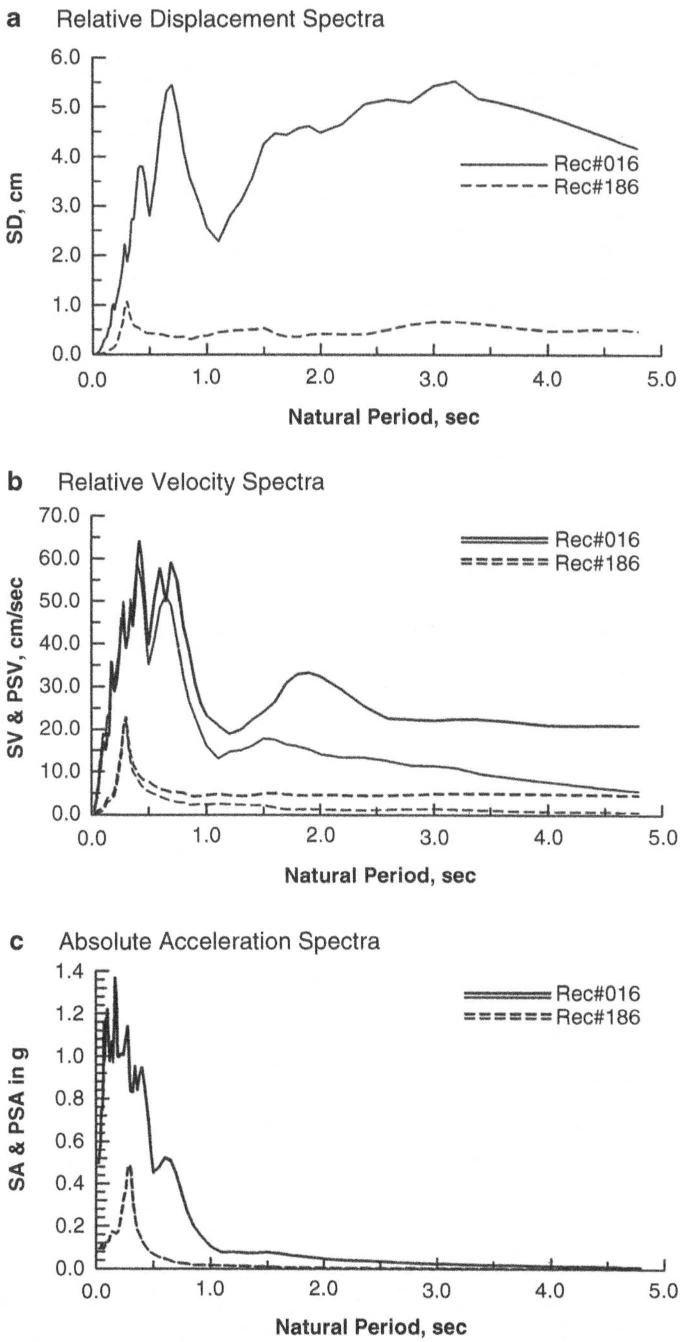

Fig. 16.3 Various types of response spectra for the two example accelerograms. In (**b**) and (**c**), thick curves represent the exact SV and SA spectra, whereas thin curves represent the PSV and PSA approximations

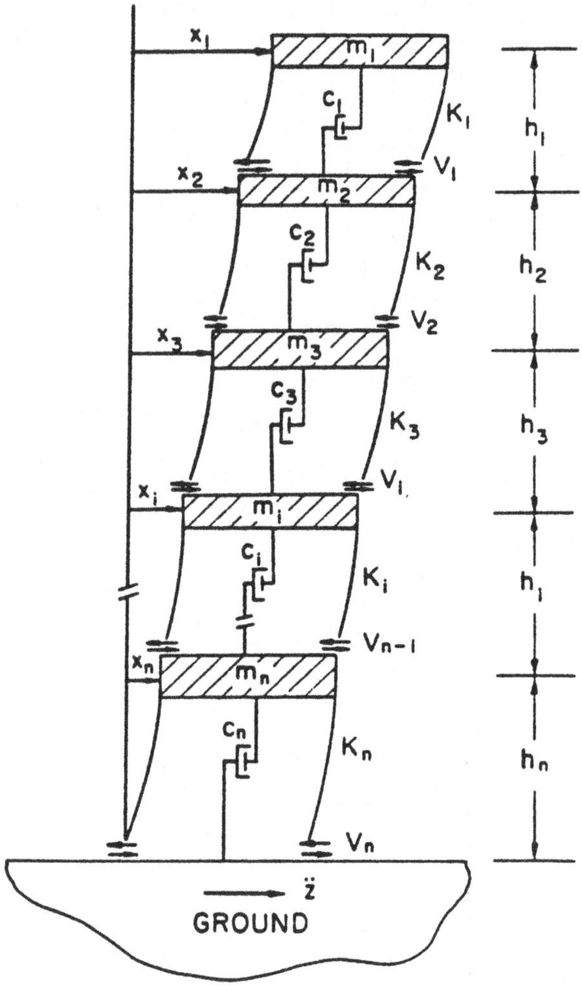

Fig. 16.4 A lumped mass, spring and dashpot, model of a multistorey building

$$[M] = \begin{bmatrix} m_1 & 0 & 0 & \dots & 0 \\ 0 & m_2 & 0 & \dots & 0 \\ 0 & 0 & m_3 & \dots & 0 \\ & & & \ddots & \vdots \\ 0 & 0 & 0 & \dots & m_n \end{bmatrix} \tag{16.6}$$

$$[K] = \begin{bmatrix} k_1 & -k_1 & 0 & \dots & 0 \\ -k_1 & k_1 + k_2 & -k_2 & \dots & 0 \\ 0 & -k_2 & k_2 + k_3 & \dots & 0 \\ & & & \ddots & \vdots \\ 0 & 0 & 0 & \dots & k_{n-1} + k_n \end{bmatrix} \tag{16.7}$$

$$[C] = \begin{bmatrix} c_1 & -c_1 & 0 & \dots & 0 \\ -c_1 & c_1 + c_2 & -c_2 & \dots & 0 \\ 0 & -c_2 & c_2 + c_3 & \dots & 0 \\ & & & \ddots & \vdots \\ 0 & 0 & 0 & \dots & c_{n-1} + c_n \end{bmatrix} \tag{16.8}$$

The systems of differential equations of motion described by Eq. (16.5) are, in general, coupled and need to be solved simultaneously, which is an extremely difficult task. However, under the assumption of classical damping, that is the damping matrix is a linear combination of the mass and the stiffness matrices (Caughey, 1959), these equations can be decoupled using the following transformation:

$$\{x\} = [A]\{\xi\} \tag{16.9}$$

In Eq. (16.9), $\{x\}$ is the column vector of the relative displacements of the various floor masses, $\{\xi\}$ is a column vector of the generalized normal coordinates, and $[A]$ is the matrix with its columns as the mode shape vectors of the building. The jth mode shape vector $\{A^j\}$ is obtained by solving the following eigenvalue problem for the undamped structure:

$$\left[[K] - \lambda_j[M]\right]\{A^j\} = 0; \quad \lambda_j = \omega_j^2 \tag{16.10}$$

where ω_j is the natural frequency of the jth normal mode. Substituting the transformation of Eq. (16.9) into Eq. (16.5) and pre-multiplying the whole expression by $[A]^T$, give

$$[A]^T[M][A]\{\ddot{\xi}\} + [A]^T[C][A]\{\dot{\xi}\} + [A]^T[K][A]\{\xi\} = -[A]^T[M][I]\{\ddot{z}\} \tag{16.11}$$

From this, the equation of motion for the jth normal coordinate can be obtained as

$$\begin{aligned} &\{A^j\}^T[M]\{A^j\}\ddot{\xi}_j + \{A^j\}^T[C]\{A^j\}\dot{\xi}_j + \{A^j\}^T[K]\{A^j\}\xi_j \\ &= \{A^j\}^T[M]\{I\}\ddot{z}(t) \end{aligned} \tag{16.12}$$

In view of Eq. (16.10), the stiffness matrix $[K]$ gets diagonalized by the transformation matrix $[A]$ as follows:

$$\{A^j\}^T[K]\{A^j\}/\{A^j\}^T[M]\{A^j\} = \omega_j^2 \tag{16.13}$$

For classical damping, it is assumed that the transformation matrix $[A]$ would also diagonalize the damping matrix $[C]$ in a similar manner as

$$\{A^j\}^T[C]\{A^j\}/\{A^j\}^T[M]\{A^j\} = 2\zeta_j\omega_j \tag{16.14}$$

Here, ζ_j and ω_j are the damping ratio and the natural frequency of jth mode of the structure. Thus, a set of uncoupled equations can be obtained from Eq. (16.12) in terms of generalized normal coordinates, ξ_j, as

$$\ddot{\xi}_j(t) + 2\zeta_j\omega_j\dot{\xi}_j(t) + \omega_j^2\xi_j(t) = -\alpha_j\ddot{z}(t); \quad j = 1, 2, \ldots, n \tag{16.15}$$

where α_j is the jth mode participation factor, defined as

$$\alpha_j = \{A^j\}^T[M]\{I\}/\{A^j\}^T[M]\{A^j\} \tag{16.16}$$

The participation factors represent the relative contributions of different modes to the total response of a MDOF structure.

The expression of Eq. (16.15) represents n independent equations for $\xi_j(t)$, each of which is of the form of Eq. (16.1) for a SDOF structure, except for a multiplication factor α_j with $\ddot{z}(t)$. Hence, the solution for modal response $\xi_j(t)$ can be obtained from the Duhamel integral of Eq. (16.3) with input excitation as $\alpha_j\ddot{z}(t)$. From a knowledge of $\xi_j(t)$ for all the n modes of a structure, the relative displacement response, $x_i(t)$, at the ith storey level of the structure can be written from the transformation of Eq. (16.9) as follows:

$$x_i(t) = \sum_{j=1}^{n} A_{ij}\,\xi_j(t) \tag{16.17}$$

The column vector of elastic forces at different storey levels can be defined as

$$\{q(t)\} = [K]\{x(t)\} = [K][A]\xi(t) \tag{16.18}$$

Using Eq. (16.10), this can further be written as

$$\{q(t)\} = [M]\,[A]\,\left[\omega^2\right]\,\{\xi(t)\} \tag{16.19}$$

Carrying out matrix multiplication, the elastic force at the lth storey level can be written as

$$q_l(t) = \sum_{j=1}^{n} m_l A_{lj}\omega_j^2\xi_j \tag{16.20}$$

The base shear response $V_i(t)$ for the ith storey level can thus be written as

$$V_i(t) = \sum_{l=1}^{i} q_l(t) = \sum_{j=1}^{n}\omega_j^2\sum_{l=1}^{i} m_l A_{lj}\xi_j \tag{16.21}$$

With h_i as the storey height, the overturning moment response at the ith storey can be defined as

$$M_i(t) = \sum_{k=1}^{i} h_k V_k(t) = \sum_{j=1}^{n} \omega_j^2 \sum_{k=1}^{i} h_k \sum_{l=1}^{k} m_l A \xi_j(t) \tag{16.22}$$

Comparison of Eqs. (16.21) and (16.22) with Eq. (16.17) indicates that the formulation for displacement response can be used to get the shear force and bending moment responses simply by replacing A_{ij} with $S_{ij}$and B_{ij}, respectively, defined as

$$S_{ij} = \omega_j^2 \sum\nolimits_{l=1}^{i} m_l A_{lj} \quad \text{and} \quad B_{ij} = \omega_j^2 \sum\nolimits_{k=1}^{i} h_k \sum\nolimits_{l=1}^{k} m_l A_{lj} \tag{16.23}$$

4 Response Spectrum Superposition Methods

From the foregoing formulation, it is seen that the computation of the exact time histories of various response quantities at different storey levels of a MDOF structure involves lengthy calculations of evaluating the time histories of modal responses. However, the earthquake-resistant design of structures is normally based on the maximum values of the displacement, base shear, and overturning moment responses at various storey levels. The maximum response amplitudes can be estimated more conveniently and efficiently using the response spectrum of the input excitation as described in the following.

A comparison of Eq. (16.1) with Eq. (16.15) indicates that the displacement response spectral amplitude, $\text{SD}(\omega_j, \zeta_j)$, for modal frequency ω_j and damping ratio ζ_j can be used to define the absolute maximum value of modal response $\xi_j(t)$ as

$$\left|\xi_j(t)\right|_{\max} = \left|\alpha_j\right| \text{SD}\left(\omega_j, \zeta_j\right) \tag{16.24}$$

Further, from Eq. (16.17), the contribution of the jth normal mode to the total displacement response at the ith storey level is given by $A_{ij}\xi_j(t)$. Thus the maximum displacement response amplitude at ith storey level due to the jth normal mode can be written as

$$R_{ij} = \left|\alpha_j A_{ij}\right| \text{SD}\left(\omega_j, \zeta_j\right) \tag{16.25}$$

As mentioned before, the maximum shear force and bending moment responses at ith story due to jth mode can also be obtained simply by replacing A_{ij} with S_{ij} and B_{ij}, defined by Eq. (16.23).

The basic philosophy of response spectrum superposition methods for estimating the maximum response is to combine the contributions R_{ij} in such a way that the total response at each storey level is in good agreement with the maximum response from the exact time-history solution. Several different spectrum superposition methods have been proposed over the years by different investigators, the earliest among which are reviewed in the next section. The more advanced later developments are described in the subsequent sections.

5 Early Spectrum Superposition Methods

The direct sum of the maximum response due to each mode is not expected to provide good results because the maximum values of the various modal responses do not all occur at the same time and in the same phase. However, it was the first modal combination method proposed by Biot (1943) to get an upper bound on the total response of MDOF structures. The maximum response at ith storey level using SUM method can be defined as

$$(x_i)_{\max} = \sum_{j=1}^{n} R_{ij} \tag{16.26}$$

where R_{ij} is the absolute maximum value of the response at the ith storey level due to jth mode and n is the total number of modes of the structure.

To get more realistic estimate of the maximum responses at different storey levels, Rosenblueth (1956) suggested to combine the modal responses by square root of the sum of the squares (SRSS). This method represents the most probable combination (Goodman et al., 1958) and gives lower values of the total response than that given by the SUM method. Mathematically, the expression for SRSS method is defined as

$$(x_i)_{\max} = \left[\sum_{j=1}^{n} R_{ij}^2\right]^{1/2} \tag{16.27}$$

Application of the above two methods indicated that the SRSS method generally underestimates the maximum storey responses for structures with closely spaced frequencies and the SUM method does not always provide the upper bound estimate of the storey responses.

An average of the SUM and SRSS results was, therefore, proposed by Jennings (1958) to get somewhat better results for structures with closely spaced modes:

$$(x_i)_{\max} = (\mathrm{SRSS} + \mathrm{SUM})/2.0 \tag{16.28}$$

To evaluate the maximum response of submarine structures to underwater explosions, O'Hara and Cunnif (1963) developed a method for the Naval Research Laboratory (NRL). The NRL method was also used for seismic structural design, which can be expressed as

$$(x_i)_{\max} = R_{ik} + \left[\sum_{j=1, j \neq k}^{n} R_{ij}^2 \right]^{1/2} \tag{16.29}$$

where R_{ik} is the maximum of all the R_{ij}, assumed to occur in the kth mode.

As an improvement over the SRSS method to account for closely spaced modes, the US Nuclear Regulatory Commission (1976) introduced the "Grouping Method," in which modes are divided into groups that include all modes having frequencies between the lowest frequency in the group and a frequency 10% higher. For each group, the representative value of the response is taken as the sum of the absolute values of modal maxima belonging to the group. The maximum response is then obtained as the SRSS of the representative group values. If the modal maxima in the qth group are represented by $R_{ij,\ q}$, the resultant of the qth group is given by

$$R_{i,q} = \sum_{j=1}^{n_q} R_{ij,q} \tag{16.30}$$

where n_q is the number of modes in the qth group. If Q is the total number of such groups, the maximum response by "Grouping Method" is given by

$$(x_i)_{\max} = \left[\sum_{q=1}^{Q} R_{i,q}^2 \right]^{1/2} \equiv \left[\sum_{j=1}^{n} R_{ij}^2 + \sum_{q=1}^{Q} \sum_{j=1}^{n_q} {}^{q} \sum_{k=1}^{n_q} R_{ij,q} \; R_{ik,q} \right]^{1/2} \tag{16.31}$$

The US Nuclear Regulatory Commission (1976) also introduced another method for closely spaced modes, known as "Ten Percent Method," which is expressed as

$$(x_i)_{\max} = \left[\sum_{j=1}^{n} \sum_{k=1}^{n} \rho_{jk} R_{ij} R_{ik} \right]^{1/2} \tag{16.32}$$

The coefficient ρ_{jk} is taken as 1.0 for $\omega_j \leq \omega_k \leq 1.1\omega_j$ and zero otherwise. This method gives similar or more conservative results compared to the Grouping Method.

6 Numerical Results From Early Methods

To illustrate the relative performance of the foregoing early response spectrum superposition methods, numerical results are computed for a five-storey hypothetical building, first with a symmetric floor plan and next by introducing asymmetry in the floor plan. The displacement, shear force, and bending moment responses for various storeys obtained by spectrum superposition methods have been compared with the exact time-history results for the two input accelerograms (Rec. # 016 and 186)

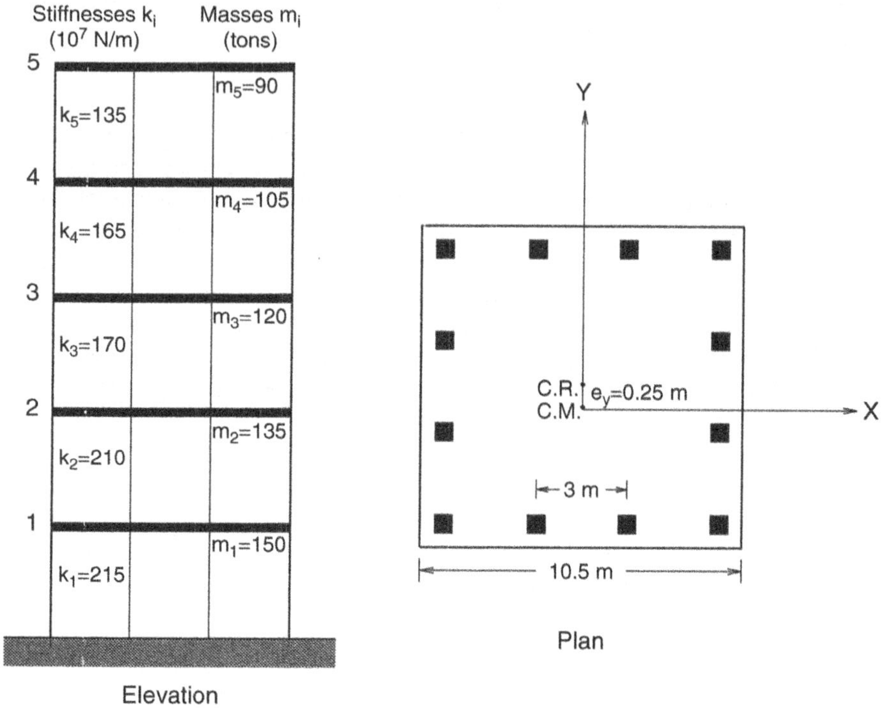

Fig. 16.5 Elevation and plan of the five-storey example building

shown in Fig. 16.1. The elevation and plan of the example building along with the storey masses and stiffnesses are shown in Fig. 16.5.

The translational stiffnesses of various storeys of the building are assumed to be identical in both x- and y-directions. Also, the total storey stiffness is assumed to be distributed equally among all the resisting elements. The center of resistance (C.R.) thus coincides with the geometrical center of the floor plan. To have the symmetric building, the center of mass (C.M.) is assumed to coincide with the C.R., whereas the asymmetric building is obtained by introducing an offset of 0.25 m between C.R. and C.M. at each floor in the y-direction only as shown in Fig. 16.5. The height of all the storeys is taken uniformly equal to 3.66 m. The floors are considered to be rigid, and the building is excited only along the x-direction.

The response of the symmetric building is characterized by one translational degree of freedom along the direction of excitation for each storey level. The mass and stiffness matrices for this building can thus be defined using Eqs. (16.6) and (16.7), respectively. The undamped modal frequencies of the symmetric structure are found to be 12.1, 31.7, 48.8, 63.8, and 72.1 rad/s, which are separated quite widely. On the other hand, the response of the asymmetric structure is also characterized by a torsional motion at each floor in addition to the translation along direction of excitation. Also, the translational and torsional degrees of freedom are coupled, which results in closely spaced modal frequencies.

If $k_{xi,\,m}$ and $k_{yi,\,m}$ are the translational stiffnesses of the mth resisting element at ith storey level along x- and y-directions, respectively, the torsional stiffness of the ith storey about a vertical axis through the center of mass is given by Gupta and Trifunac (1987a):

$$k_{ti} = \sum_{m=1}^{M} \left(k_{xi,m} b_{i,m}^2 + k_{yi,m} a_{i,m}^2 \right) \tag{16.33}$$

where $a_{i,\,m}$ and $b_{i,\,m}$ are the x and y coordinates of the mth element with the origin at center of mass. The torsional stiffnesses, k_{ti}, of various storeys of the example building from top are thus obtained as 3855.94, 4712.81, 4855.63, 5998.13, and 6140.94 × 10^6 Nm/rad. To obtain the mass moment of inertia of various floors of the building, the radius of gyration about the center of mass is taken as 3.68 m for all the floors, which gives the floor mass moment of inertia, J_i, as 1220.6, 1424.1, 1627.5, 1831.0, and 2034.4 ton m^2.

The equation of motion for the coupled translation-torsional response of the building under excitation along x-direction can be written in the following generalized form (Kan and Chopra 1977):

$$\begin{bmatrix} [M] & \\ & [J] \end{bmatrix} \begin{Bmatrix} \{\ddot{x}\} \\ \{\ddot{\psi}\} \end{Bmatrix} + \begin{bmatrix} [C_{xx}] & [C_{xt}] \\ [C_{xt}]^T & [C_{tt}] \end{bmatrix} \begin{Bmatrix} \{\dot{x}\} \\ \{\dot{\psi}\} \end{Bmatrix} + \begin{bmatrix} [K_{xx}] & [K_{xt}] \\ [K_{xt}]^T & [K_{tt}] \end{bmatrix} \begin{Bmatrix} \{x\} \\ \{\psi\} \end{Bmatrix} = \begin{bmatrix} [M] & \\ & [J] \end{bmatrix} \begin{Bmatrix} \ddot{z}(t)\{I\} \\ 0 \end{Bmatrix} \tag{16.34}$$

In this equation, $\{x\}$ and $\{\psi\}$ are the vectors of relative translational and torsional displacements at various storeys of the building, $[M]$ and $[J]$ are the diagonal mass and moment-of-inertia matrices, $[K_{xx}]$ and $[K_{tt}]$ are the stiffness matrices for pure translational and pure torsional responses, and $[K_{xt}]$ is the stiffness matrix related to the coupling between translational and torsional responses. Matrices $[K_{xx}]$ and $[K_{tt}]$ are defined by using Eq. (16.7), whereas $[K_{xt}]$ is defined in terms of the static eccentricity, e_{yi}, and translational stiffness, k_{xi}, of the ith storey as

$$[K_{xt}] = \begin{bmatrix} -e_{y1}k_{x1} & e_{y1}k_{x1} & 0 & & \\ e_{y1}k_{x1} & -(e_{y1}k_{x1}+e_{y2}k_{x2}) & e_{y2}k_{x2} & & \\ 0 & e_{y2}k_{x2} & -(e_{y2}k_{x2}+e_{y3}k_{x3}) & & \\ & \ddots & \ddots & & \\ & & & & e_{y,n-1}k_{x,n-1} \\ & & & e_{y,n-1}k_{x,n-1} & -(e_{y,n-1}k_{x,n-1}+e_{yn}k_{xn}) \end{bmatrix} \tag{16.35}$$

Similar to the case of symmetric building, the damping matrix in Eq. (16.34) is not defined explicitly, and the total damping is later replaced by its diagonal form under the assumption of classical damping to obtain the decoupled modal equations of the form of Eq. (16.15). Thus the various spectrum superposition methods can be applied to the asymmetric example structure also. The undamped modal frequencies

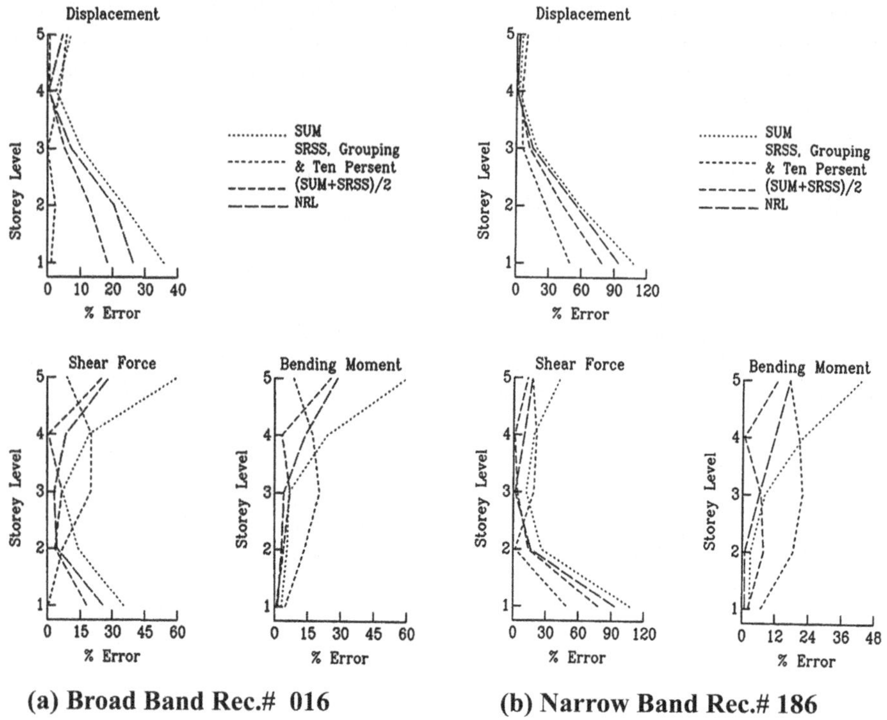

Fig. 16.6 Comparison of the errors in various early spectrum superposition methods as percentage of exact time-history solution for displacement, base shear, and bending moment responses of the symmetric example building

of the asymmetric structure are found to be 12.1, 12.5, 31.5, 32.5, 48.7, 50.1, 63.4, 65.8, 71.3, and 74.8 rad/s. The natural modes of the structure are seen to occur in pairs of closely spaced frequencies, and hence there will be significant contribution from the modal interaction terms. All the modes for the symmetric as well as asymmetric building are assumed to have a constant damping ratio of 0.05.

Figure 16.6a, b show the results in terms of the percentage errors for the displacement, shear force, and bending moment responses for the symmetric building excited by the broadband input excitation of Rec. # 016 and the narrowband excitation of Rec. # 186, respectively. The results in Fig. 16.6a for broadband excitation show almost similar errors for the SRSS, Grouping, and Ten Percent methods. For the displacement response, average of SUM and SRSS gives good results at the top storey; whereas for the shear force response, SRSS gives the minimum error at the bottom. In case of bending moment response at the base, all the methods give errors within 5% of the exact values.

For the case of narrowband excitation (Fig. 16.6b), the modal frequencies of the building lie almost outside the predominant frequency band of the ground motion. The various modes thus do not act much out of phase, and hence NRL and SUM

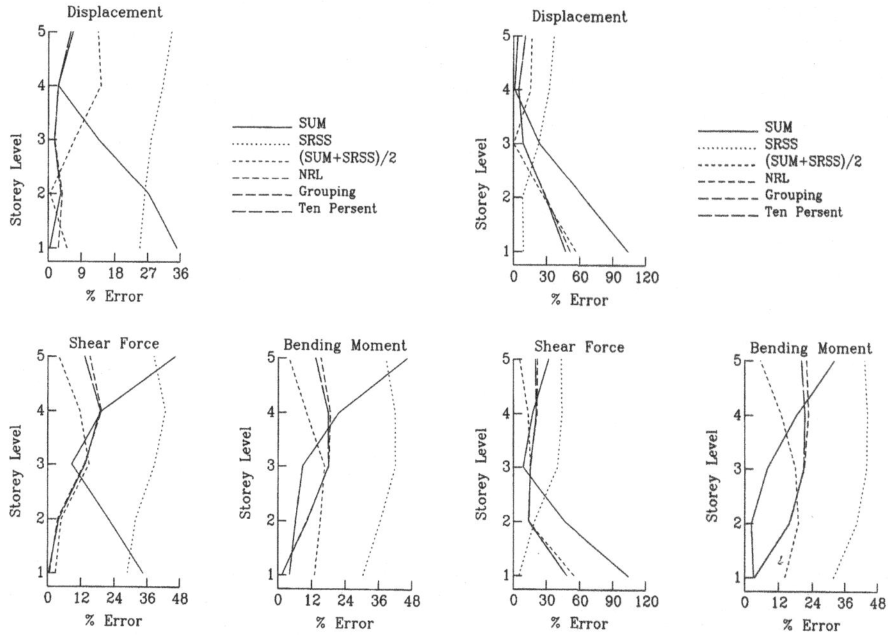

(a) Broad Band Rec.# 016 **(b) Narrow Band Rec.# 186**

Fig. 16.7 Comparison of the errors in various early spectrum superposition methods as percentage of exact time-history solution for displacement, base shear and bending moment responses of the asymmetric example building

methods give minimum errors for the displacement response at the top storey and the bending moment response at the bottom. Whereas for the shear force response, averages of SUM and SRSS show minimum errors at the bottom. As the modal frequencies of the symmetric structure are spaced widely, error in the SRSS method is almost identical to that in the Grouping and Ten Percent methods.

Figure 16.7a, b show the percentage errors in the results for the asymmetric building excited by broadband and narrowband excitations, respectively. Due to closely spaced modal frequencies, the results in Fig. 16.7a, b show the SRSS method to be in larger errors compared to the other methods like Grouping, Ten Percent, and NRL methods, which take the modal interaction effects into account to some extent. The other methods show errors within 5% for the displacement response at the top and shear force and bending moment responses at the bottom. Thus, for structures with closely spaced modal frequencies, the SRSS method does not provide accurate results, and it normally underestimates the exact response. For the case in Fig. 16.7a, the SUM method is also seen to give good results at some of the storey levels, and the assumption that this method gives an upper bound to the response does not seen always to be true.

7 Improved Spectrum Superposition Methods

The use of the early response spectrum superposition methods has shown very wide range of errors in the results for the asymmetric example building with closely spaced modes. This is because some of them do not take into account the modal interaction effects, and the others consider it in an ad hoc manner. In reality, the modal interaction is a complicated phenomenon depending on the closeness of the modal frequencies and the mutual disposition of the structural frequencies with respect to the frequency band of input excitation. Therefore, several later investigators have proposed improved modal combination methods based on more rigorous theoretical background. Most of these methods can be expressed in the following generalized form:

$$(x_i)_{\max} = \sum_{j=1}^{n}\sum_{k=1}^{n} \alpha_j A_{ij} \alpha_k A_{ik} \mathrm{SD}_j \rho_{jk} \mathrm{SD}_k \tag{16.36}$$

where $\mathrm{SD}_j = \mathrm{SD}(\omega_j, \zeta_j)$ is the relative displacement spectrum amplitude for the *j*th mode and ρ_{jk} is the cross-correlation coefficient between *j*th and *k*th modal responses $\xi_j(t)$ and $\xi_k(t)$, defined by

$$\rho_{jk} = \frac{\lambda_{0,jk}}{\sqrt{\lambda_{0,jj}\lambda_{0,kk}}} \tag{16.37}$$

In this expression, $\lambda_{0,\,jj}$ and $\lambda_{0,\,jk}$ are respectively, the autocorrelation and cross-correlation functions of $\xi_j(t)$ and $\xi_k(t)$ with zero-time delay. If the ground acceleration is defined by a stationary power spectral density function (PSDF), $G(\omega)$, and the modal responses are also assumed to be stationary, these correlation functions can be defined as

$$\lambda_{0,jk} = \int_0^{\infty} G(\omega) \mathrm{Re}\left[H_j(\omega) H_k^*(\omega)\right] d\omega \tag{16.38}$$

where $H_j(\omega) = -\left(\omega_j^2 - \omega^2 + 2i\zeta_j\omega_j\omega\right)^{-1}$ is the complex frequency transfer function for the *j*th modal response. The various methods basically differ only in the definition of the modal cross-correlation coefficients, which are based on different idealizations and approximations.

Approximating the earthquake ground motion by an equivalent stationary white-noise segment of duration *S* sec, and the modal response to be of damped periodic form, $\mathrm{e}^{-\zeta_j\omega_j t} \sin\left(\sqrt{1-\zeta_j^2}\omega_j t\right)$, Rosenblueth and Elorduy (1969) developed the following expression for the correlation coefficient:

$$\rho_{jk} = \left[1 + \left(\frac{\sqrt{1-\zeta_j^2}\,\omega_j - \sqrt{1-\zeta_k^2}\,\omega_k}{\zeta_j\omega_j + \zeta_k\omega_k + 4/S}\right)^2\right]^{-1} \tag{16.39}$$

This expression approximately accounts for the fact that for very-low-frequency modes, if the duration S is not sufficiently long, the modal responses do not attain their steady-state value. However, Rosenblueth and Elorduy (1969) have not proposed any scheme to define duration S for a given accelerogram. To compute the example results in this paper, duration S has been taken equal to the strong motion duration defined by Trifunac and Brady (1975). To obviate the need for computing duration S, Gupta and Cordero (1981) have replaced the term $4/S$ in the expression of Eq. (16.39) by a coefficient C_{jk}, defined empirically using strong motion durations of ten real accelerograms, as

$$C_{jk} = \left(0.16 - \frac{\zeta_j + \zeta_k}{4}\right)\left(1.4 - \left|\omega_j^2 - \omega_k^2\right|\right) \quad \geq \quad 0 \tag{16.40}$$

Further, for the cases when damping ratios ζ_j and ζ_k differ widely, Gupta and Cordero (1981) suggested to multiply the expression of Eq. (16.39) by a factor of $2\sqrt{\zeta_j\zeta_k}/(\zeta_j + \zeta_k)$.

Representing the input excitation by a white-noise or a smoothly varying broadband stationary PSDF covering all the modal frequencies, and assuming that the modal responses are also stationary in nature, DerKiureghian (1980, 1981) and Wilson et al. (1981) have presented expressions for ρ_{jk} in terms of the modal frequencies and damping ratios only. Because all these expressions give almost identical results, without any special preference, it is proposed to use the following expression due to Wilson et al. (1981) to compute the correlation coefficients for the case of white-noise stationary excitation:

$$\rho_{jk} = \frac{8\sqrt{\zeta_j\zeta_k}(\zeta_j + r\zeta_k)r^{3/2}}{(1-r^2)^2 + 4\zeta_j\zeta_k r(1+r^2) + 4\left(\zeta_j^2 + \zeta_k^2\right)r^2} \quad ; \quad r = \frac{\omega_k}{\omega_j} \tag{16.41}$$

The foregoing expressions for the correlation coefficient give significant correlation between modes with closely spaced frequencies only. However, if the input excitation is defined by a narrowband PSDF, all the modes lying outside the frequency range of input excitation are characterized by strong correlation among themselves as well as with the other modes, irrespective of their frequency separations. If such modes contribute significantly to the total response, the white-noise approximation is not able to provide accurate results. To consider the effect of the narrow-bandedness of input excitation, DerKiureghian and Nakamura (1993) suggested to obtain the correlation coefficients using the expressions of Eqs. (16.37) and (16.38) with the PSDF of ground acceleration defined approximately from the given design displacement response spectrum amplitudes, $\mathrm{SD}(\omega, \zeta)$, as

$$G(\omega) = \frac{4\zeta\omega^3}{\pi p^2} \mathrm{SD}^2(\omega, \zeta) \tag{16.42}$$

In this expression, p is the peak factor, defined in terms of the frequency ω, damping ratio ζ of the oscillator, and stationary duration of the input excitation (DerKiureghian 1980). The method of DerKiureghian and Nakamura, however, doesn't consider the effect of transient nature of the response on account of the fact that low-frequency modes may not attain their steady-state response.

Another class of response spectrum superposition methods is based on defining an expression for the PSDF of the displacement response $r_i(t)$ at the ith storey level in terms of the PSDF of the ground acceleration. Such an expression due to Singh and Chu (1976) can be written in the following generalized form:

$$G_{x_i}(\omega) = \sum_{j=1}^{n} \sum_{k=1}^{n} \alpha_j A_{ij} \alpha_k A_{ik} \left(B_{jk} + \frac{\omega^2}{\omega_j^2} C_{jk} \right) G(\omega) \left| H_j(\omega) \right|^2 \tag{16.43}$$

where B_{jk} and C_{jk} are the coefficients defined in terms of the modal frequency ratio $r = \omega_k/\omega_j$ and the modal dampings as follows:

$$B_{jk} = \left[\left\{ 8\left(\zeta_j^2 - \zeta_k^2 r^2\right) - 4(1 - r^2) \right\} \left\{ 4\zeta_j r(\zeta_k - \zeta_j r) - (1 - r^2) \right\} - 2(1 - r^2)(1 - r^4) \right] \times / D_{jk} \tag{16.44a}$$

$$C_{jk} = 2(1 - r^2)\left\{ 4r(\zeta_j - \zeta_k r)(\zeta_k - \zeta_j r) - (1 - r^2)^2 \right\} / D_{jk} \tag{16.44b}$$

$$D_{jk} = 8r^2 \left[2\left(\zeta_k^2 - \zeta_j^2 r^2\right)\left(\zeta_j^2 - \zeta_k^2 r^2\right) - \left(\zeta_j^2 + \zeta_k^2\right)(1 - r^2)^2 \right] - (1 - r^2)^4 \tag{16.44c}$$

Assuming that the peak factors for the modal displacement and velocity responses and that for the complete structural responses at different storey levels are all identical, Singh and Chu (1976) have proposed a response spectrum superposition method as follows:

$$(x_i)_{\max} = \left[\sum_{j=1}^{n} \sum_{k=1}^{n} \alpha_j A_{ij} \alpha_k A_{ik} \left\{ B_{jk} \mathrm{SD}_j^2 + \frac{C_{jk}}{\omega_j^2} \mathrm{SV}_j^2 \right\} \right]^{1/2} \tag{16.45}$$

In this expression, SD_j and SV_j are the relative displacement and the relative velocity spectrum amplitudes for the frequency and damping ratio of the jth mode. Vanmarcke (1972) has also presented an expression similar to that of Eq. (16.43) for the PSDF of the response of MDOF structures but he did not propose any spectrum superposition method based on that.

All the foregoing response spectrum superposition methods are based on the assumption that the peak factors for the modal responses are all equal to the peak factors for the total structural responses at various storey levels. Depending upon the structural properties and frequency contents of the input excitation, this may be violated to different degrees leading to erroneous results. The method of Singh and Chu (1976) also has the additional requirement of exact relative velocity response spectrum of the input excitation, which is generally not available readily. The use of PSV in place of SV may lead to errors. Therefore, as an improvement, Gupta and Trifunac (1987b, 1998a) have proposed a stochastic spectrum superposition method based on computing the statistics of the various orders of peaks in the structural response, when the peaks are arranged in decreasing order of amplitudes. The probability distributions for the ordered response peaks are defined in terms of the parameters estimated from the first few moments of the PSDF of the response as defined by Eq. (16.43). The PSDF of ground acceleration, $G(\omega)$, needed for this purpose has been taken to be compatible with the input displacement response spectrum, as obtained from the iterative method due to Gupta and Trifunac (1998b).

7.1 Stochastic Spectrum Superposition Method

If the N local peaks of a stochastic response time-history $x_i(t)$ at the ith storey level of a MDOF structure are arranged in decreasing order of amplitudes, the probability that the nth ordered peak of the response time history normalized by its root mean-square (rms) value a_{rms} will exceed a specified value η is given by (Gupta and Trifunac, 1988)

$$F_{(n)}(\eta) = \sum_{i=n}^{N} \binom{N}{i} [P(\eta)]^i [1 - P(\eta)]^{N-1}; \quad \eta(t) = \frac{x_i(t)}{a_{\text{rms}}} \tag{16.46}$$

From this, the probability density function of the nth order peak can be obtained as

$$f_{(n)}(\eta) = -\frac{dF_{(n)}(\eta)}{d\eta} = n\binom{N}{n}(P(\eta))^{n-1}(1 - P(\eta))^{N-n}p(\eta) \tag{16.47}$$

The functions $p(\eta)$ and $P(\eta)$ in Eqs. (16.46) and (16.47) represent the probability density and the cumulative probability of local maxima of $\eta(t)$, defined as (Rice 1944, 1945)

$$p(\eta) = \frac{1}{\sqrt{2\pi}}\left[\varepsilon e^{-\frac{1\eta^2}{2\varepsilon^2}} + (1 - \varepsilon^2)^{\frac{1}{2}}\eta e^{-\frac{1}{2}\eta^2}\int_{-\infty}^{\frac{\eta(1-\varepsilon^2)^{1/2}}{\varepsilon}} e^{-\frac{1}{2}t^2}\,dt\right] \tag{16.48}$$

$$P(\eta) = \int_{\eta}^{\infty} p(u)du = \frac{1}{\sqrt{2\pi}} \left[\int_{\frac{\eta}{e}}^{\infty} \mathrm{e}^{-\frac{1}{2}t^2} dt + \left(1 - e^2\right)^{\frac{1}{2}} \mathrm{e}^{-\frac{1}{2}\eta^2} \int_{-\infty}^{\frac{\eta(1-e^2)}{e}} \mathrm{e}^{-\frac{1}{2}t^2} dt \right] \quad (16.49)$$

From the probability density function of Eq. (16.47), the expected value of the *n*th order peak of $\eta(t)$ can be evaluated by numerical integration. Also, the expression of Eq. (16.46) can be used to obtain the *n*th order peak amplitude with any desired confidence level.

The statistical parameter a_{rms}, ε, and N in the above formulation are defined in terms of the zero-, second-, and fourth-order moments m_{0i}, m_{2i}, and m_{4i} of PSDF of $r_i(t)$ as given by Eq. (16.43) and the total duration, T, of the response taken equal to the duration of input excitation:

$$a_{\mathrm{rms}} = \sqrt{m_{0i}}, \quad \varepsilon = \sqrt{1 - \frac{m_{2i}^2}{m_{0i} m_{4i}}}, \quad \text{and} \quad N = \frac{T}{\pi} \sqrt{\frac{m_{4i}}{m_{2i}}} \quad (16.50)$$

Thus, the stochastic method can be used to compute amplitudes of several significant peaks of the structural response, but only the maximum values (first-order peaks) of the displacement, shear force, and bending moment responses at various storey levels of the example buildings have been computed in this chapter for the purpose of comparison with the other response spectrum superposition methods.

8 Numerical Results and Discussion on Improved Methods

To investigate the relative performance of the various advanced spectrum superposition methods described in the previous section, the percentage errors with respect to the exact time-history solutions have been evaluated for the displacement, shear force, and bending moment responses at various storey levels of the asymmetric five-storey example buildings. Figures 16.8 and 16.9 show the plots of these errors for the selected broadband (Rec. # 016) and narrowband (Rec. # 186) excitations, respectively.

Figures 16.8 and 16.9 present the comparison of results obtained from three different response spectrum superposition methods due to Rosenblueth and Elorduy (1969), Wilson et al. (1981), and DerKiureghian and Nakamura (1993) based on the generalized Eq. (16.36), the spectrum superposition method of Singh and Chu (1976) defined by Eq. (16.45), and the stochastic method due to Gupta and Trifunac (1998a) based on the response spectrum compatible PSDF of input ground motion. These figures also show the PSDF used for stochastic method along with the location of the modal frequencies of the example building.

From the results in Figs. 16.8 and 16.9 and similar results for many other input excitations with widely differing characteristics, none of the methods are seen to provide the same quality of agreement with the exact time-history solutions for all the response quantities and the input excitations. However, the stochastic method is,

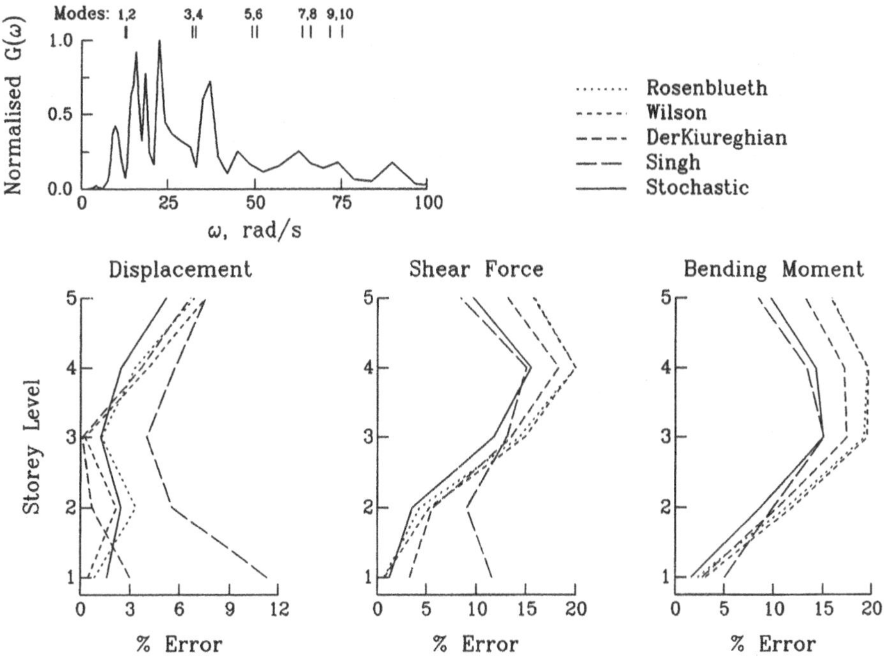

Fig. 16.8 Comparison of errors in various improved spectrum superposition methods for the response amplitudes of asymmetric building under broadband excitation of Rec. # 016

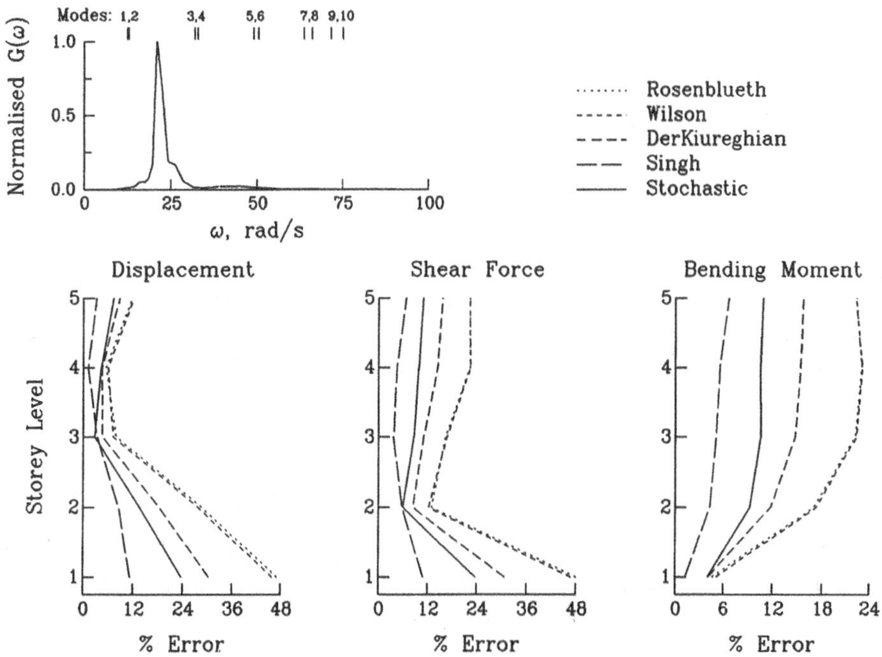

Fig. 16.9 Same as Fig. 16.7 for narrowband excitation of Rec. # 186

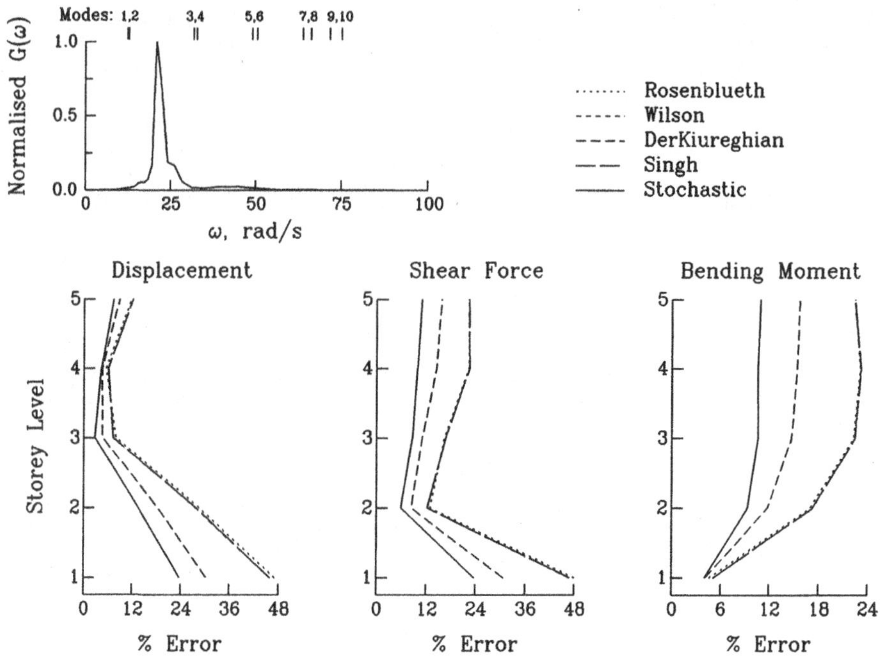

Fig. 16.10 Comparison of the performance of Singh and Chu's method with other improved methods for the narrow band Rec. # 186 when SV is approximated by PSV

in general, seen to provide consistently good results. For example, this method can be considered the best for the case of broadband excitation in Fig. 16.8 and the second best for the narrowband excitation in Fig. 16.9. However, in Fig. 16.9 also, it is not much different from the best one due to the method of Singh and Chu (1976).

However, the method of Singh and Chu (1976) has the additional requirement of the exact SV spectrum, which is difficult to meet in practical applications. As seen from the results in Fig. 16.10 for the case of narrowband excitation of Rec. # 186, the method of Singh and Chu gives very larger errors if SV is approximated by PSV. The larger errors observed in the results from the three methods based on the generalized Eq. (16.36) compared to the stochastic method may be related mainly to the assumptions and idealizations made in estimating the correlation coefficient ρ_{jk}, which are violated to different extents for different response quantities and input excitations. Further, all the other methods are based on the assumption that the peak factors for the total storey responses and the modal responses are all identical, which is not true.

It may be noted that correlation coefficient ρ_{jk} is not always high and positive for closely spaced modes and negligible for widely spaced modes. Joshi and Gupta (1998) have illustrated that for narrow band excitation, interaction among all the modes with frequencies outside the range of input excitation is large and positive. Interaction of modes above the input frequency range with those below the significant range of input frequency is also large but negative. Thus, the correlation factor for narrowband excitations may be large with both positive and negative signs, even

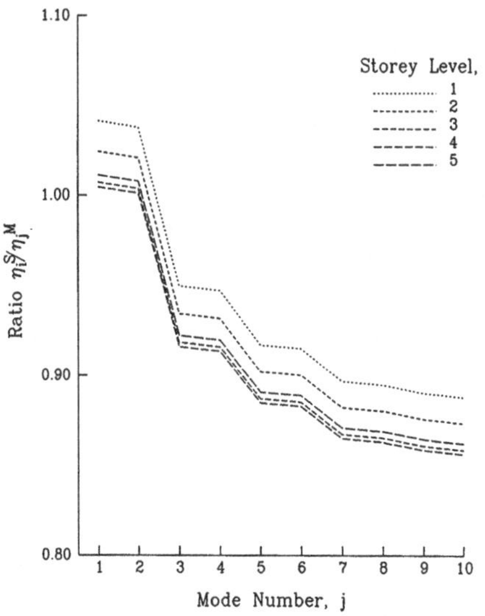

(a) Broad Band Rec.# 016

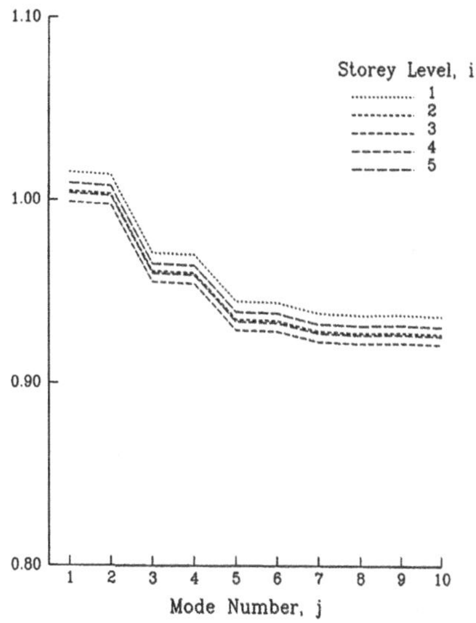

(b) Narrow Band Rec.# 186

Fig. 16.11 Variation in the ratio of peak factor η_i^S for total structural response at ith storey and η_j^M for jth modal response

for widely separated modes. This aspect is taken care of well in the stochastic method only.

Joshi and Gupta (1998) have indicated that the ratio, η_i^S/η_j^M, of the peak factors for the ith storey and the jth modal responses may be quite different from unity as illustrated in Fig. 16.11. This ratio is seen to increases from base (storey-5) to top (storey-1), and it is greater than unity for all the storey levels for the first few domination modes. It then decreases very fast and becomes much smaller than unity for higher modes. Thus, depending upon the relative contributions of various modes to the total structural response at a particular storey level and the signature of the correlation coefficient ρ_{jk}, the assumption $\eta_i^S/\eta_j^M \approx 1.0$ made in various spectrum superposition methods may lead to over or under estimation of the true response amplitudes. The stochastic method being free from these approximations is able to give equally good results in all the cases.

9 Response Spectrum Superposition Methods with Limited Modes

The exact time-history solution as well as the various response spectrum superposition methods described so far needs to carry out detailed computation for all the modes of MDOF structure, which becomes impractical for real structures having very large number modes. Also, it is difficult to evaluate accurately the frequencies and mode shapes of the high-frequency modes. Therefore, in practical applications, only a limited number of lowest modes are included in the analysis, and the contribution of higher modes is generally neglected. However, the truncation of higher modes may cause significant error for structures with significant modes outside the range of input frequencies, which are termed as rigid modes. To get accurate results in such cases by using only a limited number of modes explicitly, Singh and Mehta (1983) proposed a spectrum superposition method in terms of the relative velocity and the relative acceleration response spectrum amplitudes. To obviate the inconvenient need of relative acceleration response spectrum, Singh and Maldonado (1991) developed a modal combination method for the case of truncated modes, which needs the relative displacement and relative velocity spectrum amplitudes. This and some other methods for structures with rigid modes are described in the following.

If all the modes of a structure are numbered in increasing order of frequency and if those beyond the *m*th mode can be treated as rigid, solution of Eq. (16.15) for modes ($m + 1$) to n can well be approximated by $-\alpha_j \ddot{z}(t)/\omega_j^2$, and the exact time history of total displacement response at the ith storey level of Eq. (16.17) can be rewritten as

$$x_i(t) = \sum_{j=1}^{m} A_{ij}\,\xi_j(t) - \sum_{j=m+1}^{n} \frac{\alpha_j A_{ij}}{\omega_j^2}\ddot{z}(t) \tag{16.51}$$

To eliminate the modal properties of the rigid modes from this expression, it is assumed that they respond to the ground acceleration as a rigid body. Their contribution can thus be accounted by finding the quasi-static response vector, $\{X_S\}$, of the structure to a constant acceleration of unit amplitude as:

$$\{X_S\} = [K]^1[M]\{I\} \tag{16.52}$$

Also, from Eqs. (16.17) and (16.15), the quasi-static response at *i*th storey level, X_{Si}, can be expressed in terms of the modal properties as

$$X_{Si} = \sum_{j=1}^{m} \frac{\alpha_j A_{ij}}{\omega_j^2} \tag{16.53}$$

Thus, the second term in Eq. (16.51) can be expressed in terms of the static solution X_{Si} and the dynamic properties of only the first m modes as

$$x_i(t) = \sum_{j=1}^{m} A_{ij}\,\xi_j(t) - C_{Si}\ddot{z}(t); \quad \text{with} \quad C_{Si} = X_{Si} - \sum_{j=1}^{m} \frac{\alpha_j A_{ij}}{\omega_j^2} \tag{16.54}$$

This expression can also be used to obtain the shear force and the bending moment responses by replacing A_{ij} with S_{ij} and B_{ij} as given by Eq. (16.23).

Similar to Eq. (16.43), by taking the complex Fourier transform of the expression of Eq. (16.54), Singh and Maldonado (1991) obtained an expression for the PSDF of the displacement response at ith storey of a MDOF building as follows:

$$G_{x_i}(\omega) = \sum_{j=1}^{m}\sum_{k=1}^{m} \alpha_j A_{ij} \alpha_k A_{ik}\left(B_{jk} + \frac{\omega^2}{\omega_j^2} C_{jk}\right) G(\omega)\left|H_j(\omega)\right|^2 + C_{Si}^2 G(\omega) \\ 2C_{Si}\sum_{j=1}^{m} \alpha_j A_{ij}\left(\omega^2 \quad \omega_j^2\right) G(\omega)\left|H_j(\omega)\right|^2 \tag{16.55}$$

with B_{jk} and C_{jk} defined by Eq. (16.44a). The zeroth moment of the PSDF of Eq. (16.55) gives the mean-square displacement response at ith storey of the structure. The absolute response amplitude can be obtained by multiplying the root-mean-square response with an appropriate peak factor. Assuming that the peak factor for the responses at various storey levels doesn't differ significantly from the peak factors for the ground acceleration and those for the relative displacement and relative velocity responses for all the modes of the structure, the PSDF of Eq. (16.55) has been used by Singh and Maldonado (1991) to write an expression for the maximum displacement response in terms of the peak ground acceleration $a_{\max}$, relative displacement spectrum $\mathrm{SD}_j = \mathrm{SD}(\omega_j, \varsigma_j)$, and relative velocity spectrum $\mathrm{SV}_j = \mathrm{SV}(\omega_j, \varsigma_j)$ as

$$(x_i)_{\max} \\ = \left[\sum_{j=1}^{m}\sum_{k=1}^{m} \alpha_j A_{ij} \alpha_k A_{ik}\left(B_{jk}\mathrm{SD}_j^2 + \frac{C_{jk}}{\omega_j^2}\mathrm{SV}_j^2\right) + C_{Si}^2 a_{\max}^2 - 2C_{Si}\sum_{j=1}^{m} \alpha_j A_{ij}\left(\mathrm{SV}_j^2 - \omega_j^2 \mathrm{SD}_j^2\right)\right]^{1/2} \tag{16.56}$$

From the PSDF of Eq. (16.55), DerKiureghian and Nakamura (1993) defined another expression for maximum displacement response as follows:

$$(x_i)_{\max} = \left[\sum_{j=1}^{m}\sum_{k=1}^{m} \alpha_j A_{ij} \alpha_k A_{ik} \mathrm{SD}_j \overline{\rho}_{jk} \mathrm{SD}_k + C_{Si}^2 a_{\max}^2 - 2C_{Si}\sum_{j=1}^{m} \alpha_j A_{ij}\left(\frac{\lambda_{2,j}}{\lambda_{0,j}} - \omega_j^2\right)\mathrm{SD}_j^2\right]^{1/2} \tag{16.57}$$

In this expression $\overline{\rho}_{jk}$ is a new correlation coefficient, defined as

$$\overline{\rho}_{jk} = \left(B_{jk}\lambda_{0,j} + C_{jk}\frac{\lambda_{2,j}}{\omega_j^2}\right) / \sqrt{\lambda_{0,j}\lambda_{0,k}} \tag{16.58}$$

with

$$\lambda_{k,j} = \int_0^{\infty} \omega^k G(\omega)\left|H_j(\omega)\right|^2 d\omega; \quad k = 0\&2 \tag{16.59}$$

To evaluate the integrals in Eq. (16.59), DerKiureghian and Nakamura have proposed to use the approximate PSDF of ground acceleration defined in terms of displacement response spectrum by Eq. (16.42).

Gupta and Chen (1984) have presented a spectrum superposition method for structures with rigid modes, which can also be written in a form very similar to that of Eqs. (16.56) and (16.57) as

$$(x_i)_{\max} = \left[\sum_{j=1}^{m}\sum_{k=1}^{m} \alpha_j A_{ij} \alpha_k A_{ik} \mathrm{SD}_j \overline{\rho}_{jk} \mathrm{SD}_k + C_{Si}^2 a_{\max}^2 - 2C_{Si} \sum_{j=1}^{m} \alpha_j A_{ij} \gamma_j \mathrm{SD}_j^2\right]^{1/2} \tag{16.60}$$

In this expression, γ_j, termed commonly as rigid response coefficient, is the correlation between jth mode response and the input acceleration:

$$\gamma_j = \frac{\ln\left(\omega_j/\omega^1\right)}{\ln\left(\omega^2/\omega^1\right)}; \quad \text{with} \quad \omega^1 = \frac{\mathrm{SA}_{\max}}{\mathrm{SV}_{\max}} \quad \text{and} \quad \omega^2 = \left(\omega^1 + 2\omega^r\right)/3 \tag{16.61}$$

Here ω^r represents the rigid frequency beyond which the spectral acceleration is equal to the maximum ground acceleration. Gupta and Chen (1984) have defined the new correlation coefficient, $\overline{\rho}_{jk}$, in terms of γ_j as

$$\overline{\rho}_{jk} = \gamma_j \gamma_k + \sqrt{1-\gamma_j^2}\sqrt{1-\gamma_k^2}\rho_{jk} \tag{16.62}$$

where ρ_{jk} is the correlation coefficient of Eq. (16.39) as modified empirically by Gupta and Cordero (1981).

All the foregoing response spectrum superposition methods considering only a limited number of modes explicitly are based on varying degrees of approximations and idealizations. They also need additional parameters like peak ground acceleration and exact SV and SA spectral amplitudes, which are generally not available readily in practical applications. To improve upon the various approximations and to obviate the need for additional parameters, Gupta and Joshi (1998) have proposed to use the stochastic approach of Sect. 7.1 for obtaining the amplitudes of various orders of response peaks using the PSDF of Eq. (16.55) considering rigid modes also. As the stochastic method is based on the response spectrum compatible PSDF, this can be considered equivalent to the other spectrum superposition methods.

To illustrate the relative performance of the various methods with rigid modes, the five-storey asymmetric building as shown in Fig. 16.5 is considered to compute the example results. However, to have the modal frequencies outside the range of the frequency range of input excitations, the building has been made rigid by increasing the storey stiffnesses by a factor of 10. The modal frequencies of this rigid structure are found to be 38.1, 39.5, 99.7, 102.7, 154.1, 158.6, 200.6, 208.1, 225.5, and 236.7 rad/s. Figure 16.12a, b shows typical results for shear force response at selected storey levels of the example building to the input excitations as Rec. #

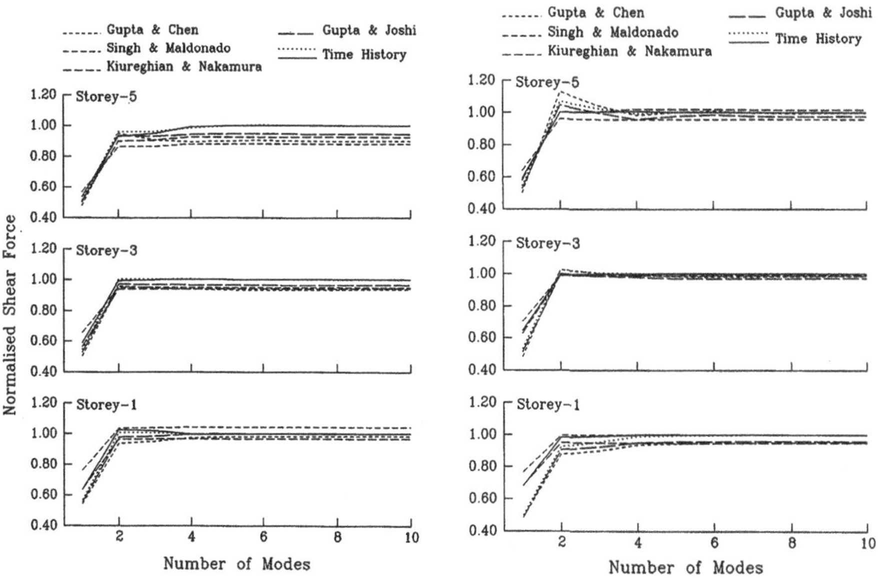

(a) Broad Band Rec.# 016 **(b) Narrow Band Rec.# 186**

Fig. 16.12 Comparison between the shear force response obtained from different response spectrum superposition methods with limited modes and the time-history solution considering the effect of higher modes by quasi-static method (solid line). The dotted line is the time-history response with the contribution of higher modes neglected

016 and 186, respectively. As the modal shear force response is proportional to the modal displacement response multiplied by square of modal frequency, the contributions of higher modes is expected to be larger in the shear force response.

Results obtained from the response spectrum superposition methods due to Gupta and Chen (1984), Singh and Maldonado (1991), DerKiureghian and Nakamura (1993), and the stochastic method are compared with the exact time-history solutions in these figures. The response amplitudes are plotted as a function of the number of modes considered explicitly and are normalized by the exact time-history solution corresponding to all the ten modes considered explicitly for detailed calculations. It is seen that the truncation of modes may cause appreciable error in the response amplitudes. The number of modes to be considered explicitly to eliminate the truncation error for the example structure is only about four, because all the higher modes lie outside the frequencies of input excitations. Also, the response amplitudes obtained from various methods are seen to underestimate the exact response by varying degrees, whereas the stochastic method shows closer agreement in maximum number of cases.

10 Conclusions

The response spectrum superposition methods are used very commonly in earthquake engineering applications to obtain the maximum response amplitudes of MDOF structures directly from response spectrum of ground acceleration without carrying out detailed time-history solution. The SUM was the earliest response spectrum superposition method proposed by Biot (1943) to get an upper bound on the maximum response amplitudes. The SRSS method was proposed by Goodman et al. (1958) as an improvement over SUM to get more realistic estimate of maximum response amplitudes. However, the SRSS method was found to underestimate significantly the response amplitudes for structures with closely spaced modal frequencies. Therefore, several ad hoc proposals were initially made by different investigators to account for the modal interaction effects to get more accurate results. Subsequently, response spectrum superposition methods for structures with closely spaced modes were evolved from more rigorous theoretical basis.

Most of the spectrum superposition methods proposed by different investigators to compute the response of multi-degree-of-freedom structures with closely spaced modes include the modal interaction effects via the correlation coefficients, defined in various approximate ways. Further, to express the final results directly in terms of the response spectrum amplitudes, it is assumed that the peak factors for the modal responses are all equal to those for the total responses at various storey levels of the structure. Thus, the various methods show varying levels of errors compared to the exact time-history solution. Further, for real structures with very large number of degrees of freedom, the contribution of only a limited number of lowest modes is included, which may lead to significant error for stiff structures with significant modes outside the range of input frequencies. Response spectrum superposition method has been also extended for such structures by considering the contribution of rigid modes by quasi-static response to a static acceleration of unit amplitude. Additional approximations and idealizations are made to propose these methods, leading to further error in the results.

The stochastic method due to Gupta and Trifunac (1996) is free from the various approximations and thus provides the results in good agreement with the exact solution in majority of cases. Further, the stochastic method can compute the response for a specified confidence level, whereas the other methods give only the expected response amplitudes. To design the structures with different importance, it may be useful to have the response estimates with different confidence levels. Also, without any significant additional computational efforts, the stochastic method can predict the amplitudes of all the significant peaks of the response (Gupta and Trifunac, 1996) and not just the highest peak as is done by the other methods. The knowledge of several highest-amplitude response peaks can provide an idea about the progress of damage in a ductile structure, as it is subjected to increasing number of excursions in the nonlinear range (Basu and Gupta, 1995).

References

Basu, B., & Gupta, V. K. (1995). A probabilistic assessment of seismic damage in ductile structures. *Earthquake Engineering and Structural Dynamics, 24*(10), 1333–1342.

Biot, M. A. (1933). Theory of elastic systems vibrating under transient impulse, with an application to earthquake proof buildings. *Proceedings of the National Academy of Sciences of the United States of America, 19*, 262–268.

Biot, M. A. (1941). A mechanical analyzer for the prediction of earthquake stresses. *Bulletin of Seismological Society of America, 31*, 151–171.

Biot, M. A. (1942). Analytical and experimental methods in engineering seismology. *Proceedings of the American Society of Civil Engineers, 68*, 49–69.

Biot, M. A. (1943). Analytical and experimental methods in engineering seismology. *Transactions of the American Society of Civil Engineers, 108*, 365–385. Paper No. 2183, 356.

Caughey, T. K. (1959). Classical normal modes in damped linear systems. *Journal of Applied Mechanics, ASME, 32*, 583–588. 59-A-62.

Chandrasekaran, A. R. & Das J. D. (1993). Strong earthquake ground motion data in EQUINFOS for India: Part 1B, In M. D. Trifunac, M. I. Todorovska, & V. W. Lee (Eds.). Central Water and Power Research Station and Univ. of Southern California, Report No. CE87-02, Pune, India and Los Angeles, California.

Clough, R. W., & Penzien, J. (1975). *Dynamics of structures*. New York, NY: McGraw-Hill.

DerKiureghian, A. (1980). Structural response to stationary excitation. *Journal of the Engineering Mechanics Division, 106*, 1195–1213.

DerKiureghian, A. (1981). A response spectrum method for random vibration analysis of MDOF systems. *Earthquake Engineering and Structural Dynamics, 9*, 419–435.

DerKiureghian, A., & Nakamura, Y. (1993). CQC modal combination rule for high-frequency modes. *Earthquake Engineering and Structural Dynamics, 22*, 943–956.

Goodman, L. E., Rosenblueth, E., & Newmark, N. M. (1958). Aseismic design of firmly founded elastic structures. *Transactions of the American Society of Civil Engineers, 120*, 782–802.

Gupta, A. K., & Chen, D. C. (1984). Comparison of modal combination methods. *Nuclear Engineering and Design, 78*, 53–68.

Gupta, A. K. & Cordero K. (1981). Combination of modal responses, Trans. of 6th Int. Conf. on SMiRT, Paper No. K7/15, Paris, August 1981.

Gupta, I. D., & Joshi, R. G. (1998). Response spectra with uncertain oscillator properties. *European Earthquake Engineering, 12*(3), 30–37.

Gupta, I. D., Rambabu V., & Joshi R. G. (1993). Strong earthquake ground motion data in EQUINFOS for India: Part 1A, In M. D. Trifunac, M. I. Todorovska, & V. W. Lee (Eds.). Central Water and Power Research Station and Univ. of Southern California, Report No. CE87-01, Pune, India and Los Angeles, California.

Gupta, I. D. & Trifunac M. D. (1987a). Statistical analysis of response spectra method in earthquake engineering, Report No. 87-03, Dept. of Civil Engineering, Univ. of Southern California, Los Angeles, USA.

Gupta, I. D., & Trifunac, M. D. (1987b). Order statistics of peaks in earthquake response of Multi-degree-of-freedom systems. *Earthquake Engineering and Engineering Vibration, 7*(4), 15–50.

Gupta, I. D., & Trifunac, M. D. (1988). Order statistics of peaks in earthquake response. *Journal of the Engineering Mechanics Division, 114*(10), 1605–1627.

Gupta, I. D. & Trifunac M.D. (1996). Investigation of nonstationarity in stochastic seismic response of structures, Report No. CE96-01, Dept. of Civil Eng., University of Southern California, Los Angeles, USA.

Gupta, I. D., & Trifunac, M. D. (1998a). An improved probabilistic spectrum superposition. *Soil Dynamics and Earthquake Engineering, 17*, 1–11.

Gupta, I. D., & Trifunac, M. D. (1998b). Defining equivalent stationary PSDF to account for nonstationarity of earthquake ground motion. *Soil Dynamics and Earthquake Engineering, 17*, 89–99.

Housner, G. W. (1941). Calculating the response of an oscillator to arbitrary ground motion. *Bulletin of Seismological Society of America, 31*, 143–149.

Jennings, R.L. (1958). *The response of multi-storied structures to strong ground motion*, M.Sc. Thesis, University of Illinois, Urbana, Illinois.

Joshi, R.G. (1997). *Investigation of Response Spectrum Superposition for Stochastic Response Analysis of Structures*, Ph.D. Thesis, Poona University, 239 pp.

Joshi, R. G., & Gupta, I. D. (1998). On the relative performance of spectrum superposition methods considering modal interaction effects. *Soil Dynamics and Earthquake Engineering, 17*, 357–369.

Kan, C. L., & Chopra, A. K. (1977). Elastic earthquake analysis of torsionally coupled multi-storey buildings. *Earthquake Engineering and Structural Dynamics, 5*, 395–412.

O'Hara, G. J., & Cunnif, P. F. (1963). *Elements of normal mode theory, NRL Report 6002*. Washington, D.C.: U.S. Naval Research Laboratory.

Rice, S. O. (1944). Mathematical analysis of random noise. *The Bell System Technical Journal, 23*, 282–332.

Rice, S. O. (1945). Mathematical analysis of random noise. *The Bell System Technical Journal, 24*, 46–156.

Rosenblueth, E. (1956). Some applications of probability theory in aseismic design, Proceedings of the World Conference on Earthquake Engineering, Earthquake Engineering Research Institute and University of California, Berkeley.

Rosenblueth, E. and J. Elorduy (1969). Response of linear systems to certain transient disturbances, Proc. 4th World Conf. on Earthq. Eng., Chile, A-1, 185–196.

Singh, M. P., & Chu, S. L. (1976). Stochastic considerations in seismic analysis of structures. *Earthquake Engineering and Structural Dynamics, 4*, 295–307.

Singh, M. P., & Maldonado, G. O. (1991). An improved response spectrum method for calculating seismic design response, part 1: classically damped structures. *Earthquake Engineering and Structural Dynamics, 20*, 621–635.

Singh, M. P., & Mehta, K. B. (1983). Seismic design response by an alternative SRSS rule. *Earthquake Engineering and Structural Dynamics, 11*, 771–783.

Trifunac, M.D. (1972). Analysis of strong motion earthquake accelerograms, Report No.EERL 72-80, Earthquake Engineering Research Laboratory at California Institute of Technology, Pasadena, California.

Trifunac, M. D., & Brady, A. G. (1975). A study on the duration of strong earthquake ground motion. *Bulletin of Seismological Society of America, 65*(3), 581–626.

U.S. Nuclear Regulatory Commission (1976). Regulatory guide 1.9.2, Revision 1: Combining modal responses and spatial components in seismic response analysis, Washington, D.C.

Vanmarcke, E. H. (1972). Properties of spectral moments with application to random vibration. *Journal of the Engineering Mechanics Division, 98*, 425–446.

Wilson, E. L., DerKiureghian, A., & Bayo, E. (1981). A replacement for the SRSS method in seismic analysis. *Earthquake Engineering and Structural Dynamics, 9*, 187–192.

Part IV
Disaster Preparedness and Mitigation

Chapter 17
Disaster Risk Reduction for Buildings

Anand S. Arya

1 Introduction

1.1 Statement of the Problem (Impacts of Hazards on the Building Sector)

Historically, India has been prone to the occurrence of damaging earthquakes, floods, cyclones and landslides in different parts of the country which have resulted into damage or total loss of more than 12 lakh housing units on the average every year. Given both the population explosion and increased need of developmental activities exposure, the scale and impact of natural hazards have increased considerably during past three to four decades.

These events, coupled with underdevelopment and changing poverty profiles in most urban centres, lead to catastrophic situations for vulnerable sections of society and the housing stock, both in urban and rural settlements. Building sector is the most commonly affected in any disaster be it earthquake, flood, cyclone, tsunami or landslides. The main reason for this sector to be most vulnerable is that majority of building stock in India comprise of non-engineered category of construction with least capacity to respond to disaster.

Specifically, the disaster risk in India may be described briefly as follows:

A. S. Arya (✉)
Department of Earthquake Engineering, IIT Roorkee, Roorkee, Uttarakhand, India

State Disaster Management Authority, Patna, Bihar, India

M. L. Sharma et al. (eds.), *Advances in Indian Earthquake Engineering and Seismology*, https://doi.org/10.1007/978-3-319-76855-7_17

1.1.1 Earthquake

India's high earthquake risk and vulnerability is clear from the fact that about 59% of India's land area could face moderate to severe earthquakes in seismic zones III, IV and V. During the period 1990–2006, more than 23,000 lives were lost in India due to 6 major earthquakes, which also caused enormous damage to buildings and public infrastructure. These earthquakes include the Uttarkashi earthquake of 1991, the Latur earthquake of 1993, the Jabalpur earthquake of 1997 and the Chamoli earthquake of 1999, followed by the Bhuj earthquake of 26 January 2001 and the Jammu & Kashmir earthquake of 8 October 2005. All these earthquakes established that the casualties were caused primarily due to the collapse of buildings. However, similar high intensity earthquakes in the United States, Japan, etc., do not lead to such enormous loss of lives, as the structures in these countries are built with structural mitigation measures. This emphasizes the need for strict compliance of town planning bylaws and earthquake-resistant building codes in India.

1.1.2 Flood

Floods have been a recurrent phenomenon in India and cause huge losses to lives, housing, properties, livelihood systems, infrastructure and public utilities. India's high risk and vulnerability is highlighted by the fact that 40 million hectares is prone to floods. On an average every year, 7.5 million hectares of land is affected, 1600 lives are lost and 12 lakh kutcha houses are destroyed due to floods. The maximum number of lives (11,316) was lost in the year 1977. The frequency of major floods is more than once in 5 years. Floods have also occurred in areas, which were earlier not considered flood prone.

1.1.3 Cyclone

A long coastline of about 7,500 km of flat coastal terrain, shallow continental shelf, high population density, geographical location and physiological features of its coastal areas makes India, extremely vulnerable to cyclones and its associated hazards like storm tides (the combined effects of storm surge and astronomical tide), high velocity wind and heavy rains. The impact on the east coast of India is relatively more devastating. This is evident from the fact that in the last 270 years, 21 of the 23 major cyclones (with a loss of about 10,000 lives or more) world wide occurred over the area surrounding the Indian subcontinent (India and Bangladesh). This is primarily due to the storm-tides effect in the area.

About 8% of the area in the country is prone to cyclone-related disasters. Recurring cyclones account for large number of deaths, loss of livelihood opportunities, loss of public and private buildings, and severe damage to infrastructure.

1.1.4 Landslide

Landslides in the Himalayan states of India constitute a major natural hazard, which accounts for considerable loss of life and damage to communication routes, human settlements, agricultural fields and forest lands. Their widespread and frequent occurrence causes great damage because of unpredictability in both space and time, innumerable types and mechanisms of slope failure involved. Removal of vegetation and toe erosion have also triggered slides. Torrential monsoon on the hill slopes, with the vegetation cover removed, has been the main causative factor in the Peninsular India namely in Western Ghats and Nilgiris. Human intervention by way of slope cutting has added to this effect.

1.2 Building

As per Census of Housing 2001, "building" is generally a single structure on the ground. Sometimes it is made up of more than one component unit which are used or likely to be used as dwellings (residences) or establishments such as shops, business houses, offices, factories, workshops, worksheds, schools, places of entertainment, places of worship, godowns, stores, etc. It is also possible that buildings which have component units may be used for a combination of purposes such as shop-cum-residence, workshop-cum-residence, office-cum-residence, etc.

1.2.1 Building Sector

We may consider building sector comprising of all buildings as per Census 2001-Housing Series, where buildings are classified in their different ways **"rural and urban"**, based on **"functional uses"**, and as **"permanent, semi-permanent or temporary"**. These are defined below:

1. *Rural–Urban Areas*

   The unit of classification is "town" for urban areas and "village" for rural areas. The definition of urban area includes: (i) All places with a municipality, corporation, cantonment board or notified town area, etc., and (ii) A place satisfying the following three criteria simultaneously: a minimum population of 5,000, at least 75% of male working population engaged in non-agricultural pursuits; and a density of population of at least 400 km^2, and (iii) Towns with population of 1,00,000 and above are called cities.
2. *Uses of Census Houses*

   The different uses of census houses were standardized and grouped into ten categories, as shown below: (i) Residence, (ii) Residence-cum-other use, (iii) Shop/office, (iv) School/college, etc., (v) Hotel/lodge/guest house, etc., (vi) Hospital/dispensary, etc., (vii) Factory/workshop/workshed, etc., (viii) Place of worship, (ix) Other nonresidential use, and (x) Vacant.

1.2.2 Type of Census Houses

These have been classified according to the types of material used in the construction of wall and roof of the house. The basis of their classification is described hereunder:

Permanent Houses: Houses, the walls and roof are made of permanent materials, namely, galvanized iron sheets or other metal sheets, asbestos sheets, burnt bricks, stones or concrete. Roof may be made of any one of the materials, namely, tiles, slate, galvanized iron sheets, metal sheets, asbestos sheets, bricks, stones or concrete.

Temporary Houses: Houses in which both walls and roof are made of materials, which have to be replaced frequently, e.g., grass, thatch, bamboo, plastic, polythene, mud, unburnt bricks or wood. Roof may be made from any one of the temporary materials, namely, grass, thatch, bamboo, wood, mud, plastic or polythene.

Semi-permanent Houses: Houses in which either the wall or the roof is made of permanent material and the other is made of temporary material.

1.3 Disaster Risk Reduction (DRR)

Disaster Risk Reduction: Technical, social or economic actions or measures used to reduce direct, indirect and intangible disaster losses. The expression "disaster risk reduction" is now widely used as a term that encompasses the two aspects of a disaster reduction strategy: *Mitigation and Preparedness.*

Mitigation: Measures aimed at reducing the risk, impact or effects of a disaster or threatening disaster situation.

Preparedness: The state of readiness to deal with a threatening disaster situation or disaster and the effects thereof.

1.4 Mainstreaming DRR

Mainstreaming: Mainstreaming risk reduction describes a process to fully incorporate disaster risk reduction into relief and development policy and practice. It means expanding and enhancing disaster risk reduction so that it becomes normal practice, fully institutionalized within an organization's development agenda. Mainstreaming has to make certain that all the development programs and projects that originate from or are funded by a public or private organization:

- Are designed with full consideration for potential disaster risks and to resist hazard impact.
- Do not inadvertently increase vulnerability to disaster in all sectors: social, physical, economic and environment.
- Are designed to contribute to developmental aims and to reduce future disaster risk.

1.5 Object and Scope of the Chapter

To include the discussion and presentation of the following issues:

1. The primary objective of vulnerability reduction is to reduce avoidable loss of life, livelihoods and property and to safeguard development gains.
2. To increase the capacity of local communities and organizations to prevent, prepare for and respond to the impact of disasters.
3. To develop performance targets and indicators to assist development organizations so as to mainstream risk reduction into development planning.
4. To identify and prioritize methods of mainstreaming risk reduction into institutional practice.

1.6 Scope of the Paper Chapter

The scope includes the following main items:

1. Framework of mainstreaming DRR in housing sector, initially at the policy decision by the authorities
2. Mainstreaming DRR in housing programs and projects undertaken by the state or other organizations

2 Housing Typology in India

2.1 House Types as per 'Census of India, 2001-Housing Series'

The Housing Series (Census of India, 2001) gives the following details of houses based on materials of construction for roofs and walls, see Table 17.1. These are adversely impacted by the natural hazards:

1. *Type of roof*:
   (a) *Pitched or sloping:* including tiles, slate, corrugated iron, zinc or other metal sheets, asbestos cement sheets, plastic polythene, thatch, grass, leaves, bamboo, etc.
   (b) *Flat:* including brick, stone and lime, reinforced brick concrete and reinforced cement concrete
2. *Type of wall*:
   (a) Mud, unburnt bricks, stone laid in mud or lime mortar
   (b) Burnt bricks laid in cement, lime or mud mortar

Table 17.1 (a) Houses by material of roof in India (Census of India, 2001—Housing Data)

		Grass, thatch, bamboo, wood, mud, etc.		Plastic, polythene		Tiles		Slate		G.I., metal, asbestos sheets		Brick		Stone		Concrete		Any other material	
	Total census houses	No. of houses	%	No. of houses	%	No. of houses	%	No. of houses	%	No. of houses	%	No. of houses	%	No. of houses	%	No. of houses	%	No. of houses	%
Rural	17,75,37,513	4,88,12,470	27.5	6,69,815	0.4	6,29,35,397	35.4	23,64,095	1.3	1,86,65,296	10.5	99,27,250	5.6	1,19,06,910	6.7	2,10,61,294	11.9	11,94,986	0.7
Urban	7,15,58,356	45,73,534	6.4	5,03,956	0.7	1,25,91,573	17.6	4,44,565	0.6	1,18,21,919	16.5	41,47,242	5.8	52,46,952	7.3	3,17,77,933	44.4	4,50,682	0.6
Total	24,90,95,869	5,33,86,004	21.4	11,73,771	0.5	7,55,26,970	30.3	28,08,660	1.1	3,04,87,215	12.2	1,40,74,492	5.7	1,71,53,862	6.9	5,28,39,227	21.2	16,45,668	0.7

(b) Houses by Material of Wall in India

		Grass, thatch, bamboo, wood, mud, etc.		Plastic, polythene		Mud, unburnt brick		Wood		G.I., metal, asbestos sheets		Burnt brick		Stone		Concrete		Any other material	
	Total census houses	No. of houses	%	No. of houses	%	No. of houses	%	No. of houses	%	No. of houses	%	No. of houses	%	No. of houses	%	No. of houses	%	No. of houses	%
Rural	17,75,37,513	2,21,62,932	12.5	4,77,498	0.3	6,58,07,212	37.1	23,63,200	1.3	8,76,677	0.5	6,27,15,919	35.3	2,03,47,899	11.5	22,53,979	1.3	5,32,197	0.3
Urban	7,15,58,356	25,74,189	3.6	2,44,278	0.3	79,91,950	11.2	8,33,792	1.2	11,22,001	1.6	4,91,75,710	68.7	51,33,918	7.2	42,86,359	6	1,96,159	0.3
Total	24,90,95,869	2,47,37,121	9.9	7,21,776	0.3	7,37,99,162	29.6	31,96,992	1.3	19,98,678	0.8	11,18,91,629	44.9	2,54,81,817	10.2	65,40,338	2.6	7,28,356	0.3

(c) Cement concrete
(d) Wood or Ekra walling
(e) Corrugated iron, zinc or other metal sheets
(f) Grass, leaves, reeds or bamboo or thatch, plastic polythene, and others

2.1.1 Wall Types

From vulnerability to hazards' impact, building may be categorized as follows (see Table 17.2):

Category—A: Buildings in field-stone, rural structures, unburnt brick houses, clay houses
Category—B: Ordinary brick building, buildings of the large block and prefabricated type, half-timbered structures, building in natural hewn stone
Category—C: Reinforced building, well-built wooden structures
Category—X: Other materials not covered in A, B and C. These are generally light shelter huts.

2.1.2 Roof Types

Category—R1: Light weight (Grass, Thatch, Bamboo, Wood, Mud, Plastic, Polythene, GI Metal, Asbestos Sheets, Other Materials)
Category—R2: Heavy Weight (Tiles, Slate)
Category—R3: Flat Roof (Brick, Stone, Concrete)

3 Framework of Mainstreaming DRR in Housing Sector

3.1 Guidelines by International Organizations

The UNI-SDR defines disaster risk reduction as "the systematic development and application of 'policies', 'strategies' and 'practices' to minimize vulnerabilities and disaster risks throughout a society, to avoid (prevention) or to limit (mitigation and preparedness) the adverse impact of hazards, within the broad context of sustainable development."

The mainstreaming framework provides a functional way to build collaboration between stakeholders in order to reduce the impact of natural disasters by integrating disaster risk reduction measures into development policies. This framework lays responsibility for reducing the impact of disasters on "policy makers", "communities", "non-governmental organizations", and "the private sector". The objective is to develop a way of capturing progress qualitatively and quantitatively, in each

Table 17.2 Distribution of houses by predominant materials of roof and wall and level of damage risk

India												
		Census houses		Level of risk under								
				EQ zone				Wind velocity (m/s)				
				V	IV	III	II	55 & 50	47	44 & 39	33	
Wall/Roof		No. of houses	%	Area in %				Area in %				Flood prone area in %
India				10.9	17.3	30.4	41.4	5.0	40.2	48.0	6.7	7.9
Wall												
A1—Mud & Unburnt Brick Wall	Rural	65,807,212	26.4									
	Urban	7,901,950	3.2									
	Total	**73,799,162**	**29.6**	*VH*	*H*	*M*	*L*	*VH*	*H*	*M*	*L*	*VH*
A2—Stone Wall	Rural	20,347,899	8.2									
	Urban	5,133,918	2.1									
	Total	**25,481,817**	**10.3**	*VH*	*HI*	*M*	*L*	*H*	*M*	*L*	*VL*	*VH*
Total—Category—A		**99,280,979**	**39.9**									
B—Burnt Bricks Wall	Rural	62,715,919	25.2									
	Urban	49,175,710	19.7									
	Total	**111,891,629**	**44.9**	*H*	*M*	*L*	*VL*	*H*	*M*	*L*	*VL*	*H/M*
Total—Category—B		**111,891,629**	**44.9**									
C1—Concrete Wall	Rural	2,253,979	0.9									
	Urban	4,286,359	1.7									
	Total	**6,540,338**	**2.6**	*M*	*L*	*VL*	*VL*	*L*	*VL*	*VL*	*VL*	*L/VL*
C2—Wood wall	Rural	2,363,200	0.9									
	Urban	833,792	0.3									
	Total	**3,196,992**	**1.2**	*M*	*L*	*VL*	*VL*	*VH*	*H*	*M*	*L*	*H*
Total—Category—C		**9,737,330**	**3.9**									
X—Other Materials	Rural	24,049.304	9.7									
	Urban	4,136,627	1.7									
	Total	**28,185,931**	**11.4**	*M*	*VL*	*VL*	*VL*	*VII*	*Π*	*M*	*L*	*VII*

Total—Category—X		**28,185,931**	**11.3**									
Total buildings		**249,095,869**										
ROOF												
R1—Light Weight Sloping Roof	Rural	69,342,567	27.8									
	Urban	17,350,091	7.0									
	Total	**86,692,658**	**34.8**	*M*	*M*	*L*	*VL*	*VH*	*VH*	*H*	*M*	*VH*
R2—Heavy Weight Sloping Roof	Rural	65,299,492	26.2									
	Urban	13,036,138	5.2									
	Total	**78,335,630**	**31.4**	*H*	*M*	*L*	*VL*	*H*	*M*	*L*	*VL*	*H*
K3—Flat Roof	Rural	42,895,454	17.2									
	Urban	41,172,127	16.5									
	Total	**84,067,581**	**33.7**	*Damage. Risk as per that for the Wall supporting it*								
Total buildings		**249,095,869**										

Housing category: Wall types
Category—A: Buildings in field stone, rural structures. unbumt brick houses, clay houses
Category—B: Ordinary brick building: buildings of the large block & prefabricated type, half-timbered structures, building in natural hewn stone
Category—C: Reinforced building, well-built wooden structures
Category—X: Other materials not covered in A.B.C. These are generally light.
Notes: (1) Flood prone area includes that protected area which may have more severe damage under failure of protection works. In some other areas the local damage may be severe under heavy rains and chocked drainage. (2) Damage Risk for wall types is indicated assuming heavy flat roof in categories A, B and C (Reinforced Concrete) building. (3) Source of Housing Data: Census of Housing GOI 2001
Housing category: Roof type
Category—R1 Light Weight (Grass, Thatch. Bamboo, Wood, Mud, Plastic, Polythene, GI Metal, Asbestos Sheets, Other Materials)
Category—R2 Heavy Weight (Tiles, Slate)
Category—R3 Flat Roof (Brick, Stone, Concrete)
EQ Zone V: Very High Damage Risk Zone (MSK > IX)
EQ Zone IV: High Damage Risk Zone (MSK VIII)
EQ Zone III: Moderate Damage Risk Zone (MSK VII)
EQ Zone II: Low Damage Risk Zone (MSK < VI)
Level of risk: *VH* = Very High, *H* = High, *M*= Moderate, *L*= Low, *VL*=Very low

thematic area that contributes to reduction of "identified risks", The thematic areas include "governance", "preparedness and emergency management", "risk identification", "knowledge management and risk management application" (vulnerability reduction).

3.2 Governance

According to the *World Disasters Report*, effective and accountable local body authorities namely; municipal corporations, development authorities, municipal councils, nagar panchayats, zila panchayats, rural panchayats, cantonment boards etc., are the most important institutions and their "governance structure" is most critical for reducing the impact of natural and human-induced disasters in human settlements whether urban or rural.

The governance model in a local body is largely responsible for policies, planning and implementation of regulatory systems which, if taken cared of properly, can help in reduction of social, physical and economic vulnerability for the communities, housing and infrastructure in the areas under their control. The local bodies can achieve this by proper planning and by ensuring adherence to approved land uses, building by-laws and regulations and promoting construction activities which address disaster safety aspects adequately. Disaster risk reduction has to be considered as a local priority for which local bodies should strengthen their institutional base and build internal capacities for properly utilizing the disaster risk information while approving project proposals submitted with them.

3.2.1 Risk Assessment/Risk Identification

Collating and validating information from past large events, loss studies and developing geographically precise risk data is a major challenge for those responsible for "risk assessment". Quality of data and information (processed or raw) is important for assessment of risks. A high level of accuracy and detail is generally possible through maps, remote sensing and GIS. In India carrying out new studies may not be necessary in every case as adequate information is available on maps obtainable from remote sensing organizations and hazard zoning maps prepared by BIS, GSI, IMD etc. available in public domain. Sometimes simulations and scenarios can also be useful in assessing how the proposed project might accentuate or mitigate hazards in the project area.

Capacity building of project planners is necessary for information collection and its analysis so that collected information leads to proper "risk assessment" and is so formatted as to guide decision making for the housing program and project.

3.3 Knowledge Management

Natural hazard information helps project planners in many ways i.e. recognize and understand hazards in the areas identify knowledge gaps, identify risks to the project from natural hazards now and in future and make decisions about how to deal with those risks.

3.3.1 Risk Management

ISDR recognizes that "Disaster Risk Management (DRM) is a set of processes, planning actions, policies and legal and institutional arrangements aimed at managing and eventually reducing the effects of hazardous events on the human and physical assets of a Community and minimizing the impacts of these hazards on the delivery of essential services to the population."

Disaster risk management should also be recognized as a professional practice, requiring its own processes, trained professionals, experience and culture. In most developing countries, it is an emerging practice often in need of experience, investment and capacity development.

The priorities for action recommended in Hyogo Framework emphasize that it is necessary to develop "sector-specific risk reduction plans" by integrating national plans and programs of every sector and area of development. In the housing sector probably area like land-use planning, the location of critical infrastructure, the management of natural resources, the protection of key assets appear to be important to ensure that risk is identified and reduced at all stages from planning to implementation.

Risk management, therefore, needs to be seen as a systematic approach and practice of managing uncertainty and potential losses, involving risk assessment and analysis and the development of strategies and specific actions to control and reduce risks and losses.

Risk from natural hazards in the housing sector most often can be addressed by preventive measures, such as avoiding settlement in flood plains and building disaster resistant buildings; monitoring, early warning and response measures to manage extreme events; and risk transfer including insurance, to cope with unavoidable impacts.

3.4 Vulnerability Reduction

It is to be seen as an important tool to achieve risk reduction targets. Vulnerability reduction has to be achieved by integrating Vulnerability and Capacity Assessment (VCA) into the project planning process and adequately factoring natural hazards and disasters into the process. VCA considers a wide range of environmental, economic, social, cultural, institutional and political pressures that create vulnerability.

3.5 Disaster Preparedness for Effective Response

The Yokohama Strategy and that of ISDR, both emphasize that strengthening local resilience and capacity to prepare and respond to natural hazard events is essential, it will need to be supported by national humanitarian assistance, such as the medical, water, shelter and other forms of technical and logistical support.

The Hyogo Framework for Action 2005–2015 provides the international foundation for reducing disaster risks as agreed by Governments at the World Conference on Disaster Reduction in Jan 2005, sets out five priorities for action; (i) promoting community participation in disaster risk reduction, (ii) recognizing the differing vulnerabilities and capacities of disadvantaged and disabled sections of population, (iii) conducting assessments of changing hazards, vulnerabilities, risk and capacities to provide priorities for intervention, (iv) strengthening of early warning systems to ensure that warnings reach all populations and enabling people's preparedness to respond to emergencies, and (v) updating emergency preparedness programs and contingency plans for effective response to disasters, supported by legislation, resources and coordination mechanisms.

4 Disaster Risk Reduction in the Project Cycle Management

There are four distinct phases considered in the management of the total project cycle stated briefly below:

Phase 1: Project identification: This phase involves the starting idea of the project, consideration of disaster proneness of the project location and conducting quick risk appraisal of the project. If the risk appraisal is found satisfactory, the project preparation is undertaken. If the quick appraisal is not found satisfactory, more detailed assessment of the risks will have to be carried out.

Phase 2: Project preparation: In this stage disaster risk reduction measures are worked out and cost thereof estimated. The steps to be taken for reducing the risk are detailed in the plan.

Phase 3: Project implementation: In this stage the project is taken up for implementation and execution. The resilience of the various project components under the stipulated hazard occurrences has to be checked continuously.

Phase 4: Project evaluation: At the completion of the project, it needs to be evaluated if the targets of disaster risk reduction have been met and the project will not have any adverse disaster impacts on the communities and the environment.

5 Awareness, Sensitization and Capacity Building

Implementation of disaster risk reduction policies, strategies and guidelines can be successfully accomplished through concerted efforts on the part of all stakeholders of housing and human settlement development sector and construction industry. The list of stakeholders, though not exhaustive, includes general public, building owners, engineers, architects, developers, builders, contractors, building managers, functionaries of concerned government department, fire and police departments, local bodies and managers of life line buildings and facilities. The other group of stakeholders includes, policy makers (elected representatives) and bureaucrats at higher levels of decision-making and governance.

5.1 Awareness

The important measures required for creating awareness among stakeholders include; (i) improving understanding of hazard status and threats in their area, (ii) preparing the general public on how to organize themselves before, during and after disasters, and (iii) equipping the professional groups on technical strategies for developing, upgrading and maintaining disaster-resistant built environment.

It is the responsibility of the national, state and local governments to continuously inform and upgrade/update information on disaster safety aspects. The city governments should also utilize print and electronic media (radio, television, Internet) to communicate with the population and strengthen the awareness campaign.

5.2 Sensitization

It is common observation that many groups in civic society do not even understand the hazard status and threats to their settlements. Thus a comprehensive awareness and sensitization campaign should be institutionalized to emphasize periodically about the prevalent hazard risk of their area and the role of individual stakeholders to address the same in his area of professional work. Widespread availability of information relating to different facets of natural hazards, their potential impact on built environment, socio-economic life of community is to be organized with the help of media. All sections of general public and professional groups in disaster prone regions and urban and rural settlements must be adequately sensitized about the problems of hazards and about the issues concerning the disaster safety.

It is also useful to undertake sensitization programs by state and local governments in schools, identifying and linking the schools with the relevant government departments (fire, police, hospitals, etc.).

Sensitizing the general public should be done carefully and with sensitivity. Services of professional media companies should be sought to develop mass media sensitization tools.

5.3 Capacity Building

Once the group of engineers and architects have been sensitized, programs need to be developed for training them in the basic concepts of safety against hazards in the design and construction. Tailor-made training programs could be developed by experts for intense training of these professionals. Such short-term training programs for practicing architects, engineers, design groups/consultants, municipal engineers and those from public sector departments and companies (PSUs) who are dealing with building construction projects, should be conducted by professional organizations like Indian Institute of Architects (IIA), Institution of Engineers (IE), Construction Industry Development Council (CIDC), National Institute of Disaster Management (NIDM), Building Materials and Technology Promotion Council (BMTPC) and academic institutions like IITs etc.

The training programmes for above mentioned professionals should be designed to cover various aspects: (*i*) *Design, construction and supervision of new structures constructed with different material*, (*ii*) *Safety evaluation and strengthening of existing buildings*, (*iii*) *Retrofitting of buildings constructed with different materials*, (*iv*) *Behavior of non-engineered dwellings* and disaster safe houses built using traditional methods and wisdom as well as (*v*) *Reinforce concrete and steel frame buildings and infrastructures.*

6 Conclusion

The process of disaster risk reduction starts with assessment of hazards and preparation of country maps followed by study of the components of the habitat in general and building types in particular. Vulnerability assessment of the physical components as well as the communities is the next important component. A combination of the hazards, the exposure and the vulnerabilities leads to assessment of risk and its mapping. The Vulnerability Atlas of India, 1997 and 2006, presents this information on macro-scale.

The frame work of mainstreaming disaster risk reduction involves: policy and strategy development by the national and state governments to be followed by the "local body authorities" through proper governance using a process of knowledge management, management of risk, creating awareness and sensitization at the various levels and in various sectors of the society. The most important component of achieving the desired results of DRR is building of capacities at various levels of the society including administration, professional engineers and architects and the building technicians through education and specialized training. The task of mainstreaming is complex and vast and could be accomplished only with concerted effort and cooperation from all state holders of the society.

References

Census of India (2001). *Housing Series, Registrar Census, Govt. of India.*

Vulnerability Atlas of India. (1997). 2nd ed., Building Materials & Technology Promotion Council. Ministry of Housing & Urban Poverty Alleviation, Govt. of India.

Vulnerability Atlas of India. (2006). 2nd ed., Building Materials & Technology Promotion Council. Ministry of Housing & Urban Poverty Alleviation, Govt. of India.

Chapter 18
Recent Trends in Disaster Mitigation and Management: *Vulnerability Atlas of India*

Shailesh Kr. Agrawal and Dalip Kumar

1 Introduction

Indian subcontinent is being visited by natural disasters of varying intensities since time immemorial, inflicting huge losses to life and property. Even a small event, turns into a disaster in India primarily because of our relief-centric approach. It was the year 2005, when the Government of India enacted Disaster Management act which empowered all agencies to change its approach toward disasters from 3Rs, i.e., response, recovery, and reconstruction, to 3Ps approach, i.e., proactive, prevent, and prepare. Building Materials and Technology Promotion Council BMTPC although mandated to transfer appropriate technologies from lab to land, took disaster mitigation and management into its stride and has been striving to mainstream disaster risk reduction in housing for the last two decades.

As per (National Disaster Management Authority NDMA report, about 58.6% land area of India is vulnerable to earthquakes of moderate to very high intensity; about 12% of land (over 40 million Ha) is prone to floods and river erosion; 5700 km of coas tline out of 7516-km-long coast line is prone to cyclones, windstorm surges, and tsunamis; 68% of cultivable area is vulnerable to drought; and hilly areas are at risk from landslides and avalanches. As such the vulnerability to these natural phenomena exists, and one cannot do much about it; however, it is the man-made habitat which poses much danger and causes irreparable losses to life and property. This is more aggravated by exponential rise in population, urbanization and hackneyed construction practices.

Occurrence of natural hazards adversely affects the existing housing and building stock, and it is estimated that about 1.2–1.5 million houses get damaged by one natural hazard or the other every year. It is high time that we recognize the risk to the existing

S. K. Agrawal (✉) · D. Kumar
Building Materials and Technology Promotion Council, New Delhi, Delhi, India
e-mail: ska@bmtpc.org

M. L. Sharma et al. (eds.), *Advances in Indian Earthquake Engineering and Seismology*, https://doi.org/10.1007/978-3-319-76855-7_18

housing and infrastructure and at the same time take preventive measures in the future constructions. The institutions who are major players in the area will have to come forward and contribute relentlessly hand in hand with other stakeholders and institutions, which is obviously not happening at this moment of time. Each institution is tackling it with its own perspective and there is no synergy among these institutions.

Under the umbrella of Ministry of Housing and Urban Poverty Alleviation, Government of India, BMTPC can boost of playing a major role toward national efforts for disaster mitigation and management. The BMTPC's role entails both in the pre-disaster and post-disaster stages. Due to a qualitative shift in the national policy for improving preparedness at all levels as a proactive strategy to reduce the impact of disasters as against the traditional approach of responding to natural disasters after the natural hazard has struck in a particular region, the role of BMTPC has further become more important in helping the state governments to evolve proactive strategies, methods, and approaches for risk and vulnerability reduction of buildings and strengthening the institutional infrastructure to mitigate the effects of natural disasters.

The chapter elaborates one of the major contributions of BMTPC, i.e., bringing out *Vulnerability Atlas of India*, which till date is the only document existing on damage risk to housing stock in India w.r.t. natural hazards, i.e., earthquake, wind and cyclone, and flood. The atlas was first published way back in 1997, and then in 2006, 2008 (CD form) based on 2001 Census data and at the time of writing this paper, the atlas is being updated based on new available data from Survey of India, GSI, and Census 2011. Also, recently, the *Vulnerability Atlas of India* (2006) is being hosted by NIC on government portal for better advocacy and outreach.

2 *Vulnerability Atlas of India*: Introduction

BMTPC in 1997 brought out the *first of its kind Vulnerability Atlas of India* covering the whole country with respect to earthquakes, cyclones, and floods. The atlas contains State- and UT-wise hazard maps showing the degree of vulnerability right up to the district levels and also district-wise tables of housing stock indicating the risk against the forces of natural hazards which may strike in a particular region.

Nearly 80% of the housing stock in the country belongs to non-engineered category, thereby leading to much higher risk of damage and destruction on account of natural hazards like earthquakes, cyclones, floods, landslides, etc.

This atlas has been a bench-mark contribution of the BMTPC and has been commended nationally and internationally. The formulation of the *Vulnerability Atlas of India* can be claimed as the first example of its type not only in India but in the whole developing world. As a result, the International Jury setup by IDNDR Secretariat ranked this work of the BMTPC at the 4th position out of the 41 international projects of such kind. The *Vulnerability Atlas of India* was also recognized as *good practice* among the cases received for best practices by the *UN-Habitat* under Dubai International Awards for the year 2006.

With increasing emphasis by the central government on proactive, preventive, and mitigation measures by all states and union territories (UTs), availability of latest information about hazards and vulnerability in a user-friendly manner has become a necessity. Since the publication of the *Vulnerability Atlas of India* in 1997, hazard scenario especially with respect to earthquakes and floods has undergone changes. At the same time, more information are available on seismotectonic feature of the country, tsunami effect of earthquakes, storm surge, rainfall data, and landslides. Housing scenario has also changed, and latest information is available through Census 2001. Politically also new States have been formed and a number of districts have been created. Keeping all these in view, under the guidance of the Ministry of Housing and Urban Poverty Alleviation, Government of India, BMTPC was entrusted the task of revising the Atlas of 1997 through a peer group.

Accordingly, the *Vulnerability Atlas of India* has been revised and published in 2006 on the basis of Census 2001 data in digitized format with latest information drawn from concerned organizations like Geological Survey of India (GSI), Central Water Commission (CWC), India Meteorological Department (IMD), Survey of India, and Department of Science and Technology (DST). It provides the following important information which makes it a useful tool for formulating proactive strategies for disaster reduction and for improving safety of houses during occurrence of future natural hazards:

- Hazard maps for all the States and UTs indicating the areas prone to earthquakes, cyclones, and floods and the likely intensities on such natural hazards
- District-wise tables showing the risk to existing housing stock as per 2001 Census in all the States and UTs with respect to forces of earthquakes cyclones and floods

Later, the revised *Vulnerability Atlas of India* was also brought out in CD version in 2008.

2.1 Issues Addressed Through the Vulnerability Atlas of India

The *Vulnerability Atlas of India* addresses the following issues:

- Availability of digitized hazard zonation maps for accuracy, clarity, and usefulness
- Making available vulnerability data for new districts and States
- Inclusion of seismotectonic details in earthquake hazard maps
- Inclusion of landslide hazard maps of major affected areas based on landslide hazard zonation atlas
- Inclusion of latest storm surge data, precipitation, and surge height
- Inclusion of flood hazard maps by incorporating latest data of flood scenario and flow of rivers
- Inclusion of information about tsunami effects of earthquakes, landslides, and volcano
- Complete modification of risk tables based on Census 2001 data

2.2 Data Collection and Analysis

In preparation of the *Vulnerability Atlas of India*, the following data was collated and presented:

- Administrative boundary of India and State/UT from Survey of India, Dehradun
- District administrative boundary from Census of India (2001)
- Earthquake hazard zone map from Bureau of Indian Standards
- Faults and thrusts map from Geological Survey of India (2000)
- Information on volcano and tsunami from Geological Survey of India
- Epicenters of earthquakes occurred since 1505 from India Meteorological Department
- Wind/cyclone hazard zone map from Bureau of India Standards
- Wind speed data from India Meteorological Department
- Surge height data from India Metrological Department
- Rainfall data from India Meteorological Department
- Flood hazard map from Central Water Commission
- River map from Central Water Commission
- Census housing data 2001 from Census of India
- Landslide hazard zone maps from BMTPC

3 The Hazard Maps

The monitoring of hazards in the country is being carried out by (a) seismic occurrence and cyclone hazard monitoring by India Meteorological Department (IMD) and (b) flood monitoring by the Central Water Commission. In addition, noteworthy contributions are made by Geological Survey of India in mapping of seismic hazard and landslide hazard prone areas and the Department of Earthquake Engineering, Indian Institute of Technology, Roorkee (DEQ), on all aspects of engineering concerning seismic risk. It is pertinent to mention here that the Bureau of Indian Standards Committees on Earthquake Engineering and Wind Engineering has already prepared a Seismic Zoning Map and the wind velocity map including cyclonic winds for the country, and the Central Water Commission has prepared a *Flood Atlas of India*. BMTPC has also published a Landslide Hazard Zonation Atlas of India.

BMTPC used these hazard maps to prepare 1:2 million scale maps by superimposing the above available data on Survey of India map of this scale as the base map. The earthquake, windstorm, and flood hazard maps are drawn for each State and Union Territory separately in which the various district boundaries are clearly shown for easy identification of the hazard risk-prone areas. The data in Landslide Hazard Zonation Atlas of India is available only on small scale and not considered appropriate to blow up the state maps on 1:2 million scale. Therefore, three regional maps giving data for Western Himalayas, North Eastern, as well as Southern India as drawn to a scale of 1:5 million are given along with other Hazard Maps of India. The seismic zones of India based on intensities of earthquakes on MSK scale and intensity of the wind hazard related with wind speed are shown on the maps clearly identifying the various intensity zones. Flood-prone areas were earlier categorized in terms of unprotected and protected areas in 1997 Atlas. However, the division using protected areas is now removed since, when the protection fails under a large flood, the devastation in the *so-called protected* area becomes even more severe since people have a false sense of protection about the calamity they may face. No information is available on low-lying areas in urban centers, which are liable to inundation during heavy rains; hence such areas could not be identified. Landslide proneness is indicated in terms of severity as very high, high, moderate to low, and unlikely occurrences.

3.1 Earthquake Hazard Maps

The Indian subcontinent has a history of recurrent occurrence of earthquakes with moderate to high intensities. The latest version of seismic zoning map of India as given in IS 1893 (Part 1): 2002 assigns four levels of seismicity for India in terms of zone factors. In other words, the earthquake zoning map of India divides India into four seismic zones (Zone II, III, IV, and V). According to the present zoning map, Zone V expects the highest level of seismicity whereas Zone II is associated with the lowest level of seismicity. The general basis of the zones as deliberated in the code is as follows:

Zone V: Covers the areas liable to seismic intensity IX and above on MSK (1964) Intensity Scale. This is the most severe seismic zone and is referred here as Very High Damage Risk Zone.

Zone IV: Gives the area liable to MSK VIII. This zone is second in severity to Zone V. This is referred here as High Damage Risk Zone.

Zone III: The associated intensity is MSK VII. This is termed here as Moderate Damage Risk Zone.

Zone II: The probable intensity is MSK VI or less. This zone is referred to as Low Damage Risk Zone.

The MSK (Medvedev–Sponheuer–Karnik) intensity broadly associated with the various seismic zones is VI (or less), VII, VIII, and IX (and above) for Zones II, III, IV, and V, respectively, corresponding to maximum considered earthquake (MCE). The IS code follows a dual design philosophy: (a) under low probability or extreme earthquake events (MCE), the structure damage should not result in total collapse, and (b) under more frequently occurring earthquake events, the structure should suffer only minor or moderate structural damage. The specifications given in the design code (IS 1893 2002) are not based on detailed assessment of maximum ground acceleration in each zone using a deterministic or probabilistic approach. Instead, each zone factor represents the effective period peak ground accelerations that may be generated during the maximum considered earthquake ground motion in that zone. Each zone indicates the effects of an earthquake at a particular place based on the observations of the affected areas and can also be described using a descriptive scale like Modified Mercalli Intensity scale (MMI) or the Medvedev–Sponheuer–Karnik scale (MSK).

The earthquake hazard map of India, states and union territories, includes the seismic zonation of India, faults and thrusts, and epicenters of earthquakes that occurred since 1505 from India Meteorological Department besides giving administrative boundary of states and districts.

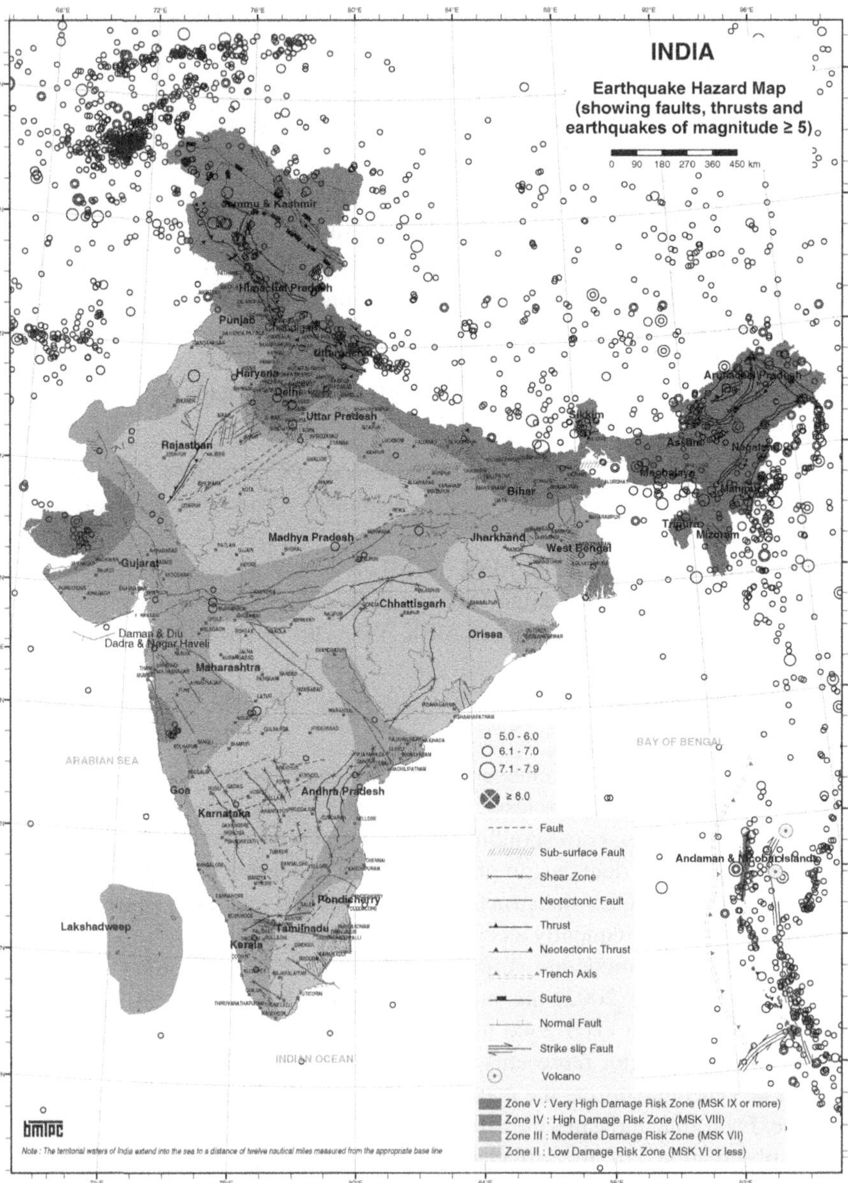

BMTPC : Vulnerability Atlas - 2nd Edition; Peer Group, MoH&UPA; Map is Based on digitised data of SOI, GOI; Seismic Zones of India Map IS:1893 - 2002, BIS, GOI, Seismotectonic Atlas of India and its Environs, GSI, GOI

3.2 Wind/Cyclone Hazard Maps

The state-wise wind hazard maps contain the following information:

Basic Wind Speed Zones: The macro-level wind speed zones of India have been formulated and published in IS 875 (Part 3) (1987) entitled "Indian Standard Code of Practice for Design Loads (other than earthquakes) for Buildings and Structures, Part 3 Wind Loads." There are six basic wind speeds *Vb* considered for zoning, namely, 55, 50, 47, 44, 39, and 33 m/s. From wind damage viewpoint, these could be described as follows:

55 m/s (198 km/h) - Very High Damage Risk Zone—A
50 m/s (180 km/h) - Very High Damage Risk Zone—B
47 m/s (169.2 km/h) - High Damage Risk Zone
44 m/s (158.4 km/h) - Moderate Damage Risk Zone—A
39 m/s (140.4 km/h) - Moderate Damage Risk Zone—B
33 m/s (118.8 km/h) - Low Damage Risk Zone

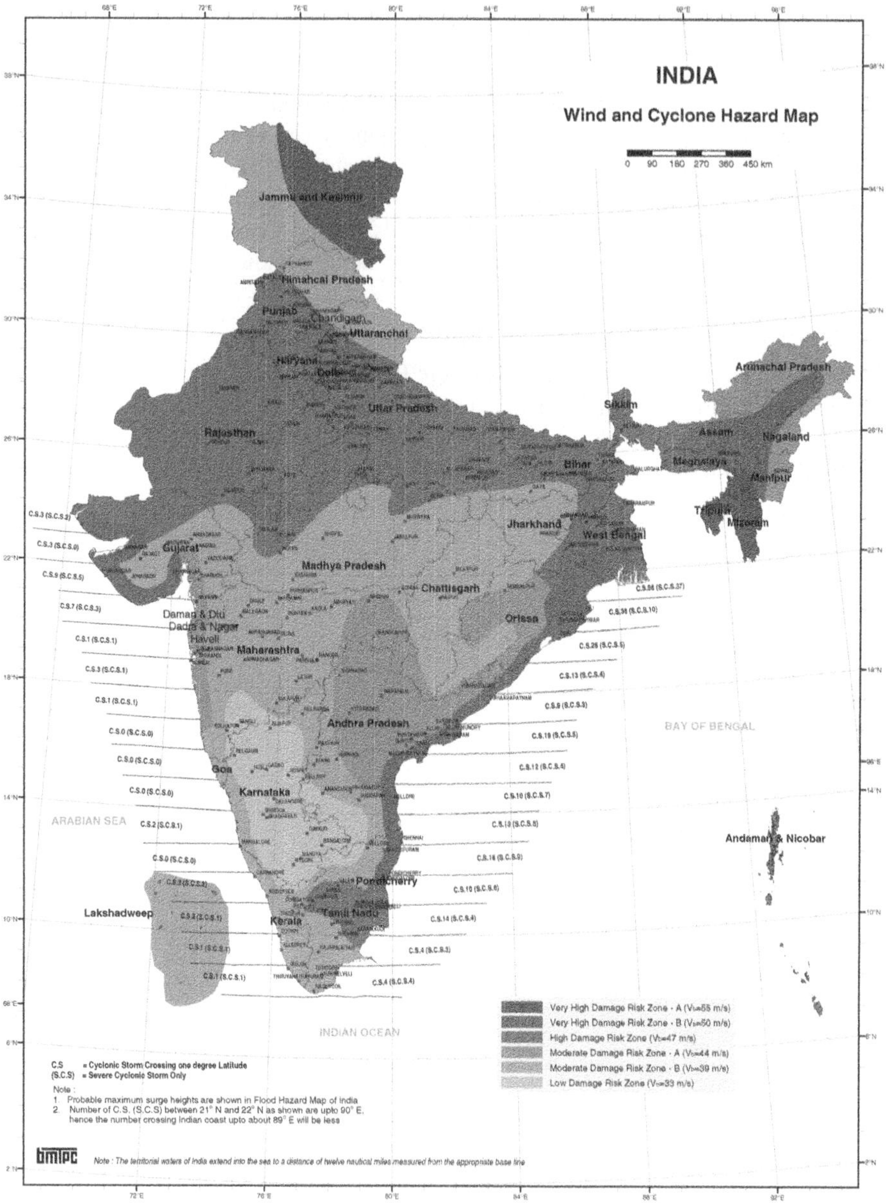

BMTPC : Vulnerability Atlas - 2nd Edition; Peer Group, MoH&UPA; Map is Based on digitised data of SOI, GOI ; Basic Wind Speed Map, IS 875(3) - 1987; Cyclone Data, 1877-2005, IMD, GOI

Infact, the cyclone-affected coastal areas of the country are classified in 50 and 55 m/s zones. Wind speeds are applicable to 10 m height above mean ground level in an open terrain. Bureau of Indian Standards also brought out two guidelines IS 15498 (2004) and IS 15499 (2004) on improving cyclonic resistance of low-rise houses and other buildings/structures and survey of housing and building typology in cyclone-prone areas, respectively. IS 15498: (2004) gives guidelines for increased wind speeds based on importance of structures in cyclone-prone areas.

The above basic maximum wind speeds in m/s represent the peak gust velocity averaged over a short time interval of about 3 s duration. The wind speeds have been worked out for 50-year return period with probability of exceedance of 63%, based on the up-to-date wind data of 43 Dines Pressure Tube (DPT) anemograph stations and study of other related works available on the subject since 1964.

The basic wind speed zones are plotted in state-wise maps which show the district boundaries as well as the district towns for their easy identification.

3.3 Flood Hazard Maps

The *Flood Atlas of India* brought out by Central Water Commission (CWC) shows pictorially the areas liable to floods, the expenditure made, and the achievement of flood protection measures. The Atlas was first published in 1962 and again published in 1977 and updated up to March 1985. A further revision is also in process. As per the information collected from CWC, a total area of 14.37 million hectares is reported to have been protected in various states out of the total flood-prone area of the country of about 40 million hectares as assessed by Rashtriya Barh Ayog (RBA) (1980). The protectable area has been considered to be of the order of 32 million hectares. The area liable to floods is the aggregate of different areas flooded in any year during the period of record. This, therefore, include the unprotected and protected areas.

The protected area is also vulnerable to floods as the flood control structures, mainly embankments, may breach during a severe flood and the so-called protected areas may also get flooded due to wrong alignment or breach of embankments. However, because of the protective measures adopted, vulnerability of houses, etc., in such areas, are considered to be comparatively less in usual circumstances. The areas outside the flood-prone areas are generally not vulnerable to flood. But experience shows that heavy rains in some of these areas can result in flood condition, and at times flooding in such areas may be very severe and create more acute problem than in the identified flood-prone areas.

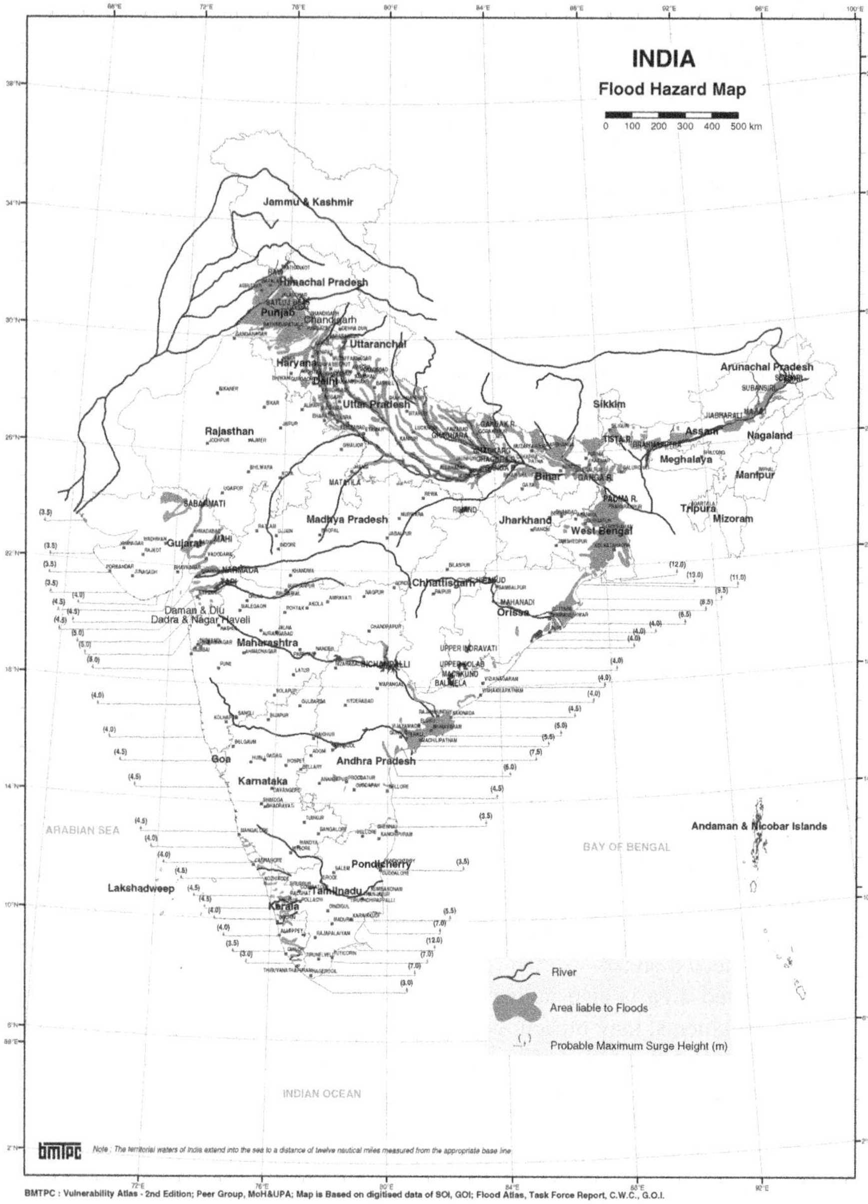

BMTPC : Vulnerability Atlas - 2nd Edition; Peer Group, MoH&UPA; Map is Based on digitised data of SOI, GOI; Flood Atlas, Task Force Report, C.W.C., G.O.I.

The state-wise flood hazard maps in *Vulnerability Atlas of India* are based on the *Flood Atlas of India* (1987) and updated flood-prone areas of Assam and other neighboring states including Bihar, West Bengal, and Eastern Uttar Pradesh included in the Task Force Report (2004). These maps mark the areas which are liable to flooding including those which have been protected. Since these maps given in *Vulnerability Atlas of India* also show the district boundaries and the location of the district towns along with the rivers, district-wise identification of the vulnerable areas will be easy.

The India Meteorological Department has compiled statistics on probable maximum precipitation (PMP) over the country considering point rainfall over a time period of 24 h. The design of drainage should consider such PMP values, the catchment areas of the drain, and the characteristics of the catchment area to avoid flooding. The district-wise PMP values are also shown on the maps for ready reference which are, however, interpolated average values of point PMP over the district.

Further, under cyclonic winds in coastal areas, the sea coast of India can be flooded due to heavy downpour on the one hand and the storm surge on the other. Whereas the PMP values give the probable intensity of raining, the probable maximum storm surge heights worked out by IMD and shown on the state-wise maps will give an idea of height of water which could flow from the sea toward the coastal plains in extreme cases and the levels to which protection will be required. The depth of inland inundation could be worked out by taking the storm surge heights, where high-resolution coastal maps with half meter contours are available.

3.4 The State-Wise Maps

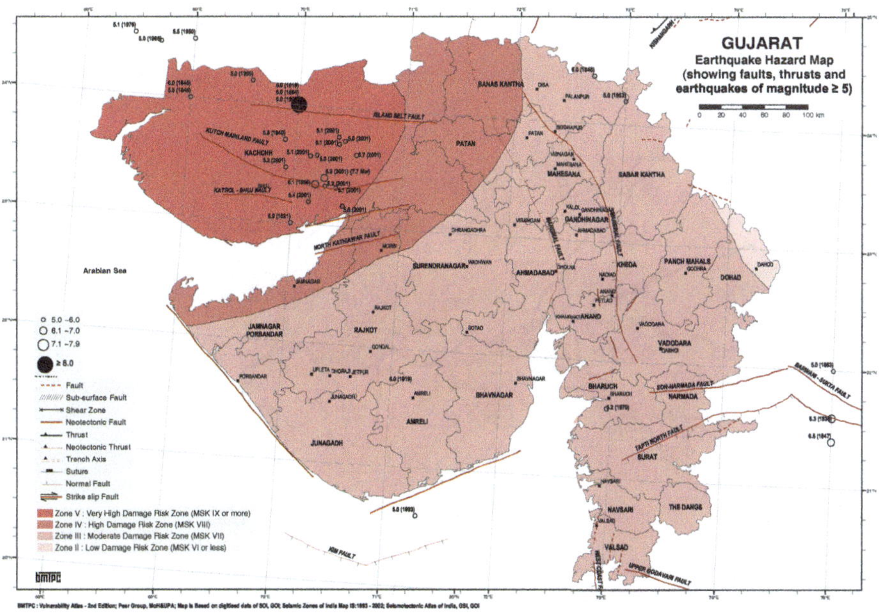

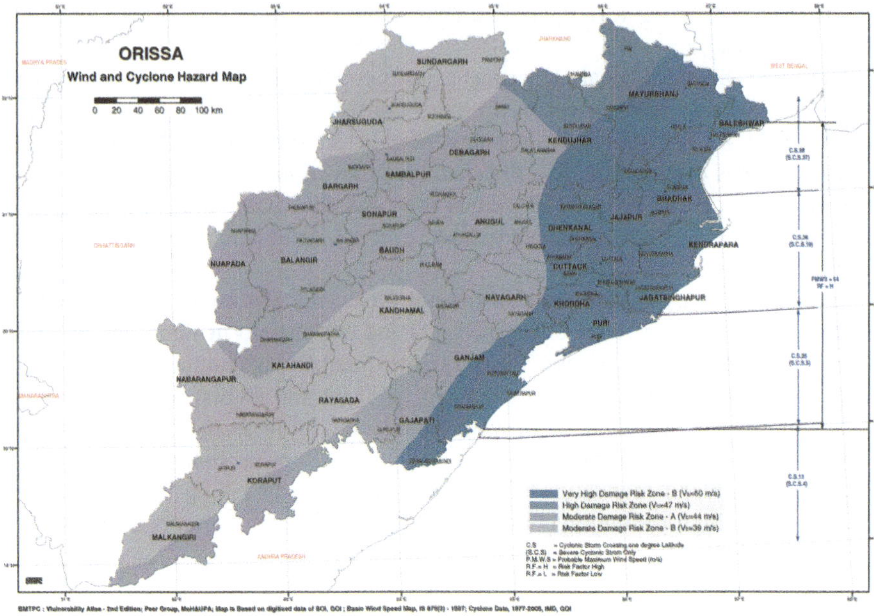
ORISSA
Wind and Cyclone Hazard Map
0 20 40 60 80 100 km
SUNDARGARH
MAYURBHANJ
BALESHWAR
JHARSUGUDA
KENDUJHAR
DEBAGARH
SAMBALPUR
BARGARH
BHADRAK
SONAPUR
ANUGUL
DHENKANAL
JAJAPUR
KENDRAPARA
NUAPADA
BALANGIR
BAUDH
CUTTACK
KANDHAMAL
NAVAGARH
KHORDHA
JAGATSINGHAPUR
PURI
KALAHANDI
GANJAM
NABARANGAPUR
RAYAGADA
GAJAPATI
KORAPUT
MALKANGIRI

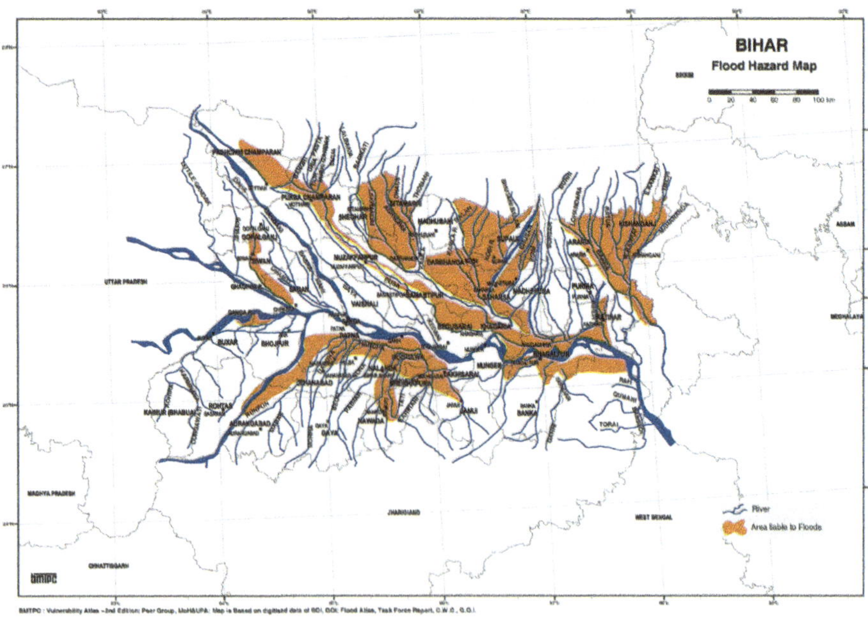
BIHAR
Flood Hazard Map
0 20 40 60 80 100 km
UTTAR PRADESH
ASSAM
MADHYA PRADESH
JHARKHAND
WEST BENGAL
CHHATTISGARH
River
Area liable to Floods

4 Vulnerability and Risk Assessment

It is noted that preliminary effort toward vulnerability assessment of buildings under seismic and cyclone intensities has been made by the Department of Earthquake Engineering, Indian Institute of Technology Roorkee, Roorkee, and Structural Engineering Research Centre (SERC), Chennai, respectively. Taking guidance from this work, the types of housing as existing in each district has been taken from the Census of India (2001) and categorized from vulnerability consideration. The vulnerability of these types to various intensities of the hazards including floods was estimated based on knowledge gained in past disaster damage surveys, and the damage risk present in each district is presented accordingly in a separate table for each district wherein the area of the district prone to various hazard intensities has also been shown.

The Census of Housing, 2001 Census of India, gives the following details of houses based on materials of construction for walls and roofs:

1. *Types of roof*:
   (a) Pitched or sloping including tiles; slate; corrugated iron, zinc, or other metal sheets; asbestos cement sheets; plastic polythene; thatch, grass, leaves, bamboo, etc.
   (b) Flat including brick, stone, and lime; reinforced brick concrete/reinforced cement concrete
2. *Types of wall*:
   (a) Mud, unburnt bricks, stone laid in mud or lime mortar
   (b) Burnt bricks laid in cement, lime, or mud mortar
   (c) Cement concrete
   (d) Wood or Ekra walling
   (e) Corrugated iron, zinc, or other metal sheets
   (f) Grass, leaves, reeds or bamboo or thatch, plastic polythene, and others
3. *Types of flooring*:
   Various types like mud, stone, concrete, wood or bamboo, mosaic floor tiles, etc.

The distribution of houses based on predominant materials of roof and wall over whole of India according to 2001 Census is shown in Table 18.1. From the point of view of vulnerability to the earthquake, wind, or flood hazards, it was seen that the type of flooring had hardly any significance, hence omitted from consideration, and that the roof types and wall types could not be grouped together. The appropriate grouping for the whole of India is shown in Table 18.2, wherein the wall and roofing groups are categorized as follows:

Table 18.1 Distribution of houses based on predominant materials of roof and wall as per 2001 Census in India

(a) Houses by material of roof in India

	Total census houses	Grass, thatch, bamboo, wood, mud, etc.		Plastic, polythene		Tiles		Slate		GI metal, asbestos sheets		Brick		Stone		Concrete		Any other material	
		No. of houses	%	No. of houses	%	No. of houses	%	No. of houses	%	No. of houses	%	No. of houses	%	No. of houses	%	No. of houses	%	No. of houses	%
Rural	177,537,513	48,812,470	27.5	669,815	0.4	62,935,397	35.4	2,364,095	1.3	18,665,296	10.5	9,927,250	5.6	11,906,910	6.7	21,061,294	11.9	1,194,986	0.7
Urban	71,558,356	4,573,534	6.4	503,956	0.7	12,591,573	17.6	444,565	0.6	11,821,919	16.5	4,147,242	5.8	5,246,952	7.3	31,777,933	44.4	450,682	0.6
Total	249,095,869	53,386,004	21.4	1,173,771	0.5	75,526,970	30.3	2,808,660	1.1	30,487,215	12.2	14,074,492	5.7	17,153,862	6.9	52,839,227	21.2	1,645,668	0.7

(b) Houses by material of wall in India

	Total census houses	Grass, thatch, bamboo, wood, mud, etc.		Plastic, polythene		Mud, unburnt brick		Wood		GI metal, asbestos sheets		Burnt brick		Stone		Concrete		Any other material	
		No. of houses	%	No. of houses	%	No. of houses	%	No. of houses	%	No. of houses	%	No. of houses	%	No. of houses	%	No. of houses	%	No. of houses	%
Total	249,095,869	24,737,121	9.9	7,21,776	0.3	73,799,162	29.6	3,196,992	1.3	1,998,678	0.8	111,891,629	44.9	25,481,817	10.2	6,540,338	2.6	728,356	0.3
Rural	177,537,513	22,162,932	12.5	4,77,498	0.3	65,807,212	37.1	2,363,200	1.3	876,677	0.5	62,715,919	35.3	20,347,899	11.5	2,253,979	1.3	532,197	0.3
Urban	71,558,356	2,574,189	3.6	2,44,278	0.3	7,991,950	11.2	833,792	1.2	1,122,001	1.6	49,175,710	68.7	5,133,918	7.2	4,286,359	6.0	196,159	0.3

(c) Houses by material of floor in India

	Total census houses	Mud		Wood, bamboo		Brick		Stone		Cement		Mosaic floor tiles		Any other material	
		No. of houses	%	No. of houses	%	No. of houses	%	No. of houses	%	No. of houses	%	No. of houses	%	No. of houses	%
Total	249,095,869	136,779,853	54.9	2,286,504	0.9	6,287,685	2.5	14,507,423	5.8	69,712,015	28.0	18,544,232	7.4	978,157	0.4
Rural	177,537,513	124,855,981	70.3	1,822,348	1.0	4,030,993	2.3	8,119,872	4.6	34,235,926	19.3	4,010,786	2.3	461,607	0.3
Urban	71,558,356	11,923,872	16.7	464,156	0.6	2,256,692	3.2	6,387,551	8.9	35,476,089	49.6	14,533,446	20.3	516,550	0.7

Note: Percentage (%) is calculated with respect to the respective total census houses given in column one

Table 18.2 Distribution of houses by predominant materials of roof and wall and level of damage risk

Wall/Roof		India		Level of risk under								
		Census houses		EQ zone				Wind velocity m/s				Flood prone area in %
		No. of houses	%	V	IV	III	II	55 & 50	47	44 & 39	33	
					Area in %			Area in %				
India				**10.9**	**17.3**	**30.4**	**41.4**	**5.0**	**40.2**	**48.0**	**6.7**	**7.9**
Wall												
A1—Mud & Unburnt Brick Wall	Rural	65,807,212	26.4									
	Urban	7,991,950	3,2									
	Total	**73,799,162**	**29.6**	*VH*	*H*	*M*	*L*	*VH*	*H*	*M*	*L*	*VH*
A2—Stone Wall	Rural	20.347,899	8.2									
	Urban	5,133,918	2.1									
	Total	**25,481,817**	**10.3**	*VH*	*H*	*M*	*L*	*H*	*M*	*L*	*VL*	*VH*
Total—Category—A		**99,280,979**	**39.9**									
B—Burnt Bricks Wall	Rural	62,715,919	25.2									
	Urban	49.175,710	19.7									
	Total	**111, 891,629**	**44.9**	*H*	*M*	*L*	*VL*	*11*	*M*	*L*	*VL*	*H/M*
Total—Category—B		**111,891,629**	**44.9**									
C1—Concrete Wall	Rural	2,253,979	0.9									
	Urban	4,286,359	1.7									
	Total	**6,540,338**	**2.6**	*M*	*L*	*VL*	*VL*	*L*	*VL*	*VL*	*VL*	*L/VL*
C2—Wood wall	Rural	2.363.200	0.9									
	Urban	833,792	0.3									
	Total	**3,196,992**	**1.2**	*M*	*L*	*VL*	*VL*	*VH*	*H*	*M*	*L*	*H*
Total—Category—C		**9,737,330**	**3.9**									
X—*Other Materials*	Rural	24,049,304	9.7									
	Urban	4,136,627	1.7									
	Total	**28,185,931**	**11.4**	*M*	*VL*	*VL*	*VL*	*VH*	*H*	*M*	*L*	*VH*

(continued)

Table 18.2 (continued)

Wall/Roof		India										
		Census houses		Level of risk under								
				EQ zone				Wind velocity m/s				
				V	IV	III	II	55 & 50	47	44 & 39	33	
		No. of houses	%		Area in %			Area in %				Flood prone area in %
Total—Category—X		**28,185,931**	**11.3**									
Total buildings		**249,095,869**										
Roof												
R1—Light weight sloping roof	Rural	69,342,567	27.8									
	Urban	17,350,091	7.0									
	Total	**86,692,658**	**34.8**	*M*	*M*	*L*	*VL*	*VH*	*VH*	*H*	*M*	*VH*
R2—Heavy weight sloping roof	Rural	65,299,492	26.2									
	Urban	13.036,138	5.2									
	Total	**78,335,630**	**31.4**	*H*	*M*	*L*	*VL*	*H*	*M*	*L*	*VL*	*H*
R3—Flat roof	Rural	42,895,454	17.2									
	Urban	41,172,127	16.5									
	Total	**84,067,581**	**33.7**	*Damage Risk as per that for the Wail supporting it*								
Total buildings		**249,095,869**										

Housing category: wall types
Category—A: Buildings in field-stone, rural structures, unburnt brick houses, clay houses
Category—B: Ordinary brick building: buildings of the large block and prefabricated type, hall-Umbered structures, budding in natural hewn stone
Category—C: Reinforced building, well built wooden structures
Category—X: Other materials not covered in A, B, C. These are generally light.
Notes: (1) Flood prone area includes that, protected area which may have more severe damage under failure of protection works. In some other areas the local damage may be severe under heavy rains and chocked drainage. (2) Damage Risk for wall types is indicated assuming heavy flat roof in categories A, B and C (Reinforced Concrete) building. (3) Source of Housing Data: Census of Housing. GOI, 2001
Housing Category: Roof Type
Category—R1—Light Weight (Grass, Thatch, Bamboo, Wood, Mud, Plastic, Polythene, GI Metal, Asbestos Sheets, Other Materials)

Category—R2—Heavy Weight (Tiles, Slate)
Category—R3—Flat Roof (Brick, Stone. Concrete)
EQ Zone V: Very High Damage Risk Zone (MSK > IX)
EQ Zone IV: High Damage Risk Zone (MSK VIII)
EQ Zone III: Moderate Damage Risk Zone (MSK VII)
EQ Zone II: Low Damage Risk Zone (MSK < VI)
Level of risk: *VH* very high, *H* high, *M* moderate, *L* low, *VL* very low
Peer Group, MoH&UPA, GOI

Wall types

Category—A: Buildings in fieldstone, rural structures, unburnt brick houses, clay houses.

Category—B: Ordinary brick building; buildings of the large block and prefabricated type, half-timbered structures, building in natural hewn stone.

Category—C: Reinforced building, well-built wooden structures.

Category—X: Other materials not covered in A, B, and C. These are generally light structures.

Roof types

Category—R1: Lightweight (grass, thatch, bamboo, wood, mud, plastic, polythene, GI metal, asbestos sheets, other materials)

Category—R2: Heavyweight (tiles, slate)

Category—R3: Flat roof (brick, stone, concrete)

With these groupings, the vulnerability of each subgroup could be defined separately for any given intensity of earthquake, wind, or flood hazard. The risk levels of the various categories of houses for the three hazards are shown in Table 18.3 and explained in the following sections.

4.1 Risk of Damage to House Types

The damage risk to various house types is based on their average performance observed during past occurrences of damaging events. In view of numerous variations in the architectural planning, structural detailing, quality of construction, and care taken in maintenance, the performance of each category of houses in a given event could vary substantially from the average observed. The intensity scales of MMI and MSK represent average observations. For example, under seismic occurrence, the following observations have been made in many cases:

1. All Masonry Houses (Categories A and B)
   (a) Quality of construction comes out as a major factor in the seismic performance particularly under intensities MSK VII to IX. Good quality constructions perform much better than poor quality constructions in any category. Appropriate maintenance increases durability and maintains original strength.
   (b) Number of storeys in the house and the storey height are other factors. Higher the storey and more the number of storeys, greater is the observed damage.
   (c) Size, location, and number of door and window openings in the walls also determine seismic performance, since the openings have weakening effect on the walls. Smaller and fewer openings and located more centrally in the walls are better from seismic performance viewpoint.
   (d) Architectural layout, particularly in large buildings, that is, shape of building in plan and elevation, presence of offsets, and extended wings, also plays an important role in initiation of damage at certain points and its propagation as

Table 18.3 Damage risk to housing under various hazard intensities

	Level of risk under								
	EQ intensity MSK				Wind velocity (miles/s)				
	$\geq$IX (Zone V)	VIII (Zone IV)	VII (Zone III)	$\leq$VI (Zone II)	55 and 50	47	44 and 39	33	Flood prone
Wall									
A1—Mud and unburnt brick	*VH*	*H*	*M*	*L*	*VH*	*H*	*M*	*L*	*VH*
A2—Stone wall	*VH*	*H*	*M*	*L*	*H*	*M*	*L*	*VL*	*VH*
B—Burnt bricks wall	*H*	*M*	*L*	*VL*	*H*	*M*	*L*	*VL*	*H/M*
C1—Concrete wall	*M*	*L*	*VL*	*VL*	*L*	*VL*	*VL*	*VL*	*L/VL*
C2—Wood wall	*M*	*L*	*VL*	*VL*	*VH*	*H*	*M*	*L*	*H*
Roof									
R1—Light-weight sloping roof	*M*	*M*	*L*	*VL*	*VH*	*VH*	*H*	*M*	*VH*
R2—Heavy-weight sloping roof	*H*	*M*	*L*	*VL*	*H*	*M*	*L*	*VL*	*H*
X—Other materials	*M*	*VL*	*VL*	*VL*	*VH*	*H*	*M*	*L*	*VH*
R3—Flat roof	Damage risk as per that for the wall supporting it								

Building category: (by wall material)
Category—A: Buildings in fieldstone, rural structures, unburnt brick houses, clay houses
Category—B: Ordinary brick building; buildings of the large block and prefabricated type, half-timbered structures, building in natural hewn stone
Category—C: Reinforced concrete building, well-built wooden structures
Category—X: Other materials not covered in A, B, and C like light sheets and biomass materials
Note: Damage risk is indicated assuming heavy flat roof in cases A, B, and C (reinforced concrete) building
Building category: (by roof material)
Category—R1: Lightweight sloping roof (grass, thatch, bamboo, wood, mud, plastic, polythene, GI metal, asbestos sheets, etc.)
Category—R2: Heavyweight sloping roof (tiles, slate)
Category—R3: Flat roof (brick, stone, concrete)
Notes: (1) Flood-prone area includes that protected area which may have more severe damage under failure of protection works. In some other areas, the local damage may be severe under heavy rains and chocked drainage. (2) Source of Housing Data: Census of Housing, GOI, 2001
EQ Zone V: Very High Damage Risk Zone (MSK > IX)
EQ Zone IV: High Damage Risk Zone (MSK VIII)
EQ Zone III: Moderate Damage Risk Zone (MSK VII)
EQ Zone II: Low Damage Risk Zone (MSK < VI)
Level of risk: *VH* very high, *H* high, *M* moderate, *L* low, *VL* very low

well. More symmetrical plans and elevations reduce damage, and unsymmetrical ones lead to greater damage.

(e) Where clay/mud mortar is used in wall construction, its wetness at the time of earthquake is very important factor in the seismic performance since the strength of fully saturated mortar can become as low as 15% of its dry strength.

2. Wooden Houses

(a) Quality of construction, that is, seasoning of wood and the joinery are important in seismic and cyclonic wind performance. Better the quality better the performance.
(b) Wood decays with time due to dry rot, insect and rodent attack, etc., therefore, the joints tend to become loose and weak. The state of maintenance of the wooden building will determine its performance during earthquake, high wind, as well as flooding.
(c) In houses with sloped roofs, a shallow angle for the roof, extended eaves projection, and reentrant corners lead to higher damage.
(d) In light roofs, pressures often exceed the dead weight leading to blowing-off of roofs.

3. Reinforced Concrete Houses

(a) Multistorey RC frame buildings resting on soft soils and having soft first storey unconnected wall panels and floating columns in the superstructure collapsed even in seismic zone III.
(b) Besides bad quality of configuration planning and structural design, poor quality of construction leads to total collapses of 5–10-storey RC frame buildings.
(c) In reinforced concrete construction, good structural design and detailing and good quality construction only would ensure excellent performance. Carelessness in any of these can lead to poor behavior both under earthquakes and cyclones.

Now the average risk levels to various categories of houses for various hazards and their intensities are defined here below for use in the house vulnerability tables.

4.2 Damage Risk Levels for Earthquakes

The damage risk to various house types is defined under various seismic intensities on MSK scale. The following damage risks are defined based on this intensity scale.

Very high damage risk (VH)

Total collapse of buildings

High damage risk (H)

Gaps in walls; parts of buildings may collapse; separate parts of the building lose their cohesion; and inner walls collapse.

Moderate damage risk (M)

Large and deep cracks in walls, fall of chimneys on roofs.

Low damage risk (L)

Small cracks in walls; fall of fairly large pieces of plaster, pantiles slip off; cracks in chimneys, part may fall down.

Very low damage risk (VL)

Fine cracks in plaster; fall of small pieces of plaster

4.3 Damage Risk Levels for Wind Storms

For damage risk to buildings from windstorms, there appears no universally accepted scale like the seismic intensity scale. The following damage risk scale has been proposed in the *Vulnerability Atlas of India* for developing the house vulnerability tables.

Very high damage risk (VH)

Generally similar to "high risk" but damage is expected to be more widespread as in the case of cyclonic storms.

High damage risk (H)

Boundary walls overturn, walls in houses and industrial structures fail; roofing sheets, and tiles or whole roofs fly; large-scale destruction of lifeline structures such as lighting and telephone poles and a few transmission line towers/communication towers may suffer damage; and non-engineered/semi constructions suffer heavy damage.

Moderate damage risk (M)

Loose tiles of clay fly, roof sheets fixed to batten fly; moderate damage to telephone and lighting poles; moderate damage to non-engineered/semi-engineered buildings

Low damage risk (L)

Loose metal or fiber cement sheets fly; a few lighting and telephone poles go out of alignment; signboards and hoardings partially damaged; well-detailed non-engineered/semi-engineered buildings suffer very little damage.

Very low damage risk (VL)

Generally similar to "low risk" but expected to be very limited in extent

4.4 Damage Risk Levels for Flood

No detailed building damage reports under flooding appear to have been worked out as yet. Also flood intensities in terms of depth of water, velocity of flow, or time duration of inundation are not yet defined. In the absence of such data, no definite recommendation about damage risk levels could be made.

The following damage risks have been drafted by the group based on understanding of material behavior under submergence.

Very high damage risk (*VH*)

Total collapse of buildings; roof and some walls collapse; floating away of sheets, thatch, etc.; erosion of foundation; severe damage to lifeline structures and systems

High damage risk (*H*)

Gaps in walls; punching of holes through wall by flowing water; parts of buildings may collapse; light roofs float away; erosion of foundation; sinking or tilting; undercutting of floors, partial roof collapse.

Moderate damage risk (*M*)

Large and deep cracks in walls; bulging of walls; loss of belongings; damage to electric fittings

Low damage risk (*L*)

Small cracks in walls; fall of fairly large pieces of plaster

Very low damage risk (*VL*)

Fine cracks in plaster; fall of small pieces of plaster

5 The Housing Vulnerability Tables

Now correlating the house types, the hazard intensities on the maps and the damage risk levels, the housing damage risk tables have been generated. For the country as a whole, for each State and Union Territory also, an overall risk table has been developed. Such tables are then prepared for each of the districts and collated state-wise.

Each table also gives at the top of each column of hazard intensities, the percent of total area of the country, state or district covered by the table, lying under the various hazard intensities. Thus the concerned administrative or professional authority can visualize the extent of damage risk existing to any hazard at one time or the other in the future.

As an example, let us refer to a District OR 10, Kendrapara of the State of Orissa (Table 18.4). It is seen that 89.2% area of the district lies in seismic intensity VII zone (Zone III) and 100% area in the 50 m/s wind zone. Also 35.5% of its area is flood-prone. The probable maximum precipitation is 600 mm, that is, quite a high figure. According to 2001 Census, there are 354,771 housing units in the district, 77.3% of which are of Category A (very weak type), 18.8% of Category B (moderate strength), and only 1.30% of Category C (the strong types). Also 2.7% houses are of bamboo, thatch, grass, and leave type indicating the poverty condition of the people. The risk of damage from earthquakes to Category A houses is "medium," and to Category B (18.8% of total) it is "low." The example district lies in the cyclone-prone area of the Orissa coastal area and has very high risk to 77.1% housing units; hence the life and property of this population living in the district is at great cyclone

Table 18.4 Distribution of houses by predominant materials of roof and wall and level of damage risk

Table No.: OR 10		State: Orissa						Kendrapara				
		Census houses		Level of risk under								
				EQ zone				Wind velocity m/s				
				V	IV	III	II	55 &50	47	44 & 39	33	
Wall/Roof		No. of Houses	%	Area in %				Area in %				Flood prone area in%
						89.2	**10.8**	**100**				**35.5**
Wall												
A1—Mud unburnt brick wall	Rural	266,645	75.2									
	Urban	6906	1.9									
	Total	**273,551**	**77.1**			*M*	*L*	*VH*				*VH*
A2—Stone wall	Rural	593	0.2									
	Urban	35	⍰									
	Total	**628**	**0.2**			*M*	*L*	*H*				*VH*
Total—Category—A		**274,179**	**77.3**									
B—Burnt bricks wall	Rural	55.783	**15.7**									
	Urban	10,826	**3.1**									
	Total	**66,609**	**16.6**			*L*	*VL*	*H*				*H/M*
Total—Category—B		**66,609**	**16.8**									
C1—Concrete wall	Rural	585	0.2									
	Urban	13	⍰									
	Total	**598**	**0.2**			*VL*	*VL*	*L*				*L/VL*
C2—Wood wall	Rural	3264	0.9									
	Urban	584	0.2									
	Total	**3648**	**1.1**			*VL*	*VL*	*VH*				*H*
Total—Category—C		**4446**	**1.3**									
X—Other materials	Rural	8889	2.5									
	Urban	648	0.2									
	Total	**9537**	**2.7**			*VL*	*VL*	*VH*				*VH*

(continued)

Table 18.4 (continued)

Table No.: OR 10		State: Orissa						Kendrapara				
		Census houses		Level of risk under								
				EQ zone				Wind velocity m/s				
				V	IV	III	II	55 &50	47	44 & 39	33	
Wall/Roof		No. of Houses	%	Area in %				Area in %				Flood prone area in%
Total—Category—X		**9537**	**2.7**									
Total buildings		**354,771**										
Roof												
R1—Light weight sloping roof	Rural	297,744	83.9									
	Urban	12,052	3.4									
	Total	**309,796**	**87.3**			*L*	*VL*	*VH*				*VH*
R2—Heavy weight sloping roof	Rural	2.201	0.6									
	Urban	465	0.1									
	Total	**2666**	**0.7**			*L*	*VL*	*H*				*H*
R3—Flat roof	Rural	35,814	10.1									
	Urban	6495	1.8									
	Total	**42,309**	**11.9**	*Damage Risk as per that for the Wall supporting it*								
Total buildings		**354,771**										

Probable maximum precipitation at a station of the district in 24 h is 600 mm
Housing category: wall types
Category—A: Buildings in field stone, rural structures, unbumt brick houses, clay houses
Category—B: Ordinary brick building: buildings of the large block and prefabricated type, half-timbered structures, building in natural hewn stone
Category—C: Reinforced building, well built wooden structures
Category—X: Other materials not covered in A, B, C. These are generally light
Notes: (1) Flood prone area includes that protected area which may have more severe damage under failure of protection works. In some other areas the local damage may be severe under heavy rains and chocked drainage. (2) Damage Risk for wall types is indicated assuming heavy flat roof in categories A, B and C (Reinforced Concrete) building. (3) Source of Housing Data: Census of Housing, GOI, 2001

bmTpc Building materials & technology promotion council
Housing category: roof type
Category—R1: Light Weight (Grass, Thatch, Bamboo, Wood, Mud, Plastic, Polythene, GI Metal, Asbestos Sheets, Other Materials)
Category—R2: Heavy Weight (Tiles, Slate)
Category—R3: Flat Roof (Brick, Stone, Concrete)
EQ Zone V: Very high damage risk zone (MSK > IX)
EQ Zone IV: High damage risk zone (MSK VIII)
EQ Zone III: Moderate damage risk zone (MSK VII)
EQ Zone II: Low damage risk zone (MSK < VI)
Level of risk: *VH* very high, *H* high, *M* moderate, *L* low, *VL* very low
Peer Group, MoH&UPA, GOI

risk. The district has also great risk of flooding, storm surges, and tsunami. Hence serious attention has to be paid to the district from cyclone, tsunami, and storm surge disaster prevention, mitigation, and preparedness points of view. Other hazards can similarly be analyzed with the help of the table.

6 Use of *Vulnerability Atlas of India*

In preparing the *Vulnerability Atlas of India*, it has been realized that the state governments have the basic mandate for management of disasters, and the executive actions are taken at the district levels with the District Collector playing the pivotal role. The Atlas provides some ready information, though at macro-level, for use of the authorities involved in the tasks of disaster mitigation, preparedness, and preventive actions. A glance at the hazard maps will bring to the notice of the district authorities, the location and percent areas of the districts most susceptible to hazard occurrence, the probable maximum hazard intensities, the type of housing, and its vulnerability and risk to the hazards. It must be realized that most of the human problems arise due to loss of the houses, deaths mostly occur in collapsed houses, and rescue, evacuation, relief, and rehabilitation become more acute when houses get lost. Houses are threatened more due to earthquakes and floods, except loss due to wind in coastal areas subjected to cyclones. Landslides and mud flows can totally erase villages and bury them under debris. Rock falls can destroy a building very badly. Knowing the extent of the problem of future disasters, the district authorities can formulate development plans for (a) preventive actions like hazard-resistant construction, retrofitting, and upgrading of existing buildings, (b) mitigating the intensity and extent of the disaster, (c) warning system installation and drills for its use, (d) instituting a hierarchical structure for preparedness down to the village level, (e) training of manpower in various tasks in the emergency, and (f) implementation of land zoning regulations in flood plains and coastal areas and building bylaws with disaster-resistant features in various towns and cities, etc.

The district authorities with the help of the peoples' representatives can create the necessary awareness leading to self-help. Also the hazard zoning can be improved at local levels by specific studies carried out in the district particularly for minimizing the flood havoc by measures such as suitable vulnerability analysis, hazard reduction measures, and risk mapping and improved resistance of buildings wherein the local technical institutions and professionals could also be involved.

The Atlas can be used to identify areas in each district of the country which are prone to high risk from more than one hazard. This information will be useful in establishing the need of developing housing designs to resist the combination of such hazards.

7 Updation of Earthquake Hazard Maps

Looking at the overall importance of seismic hazard in Indian context and associated risks involved, the National Disaster Management Authority, Government of India, entrusted BMTPC the task of preparing updated earthquake hazard maps up to district level incorporating latest data as available from Survey of India, Census and Geological Survey of India, India Metrological Department, etc. The memorandum of understanding (MoU) has also been signed with NDMA in this regard. The BMTPC is at present preparing the earthquake hazard maps for India, all States/UTs, and districts. The typical provisional district map is shown below:

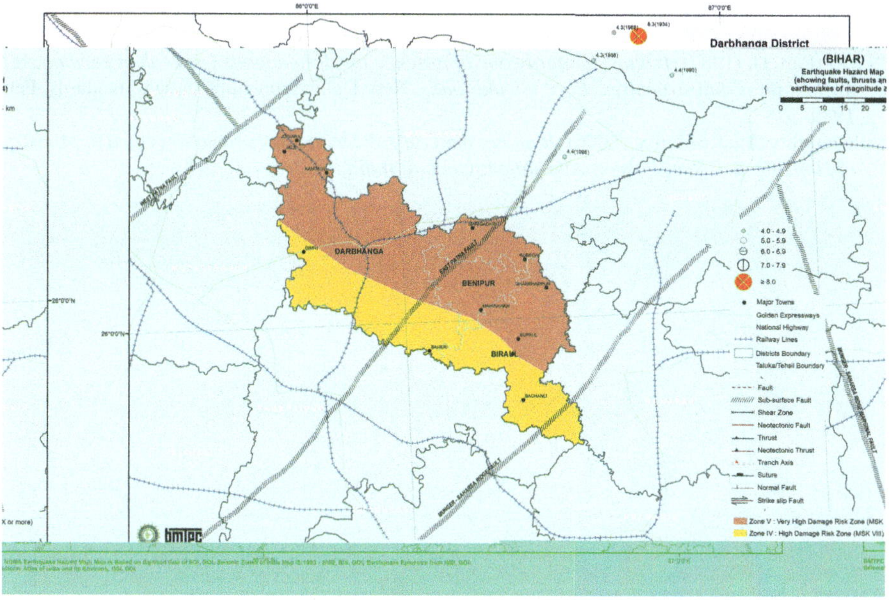

8 Conclusion

The *Vulnerability Atlas of India* presented through this chapter is a tool to understand the overall hazard scenario and risk to the existing building stock for earthquake, wind and cyclone, and floods at macro-level. It is unique in a sense that no other document is available for projecting risk to existing building stock till date. However, it is just an initiative and for better preparedness and mitigation; the studies at microlevel are required. There have been number of studies by academic and other institutions, but still, it is a long way to go before the hazard and risk maps at microlevel are published officially for public use.

References

Census of India. (2001). *Distribution of census houses by predominant material of roof, wall and floor of census houses.*

Flood Atlas of India. (1987). *Central Water Commission, Ministry of Water Resources, Govt. of India, April 1987.*

Geological Survey of India. (2000). Seismotectonic atlas of India and its environs, Special 519 Publication No. 59.

IS 15498. (2004). *Guidelines for improving the cyclonic resistance of low rise houses and other buildings/structures.* New Delhi: Bureau of Indian Standards.

IS 15499. (2004). *Guidelines for survey of housing and building typology in cyclone prone areas for assessment of vulnerability of regions and post-cyclone damage estimation.* New Delhi: Bureau of Indian Standards.

IS 1893 (Part 1). (2002). *Indian standard criteria for earthquake resistant design of structures, 5th revision.* New Delhi: Bureau of Indian Standards, 1986.

IS 875 (Part 3). (1987). *Indian standard code of practice for design loads (other than earthquake) for buildings and structures, Part 3, wind loads.* New Delhi: Bureau of Indian Standards, Feb 1989.

Vulnerability Atlas of India. (2006). *Building materials & technology promotion council, Ministry of Housing & Urban Poverty Alleviation, Govt. of India*, 2nd ed.

Chapter 19
Socioeconomic Issues in Disaster Risk Reduction in Local Buildings

Rajendra Desai

1 Introduction

India is a country that is frequented by natural disasters almost every year in one part or another. This includes many moderate to high intensity earthquakes and cyclones during the past two decades. It witnesses much damage, destruction, injury, and loss of life. A resource-starved nation like India loses much of its scarce valuable resources on post-disaster rehabilitation.

Such type of natural phenomena cannot be prevented but it is possible to prevent them from turning into disasters or emergencies. In most disasters resulting from earthquakes, cyclones, and floods, the inadequacy of the buildings, especially the houses, to stand up against such natural phenomena is the single most cause of the loss of life and loss of property. In Latur earthquake 1993, nearly 9000 lives were lost under the debris of the houses that collapsed in the quake that was of moderate magnitude. After this many more people have lost their lives in various earthquakes in the last two decades. For this reason it is important to ensure that the *new structures are built adequately strong and the existing ones*, especially the houses, schools, and places of work, *are strengthened or retrofitted to reduce their vulnerability*. In reality the existing structures are likely to remain in use for the next several decades; many of them may not be strong enough to withstand the rigors of earthquake forces and, hence, may succumb to them in future disasters.

R. Desai (✉)
National Centre for Peoples' Action in Disaster Preparedness (NCPDP), 103 Antariksh, Panjarapol Char Rasta, Near Polytechnic, Ahmedabad, Gujarat, India
e-mail: mitigation@ncpdpindia.org

M. L. Sharma et al. (eds.), *Advances in Indian Earthquake Engineering and Seismology*, https://doi.org/10.1007/978-3-319-76855-7_19

2 Non-engineered Structures in Towns and Villages

In India, as mentioned above, an *overwhelming majority* of buildings, in towns and villages, are of *non-engineered load-bearing* type. These structures have been *traditionally built* over the past century or longer, using the *locally available materials* and the *locally known practiced technologies* that have been most common. In the recently built structures, one finds a mix of the traditional and new materials/technology such as cement, concrete, and steel. It is important to note that in all this there is no input from engineers (barring buildings in the big cities). The people who build using new materials and technologies have no technical knowledge of them. For various reasons during the past decades, most of the new buildings have simply not been built to withstand the forces of an earthquake anticipated in the region.

3 Significance of Non-engineered Construction

Non-engineered structures like these *play a major role in meeting the basic but crucial needs of the society*. This could be better understood through looking at aspects like (a) cost, (b) climate, (c) lifestyle, and (d) economy. One could say with fair amount of certainty that a majority of building construction will continue to be done this way in a foreseeable future because of the eternal resource crunch and country's huge population. Other technical options have become available in the past few decades, and more will become available with time. But *few will be as viable and optimized for the rural and semi-urban context as the options that have been in use from generations.*

3.1 Large Scale of Vulnerability

In a study conducted by BMTPC based on the *Vulnerability Atlas of India,* 1997, in the Seismic Zone V alone, there are over 11 million housing units that are vulnerable. Similarly in Zone IV this number could be as much as 50 million. By another counts there are over 80 million houses in the country that are vulnerable. These counts, however, do not include the *non-engineered infrastructure buildings in towns and villages as also those built with modern technologies but without any engineering input*. In other words the number of structures that are vulnerable is extremely large, and *the people that are at risk may amount to as much as 50% of the country's population.*

3.2 *Performance by Non-engineered Structures in Recent Disasters*

The performance of non-engineered structures with a variety of building systems and materials has been studied extensively in the past decade. All these types of buildings that were built without taking into account the possible earthquakes in the area performed badly. The basic reasons for the bad performance are well known. These are: (a) *poor quality of construction*, which reflects the violation of the basic rules of construction, and (b) *absence of disaster-resistant features*. But unfortunately, more often than not, it is the materials that are used in construction that are blamed for bad performance.

4 Problems on Hand

There are two major problems in regard to the disaster risk reduction in our country.

- A large number of buildings are being constructed that are vulnerable to future disasters.
- A very large number of buildings already exist that are vulnerable to future disasters.

In order to reduce the future disaster risk, it is important to focus on finding ways to tackle both the problems, or identifying the hurdles in tackling the problems so that in a foreseeable future we are able to make our country disaster resistant. As far as the technologies are concerned, there is ample know-how in the country that is easily accessible from a wide variety of sources.

5 Disaster-Resistant New Construction

For new construction there are IS codes as well as a variety of regional guidelines that have been developed in the past two decades. These provide rules and recommendations for building new structures, using local as well as the modern materials. To a great extent, they also cover the combination of the local and modern materials, where viable. There are still many blank areas where the local technologies have not been studied adequately to bring them in to such guidelines but could be developed in near future. In the aftermath of Kutch earthquake, guidelines were evolved for bricks and stone buildings and even for the use of cement stabilized compressed soil blocks. Similarly, in the aftermath of Kosi floods in Bihar, guidelines were evolved for flood-resistant as well as earthquake-resistant brick masonry construction and also bamboo construction. Early this year guidelines were compiled for earthquake-resistant stone masonry construction with pitched stone slab (Pathal) buildings for

Uttarakhand. Typically the abovementioned documents provide rules for designing the buildings pertaining to things like wall thickness, length, and height, number of storeys, door and window openings, seismic bands, seismic vertical reinforcement within the wall masonry, encasing of the openings, etc. The last two documents also provide a large number of construction details that are needed to actually undertake the improved local non-engineered construction. These include the timber floors, timber understructure for the pitched roof, anchoring of floors and roofs to the walls, roof and floor diaphragms, etc. The additional cost of disaster-resisting features in new construction is approximately 10% of the total cost of construction of the building.

5.1 Reducing Vulnerability of Existing Buildings by Retrofitting

In regard to the vulnerability reduction of the existing buildings, retrofitting or strengthening is the most viable option. Retrofitting is a preventive measure like polio vaccine. It is a process by which the existing building is made stronger to resist the forces of earthquake or cyclone, particularly where these were not taken into account while the building was constructed. In other words, this involves the identification of the weaknesses in construction which has been observed in the buildings in the past earthquakes. And for each weakness there is a remedy in the form of a retrofitting measure. These remedies have been evolved on the basis of the information available in a number of books and the IS guidelines in India. As mentioned above, in the aftermath of various disasters, a number of guidelines have been prepared for repair and retrofitting. Once the building is retrofitted, it is not likely to collapse, and the level of damage that it may suffer in the event of a major disaster would be significantly less than what it would be otherwise. This has been effectively demonstrated through a large number of shock table tests in the country. Internationally also, there is ample demonstrative proof that the concept is very effective. This option is much cheaper (10–15% of cost of new construction) and more than five times faster compared to the total demolition of the building followed by reconstruction for disaster safety.

6 Economics and Advantages of Retrofitting Versus New Construction

The economics of retrofitting is well known around the world. Typically the *retrofitting index*, or the ratio of the cost of retrofitting to the cost of rebuilding, could be *as low as 5%*. For retrofitting it is best if done before a disaster strikes. This would help save lives and losses and prevent suffering. Instead, the option of repair and retrofitting is *faster, economical and helps retain the conveniences that were created* in the existing structure, be it a house or an infrastructure building. It is indeed the most economical option that one could exercise for ensuring one's long-term safety. The option of *building a new*, although effective, involves the cost of (a) dismantling the existing structure, (b) carting of debris, (c) the cost of reconstruction, and, finally, (d) the cost of all the *conveniences that will have to be recreated*. With the rapidly escalating costs, the expenditure involved may be several times that of what was spent when the structure was first built. On the other hand, retrofitting costs a tiny fraction of it. Yet another important advantage of this option is its *inbuilt flexibility*. Of the several different measures that need to be applied, one has a choice to apply:

- One or a few measures at a time to a part of the structure
- One or a few measures to the whole structure
- All the options to the whole structure in one go

Thus one could take a decision based on the (a) availability of funds, (b) availability of time, and finally (c) the convenience, while ensuring the technical soundness of the overall scheme.

7 Hurdles in Reducing the Vulnerability

There are a large number of hurdles in reducing the vulnerability, be it for the new buildings or for the existing ones. The principle hurdles as most commonly observed and perceived are:

- Lack of know-how and confidence with the engineers
- Lack of know-how and right skills with the building artisans

But there are other hurdles that cause even bigger impediment. These are hurdles that have little to do with the issues of technical know-how and more to do with the peoples' psyche and their beliefs. Some of these issues could be described as follows:

- *Inability to connect technical matters with the safety*: People may be scared and seeking safety. But since they do not understand as to how exactly a particular technology or an element will work in saving their houses, and probably their lives, one finds little reason to spend a little extra or put in extra effort to implement it. For example, in Latur after the earthquake, in the repair and retrofitting component of the government rehabilitation program, those who were rebuilding their houses were supposed to install the lintel band to make their houses earthquake resistant. But there were individuals who, instead of installing the RC band, made plastered patch 100 mm wide protruding out about 10 mm, all around the building at lintel level, a make-believe RC band. The government engineer saw the "band" approved it, without knowing if there was any concrete and steel behind that patch. By doing this house owners saved 20 bags out of the 40 that they had received from the government, which they sold off in gray market.

- *Preconceived notions with no scientific basis dictate the actions*: Since after the Kutch earthquake, across the country, reinforced concrete has become the "preconceived notion" for building construction for achieving the disaster safety, even for a single storey one. Unfortunately, most people including contractors and artisans do not understand what a frame structure means. Hence, in trying to make it affordable, one major trend that has become visible across the country is the use of column like RC elements in the corners of the building in the absence of beams. The walls in such structures are load bearing. Some technologists may call this construction "confined masonry," the latest fad among technologists. But unlike as required by such construction, these structures lack adequate bands as also the bond between the RC "columns" and the masonry through the toothing projected from the end of masonry at the interface with RC. If one asks the mason or the building owner, they have no scientific explanation. And not surprisingly, everyone questions the utility of the "one rod column" common term used for a single reinforcing bar within the masonry at the wall junctions—"What will a column with one rod do." The status that RC enjoys in the society, everyone including those who cannot afford it, wants to have RC elements in the construction rather than go with vertical reinforcement within masonry.
- *House owner dictating the artisan about how to construct, even if wrong*: In Uttarakhand after Chamoli earthquake of 1997, it was observed that house owners began shifting to brick masonry in cement mortar. But when the Bihari masons who were experienced in brick construction advised house owners to appoint laborers for curing of the masonry, the house owners paid little attention to them. In one case the house owner refused to allow the mason to install RC lintel band on the wall that was against the hillside. In Kashmir after 2005 earthquake, there was an unprecedented influx of cement for masonry mortar. When in one village the villagers learned in one of our meetings about the initial setting of cement mortar, and the need to use up all the mortar before that, a house owner confessed about sacking a Bihari mason who, unlike all others, was making only a little mortar at a time. The house owner thought that he was getting cheated because every hour or so the new mortar had to be made while the mason relaxed.
- *Misplaced priorities due to social and peer pressure puts assigns personal safety a lower priority:* In Latur, it was observed that many households in conformance with the fashion used the cement that was given to them to build a strong earthquake-resistant house, for plastering the walls. Many used the funds to install stone flooring.
- *Vulnerable building shared by individuals with varying priority and perceptions*: During the Latur earthquake rehabilitation, many who wanted to retrofit their houses to reduce their vulnerability were faced by their hostile siblings who either did not understand retrofitting or had some other priorities about the house. As a

result it became possible to retrofit only one half of the building and that too without going through the common wall for anchoring the seismic belt.

- *Money is placed above everything else, at the cost of personal and family security*: In Latur, it was observed that in most houses built under government-funded rehabilitation after the earthquake, the lintel band was visible. But later when the same house owners added a room adjacent to that structure, with their personal funds, they did not install the lintel band, since there was no compulsion. These are the same people who had witnessed a big earthquake not too long ago. But now the money that would be saved by not installing RC band was more important than the safety of family.
- *Common trends all around tend to prevail over scientific logic*: Many people think that disasters are caused by God, and all the damage and deaths are God's will. They cannot relate the strength of the buildings with the damage and destruction. Very few people understand the reasons behind the damage. Even the engineers fail to relate the damage to weaknesses in the building. Most people do not want to think and have lost the understanding of logic and follow what other people are doing, and in this way even if one person does something to improve his house, everybody will follow without assessing if that measure is good or bad, and in long run what will be its impact. An excellent example is the trend of dismantling of the first and second storeys that had evolved in the aftermath of Chamoli earthquake of 1999; when upon seeing that upper storeys had cracks in the corners, someone concluded that those storeys are unsafe and hence should be dismantled. People could have kept the first and second storeys after repairing and retrofitting them, so that it could have given them the much needed floor space, which is scarce in the mountainous areas, and at the same time could have given them the safety as well.
- *"Retrofitting of a building" to reduce vulnerability is too abstract: a concept difficult to digest*: But this is a time-tested concept commonly used in our day-to-day lives. One retrofits an old worn-out cardboard carton with the help of tapes along the corners or with strings wrapped around it, instead of throwing it away and buying a new one, so that it would not fall apart when loaded heavily. Or one provides extra stitching on the handles of a cloth carry bag so that handles would not separate from the bag while carrying heavy loads. One even adds a beam when the old beam has begun to sag. But when it comes to a masonry structure, people find it difficult to comprehend how retrofitting can really strengthen it.
- *Absence of fear results into misplaced priorities*: What people are not willing to do immediately after an earthquake or for a few months after, people will do very casually a year later, when the fear is gone or subsided. In Ahmedabad, soon after Kutch earthquake that saw over 100 multistorey structures topple, the multistorey building market had collapsed. But within couple of years or so, people once again started buying apartments in them and even started moving in to them, without any concern about whether the structure was really earthquake resistant.
- *Race for modernization eliminates retrofitting of masonry structure*: With RC frame construction becoming the ultimate sign of being modern, house owners

and engineers alike shy away from retrofitting a masonry structure that is already out of fashion. Instead, they prefer to replace it with new RC frame structure.

- *Engineers eliminate economically unattractive options*: Since retrofitting means major saving in cost due to very small budget, many engineers do not prefer to it since it simply means smaller fees/monetary gain.

7.1 Corrective Actions that Need to be Taken for Vulnerability Reduction

In a developing country like India, bulk of the houses and even infrastructure buildings are non-engineered and are executed in the informal sector; the focus of promoting this option must be aimed primarily toward the people at large including building artisans. The technical community, needless to say, can have a marked influence on the psyche of the people and the artisans. Hence, the efforts of promotion must also focus on the technical community. The following is a list of things that ought to be taken up for the large-scale promotion of the option of *vulnerability reduction*.

- *Public Awareness* campaign for creating *felt need* among the people is the most crucial step in evolving the delivery system for disaster-resistant construction and vulnerability reduction through retrofitting. There is a need to instill fear of the most common natural hazards of the area that are potentially destructive. *Unless there is fear, there will be neither the desire nor the demand for vulnerability reduction*. In the age of electronic media, it is not difficult to do this. For any hazard-prone area, a disaster scenario could be prepared with powerful visuals, videos, and simulations about the impact of any disaster. Frequent showing of this could be *effective in bringing fear and, hence, the felt need*.
- *Confidence Building* program that could be organized for different audiences such as common man, children, construction industry, policy makers, etc. needs to be taken up. It is important that people have the information to satisfy their felt need. For this people have to be able to relate the technical aspects such as good construction quality, or disaster-resistant features, or retrofitting measures to the desired end result, namely, the vulnerability reduction. In other words, they must be able to understand that all these steps would be effective in helping the building perform better in the face of disaster. For this people need *simple booklets that are technically correct, relevant, reader friendly, and attractive*. Here, people not only have to understand what each step does to the building performance, they must have full confidence in those steps. Only then will the people be wanting and willing to invest their hard-earned money in to those steps/measures. For this, once again, the electronic media could effectively bring to people simple shock table videos, simulations, etc. that are very convincing.
- *Impart Visibility* through full-scale demonstration of technologies in various locations. In case of retrofitting, this would have to be done on a variety of

existing buildings, public and private, infrastructural and residential, etc. Such demonstration must be most prominently visible and most easily accessible for common person. They should be self-explanatory so that the visitors could educate themselves without any one's help. But there could be guided tours also through such buildings. Once again, such demonstrations could be given even more visibility through the electronic media.

- *Training of Petty Contractors and Building Artisan*: This forms the most important link in the whole delivery system to bring the vulnerability reduction technologies to one's doorstep. After all for a common man, the mason and the local petty contractor are the "engineers" that help him build his house or repair or modify his house. The same individuals would retrofit his house or the local infrastructure building. So unless the capacity of these individuals is built for vulnerability reduction, it would be extremely difficult to move toward eliminating or reducing the disaster impact.
- *Education of Engineers*:
  - Engineering curriculum addition/revision: Since the education curriculum lacks the vernacular systems of construction which are most commonly used for building non-engineered buildings, the engineers coming out of the colleges are totally ignorant about them. Load-bearing building systems are also absent from curriculum of most colleges. On the other hand, the education over emphasizes the RCC and steel construction. As a result for most engineers, these are the only legitimate construction systems worth using. Everything else is useless and must be discarded. This trend must be reversed.
  - Continuing education: The practicing engineers also need to be exposed to the vernacular systems and ways to retrofit them.
- *Clear and Positive Government Policy*: The government invests the most amount of money in construction, as well as in the post-disaster relief and rehabilitation work. Hence, the government must have a policy that promotes improved vernacular building systems and retrofitting to minimize future losses.

Only the synergistic effect of all these programs on a sustained basis could help in the process of vulnerability reduction of houses which also will be socio-economically beneficial.

Interaction with town people to explain safer construction

Poster exhibition at a village bust stop on earthquake safety

Poster exhibition at a village market on earthquake safety

Meeting with villagers on earthquake safety

Artisan training on retrofitting in village for earthquake safety

Artisan training on safer construction in a village for earthquake safety

Index

M. L. Sharma et al. (eds.), *Advances in Indian Earthquake Engineering and Seismology*, https://doi.org/10.1007/978-3-319-76855-7

F

T

The manufacturer's authorised representative in the EU is Springer Nature Customer Service Centre GmbH, Europaplatz 3, 69115 Heidelberg, Germany. If you have any concerns regarding our products, please contact ProductSafety@springernature.com

Printed and bound by CPI Group (UK) Ltd, Croydon, CR0 4YY
30/09/2026
02225238-0014